Effective Standardization Management in Corporate Settings

Kai Jakobs
RWTH Aachen University, Germany

A volume in the Advances in IT Standards and Standardization Research (AITSSR) Book Series

BUSINESS SCIENCE Reference
An Imprint of IGI Global

Published in the United States of America by
Business Science Reference (an imprint of IGI Global)
701 E. Chocolate Avenue
Hershey PA, USA 17033
Tel: 717-533-8845
Fax: 717-533-8661
E-mail: cust@igi-global.com
Web site: http://www.igi-global.com

Library of Congress Cataloging-in-Publication Data

Names: Jakobs, Kai, 1957- editor.
Title: Effective standardization management in corporate settings / Kai Jakobs, editor.
Description: Hershey, PA : Business Science Reference, [2016] | Includes bibliographical references and index.
Identifiers: LCCN 2015042002| ISBN 9781466697379 (hardcover) | ISBN 9781466697386 (ebook)
Subjects: LCSH: Standardization. | Industrial management.
Classification: LCC HD62 .E44 2016 | DDC 658.5/62--dc23 LC record available at http://lccn.loc.gov/2015042002

This book is published in the IGI Global book series Advances in IT Standards and Standardization Research (AITSSR) (ISSN: 1935-3391; eISSN: 1935-3405)

British Cataloguing in Publication Data
A Cataloguing in Publication record for this book is available from the British Library.

All work contributed to this book is new, previously-unpublished material. The views expressed in this book are those of the authors, but not necessarily of the publisher.

For electronic access to this publication, please contact: eresources@igi-global.com.

Advances in IT Standards and Standardization Research (AITSSR) Book Series

Kai Jakobs
RWTH Aachen University, Germany

ISSN: 1935-3391
EISSN: 1935-3405

Mission

IT standards and standardization are a necessary part of effectively delivering IT and IT services to organizations and individuals as well as streamlining IT processes and minimizing organizational cost. In implementing IT standards, it is necessary to take into account not only the technical aspects, but also the characteristics of the specific environment where these standards will have to function.

The **Advances in IT Standards and Standardization Research (AITSSR) Book Series** seeks to advance the available literature on the use and value of IT standards and standardization. This research provides insight into the use of standards for the improvement of organizational processes and development in both private and public sectors.

Coverage

- Tools and Services Supporting Improved Standardization
- Standards in the Public Sector
- Impacts of Market-Driven Standardization and Emerging Players
- Multinational and Transnational Perspectives and Impacts
- Prescriptive Theory of Standardization
- Standards for information infrastructures
- Standardization and economic development
- Future of Standardization
- Standardization for Organizational Development
- Technological innovation and standardization

IGI Global is currently accepting manuscripts for publication within this series. To submit a proposal for a volume in this series, please contact our Acquisition Editors at Acquisitions@igi-global.com or visit: http://www.igi-global.com/publish/.

The Advances in IT Standards and Standardization Research (AITSSR) Book Series (ISSN 1935-3391) is published by IGI Global, 701 E. Chocolate Avenue, Hershey, PA 17033-1240, USA, www.igi-global.com. This series is composed of titles available for purchase individually; each title is edited to be contextually exclusive from any other title within the series. For pricing and ordering information please visit http://www.igi-global.com/book-series/advances-standards-standardization-research/37142. Postmaster: Send all address changes to above address.

Titles in this Series

For a list of additional titles in this series, please visit: www.igi-global.com

Business Process Standardization A Multi-Methodological Analysis of Drivers and Consequences
Björn Münstermann (University of Bamberg, Germany)
Business Science Reference • copyright 2015 • 448pp • H/C (ISBN: 9781466672369) • US $245.00 (our price)

Modern Trends Surrounding Information Technology Standards and Standardization within Organizations
Kai Jakobs (RWTH Aachen University, Germany)
Information Science Reference • copyright 2015 • 339pp • H/C (ISBN: 9781466663329) • US $225.00 (our price)

Evolution and Standardization of Mobile Communications Technology
DongBack Seo (University of Groningen, The Netherlands and Hansung University, South Korea)
Information Science Reference • copyright 2013 • 328pp • H/C (ISBN: 9781466640740) • US $195.00 (our price)

Information Technology for Intellectual Property Protection Interdisciplinary Advancements
Hideyasu Sasaki (Ritsumeikan University, Japan)
Information Science Reference • copyright 2012 • 367pp • H/C (ISBN: 9781613501351) • US $195.00 (our price)

Frameworks for ICT Policy Government, Social and Legal Issues
Esharenana E. Adomi (Delta State University, Nigeria)
Information Science Reference • copyright 2011 • 352pp • H/C (ISBN: 9781616920128) • US $180.00 (our price)

Toward Corporate IT Standardization Management Frameworks and Solutions
Robert van Wessel (Tilburg University, Netherlands)
Information Science Reference • copyright 2010 • 307pp • H/C (ISBN: 9781615207596) • US $180.00 (our price)

Data-Exchange Standards and International Organizations Adoption and Diffusion
Josephine Wapakabulo Thomas (Rolls-Royce, UK)
Information Science Reference • copyright 2010 • 337pp • H/C (ISBN: 9781605668321) • US $180.00 (our price)

Information Communication Technology Standardization for E-Business Sectors Integrating Supply and Demand Factors
Kai Jakobs (Aachen University, Germany)
Information Science Reference • copyright 2009 • 315pp • H/C (ISBN: 9781605663203) • US $195.00 (our price)

Standardization and Digital Enclosure The Privatization of Standards, Knowledge, and Policy in the Age of Global Information Technology
Timothy Schoechle (University of Colorado, USA)
Information Science Reference • copyright 2009 • 384pp • H/C (ISBN: 9781605663340) • US $165.00 (our price)

Editorial Advisory Board

Table of Contents

Detailed Table of Contents

of the securitization supply chain via vertical disintegration, which undermined the unity of the risk analysis process. This chapter argues that the effects of technological standardization on innovation in the mortgage industry should be explored beyond a narrow focus on efficiency and profit in relation to an individual organization's business strategy; but rather within an extended scope that includes broader social and policy contexts that guide innovation.

Section 7
Two Sides of the Same Coin

In standards wars, FUD (Fear, Uncertainty and Doubt) is sometimes created to weaken the opponent's market position. Little is known about these strategies, their use in committee standardization settings and how to respond to them. This chapter explores this phenomenon. It (1) identifies various FUD strategies, (2) their context of emergence, and (3) their effect on the dynamics of a standards war in a historical case study: the European standards war on digital mobile radio communication in the 1990s. The study highlights the need to distinguish 'FUD as perceived' from 'FUD as intended'. FUD strategies and case-specific characteristics of their emergence are illustrated. The chapter shows that perceived FUD polarizes and entrenches positions of warring parties thereby affecting the course of the standards war. The authors conclude that, given its impact, reflection by corporate standardization managers on (perceived) FUD, preclusion, counter-strategies and the downscaling of standards wars is warranted.

Extremely complex systems like the smart grid or autonomous cars need to meet society's high expectations regarding their safe operation. The human designer and operator becomes a "system component" as soon as responsible decision making is needed. Tacit knowledge and other human properties are of crucial relevance for situation-dependent decisions. The uniform modeling of technical systems and humans will benefit from ethical reflection. In this chapter, we describe human action with technical means and ask, on the one hand, for a comprehensive multidisciplinary technology assessment in order to produce supporting knowledge and methods for technical and societal decision making. On the other hand—and here is the focus—we propose a system life cycle approach which integrates the human in the loop and argue that it can be worthwhile to describe humans in a technical way in order to implement human decision making by means of the use case method. Ethical reflection and even ethically based technical decision making can support the effective control of convergent technology systems.

Foreword

I am honored to be asked by Kai Jakobs to provide this forward to this work.

Standardization is a methodology for the definition, design, manufacture and creation of product and component parts, and also is a long recognized approach for efficient business processes. Standardization is achieved through careful consideration, gaining relevant information, consensus building and effective influence, top of field technical competencies, and carefully monitoring and managing initiatives and achievements against key product and process goals.

Many of the most thoughtful and successful global businesses recognize the value of standardization within both their products and business processes and credit it as an element in helping them achieve industry leadership. By leveraging component and product standards, and carefully standardizing particular aspects of their products, these businesses build and sustain a competitive advantage. By standardizing processes, business implement flexible processes, design products anywhere, and manufacture at any plant facility internationally, providing the utmost flexibility in maximizing efficiencies.

Beyond recognizing the need for internal standardization, businesses credit the strategic adoption of standards in processes and products, or *strategic standardization*, with helping them achieve industry leadership. Through strategic standardization, business leaders leverage standards to build and sustain a competitive advantage or at least avoid a competitive disadvantage.

Strategic Standardization Management (SSM) is "A management discipline and methodology that investigates all aspects of standardization across a business and/or industry, then defines, recommends, and implements appropriate strategies and policies to leverage standardization so that a firm can gain competitive advantage and avoid disadvantage" [Betancourt & Walsh, 1995]. I had the privilege of working closely with Messrs. Betancourt and Walsh through ANSI throughout the 1990's in refining the early understanding of this emerging management discipline. This understanding was further researched and surveyed in the subsequent time period and culminated in a doctoral dissertation, follow-on journal articles and the development of educational programs and influence in academia and programs and courses which continue today.

A system of management focused on standardization is required to ensure that activities are adequately performed. SSM is an enterprise-wide, systemic competitiveness strategy and ongoing process that requires careful management of subject matter expert committee representation, which standards to affect and adopt, and alignment with the organizations' technical, business and strategic plans. Management of this can be improved if the organization is considered a series of interrelated systems. Additionally, management of standardization requires ongoing oversight including return on investment and where-represented gap analyses being performed periodically.

Manufacturers understand the need for the interchangeability, and standardization assures purchasers that products are compatible and make a purchase decision based on cost, quality, and level of performance. Also, assessments of conformity with the standard must occur to further ensure quality and interoperability.

It is impossible for an organization not to comply with the requirements of some standards as standard pervade and influence our everyday lives. Considering global pressures to standardize products, it is a superior business strategy to explicitly address standardization. Standardization tends to be diffused within the organization that focuses on its core competencies; therefore it requires someone to continuously monitor standards work and cannot be left to chance. Top leadership must provide support for this level of organizational commitment.

There is also growing awareness that the term "standards" connotes much more than just the documents that prescribe technical requirements and specialized test methods that demonstrate compliance. Beyond the advantages for product or service, standards, if properly addressed, are one of the best marketing tools ever conceived.

This book contains a wealth of updated information on the value of standardization as understood by some of today's top international experts. Papers include topics related to investing now in standardization to add future value, clearly and distinctly defining the difference between standardization and standards, business benefits and the role of internal standards. The many processes for standards development are discussed as well as understanding open standards. The human resource side of standards development is addressed recognizing resistance and process. Emerging issues such as intellectual property rights and legal consequences are also discussed. These are major issues being addressed internationally by the standards developing organizations. Finally, standardization benefits within design innovation and the essential need for focus on ethics are presented. There is something here for everyone.

Richard Forselius
Northwestern University, USA

Preface

In the field of Information and Communication Technologies (ICT) international standards are the one major mechanism to ensure interoperability between systems. Accordingly, standards that suit a company's needs, for example by incorporating its Intellectual Property (IP), by anticipating its implementation activities or by helping create a new interesting market represent a considerable competitive advantage. Thus, participation in standards setting and influencing the process to one's own benefit has become a major strategic tool, utilised especially by the large players (see e.g. Bird, 1998; Updegrove, 2006). But standards also have other characteristics that make them interesting for many companies (not necessarily only for the big ones). For example, they are a valuable means of technology transfer, which makes them relevant also for academia, research institutions and corporate R&D departments (see e.g. Jakobs, 2011a). Moreover, the process of standardisation is not that dissimilar to pre-competitive joint research. That is, standardisation also provides a legal platform for co-operation with competitors and (potential) customers.

Close links between standardisation and innovation may generally be identified. The times when statements along the lines of 'standardisation hampers innovation' where believed to adequately describe this link are (mostly) gone by now. However, the unqualified 'standards foster innovation' does not fully reflect reality either. Swann and Lambert (2010) observe that standards do both – enable and constrain innovation – but that the enabling aspect is much more important. Specifically, they note that "… standardization does constrain activities but in doing so creates an infrastructure to help trade and subsequent innovation. Standardization is not just about limiting variety by defining norms for given technologies in given markets. Standardization helps to achieve credibility, focus and critical mass in markets for new technologies" (p. 370).

What's more, "Standards are not only technical questions. They determine the technology that will implement the Information Society, and consequently the way in which industry, users, consumers and administrations will benefit from it" (CEC, 1996). That is, those that develop ICT standards today at the same time shape much of the ICT environment we all will use in the future. And if they do their job properly, i.e. if they develop systems that meet the needs of all stakeholders (including e.g. individuals, communities, businesses, and governments) society at large stands to benefit.

Also, the macro-economic importance of standards is no longer questioned. Swann (2010) provides a very thorough review of the relevant literature. The reported findings include, among many others, that e.g. "Standards contribute at least as much as patents to economic growth" (according to DIN, 2000). In line with a number of similar studies[1], DIN (2004) finds that "standardization contributes to GDP growth at the rate of about one percentage point per annum."

From a firm's perspective, standardisation is perhaps not always a matter of life or death, but success in standards setting may well have a significant impact on the economic well-being of a company. This impact may materialise through different channels. For example, a proprietary technology may be 'ennobled' by becoming a standard. This, in turn, may imply increased revenues due to, for instance, a faster diffusion of the technology (see e.g. Stoneman & Diederen, 1994). Moreover, a standards-based technology supports the emergence of complementing products or services. This, in turn, increases the value of this technology (see e.g. Andersen & Fjeldstad, 2003).

Alternatively (or additionally), an organisation may capitalise on IPR that has been incorporated into a standard and for which licensing agreements may be made. Perhaps less obvious, but in many cases at least as important, a standard may help extend existing markets or even open up new ones (see e.g. den Uijl et al., 2013). Along similar lines, a company may use standards setting to change the rules of the game and to get a competitive advantage. Lemstra and Hayes (2009) report a case where NCR successfully did that by initiating the development of a wireless local are network standard (to become known as WiFi or IEEE 802.11) rather than emulating the market leader's (IBM's) technology.

Unfortunately, there is a downside: Even twenty years ago the costs for the development of an average IT standard amounted to about $ 10,000,000 (Spring & Weiss, 1995). Another estimation puts development costs for a 'major international telecommunications standard' to some 1,000 person-years of experience, twenty person-years of actual effort, plus $3 million (OTA, 1992). Thus, a company's involvement in standardisation requires careful consideration to avoid costly and perhaps ruinous mistakes.

The observations above might lead to the assumption that companies take standardisation very serious. One might also be tempted to assume that the ability to be successful in standards setting is about as important as the ability to innovate. One could also believe that sophisticated management mechanisms are in place (at least in large companies) to help make sure that a company's interests are adequately represented in the international standardisation arena. Yet, at least anecdotal evidence suggests that this is not necessarily the case.

In fact, some companies (e.g. Intel, Qualcomm, Ericsson) indeed seem to have a well-defined standardisation strategy and associated management structures. Yet, apparently quite a few others seem to have adopted more of a laissez-faire approach. Some have a central standardisation department, for others it is more of an un-coordinated and distributed activity. While some manage, monitor and track the standardisation activities of their staff, many others do not seem to have a clue of who is doing what in their name in which standards bodies. To make things worse, earlier research (see e.g. Jakobs et al., 2001; Jakobs, 2011) suggests that those active in standards working groups do net necessarily always represent their employers' interests.

Literature dedicated to the topic has been thin on the ground. In the 1990s, the topic received quite a bit of attention under the label of 'Strategic Standardisation Management' (SSM). SSM is "A management discipline and methodology that investigates all aspects of standardization across a business and/or industry, then defines, recommends, and implements appropriate strategies and policies to leverage standardization so that a firm can gain competitive advantage and avoid disadvantage" (Betancourt & Walsh, 1995). Also in the 1990s, the American National Standards Institute (ANSI) ran the 'Center for Strategic Standardization Management' to assist companies that aimed at adopting this approach.

Basically, SSM argues that standards may have a considerable impact on a company's business. Therefore, active participation in their development should be seen as a strategic tool that needs to be carefully managed and to be aligned with the company's overall business strategy in the relevant sector (see e.g. Forselius, 1998). Accordingly, a dedicated central entity needs to be charged with the man-

Figure 1. A very simple view of the entities, factors, and processes that influence a standard (adapted from Jakobs, 2015)

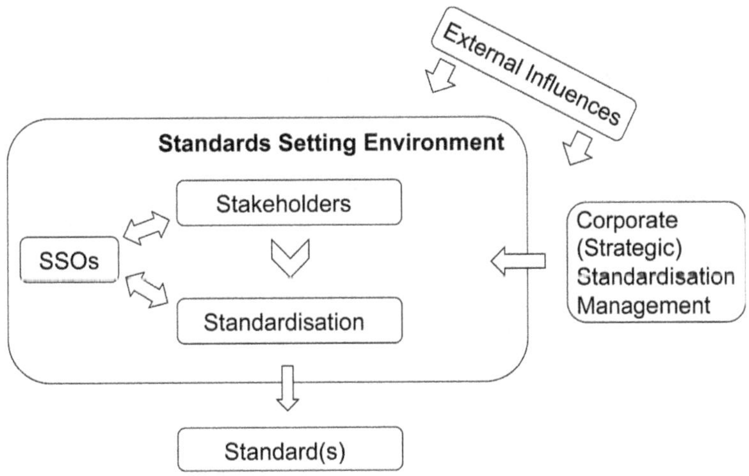

agement of standards and standardisation. Specifically, standardisation must not be considered just a technical issue, neither should it be associated only with the engineering domain. For example, process standards like the ISO 9000 series define affect all departments.

Along similar lines, Bousquet (2003, p. 52) notes that "The 'standardization in the enterprise' function, like quality must clearly take place at a management level and the experts who will participate in the standardization work will he the true spokesmen for their firm's strategies".

More recently, van Wessel (2010) developed a framework for Information Technology (IT) standardisation management. This framework, however, focuses exclusively on company-internal standards, i.e. either those that have been developed in-house or that have been selected from a set of candidates. Yet, in general standardisation management covers a much broader range of tasks than that, most notably perhaps the selection of those Standards Setting Organisations (SSOs) in whose activities a firm wants to participate, the definition and application of the criteria by which SSOs are selected and the guidance of these activities.

Jakobs (2015) came up a simple model of the entities, factors and processes that influence the development of a standard (see Figure 1). While hardly the most important factor overall, an adequate standardisation management is the most crucial success factor for each firm.

All in all, the situation regarding standardisation management in both industry and the academic literature is not exactly satisfying. After the 1990s, only very sporadic references may be found. This is a bit alarming given the overriding importance of (interoperability) standards, especially in the ICT sector.

This book sets out to improve this situation. Its 16 papers (14 contributed, 2 invited) shed some light on a vast array of topics ranging from business benefits via relevant IPR issues and resistance to change to ethical aspects. It will—hopefully—be of interest and help to both (prospective) scholars in standardisation research and management/organisation studies (and probably a few other subjects) as well as practitioners in the corporate standards departments and for everyone who would like to establish such a department in his/her company.

The first invited paper is entitled "How Corporate Standardization Shapes Tomorrow's Business" and was written by Alice de Casanove and Isabelle Lambert. As Telecom Faculty Leader at the Airbus

Defence and Space Engineering Academy and chairperson of ISO Technical Committee 279 on Innovation Management and TC 279 secretary, respectively, Alice and Isabelle are extremely well placed to address the many aspects of standardisation management. Their paper illustrate how a company can embrace standardisation and leverage it to increase its position in a competitive business environment. Inter alia, they address questions like "How to establish a standards department" and "How to benefit from standardisation". And that's precisely what standardisation management is all about.

The other invited paper was written by Carl Cargill, a well-known veteran of both corporate standardisation management and academic standardisation research. These days, he is Principal Scientist, Standards, at Adobe Systems. The paper's title "Standardization, Not Standards, Matters" shows that for him the process of standardisation is the important bit – "Standardization is one method of controlling a market". In particular, Carl argues, it is much more important than the resulting standard, which he perceives as "relatively sterile documents." Against this background, he also bemourns the lack of an adequate education in this field.

The above papers were written from a practitioner's perspective. The other papers, briefly outlined below, approach various facets of standardisation management from a slightly more academic perspective. That said, when reading the papers you will probably find out that this distinction is more or less artificial. This is not least reflected in the fact that many papers are co-authored by practitioners and academics.

The contributed papers are subdivided into six sections. The first one comprises two papers on "The Business View." Robert V. Wessel, Henk de Vries, and Piet Ribbers look at "Business Benefits through Company IT Standardization." Their study highlights that good governance and management of each of the phases of the life cycle of such an internal a standard (i.e. selection, implementation and use) will contribute the most to the success of a company-internal IT standardisation project. In the second paper, "The Role of Internal Standardization in Business Models: An Activity Configurations Perspective," Magnus Johansson, Amalia Foukaki and Matts Kärreman look at the role of internal standardisation in business models. It highlights those areas in which company-internal standardisation is prominently associated with activity configurations. Using a set of value configurations as representations of generic business models the authors develop a typology that shows the relevance of internal standardization in this context.

Section 2 is on "Developing Standards and Processes." Here, Anne-Marie Großmann's, Vincent von Gruben's, and Luisa Kim Lazina's paper entitled "Strategic Development and Implementation of Company Standards" provides companies with a toolbox of possible strategies for the development and implementation of company standards and identifies the differences in the economic effects of company-internal standards and formal ones. This should help supply-chain partners to best deploy company standards.

Looking at standards development from an entirely different angle, Erwin Folmer and Jasper Roes describe "The Pressure Cooker Approach for Open Standards Development." This approach aims to reduce the development time for open data standards while maintaining a level of quality adequate for practical (pilot) implementations? The paper shows that this can be done. Yet, it also shows that there are disadvantages and that the pressure cooker will not always yield immediately useful results.

The introduction and deployment of project management processes for an engineering consulting company's small- and medium-scale projects, based on the new ISO/IEC 29110 standard, are discussed in the paper "An Innovative Approach to the Development of Project Management Processes for Small-Scale Projects in a Large Engineering Company," by Claude Y. Laporte and Frédéric Chevalier. This standard specifically targets Very Small Entities (VSEs) and thus aims to minimise the overhead that keeps the vast majority of VSEs from using process management standards. The development and implementation of project management processes based on ISO/IEC 29110 has proven highly successful and will be spread further throughout the company.

All that glitters is not gold. Despite the many potential benefits standards may have there may also be some resistance against their introduction. Thus, the two papers of the Section "Dealing with Resistance" try to show up ways how such resistance may be avoided.

In his paper "Standardizing Knowledge Work: How 'Home Grown' Standards Overcome Resistance among Professional Employees," Kurt Sandholtz analyses the views of highly autonomous knowledge workers on attempts to standardise their work. Using a case study approach he shows that the attempt to impose ISO 9000 processes in a top-down fashion on an engineering division led to stiff resistance on the side of the engineers. In contrast, engineers in another division internalised the standard practices before being ordered to do so and have willingly, in cases even enthusiastically, adopted the processes

Ries Haverkamp and Henk de Vries address the same problem using a more philosophical approach. In their paper entitled "Managing In-Company Standardisation while Avoiding Resistance: A Philosophical-Empirical Approach," they apply Dooyeweerd's theory of modal aspects to get a better understanding why employees may resist company standardisation initiatives. This enabled the authors to identify the set of causes for resistance to standardisation and to develop an approach to overcome this resistance.

The environment within which both businesses and standards function is governed by a huge set of laws, regulations and directives. In the context of ICT standards and standardisation management, Intellectual Property (IP) laws and policies are most prominent. While other legal aspects obviously also play a role, all three papers of this section on 'Legal Aspects' discuss IPTR-related aspects, albeit to different degrees and from different angles.

For starters, Jorge L. Contreras and Andrew Updegrove offer "A Primer on Intellectual Property Policies of Standards Bodies." The paper summarise the key IPR issues that are likely to appear in standards bodies' IPR policies. The authors argue that company managers need to develop a solid understanding of these issues in order to make informed decisions regarding their organisation's participation in standards bodies

The next paper, "IP and Electric Vehicles Standards: Local Policies vs. Global Standards? Standardisation Management in a Multi-Stakeholder Environment in China," authored by Martina Gerst and Gao Xudong illustrates the current status quo in Europe and in China in the Electric Vehicle (EV) sector. It also outlines if and how IP policies influence standards management in this filed and discusses the policy, legal, and socio-economic factors that shape EV standardisation and its management.

Looking at EU competition law as well as at IPR issues, Justin Pierce's paper "Standardization Commercial Strategy: An EU Antitrust Perspective." Taking four recent cases as examples, the paper shows that commercial success in the context of standards and standardisation may well have very negative legal ramifications for a firm. The author argues that there is a clear need, for legal standards and enforcement policies to be formalised and structured to provide legal certainty how far a firm's freedom to structure commercial strategy and business practices goes especially where standards are a factor.

The link between standardisation and innovations remains to be not fully understood. Today, the view that standards may both foster and hamper innovation and that the 'fostering' element is more important seems to have reached some consensus. Yet, many issues continue to be debated. The two papers in the section "Looking at Innovation" contribute to that debate.

The paper "Standardization, Innovation, and Organization: A Contingency Perspective," by Nizar Abdelkafi and Sergiy Makhotin shows that internal organisation of a firm's standardisation function may differ considerably. The analysis of four successful companies with an organic, bureaucratic, mechanistic, or simple organisational structure suggests that the organisation of the standardisation function depends on three contingency factors: competition, size and strategy. The paper also argues that innovation is a by-product of the fit between the internal organisational structure for standardisation and the environment. Still, it remains difficult to postulate a general rule that defines when and how companies benefit from standardization as a driver for innovation.

Looking standardisation and innovation from a completely different vantage point, Antonios Kaniadakis's paper on "Standardisation of Information and Financial Innovation: Lessons from Mortgage Securitisation" argues that the effects of technological standardisation of innovation in the mortgage industry should be explored beyond the usual narrow focus on efficiency and profit. Rather, the broader social and policy context should also be considered.

The latter paper already suggests that a broader view should be adopted in the realm of standards, standardisation management and innovation. Taking this insight one step further and one (or rather, several) steps back the two papers in the this section, which is entitled "Two Sides of the Same Coin," look at ethics in standards setting (the step forward) and at standards wars (the steps back), respectively.

Starting with the latter, Tineke Egyedy and Anique Hommel analyse "Predatory Strategies in Standards Wars: On Creating Fear, Uncertainty, and Doubt." The use of predatory strategies by firms is not completely unheard of. Typically, they are deployed by dominant firms to fend of potential new marker entrants. The authors show that such strategies have also been deployed in standardisation. Specifically, behaviour like discrediting, misrepresentation and legal actions. To do so, they use a case study on the standardisation of public safety networks in Europe.

The last paper, entitled "Ethics and Standardization: A Promising Team for Designing and Operating Human-Centered Systems" and co-authored by Hartwig Steusloff and Michael Decker, represents a first attempt to combine the algorithmic view at technical systems with an ethics-oriented view at the human being as part of complex systems (comprising both technical systems and people). The basic difficulty of this approach results from the dichotomy between the fuzzy and highly individual features that inform and guide human behaviour and the predictable algorithms of the technical world. The paper provides some ideas for research and for standardization in such a heterogeneous environment (e.g. a smart city).

REFERENCES

Andersen, E., & Fjeldstad, Ø. D. (2003). Understanding interfirm relations in mediation industries with special reference to the Nordic mobile communication industry. *Industrial Marketing Management*, *32*(5), 397–408. doi:10.1016/S0019-8501(03)00013-0

Betancourt, D., & Walsh, R. (1995). The evolution of strategic standardization management. *StandardView*, *3*(3), 117–126. doi:10.1145/226191.226211

Bird, G. B. (1998). The business benefit of standards. *StandardView*, *5*(2), 76–80. doi:10.1145/301688.301691

Bousquet, F. (2003). Dealing with standardization: Do you need a guru? Paradoxes and tricks of standardization management. In *Proceedings of the 3rd Conference on Standardization and Innovation in Information Technology*. Piscataway, NJ: IEEE. doi:10.1109/SIIT.2003.1251194

CEC. (1996). *Communication from the commission to the council and the parliament on 'standardization and the global information society: The European approach'*. COM(96) 359 final. Retrieved from http://eur-lex.europa.eu/LexUriServ/LexUriServ.do?uri=COM:1996:0359:FIN:EN:PDF

den Uijl, S., de Vries, H. J., & Bayramoglu, D. (2013). The rise of MP3 as the market standard: How compressed audio files became the dominant music format. *International Journal of IT Standards and Standardization Research, 12*(1), 1–26. doi:10.4018/jitsr.2013010101

DIN. (2000). *Economic benefits of standardization: Summary of results*. Beuth Verlag GmbH and Deutsches Institut für Normung. Retrieved from http://www.din.de/sixcms_upload/media/2896/economic_benefits_standardization.pdf

DIN. (2004). *Economic benefits of standardization: Summary of results*. Berlin: Beuth Verlag GmbH and Deutsches Institut für Normung.

Forselius, R. M. (1998). *Strategic standardization management and competitive advantage*. (Unpublished Doctoral Dissertation). University of New Haven, New Haven, CT.

Jakobs, K. (2011). How people and stakeholders shape standards – The case of IEEE 802.11. In J. Filipe & J. Cordeiro (Eds.), *Web information systems and technologies 2010* (pp. 1–13). Heidelberg, Germany: Springer. doi:10.1007/978-3-642-22810-0_1

Jakobs, K. (2015). Corporate ICT standardisation management – Lessons from the literature and from case studies. In *Proceedings of the 7th Annual Conference of the EuroMed Academy of Business*. Nicosis, Cyprus: EuroMed Press.

Jakobs, K., Procter, R., & Williams, R. (2001). The making of standards. *IEEE Communications Magazine, 39*(4), 102–107. doi:10.1109/35.917511

OTA. (1992). *Global standards: Building blocks for the future*. TCT-512. Washington, DC: Congress of the United States, Office of Technology Assessment. Retrieved from http://www.princeton.edu/~ota/disk1/1992/9220/9220.PDF

Spring, M. B., & Weiss, M. B. (1995). Financing the standards development process. In B. Kahin & J. Abbate (Eds.), *Standards policy for information infrastructure* (pp. 289–320). Boston: MIT Press.

Stoneman, P., & Diederen, P. (1994). Technology diffusion and public policy. *The Economic Journal, 105*(425), 918–930. doi:10.2307/2234987

Swann, G. M. P., & Lambert, R. (2010). Why do standards enable and constrain innovation? In J. C. Gratz & K. Jakobs (Eds.), *Services standardisation: EURAS contributions to standardisation research* (Vol. 2, pp. 357–376). Aachen, Germany: Mainz Academic Publishers.

Updegrove, A. (2006). Standards stakeholders: Who should, and who does, set standards? *Consortium Standards Bulletin, 5*(4). Retrieved from http://www.consortiuminfo.org/bulletins/apr06.php#feature

van Wessel, R. (2010). *Toward corporate IT standardization management: Frameworks and solutions*. Hershey, PA: IGI Global. doi:10.4018/978-1-61520-759-6

ENDNOTE

1 A collection of such studies may be found at http://www.iso.org/iso/home/standards/benefitsof-standards/benefits_repository.htm?type=EBS-MS.

Section 1
Invited Papers

Chapter 1
How Corporate Standardisation Shapes Tomorrow's Business

Alice de Casanove
Airbus Defence and Space, France

Isabelle Lambert
AFNOR, France

ABSTRACT

Corporate standardisation function coordinates and supports experts involved in standardisation bodies. It sets up Standardisation strategy in Alignment with global strategy of the company. In this chapter, the authors illustrate how a company can embrace standardisation and leverage it to increase its position in a competitive business environment. The authors first present an overview of corporate standardization strategy: how is standardization considered in a private company, how to capture the ins and outs of a corporate standardisation, how to define, implement and manage standardization strategy, etc. Then the authors present the benefits of standardization development for a company. Finally an example of commitment in ISO TC 279 innovation management is presented.

INTRODUCTION

Because standards frame our daily life, because standards make the world safer and more interoperable, because standards try to solve babel tower issues, they Impact strongly businesses.

When the word "standard" is pronounced in a company, a lot of questions arise on the actual scope and acceptation of the word standards. Actually, a lot of concepts are covered under this concept: does "standard" make reference to internal standards, policies or a regulation or a certification, product standardisation (especially in lean management), interfaces standards (technical specifications…). Standard can also be the assessment of the customer perception. We note there are common and usual thinking about standards: about their usefulness, representatively or applicability. Henceforth, it is relevant to report on what is perceived behind each of these notions; and analyse how corporate standardisation manage these (sometimes wrong) perceptions of standardisation purpose.

DOI: 10.4018/978-1-4666-9737-9.ch001

Defining every used term can be considered as an obsession of standardisation staff but understanding how the word "standard" is used in business is mandatory to develop a standardisation development strategy. In the first part, the authors sum up typical perception of standard and standardization activities in a company. Then they recall actual objectives of standardisations, how to set up a standardization corporate function, how to efficiently benefit from standards to implement a company strategic plan. Finally, the authors introduce the current work on the standardization on innovation management. In each section, a set of questions are raised and answered to guide the reader through all the reasoning and questioning associated with the set up of a standardization activity.

HOW IS STANDARDISATION PERCEIVED IN A COMPANY?

"Standards" have different declination in a company. They all revolve on the concept of defining and agreeing upon a set of practice so as to improve business or facilitate communication. Here is a collection of the notion of "standards", ranging from internal practice up to customer-oriented needs.

Internal Standards or Policies

Companies use internal policies to harmonise their internal practices, so as to facilitates interaction across and among their departments. Even if these policies are internal, they have a huge impact on company's external image. Indeed, as these policies regulate the whole company life, they create the values of the company. The values conveyed by a company can become only marketing words to create an image if they do not have any footprints on employees and their internal processes; or become iconic and representative of a set of values customers want to share (and buy).

Setting an internal policy has many impacts on other policies up to the whole organization of a company. For example if one of the company values is green compliancy (ecology, preservation or resources, etc.), then one can expect there will be an internal policy specifically for this topic. But all other policies gravitating around this topic will be impacted (e.g. general procurement, company building) and an emphasis on this topic will be put. Generally the creation or the update of new policies is accompanied with a communication plan to aware employees. Hence, company policies are a bridge between management standards and their implementation in the company.

Certification

Certification moves internal policies one step further and confirm to external parties – customer or business partners – which particular policies are effectively in place in a company. A typical example is the ISO9001 standard that is being associated with certification activities. As a matter of fact, since the 1980's most of the company wants or has to be ISO9001 certified given its importance in quality assessment process. As the needs for quality were huge, the ISO9001 certification has been an amazing success. CQO (Chief Quality Officer) function has been created to manage this topic and create a quality culture across companies.

However, this strong link between a standard and its certification process blurred the lines on the actual intent of standards. They are now mostly considered as a topic for quality department, and induce regular audits. But only few standards generate a certification process, and standards use is always a deliberate process except for the ones references in regulations.

Regulation

The general public connect standards to regulations; they prescribe acceptable or allowed practices in an industry. These are a part of "de jure" standards, enforced by national laws. They are usually transcribed from general agreement at various scales. They are defined and enforced so as to increase the safety of goods and services, as it is the case for instance in the food industry and the management of temperature for dairy products, or safety in airplane and airport.

Yet, they are perceived as a set of complex norms limiting the business development. People are generally complaining about these regulations, stating that "everything is standardised" like if originality or out of the box thinking was killed by a unified and harmonised world. But most of the time this unification is a fertile soil for innovation development: it eases testing and rapid prototyping; it generates new ecosystem and increase the interaction between different organisations by creating a common set of knowledge and objectives. Regulation is also beneficial to the final customer: it increases trust in a product, e.g. safety regulations for food or transport.

Product Standardisation

Standard applied to goods and products complement industry-level regulation. It is a way to rationalise product life management: efficient management of configuration instead of "make-to-the-measures", savings on big volumes of material to assemble products, communication rationalization based on standardised family products, etc. A company with an efficient standardisation development strategy can make these products become the reference for the market. Large Internet-based companies demonstrated how to enforce such product standardization for smartphones or social networks, and have it accepted by its customer base.

Customers

Conversely to product standardization, customers may also favour some products. Again, the telecommunication industry showed how products could be massively adopted. When a product is successful and adopted by a large part of the population, it becomes a reference, a standard "de facto". If a new product does not fulfil the same needs as the "standards" it would be disregarded by the market and considered as below the expectations. These standards are not necessary "open" but they are at least product specifications than can be translated in a performances standard. Products or services compliant with a service are considered as safe and secure and with up to date technologies.

From Informal Standards to Standards on Standards

In the previous list, we summarized regular occurrences of standards in daily or professional life. They arise in various aspects. For instance, business standards are also key elements of request for proposal or call for tender, they speed up the creation of these Request for Proposals (RFPs) and guarantee a good communication with the potential providers. These are in place to establish customer-provider relationships across most industries, but also between state operators and the industry

Similarly, interface standards provide technical specification, creating interoperability among different systems. Hence, cars can be refuelled thanks to standards in oil industries defining characteristics of gasoline. When standards are not widely adopted they bring inconvenience, for example the electric outlet

International standardization that created multiple heterogeneous form factors.

We have presented several aspects of standards. Let us note that most of the standards are used in a deliberate way and are not supporting any regulation, but instead increase business value. If they are used wisely, they may represent a strategic asset in a company roadmap. In the following sections, the authors will discuss international standardisation and the management of this standardisation process. The corporate Standardisation department usually operates this function. In the following, the authors do no discuss internal company standards, as these are not externally visible.

WHAT IS THE REAL PURPOSE OF STANDARDISATION?

Standardisation is a voluntary and deliberate process of developing specifications based on consensus among all interested parties (industry including SMEs, consumers, trade unions, public authorities, etc.). They shape our daily live by providing minimal performance levels for services, such as safety or energy; management policies for organisation; or support for interoperability like in telecommunication.

Standardisation process is regulated by independent standard bodies, acting at national (ANSI – US, BSI – UK, AFNOR – France, DIN – Germany, etc.), European (CEN) or international (ISO, ITU, etc.) levels. They provide standards by law, which can support regulations (example in European market with CE marking).

As standardisation is based on consensus principle, it involves several levels of discussions at national, regional and then international levels. National standardisation bodies harmonies position of national stakeholders. Later, this position has to be defended at international level (regional or intergovernmental levels). Corporate standardisation has to consider this geopolitical landscape in Standardisation strategy.

Figure 1. Organization of an international standardization body

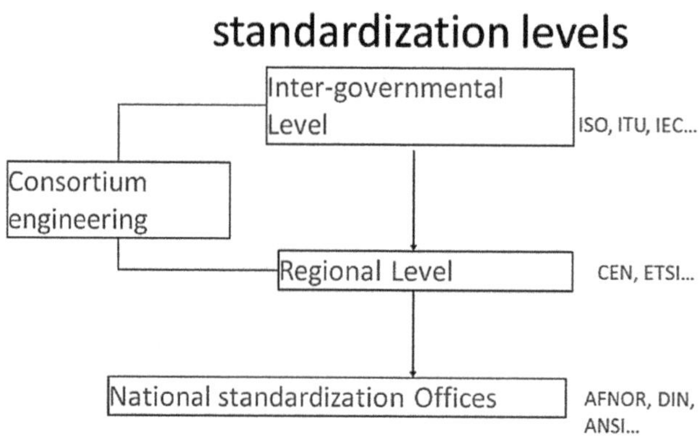

The organization of a standardization body follows the following pattern: the steering committee gives the long-term strategy of the standardisation organism and represents authority of standardisation process. All the technical committees report to the steering committee and are focused on a given topic. A technical committee sets the global vision, defines the framework and coordinates the different working groups that pertain to its area of interest. Finally, the working group gathers the experts and develops standards – its technical content.

It can be also be carried out by an industrial or domain-focused consortium. In this case, such standards would not be recognised by the states but if they are widely used they are become de facto standards adopted by the industry, like ECSS standards for the European Space standards; IETF or ECMA for the computer and network industry.

What are the Different Kinds of Industrial Standards?

There are different Kinds of standards, with different objectives:

- Fundamental standards are terminology standards, their primary objective is to build a common understanding for a given domain;
- Specification standards are technical specifications. They define interfaces standards, or performances standards; and the associated requirements for interoperability;
- Test and methods standards focus on ensuring a minimum level of quality or safety. These are achieved through defined testing methods;
- Management standards provide guidelines for organization, they precise how decision chains are managed through recommendations or requirements.

These terms are defined below and the impacts on standardisation strategies are described.

- Terminology standards are generally the first step of standardization. Sharing the same vocabulary is mandatory prior to any works. It clarifies the concept behind each word and then obviously limits misunderstanding behind two parties having different backgrounds. This first step is also necessary to define the standardisation perimeter: it will support the development of the standards documents but also mobilise the stakeholder. Indeed as the terminology is one of the first documents published, this document gives visibility to the working group and then generally new members are joining.
- Technical specifications like interfaces standards concentrate on technologies and technical works. All the stakeholders have to be around the table to debate, share, and establish relevant specifications, which would be widely adopted. This family of document requires the involvement of an efficient rapporteur to drive the agenda, moderate the discussions and coordinate the writing activities.

They are big stakes behind this kind of document, each company or consortium of companies wanted to see his in-house technology adopted by the community to capture any potential market, including significant portion of key patents necessary to implement the standard.

An efficient rapporteur will be able to keep the fundamental direction and make a standard that can be widely adopted by the community. Its key mission is to maintain the energy and commitment of all standards developers. Hence, there is a difficult but mandatory balance to find.

- Performance standards describe the minimum set of performance that a product or a service may fulfil to be compliant. This family of standards can also crystallise some market frictions. A company wanting to penalise a competitor may propose (and see enforced) unreachable, non-relevant or out of scope performances.
- Management standards are nowadays often known as management system, and support companies in the creation of their internal policies. They facilitate the relationships between different partners for co-development, research (common technical interfaces), procurement relationship (decision chain clearly identified), or supply chain management (common quality control) etc. They are also a powerful tool to disseminate good practices, and to improve the efficiency in organisations. Let us note they are also mandatory for consultants: these management standards support and legitimate their activity

How to Select the Kind of Standard to Develop?

As the authors underlined in the previous section, standard can have a dual role of market enabler, and competitor neutralizer. Hence, corporate Standardisation strategy has to define the type of standard to develop. This mostly depends on the market maturity: new technology or new management method requires a clear terminology to guarantees a efficient development market position; to become leader on a market a performance standard can eliminate less valuable competing solution.

Most of standard practitioners acknowledge that standards are nothing else than tools in a toolbox. Hence, one has to understand the purpose of each standard element, and how they can be components of a company toolbox. Furthermore, standards are embraced by the community only if they are efficient. The screwdriver with a star pattern is generally less used than the flat or cross screw because most of the screws have a flat or cross pattern. That is the same for standards, if the document does not fulfil a need efficiently; it is forgotten in the history oubliettes.

A standard is worthless without a community promoting it, using it, and maintaining it. The core of the community has generally created the seed of the standards and has grown in parallel of the development of the standard. But assembling a set of high-level experts around the table is not a guarantee of success: all the stakeholders have to be involved in the standardisation process. Generally the less represented party is the end users, even if at the end they contribute to the adoption (or not) of a standard.

Standards require also a lot of resources to be developed. Yet, one can note that standards are emerging from everywhere. The standardization landscape has become some kind of a jungle. This creates risks for a company to miss new business opportunities emerging from standards (think USB stack taking over most traditional communication buses), or updates that would Impact performances of new products (e.g. certification authorities changing regulation, for instance the FAA allowing twin-jets for long haul flight across oceans), or disrupt an established market.

Hence, it might be difficult for companies to identify where to put some efforts, resources and how to get involved. So as company management resources cannot be deployed on every standardisation field, a selection has to be done and effort put on standardisation have to be coordinated and optimized.

As a matter of fact, big companies have set up during the past decades a department for "Corporate standardisation". This Corporate Standardisation department ensures consistency in company participation to Standardisation bodies. It has to ensure that the company business and technology roadmap benefit from external standardisation works, and the return on investment of these activities is acceptable to compensate fees, delegated personnel, etc. Nowadays, this function exists in big group as well as in start-ups and SMEs. In the latter case, it can be merged with innovation CIO or CTO functions.

In the next section, the author emphasises the main activities and role of the stakeholders in the construction of a Standardisation department in a large group.

FIRST ACTIONS FOR A NEW CORPORATE STANDARDISATION DEPARTMENT

When a corporate Standardisation department is created, there are two key actions to lead: 1) depict the landscape, e.g. the standardisation bodies impacting company business, relevant standard in preparation or already in place; 2) and know who goes where: who are the experts already involved in standardisation activities. Generally this department reports to CEO in the frame of institutional relationships functions or report to CTO in the frame of engineering functions.

Several inputs are required to map relevantly the standardisation landscape: company global strategy, technology roadmap, competitive landscape, and product portfolio. Based on these data, the corporate standardisation team can identify relevant standardisation bodies, technical committees and working groups to monitor or to contribute to. This identification can be supported either by the company experts who are aware of the well know standards, or by the national standardisation bodies who are in charge of awareness actions.

Awareness and networking are key elements in the definition of this strategy. Hence, national standardisation bodies are easy to identify, but private and consortium initiatives may be less visible. It is highly advised to follow newsletters from national, regional and international standardisation offices, and also attend awareness events organised by actors and supports of standardisation. Also, standardisation coordination is also a matter of networking. The more your network across standardisation body is spread, the easier standardisation watch is.

As the landscape is complex, it may be interesting to have a visual representation. A chart or mind map assessing by influence zones the standardisation groups may help. The following picture provides an example of standardisation landscape of the European space industry. It is built across formal standardization bodies (ISO, IEC, ARINC), industry consortium (Eurospace, ECSS) in connection with other consortia like ETSI or CENELEC. All work is closely monitored by space agencies at the national or European levels, and the corresponding standardization bodies (BSI, BNAE on behalf of AFNOR, DIN, etc.). Given the various interaction levels, such mind map is crucial to share a common vision of all stakeholders, their role and prerogative in the decision taking process.

The second action consists in identifying experts among the company who are already involved in standardisation activities, or are willing to join one. This data collection is not easy and has to be always on going. Experts can be involved in multiple ways, most of the time as part of opportunities in a project. It is thus important to closely monitor any new participation, and to evaluate its potential benefits in the long term.

These data can be collected trough company internal networking. Generally top experts in a technical area are recognised and well known. Attending standardisation meetings is a way to confirm, have

Figure 2. Relevant standards setting organizations in the space industry

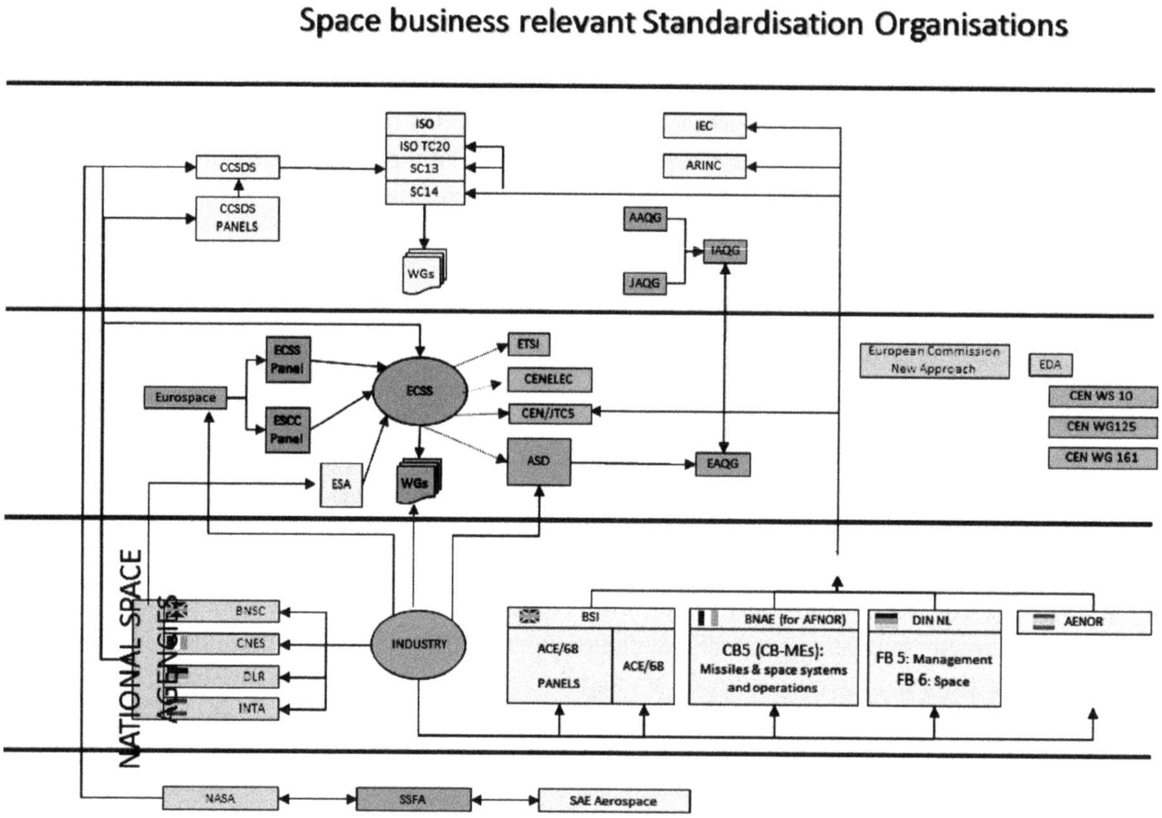

recognised and maintain this expertise. This identification work can also be made in contacting standardisation body offices; they can provide a list of people registered in their committees. At last finance department or procurement that is fully aware of the expenses related to standardisation activities can support this work.

This information is fundamental to coordinate standards development: from a management point of view several aspects have to be considered in corporate standardisation such as budget management, optimised attendance in standardisation committees, valorisation of experts (careers, training, etc.). The authors review each of these activities in the forthcoming paragraphs.

To Control the Budget Dedicated to Standards Development

There are three main approaches to manage standardization budget: centralized or decentralized budget, or a balanced mix. The three approached have advantages and drawbacks but depending on the company structure, there are preferred approaches:

- A centralized budget owned by the corporate structure will create a strict control of the expenses and the attendance strategy will be intimately bounded to the corporate strategy and will support the long-term vision of the company. Nevertheless it creates a bottleneck in the management of experts' attendance and slows down new membership in emerging consortium where it might be key to be in particular for innovation dissemination.

- A decentralized budget, which is owned by the operational level of the company, will ensure an attendance to standardization bodies developing standards that Impact today business. This is connected to the fact that operational teams generally adopt a short-term approach. The expected benefits from standards development are mainly networking (with potential customers, suppliers, or partners), market intelligence (competitors watch) and notoriety (especially for the introduction of new product into the market). Nevertheless, operational teams may cut first standards development budget in case of finance issues.
- A balanced approach, with a part of the budget in both the operational and the corporate levels, may seem a way to mitigate the pros and cons detailed previously. Yet, the main issue with this approach consists in the coordination of the experts. There is nothing worst than having two experts from the same company having opposite opinion in a standardization meeting: it duplicates company expenses for standardization activities, and most of all it creates a bad image of the company.

To Optimize Attendance in Standardization Committees

Optimizing experts' attendance to the meeting means having a clear view of the famous "who goes where?" and maintaining this mapping consistent and accurate over time. A subsequent activity is to keep it updated with a network of focal points at different operational levels (business lines, project teams, programme teams).

Generally, the corporate standardization function is in charge of the animation of this community: it has to organize regularly coordination meetings where focal point exchange information from standardization bodies works, new bodies created, experts leaving, new comers, etc. These internal communities uses also company social network tools (or intranet) to share, keep and track standardisation works.

To Valorise Internal Expertise and Develop International Influence

From a technical point of view other aspects have to be considered in corporate standardization such as company notoriety, expertise development inside and outside of the company; maintain the knowledge of company top experts, spread of technical innovation and the backward and forward compliance of company products with future standards

The best way to guarantee the compliance of today product with future standards is to turn internal standards into external standards. However, this is possible if the company has an efficient footprint on both state-of-the-art technology and standards development.

Let us recall that the standardization landscape is a kind of map with isolated islands (young consortia gathering some stakeholders of a given business) and hyper linked land. (Well-established standardization offices). If we keep the geography analogy, the role of corporate standardization is to identify relevant soil for the development of standards promoting business. Standardization coordinators are like pioneers who test the value of the soil before sending the experts who will develop the standards.

Identify Where the Company Should be

Several aspects have to be taken into account: the strength of the standardization body, the level of relevance of the standardization body with the company business and technology, the involvement of other stakeholders in this standardization bodies, the legal landscape (which kind of standards are produced: consortium standards, de jure standards, standards supporting a regulation etc.).

Standards consortia are perceived as efficient bodies but with too flexible rules to guarantee a strong consensus in the community (One usually says that "he, who has the pencil has the power"). If they gather a representative community of stakeholders they produce powerful documents, superseding de jure standards. So they are agile but might be also fragile.

On the other hand, formal Standardisation bodies (such as ISO, CEN, ITU-T…) are perceived as huge machinery, difficult to penetrate with complicated rules. Yet, they are strong enough and reliable, trying to reach a consensus, with all points of view being respected. The consensus is key to guarantee a wide adoption of the standards. It is also part of the manifesto of these bodies.

As there are several levels of standardisation development (national, regional, international), the right levels has to be identified in considering several aspects: working group dynamism (if for instance all activities are at the regional levels and no more at national ones); company strategy to identify whether company stakes are at national or international levels and then to decide commitment at the corresponding levels either as basic participants, or as a leader.

Once relevant standardization bodies are identified, working groups have to be assessed. Company generally builds a grid with its own assessment criteria such as competitiveness, working group's reactivity, works relevance, required resources, and competitor position. Nevertheless, the involvement in some working groups can benefit from a "golden token", in particular when the said working groups publish standards supporting a regulation impacting business.

To Decide Whether to Develop Proprietary Solution or Implement an External Standard

There are different degrees of involvement in standardization working groups: no involvement (proprietary solution), follower (undergo a standard), contributor (contribute to the definition of standards), and lead (contribute, create new standards, involved in steering committees).

R&D units generally consider that developing an in-house technology is three time more expensive than implementing a standard. Furthermore, standard solutions are considered safe and secure by the customers, who can considerer the technology as being open, with binding to a given provider. Finally, the manufacturer is reassured that a community shares the technical solutions; and these solutions represent the state of the art.

In-house solutions are preferred especially when they are disruptive technical choices. These solutions are generally standardised later in order to disseminate and make these solutions adopted once early adopters have recognized the value of the solution.

Hence, the degree of involvement depends on both the capability to support the development of a standard (budget for the development), and its strategic interest for the company.

Standardisation Development Budget

In most of the cases, the company reviews yearly the benefits of standardization activities during membership renewal. That's why it's important to have a clear vision of the entities in the company in charge of membership fees, but also the delegation of experts to plenary sessions.

The mission of corporate standardization is to

1. Analyse company engagements in standardization bodies, review value for business, coordinate and align contributors
2. Monitor and assess standardization bodies to identify spots where further engagement from company may be beneficial to business
3. Align company standardization strategy with company technology road map and global strategy
4. Influence key standardization bodies trough steering committees

How to Develop and Maintain a Standardisation Strategy?

Standardisation strategy is linked with the R&D roadmap and company strategy. Hence, the Standardisation strategy is the mapping of these two strategies on the future standardization landscape. The strategy elaboration consists then in 3 phases:

- Identify the future standardisation landscape, in anticipating which are the working groups or committees to be closed, the ones to be created by third parties. This identification requires being aware of the standardization Working Group activities and also steering group decisions.
- Identify the new topics of standardization that emerges in the company; and check if there is a working group, which can host this activity. If no working group exists then it is necessary to find a technical committee able to host this new working group.
- Define the level of involvement of each working groups. This is done using several criteria to evaluate adequacy to the company roadmap, such as interest for the business, competition position, in-house development, etc.

Hence, the efficiency of the corporate Standardisation strategy reflects the effectiveness of standards culture by company stakeholders, their commitment in standardisation bodies or their support in this participation, and the associated RoI. The RoI of the development of standards -- and not the usage of standards -- will be detailed in the next section.

HOW TO BENEFIT FROM STANDARDS DEVELOPMENT?

There are two main obvious motivations in standards development: support the spread of new technology or working methods, and guarantee the compliance of company products with the standards through active monitoring of forthcoming revisions. The authors want to emphasize the many other benefits: such as develop standards to enable new markets, to give visibility to existing or emerging topics, to build lifelong training, to build technology watch and market intelligence, to foster company expertise, to build company social network and branding, etc.

Business Enabler or Trigger

However, one of the key issues is to demonstrate the relevance of standardization to the company roadmap, and have it well understood by the top management so as to support it. The Standardization corporate function has to demonstrate the importance of standards to CEO and CTO top management.

The authors already presented that standards development has several added values to the company. One has to turn them into strategic elements to the top management. This is performed each time a new standard is in development: the ecosystem behind this standard has to be presented and the Impact of this standard on the business has also to be highlighted. Standards can have multiple impacts:

- Standardization can create or trigger new markets. For instance, when they introduce a new regulation. Such standard has to specify how to be compliant with this regulation. In turn, customers will update or renew their product base or systems, organisation to become compliant to the new regulation. Hence, these new constraints can be source of innovation, e.g. to meet new performance standards, and then create or renew markets.
- Standards can also enable a new market. For example Interface standards define interconnection mechanisms between systems. In making integration simpler, new business can be developed by economic actors who can enlarge their activity scopes. Standards create interfaces among different technologies and create even a new ecosystem.
- Standards also facilitate the relationship among partners. For example, a terminology standard will allow for a partnership between two companies, as sharing the same vocabulary is mandatory for an efficient collaboration. Terminology standard also favours innovation: one usually acknowledges that innovation comes from the collision of two different worlds. Should these companies have different backgrounds, these partnerships would produce new solution and lead to innovation at large.
- Performances standards can also purge a market. It gives customers a grid to assess the cost/performances ratio. If this ratio is not acceptable, a company may decide to exit this market, or on the contrary invest on it. Similarly, performances standards are scrutinized by customers to define their baseline for requirements

In the following, the authors report on three examples where standards were essential for the development of or participation in a market, and how the authors initiated the development of new standards.

- The first one is an application market place, in which images from Spot images are available on for map application, e.g. a GPS or an imaginary application. To bring updated images, we need interface standards to connect to the application provider image repository. Such general interoperability standard is in the portfolio of the OGC, the Open Geospatial Consortium, but also at ISO in the technical committee ISO/TC 211.
- Our second example is attached to regulation standards. Power plants are required to have communication means capable of working whatever the situation is. This is necessary to keep the distribution grid running, but also in case of disaster to know the situation in remote sites. In case of major disasters, part of the general infrastructure is destroyed, including land communications. Satellite-based communication is resilient to such default. They could provide an alternative solution to land lines. Yet, the industry has to show that using regular communication means and sat-

ellite communication means provide the same quality of service. Hence, a performance standard and a regulation standard have been defined: the first one describes how the performance required to support requirements for bandwidth, availability, etc.; the second one indicates how to combine satellite and land-line communication infrastructure to increase safety in the operation of power plants, and how to be compliant with these new rules of operation. This is done as part of the French BNAE standard body on behalf of the French electricity providers.

- The third example concerns certification. Let us consider issues faced by helicopters during rescue missions under bad weather condition: fog, snow, wind, etc. In such situation, the pilot requires assistance to land, but also to avoid obstacles (tree, power line, building, etc.). Using a helicopter-mounted radar, one can build in real-time a 3-D model of the local environment, and combine it with a more general 3-D map of the local area from an external database. This information can be provided through an enhanced, synthetic vision of the environment to increase the safety of the mission. In order to move forward such services, one needs to define performance standards, and associated certification process that that both the radar and the database can be used on-board an aircraft. Let us recall that the operation of aircraft is one of the most stringent standardization domains, whose main purpose is to reach the maximum level of safety in all operations. The EUROCAE and RTCA standardization bodies, in charge of the whole aeronautics industry standards, perform this work.

Hence, standards are closely related to product definition and innovation, providing various insights. First, standards can be considered as the state-of-the-art, the soil of your seed for your project. Then, when you are in development phase, you need to establish partnerships and thus you need agreement on some terminology or interfaces. Again, standards provide some elements on these topics. Finally, the authors illustrated in the previous examples how standards help organizing or shaping a market, in line with a valorisation strategy. The later usage of standard can be a strong competitive asset for a company.

Conditions to Use Standards Development as Business Trigger

A company can benefit from standardization activities only if corporate standards strategy and is aligned with the global strategy of the company, in particular with the Business development strategy and the Intellectual property strategy. It is the reason why the standards department is integrated to corporate functions. However, being in corporate functions should never imply being disconnected from operational teams. Standardization strategy has to be fed and led by company business. Market and business indicators provide insights on the degree of involvement required in standardization working group. Conversely, the attendance or involvement to standardization activities by experts from operational or business entities depends on the expected value added from operational teams.

Hence, tools and methods to capture and valorise information caught during standardization meeting have to be put in place. Information captured during standardization meeting (informal talking, or early announcement) can be shared with marketing intelligence department through regular report from standardization experts. The most relevant information for top management has also to be gathered and spread in quarterly report. It shows the importance of standardization activities but also interrogate Top management on strategy (impact of company strategy on standardization strategy).

Participation to a standardization activity is tangible evidence of company values and engagement. Also, since standardization of a topic creates buzz around a particular domain, it can be used as a communication vector to increase employees' commitment at large. They can witness the national, regional or international influence of their company. Standardization development also impacts company branding towards external customers or the general public.

Standardisation Activities Alignment with Overall Business Strategy

To assess the Alignment of standards development with the Business strategy, it is mandatory to set up in advance an internal community of practitioners, to educate this community so that its members are aware of the standards development, and can react rapidly to comment or to reorient the standards.

This internal communication is significant at key point, for instance when a milestone of the standards development is passed: the community may provide valuable feedbacks and comments. Agility is also mandatory to be an efficient actor in the standards development.

There are three main kinds of actors to mobilize to coordinate across all companies division:

- The experts in a given area, who generally attend standardization meetings, or follow remotely development of standard documents,
- The business team, capturing customers' requirements, partners' participation in standardization bodies, competitors' watch,
- The standardization coordinators, who are in contact with standardization bodies' management and steering committees.

There are different period of time for communicating with this community:

- When the community is assembled, confirming the commitments of the members is mandatory. It can be obtained through their management or the natural curiosity of experts.
- To convene this community, regular activity reports have to be sent as newsletter or update on internal corporate sites, and regular meetings to develop the commitments have to be organized.

Given the wealth of information exchange, these reports and meetings must provide minimal yet relevant information, such as: main standardisation steps, impacts of the company involvements (business and technical point of view), marketing intelligence, etc.

How to Use Standardisation to Develop Employees' Competences?

As standards address various topics most functions are impacted by standards. It turns out standardisation development can be used to develop employee's competences. As a matter of fact, standardization development and adoption impacts training for all kind of employees and in particular experts.

Experts who attend standardization committees can meet their peers in a non-competitive atmosphere. They can have open discussions on their preferred topic and challenge their views on their core expertise. The informal discussions with the other experts make their expertise grow and support technology watch

Generally company experts participating to standard development is also seen as an internal expert of the topic. Its position and role is thus increased. Hence, when a company decides to develop training in this area, this expert is a natural stakeholder, to be mobilized to participate in the realization of the learning solution. Hence, standards become a basis, frame and reference to design training towards all teams: business, legal, marketing, operational, etc.

A key issue remains: the selection of the right expert to participate actively in standards development. This choice has to be made after the decision has been taken to join a working group, and not before so as to maximize the efficiency of the decision. Actually, the following points should be considered: efficiency, visibility, and availability. One may consider the following points for selecting the right profile:

- Efficiency: working experience of 5 to 10 years or more depending on the technical domain,
- Visibility: a woman may be more visible, especially if the area is generally lead by men. Women are better remembered by WG attendees, and may have more impact in decisions,
- Availability: someone connected to both operational activities (to have enough availabilities), and to the business and company strategy.

A mission letter must be established to precise an expert code of conduct. This mission letter will frame the work of the experts in the working group. It must precisely define the limits in the topics to be discussed, recall basic principles of IP protection when considering the disclosure of technical details to be included in a standard, etc.

One of the other issues that might be solved by this mission letter is the commitment of the expert's manager. Indeed experts mission in standardisation may not be properly recognized: standardisation is generally a long-term action, time to standardize is longer than time to market with an uncertain return on investment. This mission letter will formalize the commitment of management in allocating time and budget for standardisation activities. Mission letter can also contain the different objectives that can be used as key performance indicator during employee yearly review. Hence, standardisation activities can be valorised during employee careers.

The population of standardisation experts has to be regularly trained to efficiently represent company interest during standardisation meetings. They have to be trained on three main aspects:

- To be aware of the main standardisation rules (consensus rules, types of standards, mode or operation of a standardization committee, etc.),
- To make a contribution, to obtain internal consensus to elaborate the contribution,
- To disseminate in the company information captured during meeting.

This training must allow for interaction with other domains. The authors already mentioned connections with the marketing, business development or operation teams. Intellectual Property (IPR) should also be an integral part of the strategy. IPR management department has a philosophy of confidentiality whereas corporate standardisation has a philosophy of disclosure. Nevertheless, they share a common goal to reinforce competitiveness and protect innovations. A close coordination of these both departments is essential to drive standard development. IPR may oppose the participation in a standard body if there is a risk of disclosing key technologies for which the company as a strong competitive asset to defend.

INNOVATION MANAGEMENT ISO TC 279 – CHAIRMANSHIP – BENEFITS FOR A BIG GROUP - INNOVATING IS DISOBEYING THE STANDARDS

As the authors reported in this chapter, standard are closely related to the creation of new products and markets. By extension, it is also connected to innovation, which is also a matter for standardization.

How to Boost Innovation with ISO 50500 Series on Innovation Management?

The ISO 50500 series provide an innovation management system plug-able on all existing management systems from the ISO family. It supports the setup, deployment and improvement of organizations and processes dedicated to innovation. For example, they provide a shared understanding of what innovation, innovation management and invention are, as well as tools and methods to open the innovation to other partners, assess the impact of a deployed innovation management system.

Innovation management standards are also used as a basis for innovation training, which are key to foster innovation culture. Standards and innovation are rarely in the same sentence, Innovation management is perceived as an oxymora then having a standard on innovation management is an impossible wager.

But all innovation practitioners agree on the fact that there is a need to use specific methodologies to manage innovations regarding aspects such as intellectual property, funding, marketing, dissemination, user education, which is a good incitation to build a standard.

About Chairing an ISO TC Committee

Since 2013 Airbus Defence and Space chairs ISO TC 279 on innovation management. The authors are involved in this undertaking. These standards will not standardize what is innovation but provide guidelines and good practices to support innovation management in organisations (from public research centres, universities to SMEs, big group, etc.).

AFNOR has proposed this chairmanship to different experts already involved in Innovation management standardization at French and European level. Before accepting or declining the proposition, the company compared the RoI because Company directly supports both personnel and travel costs. Strong benefits have been identified such as the image of an innovative company, the lead of the standards development. Similarly to mandate given to experts, the company prepared an internal mandate for his representative. It contains the mission to develop pragmatic guidelines to support innovation management and share the results not only in the standardization community but also in the innovation community of the group.

Which are the First Benefits?

After the 2 first years of chairmanship of his technical committee Airbus D&S confirmed some of its expected benefits. First in term of image, Airbus Defence and Space has been invited to several conferences about innovation and/or standardization and has extended his network and his notoriety. This image facilitates the recruitment of employees with an innovation mind-set and the creation of partnership with innovative start-ups. It's also a way to network and develop partnership with world top experts and also with public policy representatives in innovation. Furthermore, this notoriety in the governmental institutions makes the company an unavoidable actor to be consulted for governmental studies on national policies impact on the industry.

Also, being part of the Technical committee allows the company to make an easy and relevant benchmark on existing tools and methods deployed to foster innovation and also to be aware of new available tools. These results have to be shared inside the company through an animation of an internal community. As for expert reports, this activity summarizes standardization activity but also list emerging topics addressed in the standards development. Hence, it helps the group to challenge its good practices and also give a benchmark for a reasonable cost with other international groups or disruptive SMEs.

CONCLUSION

In this chapter, the authors provided an overview on corporate standardization activities, the main stakeholders and how standardization activities are internally connected to other corporate functions, and could lead to the emergence of new markets, new products but also innovation. The definition of involvement in standardization activities goes in high-level of management, in line with strategic roadmap; but also deeply connected to the operational activities. Such level of connectivity makes standardisation corporate functions central in multiple decision paths. Authors illustrated how to leverage this position through various examples: leveraging standards to keep or create a position; to gather experts to participate in standard development. In this chapter, the authors did not aim at providing definite answers, but rather a toolbox for setting standard corporate functions. Actually, they have to be adapted to each company culture.

KEY TERMS AND DEFINITIONS

Alignment (of Business and Standardization Strategy): A firm's standardization strategy should be designed to support its overall Business strategy – they need to be aligned.

Business Strategy: A long term plan aiming to achieve a particular goal or set of goals (in a specific sector).

Impact (of a Standard): A standard may impact a company and its business in different ways. For example, it may establish the basis for a new technology and eventually create a new market.

ISO Technical Committee (TC) 279: The International Organization for Standardization's entity responsible for setting standards in the field of innovation management.

Kinds of Standards: Standards may serve very different purposes. These include e.g. terminology standards, technical specifications, test and methods standards and management standards.

Standardisation Department: A central corporate entity in charge of managing a firm's activities in the standardisation arena. The department should have close links to Research & Development and to the Legal Department.

Standardisation Strategy: A long term plan aiming to achieve a particular goal or set of goals in the development and deployment of standards.

Chapter 2
Standardization, Not Standards, Matters

Carl Frederick Cargill
Adobe Inc., USA

ABSTRACT

This chapter looks at ways to actually use the art of Standardization (since there is not enough data to make this a discipline yet) within the Information Technology sector and within a commercial organization to effect change of some type. Note that this is not about Standards - which are relatively sterile documents - but rather a description of how to manage (some might say manipulate) cooperative action which will result, in some manner, in a Standard which can be used to create Standardization to further a defined policy, legal, social, or business management goal.

PROLOGUE

Standardization is one method of controlling a market, either politically, economically, legally, technically or in any combination of these areas. Generally, standardization is an unappreciated art because the signal product of standardization - a standard - is usually a boring document and as such, suffers from business, academic, and policy neglect. It has been observed that standards (and implicitly, standardization) are a hallmark of an industrialized society, since standards provide the necessary interconnections to make things interoperate.

Standardization can be seen and used as a social policy tool, a public policy tool, and as a business management tool. There is very little literature (academic, legal, or otherwise) that describes how to pursue standardization within a corporation or other commercial organization to accomplish these activities (policy, social control, or business management) except in a retrospective view.

The author is, and has been, involved in standardization in a multitude of software based IT companies, and this biases my view of standardization. The first and most critical bias is that, unlike many other industries, the software industry is characterized as a low capital, high intellectual property activity. It is an industry that has been characterized as "IP Intensive". Facebook, Alibaba, e-Bay and Amazon and multiple other multi-billion organizations are all examples of software based firms that started with

DOI: 10.4018/978-1-4666-9737-9.ch002

little or low capital investments and continue to operate in high Intellectual Property, as opposed to real property, investment arenas. As an example, when Alibaba wants to add another 100,000 customers, it adds a server and possibly several more systems administrators; when a brick-and-mortar store seeks to add a 100,000 customers, it needs warehouses, property, stores, and retail associates. This difference is especially critical in IT standardization, since it drives a transient standard, less influenced by process than it is by immediacy and deployment.

INTRODUCTION

Standards, and standardization, are now a fact of life in all industries globally. With the growth of the linked global economy, and with the dependence of industry on the World Wide Web and the Internet for information, the importance of standards has increased substantially over the last 20 years. While it can be argued that telecommunications (with their implicit standards and standardization) have been present since the early 1900s I would argue that the use of the web for everything from retail to supply chain management to personal entertainment devices has dwarfed the importance of the earlier telecommunications standardization activity.

Most importantly, nearly all of the Information Technology standards which are the basis of current computing are voluntary standards created by the IT industry. There is a deliberate exclusion of telecommunications standards here, since many of them, although created by a voluntary process, end up being utilized in a regulatory (or at least highly regulated) telecommunications environment. It is this distinction – the use and creation of voluntary, industry led, consensus standards as opposed to regulated or governmentally influenced standards – that has made standards less than a discipline and more of an art form, leading, in turn, to a lack of both predictability and serious academic study.

To understand the whole issue, it is first necessary to look at the document called a "standard". In this article, I will ignore "management standards" and similar types of documents, since they tend to describe best practices garnered from participants in a singular practice (quality, environment, security). What I'd like to look at is "technical specifications" – that is, documents that describe some specific technology. As an example, the following is part of a technical specification (W3C HTML 5 specification):

3.2.5.8 The Style Attribute

All HTML elements may have the style content attribute set. This is a CSS styling attribute as defined by the CSS Styling Attribute Syntax specification. [CSSATTR]

In user agents that support CSS, the attribute's value must be parsed when the attribute is added or has its value changed, according to the rules given for CSS styling attributes. [CSSATTR]

Documents that use style attributes on any of their elements must still be comprehensible and usable if those attributes were removed.

In particular, using the style attribute to hide and show content, or to convey meaning that is otherwise not included in the document, is non-conforming. (To hide and show content, use the hidden attribute.)

Element .Style

Returns a CSSStyleDeclaration object for the element's style attribute.

The style IDL attribute is defined in the CSS Object Model (CSSOM) specification. [CSSOM]

In the following example, the words that refer to colors are marked up using the span element and the style attribute to make those words show up in the relevant colors in visual media.

<p>My sweat suit is <span style="color: green; background:
transparent">green</span> and my eyes are <span style="color: blue;
background: transparent">blue</span>.</p>[1]

This is a part of a technical specification. It is meant to give instructions to software designers on how to manage the style element in the global attributes section of HTML5. It is the result of the activities of a group of technical experts who worked for several years putting together an entire specification on how the HTML 5 language was to be encoded to that it would work across multiple platforms and multiple languages.

I have chosen the HTML 5 specification for a reason. Unlike many specifications referred to by academics and others who research standards, the HTML5 spec was created by the World Wide Web consortium (W3C). W3C operates outside of the aegis of the International Organization for standardization (ISO) and the International Electrotechnical Committee's (IEC) Joint Technical Committee 1 (JTC1). It is a consortia that produces specifications at the behest of its members. It is focused on specifications and technologies that are web-centric. The consortium is of interest to and populated by developers, providers, and organizations which have an interest in the World Wide Web, whether technical, policy, or legal. And it was created and funded extensively by companies who are willing to commit significant human and monetary resources to producing web standardization.

Notice that the majority of executives in these companies never see – or even want to see – a "standard". A standard, and the arcane trivia about the process to create one, are of little or no interest to a tremendously large majority of corporate executives. Most don't care about the creation of a standard, the complexities and checks and balances of the process, the many ways to resolve conflict, or any of the other issues which so fascinate people who study standards. They don't understand the time honored process that is so beloved by standards people and standards researchers.

But, what they do understand (mostly) is their business. They understand products. They understand the concept of path dependence which is a core rationale for business to engage in standardization (see (Puffert, 2009) for a superlative discussion of the use of standards to establish path dependence). They understand interoperation, and they understand the power of joint development activities which are at the heart of standardization – in the business world. And here is where there is the tremendous disconnect between the "world of standards" and the "world of standardization".

THE PRACTICE OF STANDARDIZATION FROM A COMMERCIAL ORGANIZATIONAL PERSPECTIVE

To begin, it is necessary to bifurcate the argument for a moment into two separate streams. The first stream is a look at the actual management of the entire standardization process by the "commercial sector", and the second is the utility of standardization in a commercial setting and how organizations manage standardization to accomplish a business goal.

Management of the Standards Setting System

One of the unspoken truths about standardization is that it is structured to serve industry – the producers of goods and services. While there is always the myth that balance exists[1], the reality is that the vast majority of members of any IT standards committee are members of the IT industry or consultants who are paid by the companies comprising the IT industry. The exception to this rule (since every strong opinion must have an exception) are those committees such as biometrics or security, where governmental representation increases slightly since some governmental organizations have a legal mandate for implementing these standards. However, generally, the bulk of the participants and the bulk of the technology are contributed by the commercial segment. This is especially true in consortia, where the membership is generally comprised primarily of producers.

Complicating the issue further (and lessening the battle for balance) is the fact that larger companies tend to predominate in these organizations – or at least they tend to be the dominant voice in many of the administrative arguments. Larger companies can generally afford to populate multiple opposing or complementary standardization activities. As an example, there are numerous standardization activities occurring in the "Cloud". ISO/IEC JTC1 has an activity, as do many national bodies (with mirror committees), and there are at least a dozen or so consortia who are creating specifications (or at least marketing) around the idea of the cloud.[2] Most of these standardization organizations have representation from the major providers of cloud services and members of the IT community (except where the provider believes that they are a market dominant and have no need to participation). However, with this number of organizations, a public sector or even a NGO would be hard pressed to track all of the activities. And it in the ability to create standardization group after standardization group that the private sector - especially the companies with deep pockets – have an advantage.

The private sector engages in venue shopping – that is the use of a standardization body that is most likely to produce the results desired by the sponsoring company/companies. To complicate the discussion further, consortia do not have to model their structures on those of formal ISO/IEC/ITU standards Developing Organizations (SDOs), and it is this ability to create a structure that meets user demands (the law of "form follows function")[3] which allow consortia to differ so radically from the formal ISO/IEC standards Developing Organization (including the ISO National Bodies). As an example, the W3C is composed not just of technical standardizers, but also a host of other groups that an SDO does not have. Because the Web has social, cultural, and political significance, W3C has a Technology and Society Domain, in which activities that have social impact are gathered, systematized, and eventually, standardized. There is a Patent and standards Interest Group that looks after the issues of patents within the W3C, and is responsible for guidance of the consortia Intellectual Property matters as they relate to the Royalty Free licensing of the consortium. There are liaisons with multiple other groups (not uncommon in SDOs), but there is also a close and deliberate liaison with governments and NGOs who have an interest in the development of the World Wide Web.

The consortium – as a standardization body - has its own unique processes for operating which adhere to the general guidelines for standardization (openness, transparence, right of appeal, and so on), but it has one advantage that ISO derived bodies do not; it can refuse to take a standardization program on. Because it is an organization which has entrance fees, and those members who pay to join have the right of governance. This gives the members the right to reject (or possibly not consider) standardization outside the interests of the members. Finally, the consortium members alone have the right to pass on the goodness of a standard; it is subject to no higher appeal (excepting judicial or legal review) than the members of the consortium. There is no requirement to subject it to a national or international review. Finally, because of their very nature, consortia tend to operate as international entities. There are no German or American or Chinese national consortia standards; consortia specifications are meant to be deployed globally, usually for use by all IT producers in the IT arena.

This ability to accept or reject standardization – driven by the paying membership – and the international nature of their specifications are two of the distinguishing characteristics of a consortium. The ability to create unique processes, the right to limit or grant membership, the right to establish and mandate Intellectual Property structures and strictures on members, and finally, the right to more completely control the results of process are all specialized attributes that make consortia a preferred venue for IT standardization.

In so far as I am aware, there have been few studies of the percentage of IT standards actually being created by as a percentage of the total number of IT standards being created. The only data that I've seen suggested that over 90 percent of all standards that are being developed for the IT industry are driven through consortia – data on this would be gratefully accepted. However, to complicate this quest for quantification, there is no exact count of the consortia that exist. Absent this data base, it is difficult to actually know how many consortia standards there are[4]. The latest authoritative list was published in 1996[5]; I've seen no "official" update since then. However, I believe that the 90% number is probably moderately accurate, given the amount of continuing consortia spending that is occurring in the industry.

The privatization of standards is a phenomena that has not gone unnoticed. It was illustrated at length by Timothy Schoechle in (Schoechle, 2009). Schoechle calls this privatization by the private sector and by governmental forces (who recognize that standards are an element of international economic policy) a reshaping of the standardization world. Rebecca McKinnon (2013) sounded a similar note of concern in when she pointed out that the standards that drive the internet – indeed, all technical standards for interconnection – are a new form of legislative mandate in which the general public has little say. "The reality is that the corporations and governments that build, operate, and govern cyberspace are not being held sufficiently accountable for the exercise of their power over the lives and identities of the people who use digital networks" (McKinnon, 2013, p. xxi). She goes on to state, citing Larry Lessig, that "… software code and technical standards are for all practical purposes a new form of law, because just like laws, they shape what people and can cannot do" (McKinnon, 2013, p. 25).

If Schoechle, McKinnon, and Lessig are correct, corporations (and lately governments) have basically taken over the creation and management of technical standards – and with it, standardization. This poses two questions – how did they get away with doing this, and why did they think it necessary?

The "How" of Standards Takeover

As H.L. Mencken observed "For every complex problem there is an answer that is clear, simple, and wrong." The simple (and probably very wrong) answer lies in conspiracy theory – it was a multi-year plot by industrialists to put themselves above the law, or to at least make the law subservient to them. In reality, the gradual takeover of standardization was a complex and multi-year happening. It occurred because of a series of intertwined events happened in the industrial and post-industrial societies of the United States and Europe. It is necessary to differentiate between the IT sector and other industrial sectors her, because the IT sector was the bellwether segment from approximately 1980 onwards – but by that time, much of the groundwork for the current standardization system had been laid in the US.

Much of the credit for the standardization system as it is currently practiced in the US goes back to the unwillingness of the US government to intervene in the making of standards outside of metrology and similar commercial infrastructure standards. In the US (as much of the industrialized world) the growth of a science and technology based society replacing the rural and agrarian economy which marked the first Industrial Revolution was in turn followed by the intense innovation and expansion of the production sector in the Second Industrial Revolution. The Second Industrial Revolution saw mass production in a supply chain environment which changed the entire nature of industry. You need to ensure that your suppliers (who may also serve your competitors) provide you material that "you" need.

So, the first standards organizations were based on trade organizations, but with a technical twist. These standards organizations were based upon a particular technology knowledge set. The "experts" - usually gatherings of professional men - who were knowledgeable in a particular field (boilers, civil engineering, mechanical engineering, electrical engineering, railroads, or other applied technical discipline) gathered to pool knowledge and expand their common knowledge. They documented their knowledge with best practices or standards, specifications embodying their wisdom for the sake of their colleagues. They created societies such as the American Society of Mechanical Engineers (ASME), the Institute of Electrical and Electronics Engineers (IEEE), and The American Society for Testing and Materials (ASTM). These groups were directly responsible for technical practices which could impact their industries as a whole, and needed to ensure that the specifications which they published were correct (or at least creditable). Peer review was not only desirable, it was necessary and expected.

Douglas Puffert, in the study of the standardization of rail gauges, describes the issues that helped determine which rail gauge would be used where. One of the common activities was the use of "technical publications" to justify the choice of a rail gauge. All gauges had their learned proponents who championed their specific gauge. And all studies were used to influence a sympathetic public and legislators into mandating a particular gauge. But standardize the rails did – but not for reasons of "science". Railroads standardized because transshipment costs where discontinuous lines met finally became too much of an economic burden, and the benefits of a single gauge outweighed the economic costs of switching gauges. And this, I believe, became more and more common as industrialization continued. For a manufacturer, a standard was no guarantee of a superior product – it was merely the guarantee of a saleable product that worked with other products. Similarly, for a consumer, a "standard" product was one that met *expectations that had been set in the market* and therefore meant that you were safe to buy it. (The setting consumer expectations will be examined later).

But this collective work of standardization had a second and less obvious purpose. These standards were part of commoditizing the supply chain. By providing their suppliers – from ballast providers to tie makers to rail forgers – with precise specifications the rail lines knew exactly what to expect and how much their lines would cost. By using standards, they began to commoditize the education of the people who built the track and who ran the trains. To return to guilds – and the Ogilvie statement, railroad management "...was an enduring corporate association, usually of practitioners of a particular occupation, which was legally endowed with the exclusive right to practice certain economic activities in a certain area by *virtue of privileges granted by the political authorities.* [Emphasis the author's]" (Ogilvie, 2004, page 1). And, like the guilds, they worked for their own interests. A good companion piece to Puffert (2009) is Walter Borneman's (2010) volume Rail Rivals, which describes the business of building a railway, where the real prize was not the successful construction of rail as much as it was getting the public financing by whatever method necessary. Standards legitimized the rails and made the Second Industrial Revolution really accelerate, but it was basic greed (and the right to nearly coin money) that the leaders of the rail industry sought.

In the case of the early rail industry, standards were used when other business methods (bribery, technology, or marketing) were either inappropriate or when a different approach was needed. These organizations were based upon a particular technology knowledge set. The "experts" - usually gatherings of professional men - who were knowledgeable in a particular field (boilers, civil engineering, mechanical engineering, electrical engineering, railroads, or other applied technical discipline) gathered to pool knowledge and expand their common knowledge. They documented their knowledge with best practices or standards, specifications embodying their wisdom for the sake of their colleagues. They created societies such as the American Society of Mechanical Engineers (ASME), the Institute of Electrical and Electronics Engineers (IEEE), and The American Society for Testing and Materials (ASTM). These groups were directly responsible for technical practices which could impact their industries as a whole, and needed to ensure that the specifications which they published were correct (or at least creditable). Peer review was not only desirable, it was necessary and expected.

Note, however, that these standards were not generated by public requirements. Rather, they were generated by a self-selected group of professionals who decided – for their companies and industry – what a good solution would be. In the case of boiler standards, there was public outcry about exploding boilers (New York Times, 2015) as early as 1860. It would be one subject of the first standards the ASME would publish (ASME, 2015). But, again, it was in response to a general outcry about the results of not having the standard, not about the need for a specific standard. The founders of ASME realized that standardization would solve significant problems for them as producers as the Second Industrial revolution continued.

While this is a simplified account of how the standardization system was changed in the US, it should be understood within the context of the US industrial experience. At the time that this was occurring, the legislative process was in the midst of probably the largest laissez faire mindset seen in the industrial era. Unlike Europe and the United Kingdom, there was no history of industrial control or social good in US industrialization. These would not arrive until the reform movements of the Progressive era (circa 1890 -1920.) But by this time, the "professional society" had become established as the source for industrial standards – and the societies were created to serve their creators, not the public's, interests.

Within the US, there have been several attempts to assert more governmental control – Hoover in the 1920s and the Department of Commerce/Department of Justice in the 1960s-1970s. In both cases, the resistance to change and the promises to "self-police" made externally imposed significant reform difficult. In both cases, the reform was a restructuring and changing of process and requirements that ostensibly opened the organizations – while the parties that were most vested lost little of their clout. Generally, the standards associations that compromise the American National standards Institute (ANSI) are very sensitive to governmental influence, and maintain a significant presence on the US Capitol Hill.

However, in the mid-1980s, the US semiconductor industry came under extreme pressure from the Japanese. In searching for a way to strengthen domestic industry (while avoiding any possible inference of industrial policy), the US legislature passed National Cooperative Research Act (NCRA) of 1984 (15 U.S.C. §§ 4301 – 4305), which allowed common research and development activities to be undertaken by US industry. By 1986, over a dozen US semiconductor manufacturers had come together (with Department of Defense and their own funding) to create an organization called Sematech.

This was all well and good for a few years. Then, in 1993, a new legislative effort created the National Cooperative Production Amendments of 1993, Pub. L. No. 103-42, which amended the National Cooperative Research Act of 1984, Pub L. No. 98-462, renamed it the National Cooperative Research and Production Act (NCRPA) of 1993, and extended its provisions to joint ventures for production. Probably, there was little thought given to this particular bill by the lobbyists of the standardization agencies – it appeared benign and reasonably irrelevant to the activities of ANSI and its members. This was to be a grave miscalculation.

The NCRPA, while ostensibly focused on joint research and production ventures, was quickly recognized as an ideal tool for the legitimization and creation of standardization organizations. The ostensible reason for the creation and use of consortia was that the formal ISO based process, requiring as it did multiple layers of approval for an international standard, was too slow and too out of phase with the Information Technology industry. An international standard, from conception to completion and publication as a national standard, took up to five years, and possibly longer if the issue was contentious or novel. At the same time, the pace of technological change was increasing in the high technology industry and new models of computing began to emerge.

The Open Systems Interconnection model – the protocol to connect all devices to a common net – was undertaken simultaneously by ISO and the CCITT, which ended up being a monstrous disaster for the IT industry. It was slow, contentious, and much given to compromise. While in and of itself it was worthy, it became the poster child for why consortia were needed. And consortia began to proliferate. There were consortia for objects, operating systems, peripherals, communications schemes, structured languages, artificial intelligence, internationalization, and almost anything else in the IT panoply of technology.

The rationale for these consortia was simple. Despite the claims that consortia were faster, they probably weren't tremendously quicker. What they were was louder, and more aggressive in their standardization activities. And they caught the attention of the business and marketing community in the US and elsewhere. Standardization became a marketing tool – and the major standards bodies in the world were ill equipped to deal with the influx of marketing hype, lawyers, marketing people, and above all, publicity. Very few standards organization could then (and can, even now) describe quickly and coherently the impact of standardization on an industry they served, let alone the contribution that their organization made to the success of products and productization in that industry. Consortia, however, could, because these were the very reasons for their creation. They could make significant claims for their specifications and how these specifications would be used by the industry – because the users of these specifications belonged to the consortia.

Setting Customer Expectations

And this field of publicity quickly took entirely new meaning when consortia realized that proposing a standardized specification (or the use of standardization) could shape the market and the user expectations. At the same time claiming benefits for the use of the specification could also shape the market – and favorably incline the market to the product that claimed to instantiate the specification. In consortia, the marketers found full reign for their skill set.

And this left traditional standards bodies at a loss. They were not composed of similarly minded organizations – noir were they necessarily able to quickly produce documents that described what they were doing. All they could do was try to prepare a consensus document that described – in vague and non-threatening terms – what they were doing.[6] And, of course, it was usually too little, too late when compared with what the consortia were doing.

At the same time, the external environment was changing. The internet, and then the World Wide Web, emerged, opening the area of software creation and deployment to entirely new audiences and developers, At the same time vendor proprietary operating systems became less and less viable, being replaced by universal proprietary systems (such as UNIX), which in turn was replaced by LINUX, more completely open than its predecessors. And this change, along with the increasing popularization of computers (thanks to Microsoft and other PC makers) led to an increasing demand for software that could "do customer things" that people could use. And these applications needed to be able to run against a certain set of standards – but there was no demand for the standards to go through other than perfunctory market validation. The emphasis – as I said earlier – was on speed and deployment. The rate of change in the market drove a different set of expectations for vendors and users.

Consortia could manage to change their processes – or they could be created anew – to meet these market expectations; formal organizations could not. And this led to the consortia gaining the upper hand in IT standardization; they were able to better adapt to market needs and requirements more quickly than could the formal organizations.

However, even consortia could not meet market requirements for the speed of change and ease of evolution. As the computing environment changed with the addition of mobile apps and walled garden apps, standards development (that is, where standards were developed in more or less structured environments) has begun to decline in favor of community (either actual open source community or informal developer communities). This development – while offensive to many who believe that standards as standards should be developed in formal organizations, represents merely another evolution of standardization within the IT industry. Speed, time to product and time to profit are the drivers of this movement; large scale (or even medium scale review) is too time consuming in this environment. Additionally, most of these products are transitory (with a limited life and impact.) However, the model of develop quickly, implement, and deploy has become a new norm, and even things that becoming standards are the result of this new standardization. Things like the Google developed SPDY protocol morphed into HTTP2 in the Internet Engineering Task Force, and elements of the Webkit Open Source development activity have been introduced into various W3C and other standards.

Hence, the title of this article. To industry, the practice of standardization – that is, collective acceptance and deployment of a technology in products – is more important than are formal codified and compiled standards. The standard (and the processes that produce them) that is written and agonized over in committees and produced by complex processes is becoming an increasingly passé artifact in the Web environment. That being said, however, fundamental infrastructure standards (documented and stable) as well as hardware (connector, media, and display, for example) standards continue to be necessary and developed. However, they too are beginning to driven by the time and competitive pressures; standardization has become a competitive activity, with venue shopping and consortia creation becoming more and more common as a response to a market, rather than a technical, need. And the push of the market for faster and more standards (whether to capture a market or to protect intellectual property of to create entry barriers) is rapidly becoming "the standard" for standardization.

Managing a Corporate Standardization Effort

To begin, commercial organizations use standardization for a host of different reasons. All too often in the literature about standardization, one gets the impression participants are creatures of good will, gathering to harmoniously craft a standard that serves the greater glory of their industry, technology, and their nation.

The reality is that most professionals who are there are there to defend their corporate or business position. Consultants are there angling for business and an exploitable niche; market leaders are there to stop or mitigate the impact of adverse technology or craft barriers to entry; some companies are there to insert technology for the purpose of garnering royalties. Sometimes participants are there to "do the right thing", although this is rare. If one participates correctly, you go armed with a corporate position of goals. The goal may be to promote your technology, intellectual property, or other corporate asset. Or the rationale for attendance might be to learn what your competition is doing so that you can either mimic or leap-frog, or begin a similar or competing product/implementation/standard. Finally, you could be attending because it is in your organizations purpose to delay, modify, or otherwise impede a standard, allowing your technology/product time to survive, gain or increase market share. The politics in a standardization effort start the minute that an activity is first proposed and continue throughout the life of the activity.

The most important "thing" that a standards participant can possess when going to a standards meeting is a knowledge of their own corporate requirement for the standardization activity. If he or she knows what their organization needs to achieve in the committee, they have a basis for planning how they will respond to others on the committee. With respect to others on the committee, it is also important to know if there are potential allies who are participating, and if you are capable of "making a deal". In many instances, knowledge of the potential partner's last activity is known (in varying degrees) by all of the members of the committee or work group, so that the willingness to form coalitions or partnerships varies depending on the company, the committee, and even the participant. The potential for intrigue (and betrayal) is significant.

Once the friend(s)/foe(s) issues are settled, the contributions and participation levels of the members clarified or disguised, and the direction of the group decided, the work starts. This is where the real politics begins. Disruption, accusation, and general other disingenuous methods are used to slow a group or to disadvantage competition. Intellectual property is submitted with or without notification.[7] The committee officer roles are crucial (as might be expected), but all too often the selection of a committee chair can become a partisan issue, leading to recrimination. As the work progresses, there are various tactics that can be used to slow or stop the work or to modify it with "creeping elegance."[8]

The important take away is that standardization is not the technical love-fest of cooperating technologists. It is a business – and, as with all complex human interactions, there are always multiple agendas, multiple personalities, and multiple opportunities for unintended consequences. It would tax the skills of the three princes of Serendip to understand how a standard managed to emerge from a standardization process.

With this as a background on the way that organizations **can** participate in standardization, the question is – how often are these behaviors seen? Based on experiential data, my response would be "nearly always". There is very little actual information on this type of activity, since most of the participants either won't admit that they are doing this or they fail to see that the activities in which they participate partake of these characteristics. It is all too easy to join a coalition to modify a spec – if that specification disrupts your product planning, or your architectural vision. The problem is that if you join a standard merely to stop or "change" it, you are in fact, engaging in a disruptive and essentially negative activity.

This type of activity gives lie to the idea that standardization is a technical cooperative activity – rather, it is a multi-billion dollar business activity couched in technical guise that is used as an extension of a corporate business or marketing strategy. And, to make clear, standardization is a strategy. It requires careful planning and (usually) good execution. Because standardization fundamentally acts as a change agent, it should involve careful planning and well thought out activities. (Since the exception proves the rule, a transient commitment to a standardization activity can be used to provide temporary relief from market pressure. The benefits can be short lived if the press and industry are interested enough to track the story, however. (Microsoft's pseudo-standardization of Active X is a classic example of such a gambit.[9])

STANDARDIZATION AS POLICY

Of course, the success of businesses in regulating a market using standards has not gone unnoticed in the geo-political world. The European Union has a long history of using standardization in a significant way to smooth the functioning of the single market, for the protection of health and safety, and as a way of increasing the competitiveness of industry and the promotion of international trade. China has an even older tradition of using standardization as a unification tool, probably starting with Li Si, who argued (successfully) for weights, coinage, and character standardization as a method of unifying the Warring Kingdoms.[10]

While the use of standardization and standards is not unusual within a country or region, its use in a "global economy" creates a bit of a complication. A standardization regime that is successful in Europe for achieving unification of divergent national entities would be anathema in the United States which has a laissez faire approach to governmental interference in standardization (unless it serves to advance someone's agenda.) At the same time, standardization can be used to push an industrial objective (i.e. capturing the secretariat or chairmanship of a standardization committee such as wool, heavy earth moving equipment, or similar committees), or imposing unique IP requirements on standardization. All of these methods are used in the world as nations strive to achieve industrial parity (or possibly hegemony) through use of "non-tariff trade activities". Using standards as a non-tariff trade barrier is passé these days; it too easily caught out. Using other methods (committee control, IPR , standardization regimes that fall outside of the WTO [such as consortia]) is more opaque and more effective, since there is no effective recourse if other nations need to interoperate/interface with you. The more economically powerful the nation is, the more likely it is to be able to dominate or manage standardization to fit its own policy objectives.

The European Union has been using standards and has passed enabling legislation to create European standards Organizations and to harmonize pan-European standards. The United States created consortia with the NCRPA, at first to strengthen its semi-conductor industry and then to allow US industries an alternative to a formal standardization system. China is rewriting its standards laws to encourage "social groups" to create voluntary industry technical specifications, with the intent of creating a more innovative and open process. All of these acts are recognition that nations (and sometimes regional bodies) are using standardization in trade policy and social engineering, much as businesses have, in the past, used them to control a market.

Unfortunately, there is very little relevant precedent to guide them in their efforts. Previously, many nations could act in an independent fashion – what one did with steel or cotton depended upon market conditions internally or externally rather than whether or not the cotton or steel could interface with another set of steel or cotton. The difference in IT standards is that they are interdependent – there is no such thing as a German Internet of Things and a Chinese HTML5 or a US Cloud standard. All of these activities – the IoT, HTML5, and Cloud Computing standards – require invisibility of borders. The value of IT standards is that they allow, even encourage, borderless computing. And this is what keeps politicians awake at night. The idea that they cannot mandate either for their producers or their citizens a safe and simple solution to these complex issues which does not disadvantage their citizens or their put their careers at risk must seem grossly unfair.

The key issue is the lack of a good understanding of what "standardization", not standards, is both in practice and theory. standardization, especially in the IT fields, is an interdisciplinary, cross functional study. It is not enough to know technology and what makes a good standard. One needs to know what makes good standardization practice. This would include understanding the legal rationale for standardization across various jurisdictions, understanding the social and policy implications of a standard, understanding the market impact of standardization, and finally, the impact of standardization in the long term. All too often, the "strategy of unplanned change" (or the doctrine of unanticipated consequences) rears its head because some aspect of policy or technology or social consequence was overlooked by the originator of the activity – when it would have been easy to predict the outcome **if** there had been someone familiar with the specific discipline. And it is in this lack of interdisciplinary study and understanding that the greatest danger to standardization resides.

CONCLUSION

The major problem that I've found with standards and standardization in the 30 some years I've been involved is the lack of education about standardization. Many commercial organizations (ANSI, DIN, AFNOR, ISO)[11] are willing to tell you about how their or their processes are superior for the production of specifications; very few can tell you why or how their specifications benefit society. From a commercial point of view, they are all selling a commodity product – like nails or glue – to hold things together. It works well for nails and glue, but it really fails when you're dealing with an interface that is trying to describe how complex information systems interoperate or function.

Compounding the problem is the fact that education in this area, when it is available, rarely recognizes the cross disciplinary necessity. It's as if marketing classes were taught without any reference to how a product was to be distributed, promoted, or priced – but only on how it was to be produced. Standardization has all the attributes of a marketing plan – there is a product, a distribution plan, a price (there's always a price for standardization), and a promotional campaign. There are also competitive issues for standardization – from the lack of standardization to an overabundance of standards. But these aren't taught; rather the focus is on how to run a committee, or how to make sure the committee is balanced, or how to create an equitable IPR policy. All are important – but all are a subset of the far larger and more fascinating arena of standardization.

And hence, I repeat – it's about standardization, not standards.

REFERENCES

ANSI. (2015). *Essential Requirements: Due process requirements for American National Standards.* Retrieved from http://publicaa.ansi.org/sites/apdl/Documents/standards%20Activities/American%20National%20standards/Procedures,%20Guides,%20and%20Forms/2015_ANSI_Essential_Requirements.pdf

Borneman, W. R. (2010). *Rival Rails: The Race to Build America's Greatest Transcontinental Railroad.* Old Saybrook, CT: Tantor.

McKinnon, R. (2013). *Consent of the Networked: The Worldwide Struggle For Internet Freedom.* New York, NY: Basic Books.

Ogilvie, S. (2004). Guilds, efficiency, and social capital: Evidence from German proto-industry. *The Economic History Review*, *57*(2), 286–333. doi:10.1111/j.1468-0289.2004.00279.x

Puffert, D. (2009). *Tracks across Continents, Paths through History: The Economic Dynamics of Standardization in Railway Gauge.* Chicago: University of Chicago Press.

Schoechle, T. (2009). *Standardization and Digital Enclosure: The Privatization of Standards, Knowledge, and Policy in the Age of Global Information Technology.* Hershey, PA: IGI Global. doi:10.4018/978-1-60566-334-0

Sullivan, L. H. (1896). The Tall Office Building Artistically Considered. *Lippincott's Magazine*, *57*, 403–409.

Toth, R. B. (1996). *Standards Activities of Organizations in the United States.* NIST Special Publication 806. Retrieved from http://gsi.nist.gov/global/docs/SP%20806.pdf

KEY TERMS AND DEFINITIONS

Intellectual Property Rights (IPR): IPR s give the creator an exclusive right over the use of his/her creation for a certain period of time. In standards setting, patents are the most prominent form of IP.

IT (Information Technology): IT refers to anything related to computing technology, including hardware and software.

Standard: A standard is a document that provides descriptions of different kind to be used to ensure that materials, products, processes and services are fit for their purpose.

Standardization: The process of setting standards.

(Standards) Consortium: A private organization that develops technical specifications, primarily in the IT sector.

Standards Setting Organization: Any kind of organization that produces standards.

W3C: The World Wide Web consortium, a private consortium to develop standards for the WWW.

ENDNOTES

1 "The standards development process should have a balance of interests. Participants from diverse interest
categories shall be sought with the objective of achieving balance." (ANSI, 2015), page 4.

2 A partial list includes: European Union Agency for Network and Information Security; European Telecommunications standards Institute; Global Inter-Cloud Technology Forum; Institute for Electrical and Electronics Engineers; Internet Engineering Task Force; International Organization for standardization; International Telecommunication Union; ITU Telecommunication standardization Sector; Organization for the Advancement of Structured Information standards; Open Data Center Alliance; Open Grid Forum; Operations Support System/Business Support System; Quality Excellence for Suppliers of Telecommunications; Storage Networking Industry Association; Telecommunications Industry Association; TeleManagement Forum; The Open Group

3 "It is the pervading law of all things organic and inorganic, of all things physical and metaphysical, of all things human and all things superhuman, of all true manifestations of the head, of the heart, of the soul, that the life is recognizable in its expression, that form ever follows function. This is the law." (Sullivan, 1896).

4 ANSI, the official voice of US standardization, makes the following absurd statement: "ANSI is often asked about the total number of standards (and standards setting bodies) in the United States. It is estimated that in the U.S. today there are hundreds of "traditional" standards developing organizations - with the 20 largest SDOs producing 90% of the standards - and hundreds more "non-traditional" standards development bodies, such as consortia." Basically, ANSI is admitting that they have no idea about the size or importance of the US standards industry (see http://www.ansi.org/about_ansi/introduction/introduction.aspx?menuid=1, retrieved 8 April 2015).

5 According to data provided in (Toth, 1996), there are more than 93,000 standards produced and nearly 700 organizations. Data shown is as of 1996; newer statistics are not available.

6 It was the author's unfortunate experience to try to write a brief (two page) pamphlet describing how ASC X3 worked and why it was significant to its members. The draft was sent to the subcommittee that was to approve it. Two years later, it emerged from the subcommittee. The group could not agree that standards changed the market (this might upset entrenched interests), that standards activities required meetings (this might upset finance officers who approved travel budgets), that standards were good (this might upset those with proprietary offerings), or that standards set strategy (this would upset market leaders.) It was a shambles and taught me a valuable lesson in consensus standards when dealing with non-technical ideas.

7 Most standardization committees require that all members commit to RAND licensing of technology – but only in the final three months (or thereabouts) of a standardization creation life cycle. This late identification of embedded technology penalizes those who implement the standard before

final call – which is common practice. You can imagine someone's shock at finding that the new interface they are designing has suddenly become burdened with a royalty which had not been considered in the original business plan.

8 Technologists are especially susceptible to the creeping elegance gambit, since nearly every technologist wants to "…add that one little feature that hadn't been thought of before but which now is absolutely vital". Needless to say, creeping elegance has slowed or destroyed more standards that it has helped.

9 "Microsoft appears to be in the full-flush of damage control brought on by what Microsoft EVP Steve Ballmer last week called a naive move to thrust ActiveX into an independent standards body, reports sister paper ClieNT Server News ." http://www.cbronline.com/news/microsoft_back_pedals_activex_standardization

10 Li Si was also the official to whom the "burning of books and burial of scholars" was also ascribed, showing that there can be an occasional nasty side to standardizers.

11 I consider all standards organizations to be commercial enterprises – they sell a product, usually a specification, receiving payment in some form or another. (The exception might be the IETF, but that's only because the organization is sort of anti-organization.) They are all devoted to ensuring their continued commercial success and employ large numbers of people. They can fail when revenue fails to meet expenditures, or when the market no longer cares for their product.

Section 2
The Business View

Chapter 3
Business Benefits through Company IT Standardization

Robert M. van Wessel
Tilburg University, The Netherlands

Henk J. de Vries
Erasmus University, The Netherlands

Piet M. A. Ribbers
Tilburg University, The Netherlands

ABSTRACT

Many companies have embarked on IT standardization initiatives with specific benefits in mind, but some projects fail dramatically whereas others are very successful. Our research suggests that successful standardization projects require good governance and management across distinct lifecycle phases: selection, implementation, and use and change. We present a case study from a financial services company to demonstrate effective practices that have led to significant financial benefits, to improved service delivery and support, and to a more stable IT environment.

INTRODUCTION

Many companies try to converge on particular IT processes and/or IT products to gain business benefits such as quality improvements, cost reductions or obtaining strategic advantage (Swaminathan, 2001; Boh and Yellin, 2007; Mueller et al., 2015). These efforts can be described as standardization activities since the parties involved "*.... have the intention and expectation that the established solutions will be used within a certain period by a substantial number of the parties for which the solutions are meant.*" (De Vries, 1999, p. 162). The result of such an initiative is called a company IT standard (Van Wessel, 2010). De Vries (1999) argues that a company standard may have the form of:

1. A reference to one or more external standards officially adopted by the company.
2. A company modification of an external standard.

DOI: 10.4018/978-1-4666-9737-9.ch003

3. A subset of an external standard (for instance, a description of the company's choice of competing possibilities offered in an external standard, or a subset of the topics covered in the external standard).
4. A standard reproduced from (parts of) other external documents, for instance, suppliers' documents.
5. A self-written standard.

We define a company IT standard as: "A specification of an IT product or process to be repeatedly and consistently used in the company" and the company IT standards in this chapter relate to category 4 in the above list.

Typically, a company standardization process encompasses a number of sequential steps: selection, implementation, and use (including changes and withdrawals) of the standard, which together comprise the lifecycle of the company's IT standards. These internal IT standards are not necessarily restricted to formal standards created by official standard setting organizations, but may also include standards set by consortia or even specifications of propriety products and processes. Some of such standardization initiatives fail dramatically whereas others are very successful, and the reasons are not clear. Companies have to make choices among numerous IT products and processes to arrive at company standards, but how should they do so effectively and efficiently? Who should be involved? How should they plan and control? How should they measure their effects? What are the pros and cons, and the costs and benefits? This paper aims to find empirical evidence of the business impact of a company's IT standard and of the effective governance and management mechanisms for successful company standardization initiatives.

Since the 1980s, scholars have studied the economic aspects of standardization, such as network effects and switching costs (Van de Kaa et al., 2007). The majority of standardization studies focus on the effects of IT standards on a macro-economic scale (Blind, 2004; WTO, 2005), on the development of standards by industry, consortia, and international standards bodies (Backhouse, 2006; Nickerson and zur Muehlen, 2006; Teichmann, 2010; Jain, 2012), and on battles between competing standards (overview of studies in Van de Kaa et al., 2011). Others apply the diffusion of innovation theory (Rogers, 2003) to the field of standardization (Poba-Nzaou and Raymond, 2011), or a combination of diffusion of innovation and economic theories (West and Dedrick, 2006; Mendoza and Ravichandran, 2011). However, the number of academic studies on standardization in companies is limited and fragmented. The professional literature on IT standards seems to have adopted an almost exclusively technical point of view.

One of the classic problems facing standardization and standards usage in companies is demonstrating its contribution to the company's total success (Hesser and Inklaar, 1997). Typically, in standardization there are significant uncertainties about the factual costs and benefits and about adequate planning and control strategies (Weitzel, 2003). Kayworth and Sambamurthy (2000) show that the organizational context in which IT infrastructure standards are used is an important success factor with respect to the satisfaction of specific local needs and the degree to which the standards are integrated in the whole company. Swaminathan (2001) describes the issues that companies face when they consider mass customization to meet the needs of their businesses. He identifies four operational strategies for standardization employed by firms to minimize the increase of variability in the operating environment. These include part standardization, process standardization, product standardization, and procurement standardization.

De Vries and Slob (2006) investigate a 'best practice' for company standardization at six chemical and petrochemical industries in the Netherlands by comparing the standardization activities and, subsequently, by choosing the best way to execute them. They define success of the company standardization process as "a standard that is known to the users and that is used in practice" and identified factors that positively influence the use of such standards. Despite the difficulty of measuring the real usage of a standard, they identify a set of best practices for company standardization, grouped under the headings Standardization Policy (at strategic, tactical and operational level), Prioritization Process, Company Standard Development Process, Company Standard Introduction Process, Distribution Process, Facility Management, and Funding.

In a survey research on the use of such standards, Boh and Yellin (2007) investigate to what extent the use of IT infrastructure standards facilitates organizations to improve the sharing and integration of IT resources across the enterprise and how different mechanisms affect their usage. Their results show that the use of enterprise architecture standards is significant in helping organizations to effectively manage IT resources. Hesser (2010) presents the effects of company standardization on competitive and functional strategies in design and development, procurement, and production with direct relevance to industrial processes. Company standardization includes: numbering systems, size range systems, unit assembly systems, modularization, quality management systems, and environmental management. However, company standards are often only introduced after cost or complexity targets have been missed. He argues that company standardization must be part of corporate and competitive strategies. Li and Chen (2012) examine company standardization within corporations and its impact on sellers' incentives to invest in IT compatibility. Their research suggests that exclusive purchase commitment by a company is dependent on the degree of horizontal differentiation among sellers, its product compatibility and the relative competitive advantages. Meuller et al. (2015) describe factors that influence the intention to accept and use company standards at an employee level, and Manders (2015) investigates the implementation and impact of ISO 9001 at the country, company, and employee level.

Because the number of scholarly articles on company standardization is so limited, our paper investigates how organizations can achieve intended business benefits from company IT standardization. This research expands on this scarce theoretical base and uses insights from practice. For this purpose, case study research is an appropriate research method (Yin, 2009). We present a case study from a financial services company, ABN AMRO. Data were gathered from ABN AMRO's headquarters, and included official company records such as project plans, project reports, presentations, policy documents, memoranda, and leaflets. We evaluated the financial, organizational, and technical objectives of the company and examined their effects on efficiency and effectiveness. We conducted semi-structured interviews, lasting an average of 90 minutes, aimed at discussing the selection, implementation, and usage of IT process and/or product standards. The interviewees ranged from senior executives to IT experts. Most interviews were taped and subsequently transcribed within 24 hours. Only a few interviewees did not give permission for the interview to be audio-taped. The interviews focused on how the bank could achieve the intended business performance from IT standardization, and in particular on the governance and management practices across the entire lifecycle of the company's IT standards.

The term *governance* (Weill and Ross, 2004) in relation to company IT standards refers to which decisions must be taken, who should take these decisions, and how they are taken and monitored to ensure the effective management of the standardized IT environment. *Management* (Boynton and Zmud, 1987) means actually taking decisions on planning, organizing, directing, and controlling the company IT standard. The results of the IT standardization project have led to significant financial benefits, but also better IT service delivery and support. For this study, we use a company IT standardization management framework, as depicted in Figure 1, which was successfully validated in a number of case studies (Van Wessel, 2010).

Figure 1. Company IT standardization management framework

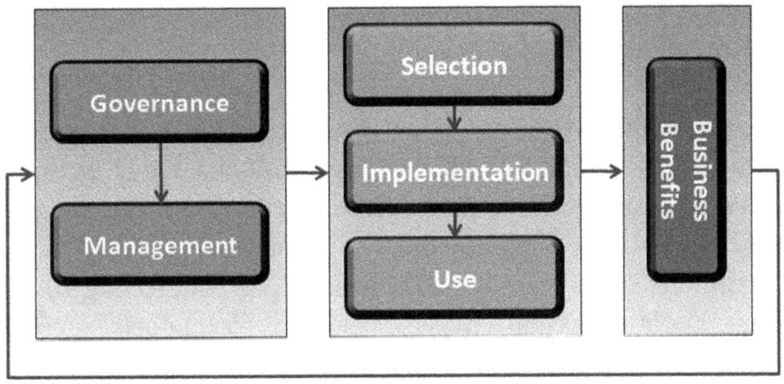

Governance and Management impact Selection, Implementation, and ultimately Use of company IT standards, which should result in Business Benefits. The arrow from Governance to Management indicates that adequate management presupposes effective governance. The feedback loop from Business Benefits to Governance and Management symbolizes the dynamic character of the framework. If Business Benefits are not achieved, changes to governance and/or management of company IT standards should be considered. (Box 1)

Box 1. "Intermezzo"

To illustrate the feedback loop in the framework, we provide an example from the same financial services company related to Enterprise Resource Planning (ERP). ERP often involves customizable, standard application software, which includes integrated business solutions for core processes (e.g., production planning and control) as well as the main administrative functions of an enterprise (e.g., accounting and human resource management).

Dey et al. (2009) discuss a number of benefits of ERP systems. However, adopting standard application software does not necessarily mean that such benefits will be achieved. They use a five-stepped risk management procedure to identify risk factors. Key success factors for ERP implementation are commitment from top management, selection of the appropriate systems, and proper management of its integration with existing systems.

In this specific case, the company was unable to reach agreements on the terminology (i.e., definitions of Human Resources (HR) terms). The corporate HR department selected an ERP HR suite that included core HR processes as the company standard. The main driver for this choice was the lack of consistency of HR information in existing HR systems, which made it impossible to carry out proper analysis and reporting. HR administration had been based on dispersed spreadsheet-like tools and often lacked historical data.

The standardization process, especially with regard to data semantics, turned out to be ineffective. The process was limited to a relatively small set of entities because no enterprise-wide agreement could be reached and because some business units did not fully support the company standard. Local HR buy-in was not stimulated, and no standard way of implementing the HR Information System was enforced, which resulted in ad-hoc deviations and a number of partly incompatible versions, Moreover, the roles and responsibilities of corporate HR and local HR units were not well defined or effectively assigned. The main issues were: 1) lack of full mandate from corporate HR; 2) lack of implementation obligation; 3) lack of ownership of HR data entities. To make matters worse, the implementation process took too long and the anticipated scope of the deliverables was not met.

After rigorous changes were made to the way the company standard was implemented, the anticipated effects of applying the company standard were achieved. The feedback loop in Figure 1 (from Business Benefits to Governance and Management) represents the changes made. These included: 1) senior management senior management fully supported the company IT standard; 2) the scope of the project was made mandatory; 3) ownership of HR data elements was clearly assigned at corporate level. This resulted in mandatory adoption of the standard, an 80% standardization of HR processes and data entities, and no possibilities of ad-hoc deviations. As a consequence, the quality of reporting improved, legacy HR systems were decommissioned, and changes could be carried more easily. A more detailed description of this case can be found in Van Wessel et al. (2006).

ESTABLISHING COMPANY IT STANDARDS AT ABN AMRO

Rationale for a New IT architecture

Our case study was carried out at ABN AMRO, a Dutch financial service company with a presence in 21 countries worldwide. At one of ABN AMRO's business units, the IT environment consisted of a plethora of different hardware and software products. Managing this environment was difficult and resulted in high support costs and long resolution times. Upgrades were difficult to implement due to lack of standardization. Therefore, a two-year IT standardization project affecting 10,000 users was carried out in this business unit. The business unit's management set project targets for cost savings and also set functional requirements, which made the project was business-oriented rather than technology-focused. The scope of the project, which included the choice and implementation of hardware and software at both the front-end and the back-end of the business unit, ranged from desktop productivity tools to cost accounting applications.

The main objectives of the standardization project were to reduce the total cost of ownership or TCO[1] by 18% (calculated against industry benchmarks) and to increase flexibility by implementing a set of company standards. Cost reductions were particularly needed in the fields of procurement and support, and were expected to be achieved by reducing the complexity of the IT environment by limiting the variety at both the hardware and software levels. Greater flexibility was needed to allow for 'hot desking', and to roll out changes seamlessly. The project was completed successfully in time and with a marginal cost overrun.

The next sub-section provides a discussion of the successful accomplishment of this company IT standardization initiative.

Selecting the Company IT Standards

Governance

The business unit's senior management initiated the standardization project. It had two main reasons to do so: increasing flexibility and reducing costs. These two goals had to be reached by rationalizing the number of IT applications used by the business unit, and by simplifying their maintenance. These objectives, in turn, were to be achieved through standardizing the IT infrastructure. The business unit and its IT department worked in close cooperation to specify and select the IT products in the new set of company IT standards. However, this rationalization and standardization process was not easy. It required much effort, and it took time to convince all stakeholders of its added value. At the start of the initiative to standardize the IT environment, the IT department asked the business unit what kind of functionalities it required (not which applications it wanted). The answer was a 300-page document with requirements that would have cost a significant amount of money to implement. To deal with this issue, the business unit asked the IT department how the IT environment could be rationalized. IT suggested to script applications for browser-based access if technically possible and financially feasible, and to introduce terminal servers for applications that cannot be ported.

Management

Management developed a three-tiered approach to devise the company's new IT standards: hardware standardization, system software standardization, and application software standardization. Alternatives were carefully considered for each IT product, and choices were based on a combination of functionality and costs. The selection process consisted of three steps: 1) request for proposal focusing on functional and non-functional requirements; 2) price negotiation by reversed auction; and 3) acceptance testing of two products, to determine the preferred one.

Hardware standardization: To decrease complexity and allow hot desking (with the aim of reducing the number of staff workplaces), an environment known as 'server-based computing' was chosen. It consisted of three main elements:

- Thin clients (PCs with a minimum of local applications)
- Web servers and terminal servers hosting applications
- Back-end servers hosting all user data

This set-up minimized dependencies among the system's hardware components. The new environment consisted of 10,000 thin client workstations, 1,000 laptops, 1,000 terminal servers, a dozen of web servers, and around 300 back-end servers. User and group data were stored on the terminal servers which were replicated daily to a central storage.

System software standardization: Managing the original desktop environment was difficult since it consisted of a large collection of locally installed applications on fat client desktops, and implementing changes could take as long as three to four months. Whenever technically possible, therefore, the new system software was not installed on desktops. Instead, users used browsers to access applications from their PCs. The browser-based applications ran on the web servers in a multi-application hosting environment on an AS/390 mainframe using products called Websphere, and Java Virtual Machine. Applications that were not browser-based were installed on terminal servers running on Wintel, also using Websphere. To minimize interdependencies between these logical nodes, no multi-application hosting was allowed on terminal servers.

Application software standardization: The 'server-based computing' approach was stretched as far as technically possible, but some bandwidth issues were encountered and it was decided to install a few applications on the workstations instead of on the servers to prevent network overload. Some of these applications were remaining legacy software, whereas some such as a word processor, spreadsheet, and presentation software and groupware were used daily by the majority of staff. After inventorying, the number of applications was reduced from more than 6,000 to 265 – more than twenty-two^2 times less!

Interestingly, the biggest savings were achieved by a reduction of license fees. In the past, several hundreds of licenses had been purchased – some for very few users or even no users at all – simply because no-one could keep track of the installed software on thousands of locally run PCs. So the guiding principle in this standardization project phase was that only one type of software was allowed – preferably the latest version – unless it considerably degraded business functionality. In a tendering process, proposals were requested from several suppliers. The key criterion for selecting applications, and subsequently listing those as preferred software, was the balance between functionality, support, and license costs. But other criteria, such as manufacturer or vendor strategy and track record, were also taken into account. The lists were drawn up per department, subject to approval of business management. The final set of specifications of IT products, the company IT standards, is listed in Table 1.

Implementing the Company IT Standards

Governance

A program organization carried out the implementation of the company standardization initiative (Figure 2). Project groups with a specific assignment reported to a program manager, who reported to a steering group that consisted of members from the business unit and the IT department. The steering group was accountable for managing project costs and progress, for approving project changes, monitoring the quality of the implementation through progress reports, deciding on organizational changes, and controlling the type and number of applications. The program manager was supported by a project office.

Table 1. Set of standard IT products (from a user perspective)

IT Products	Remarks
Desktops and Laptops	Multi-language MS Windows and Office; Web browser; PDF Reader; Winzip; Lotus Notes; Anti-Virus Software
Monitors	17" LCD
Printers	Network printers (no personal printers allowed)
Work-at-home facilities	Browser via Internet/SSL, replacing most company laptops
RSI prevention tools	Special computer mouses and software
Smartcard readers	Integrated with keyboard
Specials	On demand, like scanners, or Braille for the poorly sighted

Figure 2. Program organization

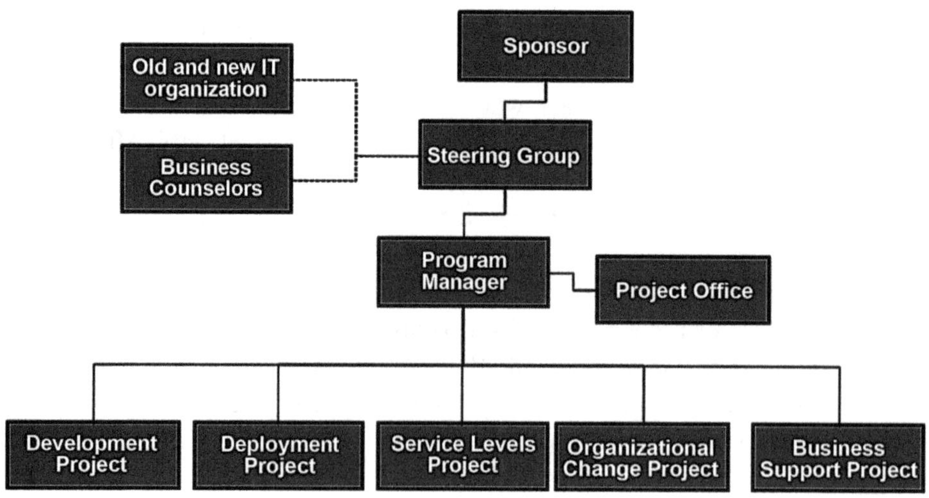

Management

To manage the complexity of the standardization initiative, the program involved many people, processes and technological changes, five separate projects were set up that each covered specific aspects of the program (Table 2). A total of 51 FTEs (Full Time Equivalent), including the program manager and the project office, were involved for 19 months.

Implementing the set of standards was vital to achieve the required flexibility of the new hardware and system software. Because of its modular structure, with only one hardware and software type allowed, the new IT environment had maximum flexibility at both client-end and server-end. Strict adherence to project management processes was another key element in the successful implementation. Moreover, the project was carried out as a joint effort by the business unit and its IT department, which increased the buy-in at both sides.

A number of potential risks were identified that could cause the project to fail in the implementation phase. These consisted of organizational and HR risks because the IT support organization had to be dismantled and its staff relocated in order to set up a new centralized IT support department. Also financial risks were taken because the estimates of operational revenues and expenditures were exactly that: estimates. The most important project risks are listed in Table 3, including the countermeasures taken to prevent these risks.

Table 2. Projects of the server-based computing program

Project	Objective	FTEs
Development	Preparing applications for terminal servers; scripting, conversion, testing	29
Deployment	Rolling out the new infrastructure in each department	8
Service Levels	Setting up service level agreements with the business unit	7
Organizational Change	Managing the IT department's organizational change caused by the project	3
Business Support	Counseling and liaising with the business departments	2

Table 3. Project risks, impact, and countermeasures

Risk	Possible Impact	Countermeasure
Organizational change is carried out without any consideration to staff.	IT staff morale deteriorates, which is reflected in lower service quality.	Pay special attention to staff involvement and open and honest communication.
The business unit's management shows no commitment to the project.	Business units do not commit to the new company IT standard. The costs reductions and flexibility required are not achieved.	Maintain close contact between the business unit's managers and the company's general management.
The reduction of the number of applications is not achieved.	Potential benefits of scale are not fully exploited. More licenses and terminal servers are needed. Support remains costly and complex.	Continuously monitor and enforce the maximum number of applications.

Use and Change of the Company IT Standard

Governance

An IT review team, (Table 4) consisting of four IT staff with clear responsibilities in governing the use of the set of standards, was appointed. The IT product coordinator was the primary point of contact for functional changes. Their core tasks were to assess requests for changes and deviations from the standards, and to make recommendations to an IT policy board.

This IT policy board approved changes and deviations to the set of IT standards. Approval was based on four policies:

1. The application level and the operating system level should be uncoupled to ensure minimum dependence between developments at these two levels.
2. Upgrades to the computing platform must be possible without large investments to update in-house developed applications.
3. Only COTS[3] products are allowed. This prevents legacy applications from remaining operational, which would cause security and stability problems, and high support and license costs.
4. No business-specific applications are allowed on workstations and workstations must be completely de-personalized to ease maintenance and increase security.

When a deviation request was granted, there were two options. The preferred one was to incorporate the specification of the requested product into the set of company IT standards. The alternative was to grant the deviation, but only on a temporary basis. After a year, the IT review team re-evaluated the deviation request.

Conformance testing of proposed IT products (e.g., by individual staff or by projects members) to the applicable IT standard by specific IT teams is a practice that was found in other companies as well (Rada & Craparo, 2000; Boh & Yellin, 2007). The review team also assessed any upgrades, replacements, or patching needed. A full review process was carried out every two years. This often resulted in a new product, the specification of which was then incorporated into the set of company standards, reflecting developments in technological and business environments.

Table 4. Staff responsible for managing the company IT standard

Function	Accountability
IT architect	Preserves IT infrastructure consistency and evaluates the overall impact of deviation requests.
IT product manager	Manages costs and charging the business departments, reduces expenses, and improves cost transparency. Calculates the financial impact of deviation requests. Accountable for all assets, budget-responsible for infrastructure deprecations, and owner of all support contracts.
IT product coordinator	Translates functional and technical requirements into specifications of IT products and assesses deviation requests from a technical point of view.
IT support coordinator	Plans and controls the IT operations and reviews change requests from an IT operations viewpoint.

Management

There are several ways in which company standards can be enforced once they are implemented (Cargill, 1989) and these can be classified into three categories: the regulatory style, the *laissez faire*, or a combination of these two. This business unit adopted the first style, which was expressed in their maxim 'each and every desktop must have the same configuration'. Consequently, the company IT standards were strictly applied. As discussed earlier, the business unit's objectives were to achieve cost reductions and increase flexibility. The business unit's IT department fully supported these goals, but realized that the costs of IT delivery and support would sometimes not be a convincing argument for business staff. In some cases, business departments were willing to pay substantially more, rather than less, for products that did not conform to company IT standards. The deviation process, therefore, had to be very strict to ensure that only requests with a sound business rationale would be made. Any such requests were then assessed by the IT review team and approved by the IT policy board, taking both business and technological considerations into account. Very serious arguments were needed to convince this group to allow deviations, and these would then have to be reported to IT management on a monthly basis.

The choice was limited to the applications and hardware that were specified in the list of company IT standards. Changes to these standards were relatively easy to make because of the modular structure of the new IT environment. But anyone requesting modifications to functionality first had to try to realize those within the company's current standards. Failing that, the IT team would only make structural changes on the basis of a sound business rationale with indicators such as added value, the number of users affected, the possibility of charging someone for it, the vendor's track record, and the total costs involved. Planning and controlling the company IT standards was the task of staff as listed in Table 4. Strict adherence to "IT service management processes", such as those specified in ISO/IEC 20000, was another key element in using the company standard successfully.

BENEFITS ACHIEVED

The new standardized IT environment was assessed by analyzing the project's financial results and by evaluating operational aspects of IT service delivery and support obtained from the Finance unit.

The Financial Results

With the introduction of the standardized IT environment, many costs were eliminated. The most important savings were achieved by reducing the number of licensed applications, by removing local installations resulting in an easier process of application support, and by eliminating hardware relocations (formerly approximately 25% of total desktop costs). As a result, the costs for IT support and development were considerably lower than before the standardization initiative. In addition, less specific staff expertise was needed, and re-use was easier. The resulting economies of scale enabled the company to negotiate significant global purchase discounts from its suppliers. Staff used their web browsers to acquire any additional applications needed on their thin client. The extent to which they used them determined the variable costs of those workstations. Authorization was obtained through an application also used for license management, which gave the business unit maximum transparency and cost control.

The project's investments ran to a total of €32 million (€17M hardware and software transitions costs and €15M project staff outlays), which was 5% more than budgeted. The direct costs of the old and new IT environments, calculated as of January 1st of the year of the project and again exactly one year later, fell from € 4,600 per desktop per year to € 2,392 (Figure 3). Calculating on the basis of 10,000 desktops and a four-year life span, the payback period[4] was 1.45 years. The return on investment rate[5] was 176% and the internal rate of return[6] 58%. Figure 4 shows the cash flow for four years. The initiative resulted in a positive cash flow of + €56M in the fourth year. The bottom line was that the new IT environment costs were reduced by about 50%.

IT Service Support Benefits

Significant organizational improvements were achieved, from decreased time to process service requests to the implementation of the standardized IT environment for a whole department. New projects showed lead time decreases of up to 75%. New applications were made operational within a few weeks and deploying the standardized IT environment typically took no more than a month. So service performance increased from both the development and the support point of view. The area in which no significant improvements were achieved related to error and rework rates, as the former organization was already responsive in this respect.

Figure 3. Direct costs per desktop per year (before and after standardization)

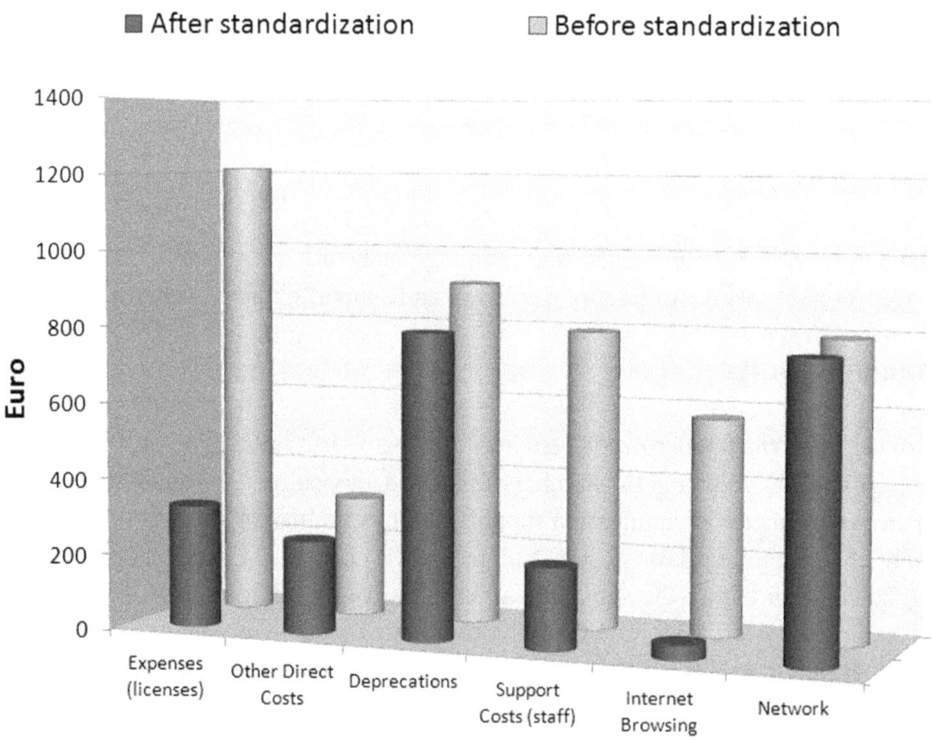

Figure 4. Cash flow following investment in the standardized IT environment

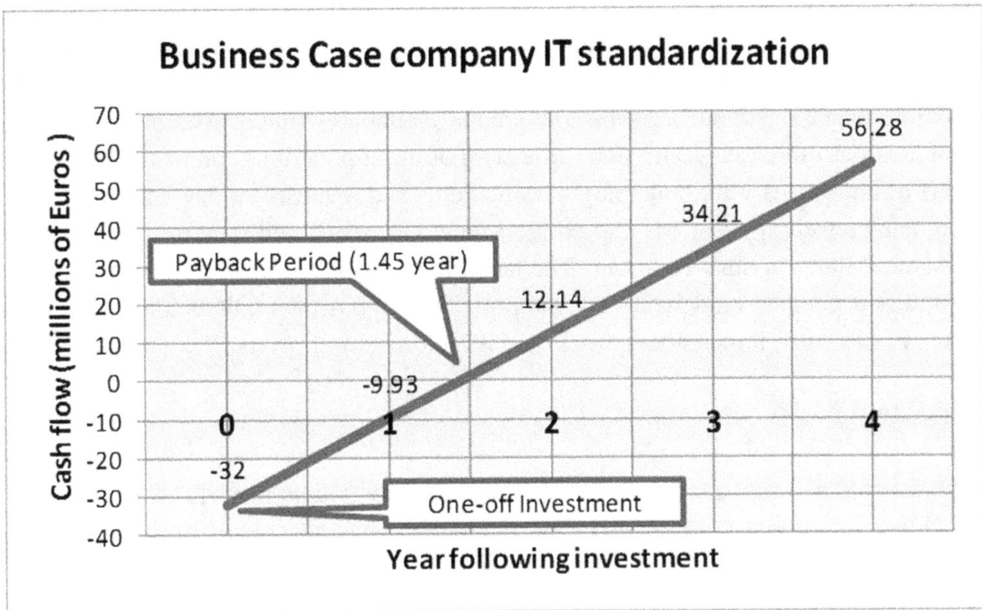

In addition, the information security was improved through automated anti-virus updates and the uniform patching of the applications and operating system. Smartcards were used for application single sign-on. Desktop downtime became negligible since instability was predominantly application-related, not hardware-related. Moreover, the impact of IT failures dropped because of the modular setup of server-based computing with one application per terminal server. Altogether, the new IT environment proved to be very robust.

The standardized IT environment facilitated improved technical, organizational, and financial flexibility. Technical flexibility was achieved in terms of adaptability, scalability, and robustness. The ATMs used by customers of ABN AMRO, for example, were incorporated in this environment as essentially just another peripheral. Furthermore, the IT environment allowed several versions of an application such as Lotus Notes 6 and 8 to run simultaneously on a single desktop, because the software was installed on separate and independent terminal servers. Organizational flexibility was exemplified by a department using a dedicated Local Area Network. The new IT environment allowed the department's staff to work anywhere, irrespective of their physical location. This meant its business functionality could be spread over several locations. Financial flexibility was achieved because application and hardware usage were charged based on the actual number of subscriptions. Users were charged for access to the system's web and terminal servers, so when a department reduced or increased its staff, the IT costs changed automatically. Of course, this kind of flexibility caused some financial uncertainty for the IT organization, but flexibility was considered necessary to achieve a more effective environment for users.

How the Users Perceived the Changes

The business unit staff needed some time to become familiar with the new IT environment. Initially, there were some negative reactions such as 'We will lose all flexibility...' but this changed for the better and most of the staff perceived the new environment as quite acceptable. The interviewees reported that the majority of business unit staff considered the change an improvement, 30% were indifferent, and some 10% were less satisfied with IT delivery and support than before – and those, interestingly, were predominantly IT staff. Generally speaking, the users were satisfied. With just a few clicks they could select from more than 250 applications that were made available within quarter of an hour. Staff could use more applications more easily, and since almost all of the applications conformed to a standard look and feel, staff training costs were only marginal. Incidents and requests for new hardware and software were logged using a web-application. The hot desking concept allowed staff to use workstations with exactly the same features at other locations. The fact that all workstations were the same also made the business unit less dependent on delivery and support. For the business unit as a whole, this meant decreased costs. For its staff, it meant less hassle and therefore an improved working environment.

Other Benefits

The organization learned that a strict way of dealing with deviations from IT standards proved to be worthwhile. First and foremost, it provided signs of changing business needs from users, and secondly, only genuine requests entered the process. In this case, the business unit chose *de facto* standard software products from Microsoft. It was aware of the danger of potential lock-in effects by being bound to a single supplier of hardware or software. But this presented no real problems, although adaptability and interoperability became more difficult when applications that deviated from Microsoft's products had to be integrated. However, these lock-in effects were considered far less important than the benefits. Hardware seemed to present almost no such risks at all. And it was considered even less important for system software, since the modular set-up of the environment allowed changing front-end or back-end software relatively easily.

LESSONS LEARNED

Governance and Management Arrangements

This case study showed us that the standardized IT environment resulted in several advantages:

- an up-to-date set of IT products, aligned with business requirements and technical developments
- significant cost reductions and improved transparency and control
- satisfied customers through improved service delivery and support
- deliverables of projects in conformance to the set of IT standards
- a stable and fully modular IT environment with better reusability and change flexibility

The following governance and management mechanisms facilitated these changes.

Governance

Selection: The business decision to standardize the IT environment was taken in collaboration with IT, with business departments taking the lead. Key objectives were cost reductions and increased flexibility. This resulted in a formalized organizational setting: an IT policy board with senior representatives from the business unit and from IT.

Implementation: A program management organization was established. Key players included a program manager and a steering group with members from both the business unit and the IT department.

Use: In the operational phase, an IT review team carries out refinement and updates of the standard with input from business and IT. This team assesses specified IT products and processes for projects, and evaluates changes and deviation requests to the set of IT standards. Decisions concerning the standardized IT environment are taken on the basis of functionality, not on specific software packages. Four key players were appointed to ensure the effective management of the standardized IT environment: an IT architect, an IT product manager, an IT product coordinator, and an IT support coordinator. In addition, an IT policy board approves or rejects changes and deviations by taking business and technological considerations into account. Ratification is based on infrastructure consistency policies. Approved requests are expected to be incorporated in the standardized IT environment.

In terms of archetypal IT governance decision-making (Weill and Ross, 2004), the arrangement described above is an 'IT Duopoly': "two party decision-making involving IT executives and one group of business leaders". This governance mechanism is in contrast to an 'IT monarchy' in which an individual or a group of IT executives are the decision makers. Our case company uses a duopoly arrangement between business units (involved mostly in the selection and implementation phase) and the IT department (involved mostly in the implementation and use phase) as shown in Figure 5.

Figure 5. IT governance arrangement in the case company: duopoly

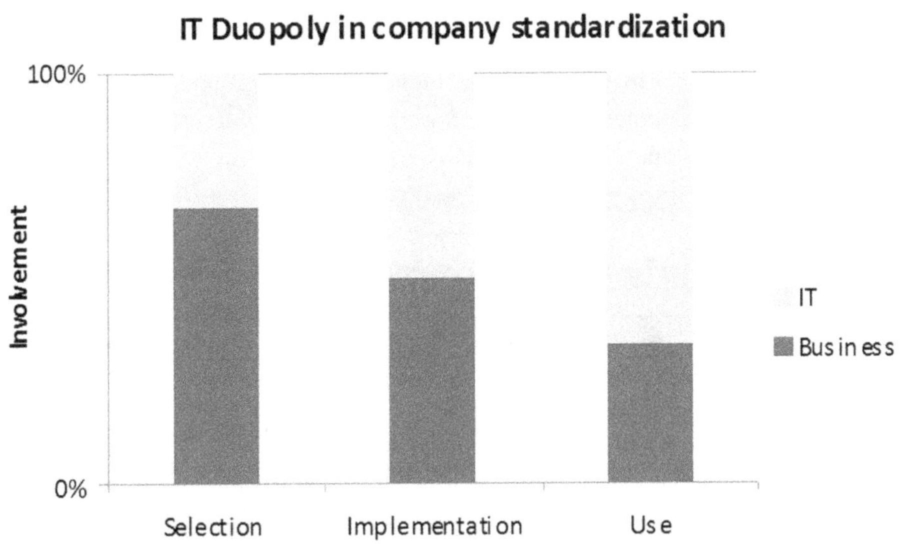

Management

Selection: The primary selection criterion was based on finding a balance between functionality and support/license costs. The selection of products was simplified by subdividing the IT infrastructure into three architectural layers (hardware, system software, and application software).

Implementation: Formalized project management was essential for the successful implementation of the standardized IT environment. This included managing progress and costs, approving project changes, and monitoring the quality of implementation. In addition, risk management was carried out, notably deciding on organizational changes, and controlling the number and type of applications.

Use: Once implemented, the company's IT standards are updated according to requests for changes and deviations from the business department (e.g., from projects) and technical considerations from the IT department. There is strict control on usage and enforcement of the IT standards via the slogan 'comply or explain', and the process of making exceptions to policy and standards is restrictive. Formalized processes for IT service management are also essential for the success of the standardized IT environment. User impact and number of users involved, the possibility to charge costs, the contribution to a reduction of expenses, and the vendor's track record are the main considerations when allowing changes to company IT standards. Table 5 summarizes the governance and management of this set of company IT standards.

The concepts of governance and management have been successfully applied in this case study. These proved to be essential in realizing the anticipated Business Benefits from company IT standardization as suggested by Weill and Ross (2004) and later investigated in detail by Van Wessel (2010). The project resulted in a low-cost and fully standardized IT environment that was successfully selected, implemented, and used.

CONCLUSION AND DISCUSSION

This study adds empirical evidence to the scarce academic literature on the effects of company IT standards on organizational performance and the influence of its governance and management arrangements. Furthermore, it provides practical insights for companies how to effectively set up governance and management in company standardization initiatives.

Table 5. Effective governance and management of company IT standards – Key elements

Phase	Governance	Management
Selection	Business and IT (evolved into IT Policy Board)	• Selection criteria • Consider three IT architectural layers (hardware, system, and application software)
Implementation	• Steering group (evolved into IT Policy Board) • Program Manager	• Project Management • Risk Management
Use (incl. changes)	• IT Policy Board • IT Review Team	• 'Comply or explain' slogan • Service management • Change criteria

Contribution to Theory

The number of studies related to company standardization and standards is still very limited. Some key success factors such as standard enforcement, institutionalized mechanisms to involve key stakeholders, commitment from top management and ensuring that company standardization is part of corporate strategy (Kayworth and Sambamurthy, 2000; De Vries and Slob, 2006; Boh and Yellin, 2007, Dey et al., 2009; Hesser, 2010) have been acknowledged in this study on product standardization. However, we have added some specific aspects to the body of knowledge which relate to the way the benefits have been assessed, and the involvement of stakeholders during the lifecycle of the company standards. Due to the explicit distinction between governance and management, as part of the company IT standardization Management Framework (Figure 1), we were able to elaborate on specific governance arrangements between Business and IT stakeholders. In addition, we have described financial, service-and-support, and user-related benefits, whereas others (such as Boh and Yellin, 2007) have not addressed these benefits from a business point of view, or have related the success of company standards only to the actual "use of the standards" (De Vries and Slob, 2006). Because of our explicit distinction between governance and management, we were able to describe an efficient and effective company standardization initiative. Although this study deals with a single case, this chapter can be used as a basis of further academic studies.

Contribution to Practice

Companies carry out numerous standardization initiatives aimed at creating value by converging on particular software, hardware, or processes, but often with mixed success. The company studied here has shown that two practices in particular have contributed to the achievement of the intended benefits of the IT product standardization project. *First*, it is important to recognize the life cycle of a company standard and its phases of selection, implementation, and use. *Second*, it is essential to set up good governance and management mechanisms throughout each of these distinct phases.

Key to the success of company standardization, as far as governance is concerned, is the manner in which both business and IT stakeholders are involved in the selection, implementation, and use phases. These business stakeholders are the internal end-users of the standards, and ultimately well-functioning company processes are beneficial for the company's customers. During the lifecycle of a company standard, involvement of stakeholders may vary depending on the type. In our case, business stakeholders took the lead. Typically, representatives from (Business and IT) management, subject matter experts and the end-user community should be involved. Furthermore, decisions regarding deviations, changes, withdrawals, and renewals of company IT standards must be dealt with seriously to ensure standards reflect up-to-date requirements, and to prevent users from bypassing these standards. Senior business and IT representatives make and monitor such decisions through program management during the implementation phase, and policy boards are involved in the use phase.

In the selection phase, each decision should be based on preserving a balance between functionality, user impact, and the costs of implementation and maintenance. To minimize costs, the chosen IT standards have to be aligned with the IT architecture of the company. During the implementation phase, adequate project management must be in place to mitigate risk, and to track that the selected company standard is implemented as intended. In the use phase, proper IT service management processes and change management procedures must be in place. This includes enforcing the use of company standards and effectively managing changes in case of changed business requirements. It is more important for an organization to repeatedly and consistently use a company standard, than to aim for the "perfect" one, but not use it consequently. The described governance and management arrangements of company IT standardization may inspire IT executives and others to successfully adopt company standards in their enterprise.

ACKNOWLEDGMENT

The authors would like to thank ABN AMRO and in particular those staff that took part in this case study. Moreover, we thank four anonymous reviewers for all improvements they suggested.

REFERENCES

Backhouse, J., Hsu, C. W., & Silva, L. (2006). Circuits of Power in Creating de jure Standards: Shaping an International Information Systems Security Standard. *Management Information Systems Quarterly*, *30*(Special Issue), 413–438.

Blind, K. (2004). *The Economics of Standards – Theory, Evidence, Policy*. Cheltenham, UK: Edward Elgar.

Boh, W. F., & Yellin, D. (2007). Using Enterprise Architecture Standards in Managing Information Technology. *Journal of Management Information Systems*, *23*(3), 163–207. doi:10.2753/MIS0742-1222230307

Boynton, A. C., & Zmud, R. W. (1987). Information Technology Planning in the 1990s. *Management Information Systems Quarterly*, *11*(1), 59–71. doi:10.2307/248826

Cargill, C. F. (1989). *Information Technology Standardization: Theory, Process, and Organizations*. Bedford, MA: Digital Press.

De Vries, H. J. (1999). *Standardization – A Business Approach to the Role of National Standardization Organisations*. Boston: Kluwer Academic Publishers. doi:10.1007/978-1-4757-3042-5

De Vries, H. J., Slob, F. J., & Zuid-Holland, V. G. (2006). Best practice in company standardization. *International Journal of IT Standards and Standardization Research*, *4*(1), 62–85. doi:10.4018/jitsr.2006010104

Dey, P. K., Clegg, B., & Bennett, D. (2009). Managing implementation issues in enterprise resource planning projects. In M. H. Sherif (Ed.), *Handbook of enterprise integration*. Boca Raton, FL: Auerbach. doi:10.1201/9781420078220-c24

Hesser, W. (2010). Standardisation within a company – A strategic perspective. In W. Hesser (Ed.), *Standardisation in companies and markets* (3rd ed.; pp. 215–258). Hamburg, Germany: Helmut Schmidt University.

Hesser, W., & Inklaar, A. (1997). *An introduction to Standards and Standardization. In DIN-Normungskunde* (Vol. 36). Berlin: Beuth Verlag GmbH.

Jain, S. (2012). Pragmatic agency in technology standards setting: The case of Ethernet. *Research Policy*, *41*(9), 1643–1654. doi:10.1016/j.respol.2012.03.025

Kayworth, T., & Sambamurthy, V. (2000). Facilitating localized exploitation and enterprise-wide integration in the use of IT infrastructures: The role of PC/LAN infrastructure standard. *The Data Base for Advances in Information Systems*, *31*(4), 54–81. doi:10.1145/506760.506765

Li, X., & Chen, Y. (2012). Corporate IT Standardization: Product Compatibility, Exclusive Purchase Commitment, and Competition Effects. *Information Systems Research*, *23*(4), 1158–1174. doi:10.1287/isre.1110.0412

Manders, N. B. (2015). Implementation and Impact of ISO 9001. ERIM Ph.D. Series Research in Management. Rotterdam, The Netherlands: Erasmus Research Institute of Management (ERIM).

Mendoza, R. A., & Ravichandran, T. (2011). An Exploratory Analysis of the Relationship Between Organizational and Institutional Factors Shaping the Assimilation of Vertical Standards. *International Journal of IT Standards and Standardization Research*, *9*(1), 24–51. doi:10.4018/jitsr.2011010102

Mueller, T., Dittes, S., Ahlemann, F., Urbach, N., & Smolnik, S. (2015). Because Everybody is Different Towards Understanding the Acceptance of Organizational IT Standards. In *Proceedings of the 48th Hawaii International Conference on System Sciences* (HICSS). Kauai, HI: HICSS. doi:10.1109/HICSS.2015.487

Nickerson, J. V., & zur Muehlen, M. (2006). The ecology of standards processes: Insights from Internet standards making. *Management Information Systems Quarterly*, *30*(1), 476–488.

Poba-Nzaou, P., & Raymond, L. (2011). Managing ERP system risk in SMEs: A multiple case study. *Journal of Information Technology*, *26*(3), 170–192. doi:10.1057/jit.2010.34

Rada, R., & Craparo, J. (2000). Standardizing Software Projects. *Communications of the ACM*, *43*(12), 21–25. doi:10.1145/355112.355117

Rogers, E. M. (2003). *Diffusion of innovations* (5th ed.). New York: Simon & Schuster International.

Swaminathan, J. M. (2001). Using Standardized Operations. *California Management Review*, *43*(3), 125 135. doi:10.2307/41166092

Teichmann, H. (2010). *Global Standardization: Coping with Multilingualism. In EURAS contributions to standardisation research 3*. Aachen, Germany: Verlag Mainz.

Van de Kaa, G., De Vries, H. J., Van Heck, E., & Van den Ende, J. (2007). The Emergence of Standards: a Meta-analysis. In *Proceedings of the 40th Hawaii International Conference on Systems Sciences*. IEEE.

Van de Kaa, G., Van den Ende, J., De Vries, H. J., & Van Heck, E. (2011). Factors for winning interface format battles: A review and synthesis of the literature. *Technological Forecasting and Social Change*, *78*(8), 1397–1411. doi:10.1016/j.techfore.2011.03.011

van Wessel, R. (2010). Toward Corporate IT Standardization Management: Frameworks and Solutions. Hershey, PA: IGI Global.

van Wessel, R., Ribbers, P., & de Vries, H. (2006). Effects of IS Standardization on Business Process Performance: A Case in HR IS Company Standardization. In *Proceedings of the 39th Annual Hawaii International Conference on System Sciences*. HICSS. doi:10.1109/HICSS.2006.141

Weill, P., & Ross, J. (2004). *IT Governance – How Top Performers Manage IT decision Rights for Superior Results*. Boston, MA: Harvard Business School Press.

Weitzel, T. (2003). *A Network ROI, MISQ Special Issue Workshop Standard Making: A Critical Research Frontier for Information Systems*. Paper presented at Pre-Conference Workshop, International Conference on Information Systems, Seattle, WA.

West, J., & Dedrick, J. (2006). Scope and Timing of Deployment: Moderators of Organizational Adoption of the Linux Server Platform. *International Journal of IT Standards and Standardization Research*, *4*(2), 1–23. doi:10.4018/jitsr.2006070101

WTO. (2005). *World Trade Report 2005: Standards, 'offshoring' and air transport*. Geneva: World Trade Organization.

Yin, R. K. (2009). *Case Study Research – Design and Methods* (4th ed.). Thousand Oaks, CA: SAGE.

KEY TERMS AND DEFINITIONS

Business Benefits: An outcome of an action or decision that contributes towards meeting one or more business objectives.

Business IT Alignment: The continuous, mutual coordination of business departments and the IT department to optimize the value that Information Technology contributes to an enterprise.

Business Performance: The efficiency and effectiveness of an organization reflected in the business objectives set by management.

Company Standard: A specification of a product or process to be repeatedly and consistently used in the company.

Company Standardization: The activity of establishing and recording a limited set of solutions to actual or potential matching problems, directed at benefits for the party or parties involved, balancing their needs, and intending and expecting that these solutions will be repeatedly or continuously used, during a certain period, by a substantial number of the parties for whom they are meant.

Company Standardization Process: A set of sequential process steps, a) selection, b) implementation, c) use (including changes and withdrawals) that comprise the lifecycle of a company standard.

Governance of Company Standards: Specifying the decision rights and accountability framework to encourage desirable behavior in the selection, implementation and use of standards within an organization.

Management of Company Standards: The decision-making efforts associated with planning, organizing, controlling, and directing the selection, implementation and use of standards within an organization.

ENDNOTES

1. Total cost of ownership is an indicator of IT efficiency, introduced by Gartner in 1987
2. Calculated on the basis of functional equivalents
3. COTS: Commercial Off The Shelf

4 Payback period = 32,000,000 / {(4,600-2,392) x 10,000} = 1.45 year

5 ROI = {(4,600-2,392) x 10,000 – 32,000,000} / 32,000,000 = 176%

6 $$32,000,000 = \sum_{j=1}^{4} \frac{(4,600 - 2,392) \text{x} 10,000}{(1+i)^j} \Rightarrow i = 0.5759$$

Chapter 4
The Role of Internal Standardization in Business Models:
An Activity Configurations Perspective

Magnus Johansson
Lund University, Sweden

Amalia Foukaki
Lund University, Sweden

Matts Kärreman
Lund University, Sweden

ABSTRACT

This chapter focuses on business models and the role of internal standardization in business models. We develop a typology that outlines the role of internal standardization for a set of value configurations that serve as representations of generic business models. The topic is of importance for both managers and researchers in relation to firm level innovation and strategy, as well as how business models relate to internal and external standardization. With regard to business model innovation, this chapter can aid in identifying shifts in the firm's internal standardization focus associated with changes in its business model.

INTRODUCTION

Interest in the Business Model concept has surged during the last twenty years (Ghaziani & Ventresca, 2005; Zott, Amit & Massa, 2011). The increased attention of management scholars in Business Models and the relation to strategy stems from the attempt to explain value creation in the arising digital economy and increasingly important networked markets (Zott et al, 2011), both in the form of markets with single-side as well as multiple-side network effects (stemming from complementarity).

DOI: 10.4018/978-1-4666-9737-9.ch004

A Business Model can be viewed as a system of activities which is not delimited to the firm itself but includes a system-level approach to how the firm interacts with its environment and does business (Zott et al, 2011). Developing a successful Business Model is often insufficient to assure competitive advantages as imitation might be easy. Business Model innovation can, however, be a pathway to competitive advantage and limit the risks of imitation (Teece, 2010). Christensen, Grossman and Hwang (2009) have addressed the topic of Business Model innovation by utilizing activity configurations (Stabell & Fjeldstad, 1998; Porter, 1985) as descriptions of generic Business Models.

The Network Configuration (Stabell & Fjeldstad, 1998) is of particular interest. It relies on Standardization to facilitate a networked market. The creation of networked markets and complementarity between them depends on the ability of Standards to provide compatibility (Gallagher, 2007; Tassey, 2000), and networked markets have been of certain interest for research on Standardization due to the fierce Standardization competition they can entail (Besen & Farrel, 1994). The Network Configuration can rely on proprietary Standards to facilitate a networked market, but also on industry Standardization to facilitate inter-network connections. Diffusion of industry Standards is critical on networked markets (Stabell & Fjeldstad, 1998) and consequently the firm's Business Model depends on its proprietary Standards as well as their relation to industry Standards.

In general, Standards enforce conformity requirements (Tassey, 2000) and serve as rules of key activities, resulting from consensus between agents (Narayanan & Chen, 2012)). The chain (Porter, 1985; Stabell & Fjeldstad, 1998) and package (Johansson & Jonsson, 2012) configurations also rely on Standardization to realize value creation, for instance by reducing transaction costs of firm products in the market (Foss, 1996).

Hence, Standardization seems to play an important role in activity configurations for value creation, but its more precise nature in association with each of generic configurations above has not been treated as a key topic in the literature. This chapter aims at elucidating the particular use of Internal Standardization in enabling different Business Models in the form of generic activity configurations. Hence, this chapter links the field of Standardization with the fields of activity configurations as Business Models. By development of a typology, the main Internal Standardization nature of the chain, network, package and Shop Configurations is outlined. The focus is on Internal Standardization, but we also recognize that internal and external Standardization are interrelated (Botzem & Dobusch, 2012; Okhomatovskiy & David, 2012; Schilling, 1999; Boiral, 2012; Narayanan & Chen, 2012), and thus our findings are of importance also in relation to external Standardization, activity configurations, and the value of external Standards (Mione, 2015).

BACKGROUND

Activity Configurations as Business Models

During the 1990s, research pursued alternatives to the value chain (Porter, 1985) by focusing on interrelationships and complementarity, but also by focusing on various types of scale effects. For instance, Normann and Ramirez (1993;1994), through their concept of value constellations, pay specific attention to co-creation of value between players. Stabell and Fjeldstad's (1998) value configurations, on the other hand, focus more on the internal activity configuration of an organization. Stabell and Fjeldstad extend

Porter's value chain by relating activity configurations to generic strategies associated with particular economic effects. This is discussed by Christensen et al. (2009) when they illustrate moves from one configuration to another as Business Model innovation paths along which firms can transform their ways of working and interacting with the market. Therefore, activity configurations are considered in this chapter as Business Models in the form of systems of activities (Zott et al, 2011).

Stabell and Fjeldstad (1998) proposed three activity configurations: the chain, shop and Network Configuration. At the center of each configuration they have placed value creation logics inspired by Thompson's (1967) technologies. These alternative configurations emphasize integration and interrelations of activities, internally and externally, and that these can be radically different from what we associate with the value chain. The value Chain Configuration, according to Stabell and Fjeldstad, utilizes long-linked technologies (Thompson, 1967), which are associated with the mass production assembly line and the transformation of input to output. The focus of mass production is the perfection of the production of a Standard output, repetitively and at a constant rate.

The value Shop Configuration utilizes intensive technology to solve customer unique problems through iterative problem solving (Stabell & Fjeldstad, 1998). Thompson also states: "The intensive technology is a custom technology. Its successful employment rests in part on the availability of all the capacities potentially needed, but equally on the appropriate custom combination of selected capacities as required by the individual case or project." (Thompson, 1967, p. 18). Typical examples of Shop Configurations are professional services firms, such as consulting firms or medical practice.

Finally, the Network Configuration relies on linking customers through mediating technologies (Thompson, 1967). Stabell and Fjeldstad (1998) describe the role of the firm as a network service supplier and a mediator between customers. The network logic utilizes network externalities (Katz & Shapiro, 1985) to create value. A classic example of a network dependent firm is a telecom network operator, bringing multiple subscribers together. Furthermore, in Johansson and Jonsson (2012) the set of three activity configurations (and logics) is extended with a fourth, the Package Configuration. The package logic is, similarly to the shop logic, centered on problem-solving. However, this is problem-solving across a range of potential customers, with the goal of making such common problem-solving efficient through underlying solutions. Typical examples are firms producing designs for licensing, such as hardware IPs for the semiconductor industry. The four activity configurations and their logics are summarized in Table 1. Despite the fact that Stabell and Fjeldstad (1998) proposed three activity configurations[1] - the chain, shop and network- they also recognized the relevance of their hybrid versions (cf. Miller, 1996).

Christensen et al. (2009) have linked these configurations to Business Model innovation by considering them to be generic descriptions of Business Models. One of the main views of Business Models in research addresses them as firm-centric but at the same time as boundary-spanning systems of activities (Zott et al, 2011). Christensen et al. emphasize the importance of innovating the firm's Business Model, and essentially the value creation logic of the firm, by moving from one configuration to another. Thus, Business Model innovation concerns innovation regarding the set of activities, their configuration, or the parties involved in performing them (cf. Amit & Zott, 2012).

Activity configurations can, in addition to being considered as Business Models (Christensen et al, 2009) and value creating entities, be understood from a governance perspective. For particular transaction attributes the costs of governing activities differ, depending on whether these activities are governed through markets or in organizations (Masten, 1991). Boundaries for activities, and consequently activity configurations, are results of governance-costs minimization attempts. Sources of governance-cost dif-

Table 1. Typology of value configurations. (Stabell & Fjeldstad, 1998, p. 415 and Johansson & Jonsson, 2012, p. 548)

	Network	Package	Chain	Shop
Value creation logic	Linking customers	Solving common problems	Transformation of inputs to products	(Re)solving customer problems
Primary technology	Mediating	Intensive	Long-linked	Intensive
Primary activity technologies	Network promotion and contract management Service provisioning Infrastructure operation	Problem gathering Problem prioritization Problem decomposition Problem solving Control of parallel development Marketing and support	Inbound logistics Operations Outbound logistics	Problem finding and acquisition Problem solving Choice Execution Control/evaluation
Main interactivity relationship logic	Simultaneous, parallel	Single to multiple point packaging funnel	Sequential	Cyclical, spiraling
Primary activity interdependence	Pooled Reciprocal	Pooled Sequential Reciprocal	Pooled Sequential	Pooled Sequential Reciprocal
Key cost drivers	Scale Capacity utilization	Scale/Development Capacity utilization	Scale Capacity utilization	Efficiency in referral Reuse of human resources
Key value drivers	Scale Capacity utilization	Quality		Reputation
Business value system structure	Layered and interconnected networks	Packaging funnels in tree structure	Interlinked chains	Referred shops

ferences can be understood from three streams of thought, those being transaction costs, measurement difficulties and knowledge differences (Santos & Eisenhart, 2005), all of which will be discussed in the next section.

From a transaction cost economics perspective, structures for defining, monitoring and enforcing activities and activity configurations are consequences of firms striving to minimize transactions costs. While measurement costs are caused by information problems (Holmstrom, 1999; Williamson, 1985), these, in turn, change over time through organizational learning. Supervision of activities, alignment of incentives and collecting information are simpler within an organization than in the market. Different activity configurations come with different challenges regarding measurement costs. Costs derived from idiosyncrasies in knowledge include coordination costs. Individuals bring and develop different knowledge and will practice or argue for different solutions to organizational tasks if not coordinated. The extent of the need for coordination more than likely varies among activity configurations, and routines developed in order to reduce costs will be aligned to the specific activity configuration.

Of particular interest during the last decades has been the Network Configuration, which provides a Business Model perspective on the deployment of network effects (Katz & Shapiro, 1985; 1994), and thus, also relates to the role of industry platforms (Eisenmann, Parker & Van Alstyne, 2006; Parker & Van Alstyne, 2003; Rochet & Tirole, 2003). A fundamental aspect of realizing network effects of an industry platform is the utilization of Standardization in order to enable compatibility across players accessing the network (Cusumano, 2010; Ulrich, 1995).

Moreover, in the activity configurations proposed by Stabell and Fjeldstad (1998), we see great consideration made to cooperative aspects, which is in line with the systemic view of Business Models (Zott et al, 2011). For instance, the Shop Configuration may include joint, iterative execution together with a customer, where the outcome may very well also depend on the efforts of the customer. This results in the firm's iterative problem-solving activity sequences. The Network Configuration also relies on cooperative actions by players participating in the network facilitated by the focal firm. In the following section, Standards and Standardization are defined; before we move on to outline the role of Standardization in the configurations.

Internal and External Standards and Standardization

Standardization codifies and diffuses state of the art technology and best practice (Swann, 2010; Krechmer 2000, 2005; Blind & Grupp, 2000), essentially establishing conformity requirements on products and processes (Naryanan and Chen, 2012; Tassey, 2000) - and thus can be seen as the outcome of consensus among agents on how to perform certain key activities according to specific rules (Naryanan & Chen, 2012). Several scholars have pointed to how agents try to influence the Standardization process in their own interests (Blind & Mangelsdorf, 2012; Dokko & Rosenkopf, 2010). Therefore, we view industry Standards as reducing variety in artefacts and activities within an industry. Also, we recognize that industry Standardization may very well encompass varying degrees of influence from the various agents that try to promote their own interests, for instance within consensus based committee work.

Consequently, we view Internal Standards as internally based variety reduction, and especially that Internal Standards have a variety reducing effect across products and production processes for instance, and as a consequence between intraorganizational entities, such as divisions, functions or teams. . This is the internal equivalent of industry codification and diffusion of technology and best practices. In many instances the development and implementation of these follow internal consensus processes, analogous to external Standardization (Grossman, 2013). These Internal Standards can be deployed internally as well as externally, i.e. utilized for the output and input markets, as is the case for private Standards (Vandemoortele & Deconinck, 2013).

External Standardization depends on the input from several participating firms. Such input may originate from Internal Standardization (of processes, products or sub-deliveries). At the same time, Internal Standardization might be the result of adoption of external Standards if they need to be accustomed to serve the firm's needs. Therefore, Standardization is not a far-away industry matter, but is interdependent on firm Internal Standardization. For instance, in Okhomatovskiy and David (2012) the firm responds to industry Standardization by creating its own Internal Standard. However, this Internal Standard does not only serve internal purposes, but also signals certain organizational traits to the external environment. Thus, the firm has several strategic options related to Standardization and these involve internal and external Standardization and their interdependence.

Since a firm centric view is applied in this chapter, the role of Standards needs to be considered from the input and output market perspectives, rather than those from the demand and supply side, as is common for external Standards (Naryanan & Chen, 2012). An Internal Standard may be applied to suppliers of the firm (as in Grossman, 2013). The use of Internal Standards on the output market, i.e. the suppliers, means that Standardization occurs through an internal process and then its outcome has consequences for other firms as well at the input or output markets of the firm. One example of Standards that are of an internal origin, but with external effects, is private Standards, a type of Standard mainly associated with minimum quality certification within the food and agricultural sector (Henson & Hooker, 2001; Vandemoortele & Deconinck, 2013). Private Standards controlled by an intermediary in a market can have an effect on the input as well as the output market of the firm but can only under certain circumstances be optimal for both buyers and suppliers (Vandemoortele & Deconinck, 2013).

Standardization is often associated with the process by which external Standards emerge (Tassey, 2000; Blind, 2004; Greenstein & Stango, 2007). For instance, Tassey (2000) distinguishes between the process of Standardization, which enables the conformity associated with Standards, and Standards themselves, which are denoted as a set of specifications. In several instances, research on Standardization has focused on the consensus processes of Standards developing entities (cf. Botzem & Dobusch, 2012), such as Standards Developing Organizations (SDOs) or Consortia. The emphasis in the aforementioned cases is primarily placed upon the inter-organizational rather than the intra-organizational processes. Also, external Standards and their economic effects are often discussed from the perspective of the impact they have on market establishment and efficiency (Blind, 2004; David, 1987). However, several of the economic effects associated with external Standardization are also relevant from a firm perspective, or from that of inter-firm relations, e.g. between a firm and its supplier network. For instance, variety reduction of components, resulting in scale effects, has been referred to as Internal Standardization (Ulrich, 1995). The firm frequently faces a situation of Standards-related alternative choices, such as keeping its Standardization internal or diffusing it among a number of other firms (through consortia or SDOs), or adopting external Standards for internal variety reduction. At its most basic level, Standardization concerns "reducing the infinite number of things in the world … to a moderate number of well-defined varieties" (March & Simon, 1958, p 181). The key question for the firm is on what level this reduction of varieties should occur, i.e. internally or externally.

A firm is not isolated from industry structures and their changes, even if outside its boundaries (Sheehan & Foss, 2007). Boundaries of the firm and its Standardization activities arguably intertwine and mutually affect organizations in terms of efficiency, competence, identity and influence (Santos & Eisenhardt, 2005). Firm boundaries are a central phenomenon that has been examined from various perspectives. Firstly, as a demarcation of the social structure that forms an organization, and as such, activities operate under a particular logic of identity that shapes how things are done in the organization and sets the rules for inclusion (Kogut, 2000). Secondly, boundaries are the demarcation of the resources controlled by the firm, thus shaping the development of the organization (Helfat, 1997). Thirdly, boundaries can be seen as the determinants for organizational influence, including the degree of industry control and power over external forces (Santos & Eisenhart, 2005). Porter (1985) points to the reciprocal relationship between industry structure and the chain of activities of the players in an industry, while claiming that "potential sources of competitive advantage are everywhere in a firm" (Porter, 1985, p. 3). Thus, Standardization influences the firm and its activities via the effects it has on industry structures and vice versa.

On the other hand, the firm can also influence Standardization and in turn the industry. Schilling (1999) presents Standards as means to gain competitive advantage, determining success or failure, depending on how a firm employs them. For instance, she stipulates how a firm could promote and influence a specific Standard, thereby affecting the evolution of the industry in a manner advantageous to the firm. Furthermore, the firm must perform variety-reducing and conformity-enabling activities while at the same time considering other players involved in a Standardization process, be it of a consensus or strictly competitive nature.

INTERNAL STANDARDIZATION AND STANDARDS

Internally and Externally Applied Standardization of Products and Processes

Viewing Standardization as the process by which Standards emerge is one way of addressing the difference between the Standards' development procedure and its outcome. However, the processes taking place within an organization might also be Standardized. That is, the process of Standardization may lead to an outcome, the Standards, being either for processes (Perez-Aleman, 2011) or for products (Warner, Fairbank, & Steensma, 2006). Brunson and Jacobsson (2000) distinguish between Standardization of processes on the one hand and services and products on the other. Similarly, Blind (2011) makes a distinction between Standardization of products, procedures and practice. Here, a distinction will be made between Standardization of products (including internal and external parts and sub-deliveries, and changes to those stemming from service execution) versus Standardization of processes (within which are included procedures and practices).

Another distinction relates to organizational boundaries. A Standard can originate internally as well as externally (Okhmatovskiy & David, 2012), i.e. it can stem from within an organization as well as from an external Standardization body. Following this line of argument, a Standard can also be applied internally or externally. An illustration is seen in Figure 1, where we also highlight the interdependence between internal and external Standards.

As an example, it is possible to distinguish between process Standards applied internally and Standards that are applied externally. Process Standards are closely linked to procedures, practices and actions (cf. Brunson & Jacobsson, 2000; Perez-Aleman, 2011) and thus the internal aspects of process Standards relate to the way that work is done in an organization. From the internal perspective, process Standardization can play a role in enabling modularity (Tu, Vonderembse, Ragu-Nathan, & Ragu-Nathan, 2004), thus representing the internal application. However, as Okhomatovskiy and David (2012) have shown; an internally developed process Standard can also be communicated externally in order to convey certain practices.

Figure 1. Origin and application of Standards from the firm perspective.

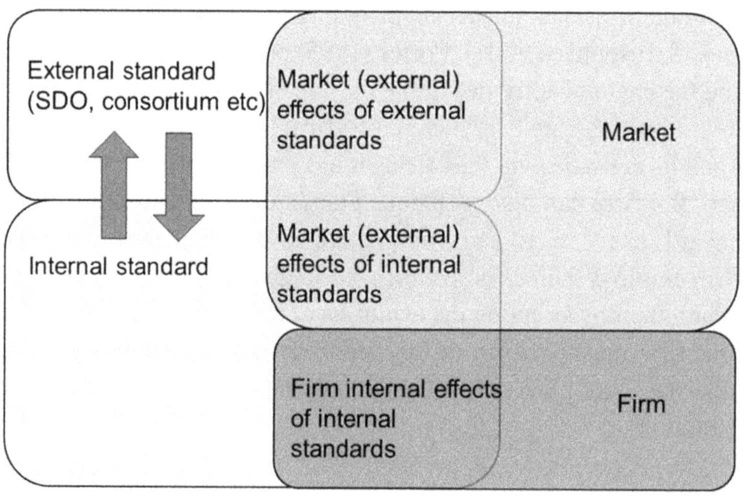

Product Standards can even be viewed from the perspective of their application, internal or external. An internally developed Standard can be *applied* internally in order to reduce variety or facilitate compatibility between components such as internal sub-deliveries. Internal application of product Standards is fundamental for enabling modularity of components (cf. Argyres & Bigelow, 2010) within the firm. An internally developed product Standard can also be *applied* externally as part of an offering in a market. As a Standardized item it can communicate certain characteristics (such as minimum quality in the case of private Standards (Vandemoortele & Deconinck, 2013)) on the output market compared to other products from the same firm or from other firms, i.e. it relies on a certain level of uniformity vis-à-vis the environment. This is important from a transaction cost perspective as it affects search and control costs related to products. (cf. Barzel, 1982; Foss, 1996). It can also be applied externally to the input market, for instance in order to control supplies from subcontractors (Ulrich, 1995).

Some Standards reduce variety resulting in, for instance, transaction cost reduction, as well as providing compatibility, leading to, for example, network effects across firms or customers (Blind, Gauch, & Hawkins, 2010; Blind, 2004). As an example, consumers might receive utility from using arrangements containing a variety of compatible components (Matutes & Regibeau, 1996) while for firms, modularity enabled by Standardization can play a role beyond the boundaries of the firm when the development and production of components is outsourced. The above-mentioned externally and internally applied Standards may relate to both processes and products.

Internal Standardization for efficiency: Here the focus is either on Standardization of a type of process, enabling a certain level of efficiency, or efficiency due to more precise interaction between several internal processes. An example would be time-critical deliveries from several sub-processes to fit into each other, thus requiring precise execution according to an Internal Standard.

Internal Standardization: This box encompasses Internal Standardization of sub-deliveries and components that often serves to facilitate efficient utilization of interfaces between modules and sub-deliveries and consequently their production processes. For instance, a product is made up of several components from separate production lines that require clearly defined and Standardized interfaces to fit together. This includes interoperability as well as minimum quality aspects.

Standard services execution: Standardization of externally oriented processes that constitute services fall under this category. One example is services execution through software systems or service execution with human interfaces fulfilling requirements of an Internal Standard.

Product Standardization: This box encompasses the Standardization and homogeneity of products to the output market resulting in transaction cost reduction among the firm's offerings. This section can also include results of services; i.e. the changes that services incur on goods and products of the customer.

Time dependent platform Standardization: Here the focus is on the timing aspect, making sure that processes are Standardized to facilitate interaction between users. For instance, online sales and auction sites need not only certify the right information in its systems but also through Standardized processes ensure that counterparts supply information and act with the timing required.

Platform Standardization: From a firm perspective this section includes Standardization to ensure compatibility between products, technologies and users on the output market. Thus, an example would be online platforms such as Facebook.

Analogously as the categories on the output market we can imagine internally developed Standards applied to the input market to facilitate certain economic effects vis-á-vis suppliers. Thus, the firm through its internally developed Standards introduces conformity requirements on its suppliers. In Table 2 they are called *Time Dependent Supply Platform, Supply Platform, Services Supply Timing* and *Supply Product and Technology*.

Platform Standardization requires additional consideration. Platforms that enable double-sided network effects between several players can stretch beyond the input and output market border through vertical Standardization of interfaces. Thus, for platforms the input and output distinction is, in some instances, not relevant as all players joining the platform can be seen as customers to some extent.

Processes are considered as internal if they predominantly concern activities that can be performed without direct interaction with the environment. That is not to say that external input is unnecessary but that direct interaction in day-to-day execution is limited. Processes of external application, however, are related to activities in direct interaction with the environment in the form of customers, suppliers etc. In the literature on Standards, *Standardization for internal efficiency* is seldom discussed in-depth. From the perspective of the strategic management and organization theory literature this area is mostly associated with terms such as practices (cf. Brown & Duguid, 1991) and routines as part of the evolutionary perspective (Nelson & Winter, 1982).

However, the Standardization literature[2] may stretch into the internal aspects of reducing variety and providing compatibility concerning *products and parts* but rarely as far as applying the concept of Standards on internal *processes*, unless they are a result of internalizing externally specified process Standards (such as Boiral, 2012; Narayanan & Chen, 2012; Sandholtz, 2012; Perez-Aleman, 2010). Here internal processes will be included in the analysis in order to point to differences between the activity configurations. Also, whereas extensive parts of the Standardization literature deal only with Standards that diffuse a substantial share of a market, this chapter also includes Standardization of output or processes from one firm, i.e. variety reduction for, and across, firm products (Foss, 1996) or services (Tether, Hipp, & Miles, 2001). This can be argued to be an important aspect since a Standard on the single firm level can provide a reduction in transaction costs (cf. Barzel, 1982; Foss, 1996).

STANDARDIZATION IN ACTIVITY CONFIGURATIONS AND ITS ANTECEDENTS

Externally Applied Standardization

Stabell and Fjeldstad (1998) make an explicit reference to Katz and Shapiro (1985) and direct network effects, i.e. single-side effects. Since then the field of network effects has advanced through the definition of indirect network effects (Katz & Shapiro, 1994), two-sided markets (Eisenmann et al, 2006; Parker

Table 2. Outline of Internal Standardization

	Input market			Output market	
	Compatibility	**Uniformity**	**Internal**	**Uniformity**	**Compatibility**
Process	Time dep Supply Platform	Services Supply Timing	Internal for efficiency	Services Execution	Time dep Platform
Product	Supply Platform	Supply Product and Tech	Internal	Output Product	Platform

& Van Alstyne, 2003; Rochet & Tirole, 2003) and the role of platforms as multi-sided market enablers (Gawer & Cusmano, 2008). In turn, industry platforms rely on Standardization. Not all Standards are platforms (Cusumano, 2010) but all platforms contain an aspect of Standardization to enable network effects. This dependence on Standards is noted by Stabell and Fjeldstad (1998) and has its origin in a fundamental aspect of the mediating technology to ensure compatibility (Thompson, 1967).

Another discussion on Standardization revolves around the level of Standardization of a firm's products from the internal perspective, and not in relation to offerings from other players in the output market. To what extent does the firm deliver similar products or services to multiple customers versus tailoring for each customer's needs? In Stabell and Fjeldstad (1998, p. 416) the value chain has Standardized output: "Consider assembly line-based manufacturing as an example of a long-linked value creation technology. The assembly line is designed to produce Standard products at low cost per unit by exploiting cost economies of scale." Stabell and Fjeldstad mention that value shop deliveries may include Standardized solutions but the value creation process is essentially designed to handle unique cases. Johansson and Jonsson (2012) deal specifically with this, claiming that the value creation associated with the Standardized solution is the basis for a unique activity configuration, the Package Configuration (see the consequences of this in Table 1, outlining four activity configurations), thus facilitating Standardized solutions that form a basis for internal and, in some cases, external output. The Package Configuration is also interesting as it emphasizes activities such as those of research and development as primary, especially in combination with value capture models built on licensing. This in turn has close connections to external Standardization in innovative industries (cf. Blind, 2002; 2004).

Internally Applied Standardization

The use of the label Standardization can, however, also have an internal character as in Thompson (1967),. Although Thompson includes an external perspective in association with defining different technologies, Standardization is in the sections related to interdependence internal, as it is inspired by the definitions in March and Simon (1958). These authors claim Standardization to concern three devices: Homogenization of semi-manufactured products for subsequent steps, interchangeability between parts, and coordination of timing between subprocesses. These three are all aspects of "reducing the infinite number of things in the world, potential and actual – to a moderate number of well-defined varieties" (March & Simon, 1958, p. 181). Standardization, as described by March and Simon. serves the purpose of enabling process specialization and in turn efficiency. Whereas in March and Simon Standardization is a general means of enabling process specialization, it is according to Thompson (1967) a distinct way of coordinating activities. Furthermore, Thompson, in the case of Standardization, stresses the aspects of pooled interdependence, i.e. what may involve several units or processes and not only successive processes. Successive processes, however, is the key example in March and Simon (1958) with regard to device three of Standardization: coordinated timing.

The tampering of labels (and the tampering to some extent of their content) by Thompson is beneficial in the sense that it enriches the typology with situations of stronger and more diverse mutual dependence but it also switches the role of Standardization from a process specialization enabler into an activity coordination mechanism. Thus, independent of March and Simon's or Thompson's take on the labels, Standardization is present from an internal perspective in all types of organizations.

Another internal application of Standardization by Thompson (1967) can be found in the repetitive nature of processes in association with the long-linked technologies and that "proportions of resources involved can be Standardized" (p. 15). Also, criteria can be established with regard to the resources involved. Thus, for the value chain, Internal Standardization takes place with regard to the process, the involvement of resources and the homogeneity of resources. Standardization of the process in the internal sense is similar to the concept of practices and to some extent routines.

Nelson and Winter (1982) argue for routines as the fundamental units of analysis in an organization, though unanimous definition of organizational routines – or agreement regarding their origins – is still lacking (Cohen, Burkhart, Dosi, Egidi, Marengo, Warglien & Winter, 1996; Felin & Foss, 2005). The issue is then how Standardization of internal processes, with limited direct external interdependence, relates to routines. The concept of routines has its roots partly in *Standard operating procedures* (Cyert & March, 1963; Felin & Foss, 2009). Thus, from the internal perspective it is worth recognizing the aspect of routines that relates to repetitive behavior in a Standardized fashion, which is also evident in the above discussion on Standardization by March and Simon (1958). When the literature focusing on routines discusses Standardization it can also do so from a design perspective where Standards set the framework for routines. Felin and Foss (2005) discuss this taking the individual level into consideration. Standard operating procedures are first set and within these individuals interact and gradually change the rules and procedures, and thus the routines. Thus, Standards and routines are different concepts and as indicated above, routines concern variety reduction across a temporal dimension. Standards, on the other hand, fulfil their function across the spatial dimension, i.e. reducing variety across a population, be it technologies, products or processes.

It is vital to acknowledge the differing views on Standardization from the perspective of routines as well as the way it is displayed by Thompson (1967). When Thompson discusses Standardization in association with technologies, he places the organization in its environmental context. On the other hand, when he discusses structure, the focus is primarily internal. Thus, whereas compatibility in the first case is inter-organizational in association with the mediating technology, it is in the latter case intra-organizational. However, with the wide definition Thompson applies to Standardization for pooled resources it pretty much encompasses any instances where an organization makes any shared commitments or shares any resources. This chapter requires a slightly more distinct use of the term Standardization and thus requires viewing Standards in relation to the process or product, as in Table 2.

Internally developed Standardization can therefore help the firm increase internal efficiency or generate a number of effects in its input or output market. Another opportunity is to diffuse an internally developed Standard to the industry (Botzem & Dobusch, 2012; Guler, Guillen & Muir Macpherson, 2002) through consensus processes such as those administered by Standardization organizations, or adopt an external Standard. In either case, Internal Standardization is interrelated with external Standardization.

STANDARDIZATION IN ACTIVITY CONFIGURATIONS

Service vs. Product Character and Processes

Before summarizing the role of internally controlled Standardization in activity configurations in detail it is relevant to consider the most common distinction of deliveries of various organizations, i.e. products versus services. The importance of recognizing this distinction here stems from the fact that aspects of it reside also in the disparity between product and process Standards. Furthermore, both of these distinctions are inherent in the activity configurations. As Rathmell (1966) points out, services are acts or processes, or as Parasurman, Zeithaml and Berry (1985) put it: performances, rather than tangible items. (This does not mean that services cannot have market characteristics that resemble those of tangibles.) Thus, service organizations will rely more strongly on processes directed at the market or individual customers, than on goods and products. And therefore, Standardization of processes will play a specific role in the case of services, as services themselves are processes.

In Parasuraman, Zeithaml and Berry (1985) two more characteristics of services besides intangibility are mentioned: heterogeneity and inseparability. These are also inherent in some of the activity configurations, and especially the Shop Configuration where deliveries are customized (and thus highly heterogeneous) and considerably iterative vis-à-vis the customer (and therefore production and consumption are inseparable). The heterogeneity aspect is difficult to relate to Standardization, as the goal of Standardization is the opposite: to reduce heterogeneity. The issue of inseparability can to some extent be captured explicitly by considering those instances where Standardization is involved in externally applied processes.

A final distinction can be made, which relates to the result of service processes. Hill (1999) argues that the output of service processes can consist of material changes in goods or persons that the service supplier works with (the client or customer). Thus, in our discussion with regard to Standardization, we separate such changes from the processes that produce them. Thus in Table 2 we see them as part of the *product* category.

By taking the nature of services versus products into account, the implications for services versus product firms are clarified. However, the prime distinction here is, as depicted in Table 2, between products and processes. Delivery characteristics are to some extent inherent in the specific activity configurations. Also, economic products are positioned somewhere along a continuum between products (goods) and services.

The following section summarizes the utilization of Internal Standardization in each of the activity configurations as defined by Stabell and Fjeldstad (1998) and with the addition by Johansson and Jonsson (2012). The summary will include externally as well as internally applied Standardization (but not diffused into the industry level) in an attempt to provide an overview of the role of Standardization in association with each activity configuration. For external application the overview will be more detailed as it will distinguish between uniformity and compatibility. Implications for services, due to their intangible nature, will be mentioned specifically in association with Standardization of processes. The concept of processes in this section refers to the way it is used in the Standardization literature (Botzem & Dobusch, 2012; Perez-Aleman, 2011). It refers to sequences that can be Standardized, for instance a specific way of executing a service in order to certify a Standardized minimum quality level. This delimited nature makes a process narrower than activities in most instances and consequently narrower than activity configurations that are systems of interlinked activities. Although a process may cut across a set of interlinked activities it most often does not encompass all the aspects of an activity configuration so that it entails the whole of a Business Model in the form of an activity system.

The Role of Standardization Types in the Network Configuration

Product Standardization for the *Network Configuration* appears primarily for external application with a compatibility dimension in order to facilitate network effects and reduce transaction costs. As an example, consider the Android operating system, including application development software, an output that is utilized by handset providers and application developers and thus enables a network effect by functioning as a platform, but still largely controlled by Google. Therefore, the prime focus of Internal Standardization for the output market in accordance with Table 2 is *Platform Standardization*. Also for the input market the prime type of Standardization will be *Supply platform Standardization*. However, when providing platforms that facilitate network effects, and especially double-sided network effects, the distinction between supplier and customer is not that necessarily clear. Players connecting to the network can be both suppliers and customers simultaneously.

Process Standardization in association with the *Network Configuration* is most prominent for service organizations as externally applied services with a compatibility dimension to enable network effects, i.e. execution follows a Standardized pattern which enables several customers and several sides of a market to access the network. The main types in accordance with Table 2 are *Time dependent platform Standardization and Time dependent supply platform.*

The Role of Standardization Types in the Package Configuration

The prime type of Standardization is product Standardization for external application with high uniformity, either in the form of a product or service concept (the Package Configuration also includes signalling of offerings, rather than actual deliveries, to the environment such as through marketing, see Johansson & Jonsson, 2012). Also, the Package Configuration may involve internal interface Standardization in order to enable modularity (cf. Sanchez & Mahoney, 1996) of a concept. In addition, an internally developed process, guiding execution, may serve a purpose of sending signals to external observers (Okhomotovskiy & David, 2012). This is the case when signalling a Standardized service delivery, i.e. what is signalled is the promise of the delivery. When it comes to diffusion, the firm may very well contribute to diffusion of its own Standards or adopt external Standards within a delimited area. However the prime outcome of the package logic, i.e. the packaged offering (or Standardized solution) will most often retain some firm uniqueness unless the firm has internal cost advantages versus its competitors (which are then a result of uniqueness from its chain logic aspects) or if uniqueness can be protected through other means, such as unique access to other sources of value capture which benefit from complementarity (cf. Teece, 1986). Thus, the prime type of Standardization is *Output Product Standardization*.

Process Standardization for the *Package Configuration* may appear through internal application to enable problem solving (although the character of activities within the configuration will always contain some uniqueness as it involves solving a problem which is new or unique). This is different from the signaling described above, as it relates to the actual execution. An example of this is the way consulting firms Standardize their approaches (Internal Standardization or adaptation stemming from external Standards) to assist client organizations (Werr, Stjernberg, & Docherty, 1997; Wright et al, 2012). Standardization may also appear in external application, in particular for services, as the package logic may facilitate development of a Standardized services execution concept. Thus, process Standardization in association with the Package Configuration is relevant for facilitation of service execution, i.e. *Services execution stadardization* in Table 2.

The Role of Standardization Types in the Chain Configuration

Product Standardization for the *Chain Configuration* occurs through sub-deliveries within the organization (the extent depends on whether execution is broken down into subsets which require uniformity and compatibility; compare Thompson, 1967 and March & Simon, 1958) to produce a single kind of output. Thus, the prime area of internally controlled Standardization of output for the Chain Configuration is: *Internal interface Standardization* and *Output Product Standardization* but also *Supply Product Standardization* to optimize input conformity with internal execution. The difference from the Package Configuration lies in the static view of the output. The Package Configuration focuses on the creation of new concepts with innovation value but Standardized over a population of customers.

Process Standardization for the Chain Configuration appears to facilitate repetitive execution and internal efficiency. High diffusion of process Standardization may be favourable if uniqueness can be protected by other means, such as unique access to other sources of value capture which benefit from complementarity (here without network effects). For instance, if a firm benefits from adhering to (or diffusing its own) process Standards in an industry while maintaining uniqueness in other aspects of its processes or its offerings, and thus maintaining cost advantages, differentiation, or advantages through complementarity, diffusion may be favourable. Therefore, the main types of Standardization are related to *Internal Standardization for efficiency* and *Services Supply Timing* to align input, and due to the single type output also *Services execution Standardization.*

The Role of Standardization Types in the Shop Configuration

The prime character of the value *Shop Configuration* is the solving of customer and temporally unique problems. The level of externally applied Standardization is by definition low and thus this section will only highlight a few items rather than go through all the sections of Table 2. Although internal processes and subdeliveries may have some common traits over time they also tend to have a high level of uniqueness from time to time and from project to project. In Stabell and Fjeldstad (1998), Standardized solutions, are included in the value shop but this chapter recognizes the central role of such as a key part of its own value creation logic and configuration in accordance with Johansson and Jonsson (2012). This is also in line with the way that Christensen et al (2009) view a Standardized execution within healthcare (although they directly categorize this as a chain logic example without recognizing the importance of creating Standardized solutions). An important exception to the above Standardized solutions in Stabell and Fjeldstad can be mentioned, though, and that is the Standardized process on the individual level, i.e. the individual professional has a Standard procedure for approaching a variety of problems through the information acquisition phase. This can be viewed as *Internal efficiency Standardization* on the individual level.

Internal and External Standardization Interrelations

To illustrate more in-depth how value configurations involve Internal Standards and how these interrelate with external Standards we provide two examples. The first one is Camfil, a developer and producer of filters and systems for air cleaning. Camfil relies on continuous development of the technology that is included in its products but which is also submitted to external Standardization in order to set minimum quality levels for systems (Freyschuss, 2015). Camfil has a long history of this way of working with Standardization. However, it is also evident that the approach relies on investments in improved or new product technology, which would correspond to a significant part of value creation relying on the Package Configuration.

Thus for Camfil, presence in external Standardization is important to influence external Standards setting to create minimum quality Standards and thus raising rivals' costs (Swann, 2000), as well as to

Table 3. Internal Standardization for each Business Model/activity configuration

		Network	Package	Chain	Shop
Process	Input Compatibility	Time Dep. Supply Platform	-	-	-
	Input Uniformity	-	Services Supply Timing	Services Supply Timing	
	Internal application	-	-	Internal for efficiency	Internal for efficency (individual)
	Output Uniformity	-	Services Execution	Services execution	-
	Output Compatibility	Time Dep. Platform	-	-	-
Product	Compatibility	Supply Platform	-	-	-
	Uniformity	-	Supply Product and Tech	Supply Product and Tech	-
	Internal application	-	-	Internal	-
	Uniformity	-	Output Product	Ouput Product	-
	Compatibility	Platform	-	-	-

align external Standards with its own Internal Standards which could potentially provide first mover advantages and temporary monopoly rents. This is based on the ability to refine or develop new technologies, concepts and products, which is the essence of the Package Configuration. This can then serve as a basis for setting the industry Standardization agenda.

Another example comes from the Volvo Group, where Internal Standardization is approached as a problem-solving mechanism, yet it is often triggered proactively and relates to early phase research and external stimulus. Here, the company is taking a product output Standard from internal to external in order to direct Standard setting within external Standardization committees and influence the external Standardization in the firm preference (proprietary innovation (Laursen & Salter, 2014; Leiponen, 2008; Rosenkopf, Metiu, & George, 2001)). That is, Volvo tries to influence external Standards and hence promote Internal Standards as private Standards towards suppliers and customers (Vandemoortele & Deconinck, 2013).

This is a case linking to Chain Configuration, where influence of external Standardization becomes crucial in order to discipline sub-deliveries, e.g. input from suppliers. In agreement with Stabell and Fjeldstad (1998) and Thompson (1967), the company's aim is to utilize long-linked technologies for the mass production assembly line, by ensuring Standardized input (through external Standards) aligned with the company's internal preferences.

The above-mentioned examples indicate that firm strategies versus industry level Standardization, depend on firm Internal Standardization in relation to the Business Model of the firm(s). Thus, to certify successful firm participation in external Standardization requires being able to develop Internal Standards that when appropriate can provide additional cost reduction or market value creation by being included in an industry Standard.

FUTURE RESEARCH DIRECTIONS

Further research should try to expand the above typology into external Standards and outline in detail how these vary with each activity configuration. As a consequence of such studies, further research could also apply the perspective of activity configurations to shed light on the relationship between internal and external Standardization from the perspective of the firm. For managers, understanding how the firm reduces variety through firm controlled action, and how this is related to value creation, should be a highly important input to firm participation in external Standardization. How do we Standardize internally? What type of value and cost effects does it have? Understanding this in detail should provide a better basis for involvement in external Standardization and an ability to decide what to share with other players, and in particular competitors, and what external Standardization the firm should try to hinder in order to protect capture of its created value (cf. Mione, 2015).

Through the discussion on Standards in association with activity configurations the use of the terms *Standard* and *Standardization* in organizational theory and in association with Business Models as activity systems is highlighted. Due to the interdependence with the concepts of practices and routines it is apparent that the basic function of reducing variety (across populations) and enabling compatibility are central to intra- as well as inter-organizational effects. Thus, in order to develop our understanding of Standards and Standardization further, more studies bridging these fields and clarifying the use of the concepts related to Standards in each setting (and determining commonalities and differences) should be pursued.

CONCLUSION

This chapter highlights the areas in which Internal Standardization is prominently associated with activity configurations (Johansson & Jonsson; 2012; Stabell & Fjeldstad, 1998) as Business Models (Christensen et al, 2008; Zott et al, 2011). This is important in the sense that it provides a map that guides Internal Standardization efforts related to value creation logic design and innovation.

For most organizations it is a matter of integrating two or several value creation logics into what Stabell and Fjeldstad (1998) call *hybrid* value configurations. Thus, from a Standardization perspective, the firm often has to combine its Standardization efforts in several areas. Similarly, firms in transition of their activity configuration dependence have to consider multiple types of Standardization. Innovating through value creation transformation, as described in Christensen et al (2009), i.e. a shift in its set of activities, their configuration or the parties involved, could be accompanied by an alteration of Standardization efforts. Christensen et al show how moves from reliance on a Shop Configuration via a chain and finally a Network Configuration can be seen as a disruptive innovation of the Business Model as opposed to disruptive innovation within the current Business Model. The typology proposed here of Internal Standardization in association with each configuration shows in what way Standardization will have to change.

ACKNOWLEDGMENT

This research was made possible through the funding provided by the Swedish Standardization Association. We would also like to thank the discussants and the reviewers at the 18th EURAS conference, Brussels 2013 and the EURAM conference, Valencia 2014 for insightful and valuable feedback.

REFERENCES

Amit, R., & Zott, C. (2012). Creating Value Through Business Model Innovation. *MIT Sloan Management Review*, *53*(3), 41–49.

Argyres, N., & Bigelow, L. (2010). Innovation, modularity, and vertical de-integration: Evidence from the early US auto industry. *Organization Science, 21*(4), 842-853. doi: 10.1287/orsc.1090.0493

Barzel, Y. (1982). Measurement Cost and the Organization of Markets. *The Journal of Law & Economics*, *25*(1), 27–48. doi:10.1086/467005

Besen, S. M., & Farrell, J. (1994). Choosing How to Compete: Strategies and Tactics in Standardization. *The Journal of Economic Perspectives*, *8*(2), 117–131. doi:10.1257/jep.8.2.117

Blind, K. (2002). Driving forces for Standardization at Standardization development organizations. *Applied Economics*, *34*(16), 1985–1998. doi:10.1080/00036840110111158

Blind, K. (2004). *The economics of Standards: Theory, Evidence, Policy*. Northampton, MA: Edward Elgar Publishing.

Blind, K. (2011). An economics analysis of Standards competition: The example of the ISO ODF and OOXML Standards. *Telecommunications Policy*, *35*(4), 373–381. doi:10.1016/j.telpol.2011.02.007

Blind, K., Gauch, S., & Hawkins, R. (2010). How stakeholders view the impacts of international ICT Standards. *Telecommunications Policy*, *34*(3), 162–174. doi:10.1016/j.telpol.2009.11.016

Blind, K., & Grupp, H. (2000). *Standards Statistics as New Indicators for the Diffusion of Technology*. Paper presented to the International JA Schumpeter Conference, Manchester, UK.

Boiral, O. (2012). ISO certificates as organizational degrees? Beyond the rational myths of the certification process. *Organization Studies*, *33*(5-6), 633–654. doi:10.1177/0170840612443622

Botzem, S., & Dobusch, L. (2012). Standardization cycles: A process perspective on the formation and diffusion of transnational Standards. *Organization Studies*, *33*(5-6), 737–762. doi:10.1177/0170840612443626

Brown, J. S., & Duguid, P. (1991). Organizational learning and communities of practice: Toward a unified view of working, learning and innovation. *Organization Science*, *2*(1), 40–56. doi:10.1287/orsc.2.1.40

Brunson, N., & Jacobsson, B. (2000). The contemporary expansion of Standardization. In N. Brunson & B. Jacobsson (Eds.), *A world of Standards*. New York: Oxford University Press.

Christensen, C. M., Grossman, J. H., & Hwang, J. (2009). *The innovator's prescription: A disruptive solution for healthcare*. New York: McGraw-Hill.

Cohen, M. D., Burkhart, R., Dosi, G., Egidi, M., Marengo, L., Warglien, M., & Winter, S. (1996). Routines and Other Recurrent Action Patterns of Organizations: Contemporary Research Issues. *Industrial and Corporate Change*, *5*(3), 653–698. doi:10.1093/icc/5.3.653

Cusumano, M. (2010). Technology Strategy and Management – The Evolution of Platform Thinking. *Communications of the ACM*, *53*(1), 32–34. doi:10.1145/1629175.1629189

Cyert, R. M., & March, J. G. (1963). *A behavioral theory of the firm*. Prentice Hall.

David, P. A. (1987). Some new Standards for the economics of Standardization in the information age. In P. Dasgupta & P. Stoneman (Eds.), Economic policy and technological performance (pp. 206-239). doi:10.1017/CBO9780511559938.010

Dokko, G., & Rosenkopf, L. (2010). Social capital for hire? Mobility of technical professionals and firm influence in wireless Standards committees. *Organization Science*, *21*(3), 677–695. doi:10.1287/orsc.1090.0470

Eisenmann, T., Parker, G., & Van Alstyne, M. W. (2006). Strategies for two-sided markets. *Harvard Business Review*, *84*(10), 92–101. PMID:16649701

Felin, T., & Foss, N. J. (2005). Strategic organization: A field in search of micro-foundations. *Strategic Organization*, *3*(4), 441–455. doi:10.1177/1476127005055796

Felin, T., & Foss, N. J. (2009). Experience and Repetition as Antecedents of Organizational Routines and Capabilities: A Critique of Behaviorist and Empiricist Approaches. *SMG WP*, *9*, 2009.

Foss, K. (1996). *A transaction cost perspective on the influence of Standards on product development: Examples for the fruit and vegetable market.* Working paper No 96-9. DRUID Working Paper Series, Copenhagen.

Freyschuss, A. (2015). Så blev en svensk koncern världsledande med hjälp av Standarder. *SIS*. Retrieved from http://www.sis.se/Global/Om%20SIS/kundseminarium%202011/camfil_ppt.pdf

Gallagher, S. (2007). The complementary role of dominant designs and industry Standards. *IEEE Transactions on Engineering Management*, *54*(2), 371–379. doi:10.1109/TEM.2007.893991

Gawer, A., & Cusumano, M. A. (2008). How Companies Become Platform Leaders. *MIT Sloan Management Review*, *49*(2), 29–30.

Ghaziani, A., & Ventresca, M. J. (2005). Keywords and Cultural Change: Frame Analysis of Business Model Public Talk, 1975–2000. *Sociological Forum*, *20*(4), 523–559. doi:10.1007/s11206-005-9057-0

Grossmann, A.-M. (2013). Suppliers' motives for applying company Standards – How company Standards can affect the buyer-supplier relationship. In *EURAS Proceedings 2013: The Euras Board Series.* Verlagshaus Mainz.

Guler, I., Guillén, M., Macpherson, J., & Guillen, M. F. (2002). Global competition, institutions, and the diffusion of organizational practices: The international spread of ISO 9000 quality certificates. *Administrative Science Quarterly*, *47*(2), 207–232. doi:10.2307/3094804

Helfat, C. E. (1997). Know-how and asset complementarity and dynamic capability accumulation: The case of R&D. *Strategic Management Journal*, *18*(5), 339–360. doi:10.1002/(SICI)1097-0266(199705)18:5<339::AID-SMJ883>3.0.CO;2-7

Henson, S. J., & Hooker, N. H. (2001). Private Sector Management of Food Safety: Public Regulation and the Role of Private Controls. *International Food and Agribusiness Management Review*, *4*(1), 7–17. doi:10.1016/S1096-7508(01)00067-2

Hill, P. (1999). Tangibles, intangibles and services: A new taxonomy for classification of output. *The Canadian Journal of Economics. Revue Canadienne d'Economique*, *32*(2), 426–447. doi:10.2307/136430

Holmstrom, B. (1999). The Firm as a Subeconomy. *Journal of Law Economics and Organization*, *15*(1), 74–102. doi:10.1093/jleo/15.1.74

Johansson, M., & Jonsson, A. (2012). The package logic: A study on value creation and knowledge flows. *European Management Journal*, *30*(6), 535–551. doi:10.1016/j.emj.2012.04.003

Katz, M., & Shapiro, C. (1985). Network Externalities, Competition, and Compatibility. *The American Economic Review*, *75*, 424–440.

Katz, M., & Shapiro, C. (1994). Systems competition and Network effects. *The Journal of Economic Perspectives*, *8*(2), 93–115. doi:10.1257/jep.8.2.93

Kogut, B. (2000). The network as knowledge: Generative rules and the emergence of structure. *Strategic Management Journal*, *21*(3), 405–425. doi:10.1002/(SICI)1097-0266(200003)21:3<405::AID-SMJ103>3.0.CO;2-5

Krechmer K. (2000). *The Fundamental Nature of Standards: Economics Perspective*. Paper presented at the International JA Schumpeter Conference, Manchester, UK.

Krechmer, K. (2005). *Standards Mark the Course of Economic Progress*. Unpublished Paper.

Laursen, K., & Salter, A. J. (2014). The paradox of openness: Appropriability, external search and collaboration. *Research Policy*, *43*(5), 867–878. doi:10.1016/j.respol.2013.10.004

Leiponen, A. E. (2008). Competing through cooperation: The organization of Standard setting in wireless telecommunications. *Management Science*, *54*(11), 1904–1919. doi:10.1287/mnsc.1080.0912

March, J. G., & Simon, H. A. (1958). *Organizations*. Cambridge, MA: Blackwell publishers.

Masten, S. E. (1991). A legal basis for the firm. In O. E. Williamson & S. G. Winter (Eds.), *The Nature of the Firm: Origins, Evolution and Development* (pp. 196–212). New York: Oxford University Press.

Matutes, C., & Regibeau, P. (1996). A selective review of the economics of Standardization. Entry deterrence, technological progress and international competition. *European Journal of Political Economy*, *12*(2), 183–209. doi:10.1016/0176-2680(95)00013-5

Miller, D. (1996). Configurations revisited. *Strategic Management Journal*, *17*(7), 505–512. doi:10.1002/(SICI)1097-0266(199607)17:7<505::AID-SMJ852>3.0.CO;2-I

Mione, A. (2015). The value of intangibles in a situation of innovation: Questions raised by the case of Standards. *Journal of Innovation Economics & Management*, *2*(17), 49–68. doi:10.3917/jie.017.0049

Narayanan, V., & Chen, T. (2012). Research on technology Standards: Accomplishment and challenges. *Research Policy*, *41*(8), 1375–1406. doi:10.1016/j.respol.2012.02.006

Nelson, R. R., & Winter, S. G. (1982). *An evolutionary theory of economic change*. Cambridge, MA: Belknap press/Harvard University Press.

Normann, R., & Ramirez, R. (1993). From value chain to value constellation: Designing interactive strategy. *Harvard Business Review*, *71*, 65–78. PMID:10127040

Normann, R., & Ramirez, R. (1994). *Designing interactive strategy – from value chain to value constellation*. Chichester, UK: Wiley & Sons.

Okhomatovskiy, I., & David, R. J. (2012). Setting your own Standards: Internal corporate governance codes as a response to institutional pressure. *Organization Science*, *23*(1), 155–176. doi:10.1287/orsc.1100.0642

Parasuraman, A., Zeithaml, V. A., & Berry, L. L. (1985). A conceptual model of service quality and its implications for future research. *Journal of Marketing*, *49*(4), 41–50. doi:10.2307/1251430

Parker, G., & Van Alstyne, M. W. (2003). Two sided networks: A theory of information product design. *Management Science*, *51*(10), 1494–1504. doi:10.1287/mnsc.1050.0400

Perez-Aleman, P. (2011). Collective learning in global diffusion: Spreading quality Standards in a developing country cluster. *Organization Science*, *22*(1), 173–189. doi:10.1287/orsc.1090.0514

Porter, M. (1980). *Competitive strategy*. New York: Free Press.

Porter, M. E. (1985). *Competitive advantage: creating and sustaining superior performance*. New York: Free Press.

Rathmell, J. M. (1966). What is meant by services? *Journal of Marketing*, *4*(4), 32–36. doi:10.2307/1249496

Ray, G., Barey, J. B., & Muhanna, W. A. (2004). Capabilities, business processes, and competitive advantage: Choosing the dependent variable in empirical tests of the resources based view. *Strategic Management Journal*, *25*(1), 23–38. doi:10.1002/smj.366

Reimers, K., & Li, M. (2005). Effectiveness of governance structures for 3G Standardization and implications for China's 3G policy. In P. Cunningham & M. Cunningham (Eds.), eAdoption and the knowledge economy: Issues, applications, case studies (pp. 256-263). IOS Press.

Rochet, J.-C., & Tirole, J. (2003). Platform competition in two-sided markets. *Journal of the European Economic Association*, *1*(4), 990–1029. doi:10.1162/154247603322493212

Rosenkopf, L., Metiu, A., & George, V. P. (2001). From the Bottom Up? Technical Committee Activity and Alliance Formation. *Administrative Science Quarterly*, *46*(4), 748–772. doi:10.2307/3094830

Salop, S.C., & Sheffman, D.T. (1983). Raising Rivals' Costs. *The American Economic Review*, *73*(2), 267-271.

Sanchez, R., & Mahoney, J. T. (1996). Modularity, flexibility, and knowledge management in product and organizational design. *Strategic Management Journal*, *17*(S2), 63–76. doi:10.1002/smj.4250171107

Sandholtz, K. W. (2012). Making Standards stick: A theory of coupled vs. Decoupled compliance. *Organization Studies*, *33*(5-6), 655–679. doi:10.1177/0170840612443623

Santos, F. M., & Eisenhardt, K. M. (2005). Organizational Boundaries and Theories of Organization. *Organization Science*, *16*(5), 491–508. doi:10.1287/orsc.1050.0152

Schilling, M. (1999). Winning the Standards race: Building installed base and the availability of complementary goods. *European Management Journal*, *17*(3), 265–274. doi:10.1016/S0263-2373(99)00005-5

Schumpeter, J. (1934). *The theory of economic development*. London: Transaction publishers.

Shapiro, C., & Varian, H. R. (1999). The art of Standards wars. *California Management Review*, *41*(2), 8–32. doi:10.2307/41165984

Sheehan, N. T., & Foss, N. J. (2007). Enhancing the prescriptiveness of the resource-based view through Porterian activity analysis. *Management Decision*, *2*(3), 240–260.

Stabell, C. B., & Fjeldstad, Ø. D. (1998). Configuring value for competitive advantage: On chains, shops, and networks. *Strategic Management Journal*, *19*(5), 413–437. doi:10.1002/(SICI)1097-0266(199805)19:5<413::AID-SMJ946>3.0.CO;2-C

Swann, G. P. (2000). *The economics of Standardization*. Manchester, UK: University of Manchester.

Swann, G. P. (2010). *The economics of Standardization: an update. Report for the UK Department of Business, Innovation and Skills*. BIS.

Tassey, G. (2000). Standardization in technology-based markets. *Research Policy*, *29*(4-5), 587–602. doi:10.1016/S0048-7333(99)00091-8

Teece, D. J. (1986). Profiting from technological innovation: Implications for integration, collaboration, licensing and public policy. *Research Policy*, *15*(6), 285–305. doi:10.1016/0048-7333(86)90027-2

Teece, D. J. (2010). Business Models, Business Strategy and Innovation. *Long Range Planning*, *43*(2-3), 172–194. doi:10.1016/j.lrp.2009.07.003

Tether, B. S., Hipp, C., & Miles, I. (2001). Standardisation and particularization in services: Evidence from Germany. *Research Policy*, *30*(7), 1115–1138. doi:10.1016/S0048-7333(00)00133-5

Thompson, J. D. (1967). *Organizations in action*. New York: McGraw-Hill.

Tu, Q., Vonderembse, M. A., Ragu-Nathan, T. S., & Ragu-Nathan, B. (2004). Measuring modularity-based manufacturing practices and their impact on mass customization capability: A customer-driven perspective. *Decision Sciences*, *35*(2), 147–168. doi:10.1111/j.00117315.2004.02663.x

Ulrich, K. (1995). The role of product architecture in the manufacturing firm. *Research Policy*, *24*(3), 419–440. doi:10.1016/0048-7333(94)00775-3

Vandemoortele, T., & Deconinck, K. (2013). When are private Standards more Stringent than Public Standards? *American Journal of Agricultural Economics*, *96*(1), 154–171. doi:10.1093/ajae/aat064

Warner, A., Fairbank, J., & Steensma, H. (2006). Managing uncertainty in a formal Standards-based industry: A real options perspective on acquisition timing. *Journal of Management*, *32*(2), 279–298. doi:10.1177/0149206305280108

Werr, A., Stjernberg, T., & Docherty, P. (1997). The functions of methods of change in management consulting. *Journal of Organizational Change Management*, *10*(4), 288–307. doi:10.1108/09534819710175839

Williamson, O. E. (1985). *The economic institutions of capitalism: firms, markets, relational contracting*. New York: Free Press.

Wright, C., Sturdy, A., & Wylie, N. (2012). Management innovation through Standardization: Consultants as Standardizers or organizational practice. *Research Policy*, *41*(3), 652–662. doi:10.1016/j.respol.2011.12.004

Zott, C., Amit, R., & Massa, L. (2011). The Business Model: Recent developments and future research. *Journal of Management*, *37*(4), 1019–1042. doi:10.1177/0149206311406265

KEY TERMS AND DEFINITIONS

Business Model: A system of activities not delimited to the firm itself but includes a system-level approach to how the firm interacts with its environment and does business.

Chain Configuration: Utilizes long-linked technologies in the transformation of input to output.

Internal Standard: Serve to reduce variety internally across products and production processes and as a consequence between intraorganizational entities. In many instances these follow internal consensus processes.

Network Configuration: Links customers through mediating technologies.

Package Configuration: Solves problems across a range of potential customers, with the goal of making such common problem-solving efficient through underlying solutions.

Shop Configuration: Utilizes intensive technology to solve customer unique problems through iterative problem solving.

Standard: Serve to reduce variety across for instance products, processes and technologies.

Standardization: The process of creating and spreading Standards.

Value/ Activity Configuration: Activity organization of a firm for value creation.

ENDNOTES

1. Stabell and Fjeldstad's original label is *value configurations* since what Stabell and Fjeldstad call the *value creation logic* is at the center of each configuration. This focus on value is in line with Porter's original idea of value created by the things that firms do, the activities (Porter, 1985; Sheehan and Foss, 2009). Here the focus is on the configurations as generic models of activity organization.
2. The Standardization literature is defined here as management oriented literature that uses the word Standard or Standardization in the title, in the keywords or the abstract of the article or book.

Section 3
Developing Standards and Processes

Chapter 5
Strategic Development and Implementation of Company Standards

Anne-Marie Christin Großmann
Technische Universität Berlin, Germany

Paul-Vincent von Gruben
Technische Universität Berlin, Germany

Luisa Kim Lazina
Technische Universität Berlin, Germany

ABSTRACT

This chapter provides a thorough introduction to the development of company standards for strategic purposes with specific focus on the use of company standards in the manufacturing sector. It gives an overview of the different types of company standards that are implemented by a company and its supply chain partners. By highlighting similarities with and distinctions from formal standards, this chapter carves out different economic effects of company standards. This allows the authors to develop different strategies a company can follow with the development of company standards. These strategies, however, are not confined for internal aspects. They also include the impacts of company standards on others, such as suppliers. This book chapter aims to aid companies and their supply chain partners to optimally use company standards for economic optimization.

INTRODUCTION

[Company standards] can de facto affect one, a few or many enterprises in a branch (Blind, 2004, p. 66)

This chapter provides companies with a toolbox of possible strategies for the development and implementation of company standards. The main aim is to aid strategic management by using company standards, to achieve economic optimization within the own company but also from passing them to supply chain partners, such as suppliers or cooperation partners. We will consider how company standards can affect

DOI: 10.4018/978-1-4666-9737-9.ch005

both the standard-setting company but also standard-receiving supply-chain partners. We outline the different types of company standards that can be implemented within a company and consider how they can support the strategic management of the company for maximum benefit.

We evaluate numerous aspects that are covered in company standards as well as their economic effects for the firm. The first aim of this chapter is to distinguish company standards from formal standards. Although often similar aspects are covered in company standards as opposed to formal standards, we will specifically highlight their differences by considering the economic effects respectively. We support this analysis with data from a survey of German companies as well as numerous specific company insights. The focus of this chapter rests on firms from the manufacturing sector, but parts of our analysis may also extend to firms within the service sector, which we discuss later.

We proceed by explaining the use of company standards within a company in detail. Although the main aim is the optimal development and application of company standards within the own company, we also consider how company standards can be passed to suppliers in order to govern their activities for mutual benefit. This chapter therefore does not only help general or standardization managers within standard-setting companies, but also suppliers to efficiently implement company standards. The joint acceptance of company standards can enable the optimization of procedures from the development onwards, so that for example cost reductions can be realized.

The chapter is built as follows: A review of the definitions as well as a short overview of the characteristics and the development process of both formal and company standards allows a clear delineation of the two types. Different types of standards are considered and we show which of these types are actually developed and implemented within exemplary companies. This sets the scene for a review of the purposes for and the economic effects of company standards. Particular attention is paid to impacts on the company-standard-setting company but also on the company-standard-receiving company, such as the supplier. The final section of this chapter reviews the strategic potential for company standards to be implemented, such as internal optimization, supply-chain optimization and strategic positioning.

BACKGROUND

Company standardization can be viewed as the positioning of a company towards external standardization (de Vries & Slob 2007; de Vries, 2007). We, however, would like to add to the definition of company standards the specific standards that a company develops independent of formal external standards. The differences between the effects of company and external (such as formal or consortium) standards to date have obtained little attention (Riillo, 2013). This section describes the difference between formal standards and company standards, starting with the definitions and theoretical descriptions that are followed by the types of standards.

Shin et al. (2015) highlight how standards allow making the specific difference in the success of innovations. Standards have an impact by establishing the rules of the game and creating a shared framework for innovation. They point out that standards define common vocabularies and set the essential characteristics of a product or service. These aspects also apply to the use of company standards with regard to company internal innovation efforts. Therefore, company standards can be a catalyst and help to ensure fruitful results (Shin et al., 2015).

Before a standard can be implemented it has to be developed. The standard DIN EN 45020 of the German Institute for Standardization (DIN) defines the standardization process as an "activity of establishing, with regard to actual or potential problems, provisions for common and repeated use, aimed at the achievement of the optimum degree of order in a given context [...]" (DIN EN 45020). We now consider how this process of standardization and therefore the agreement on a standard differs between the company level and a superior external level.

We differentiate between formal standardization that can result in national, international and European standards (DIN 820-1), for example DIN, EN, ISO/IEC standards and standards that are developed within a company, which we refer to as company standards (see the formal definition below). These company standards are used within the company or passed on to cooperating or supply-chain partners, and can be developed and implemented as a result of different motivations. In this chapter the authors' main focus is on the result of the standardization process, i.e. the standard. On a formal level, this process means consensually setting standards within technical committees with the outcome of consensus-based formal standards. On a company level, in contrast, this process is defined internally which is generally done independent of other companies. Usually during this process of the company standard development, first, an internal department hast to make a proposal for a standard or a standardization-relevant topic. The standardization department hast to consider whether the topic is economically viable for becoming a standard. If this test has been passed, it has to be put in the form and structure of a company standard. Additionally it must be accepted by all affected relevant departments. By considering the purposes and effects that the standards will have, which we will develop in more detail in this chapter, companies can also enhance the process of company standardization.

The Fundamental Differences of Formal and Company Standards

Jakobs defines standards as "a publicly available definitive specification of procedures, rules and requirements, issued by a legitimate and recognized authority..." (Jakobs, 2000, p. 11). This is applicable for formal and company standards, but there are also significant differences between them. By stating the formal, "standardized" definition of a formal and a definition of a company standard as used by the industry, we want to indicate these differences between company and formal standards.

- **Definition of Formals Standard:** "[Formal standards are] document[s], established by consensus and approved by a recognized body, that provides, for common and repeated use, rules, guidelines or characteristics for activities or their results, aimed at the achievement of the optimum degree of order in a given context.[...]" (DIN EN 45020)
- **Definition of Company Standards:** Company standards are documents with generally basic content, wide application and repeated use to set rules or characteristics for objects and operations. These are established by consensus within the company and accepted by applying or competent actors that provide, for common and repeated use, rules, guidelines or characteristics for activities or their results, aimed the maximum benefit for the company.

The above definition of company standards is based on information from the internal documentation of two German original equipment manufacturer as well as on the DIN EN 45020. It should be highlighted that the standard developing actors are employees of different departments of the respective company and also suppliers of the companies. To be able to distinguish between the different influences of company standards both on the standard-developing company as well as the standard-receiving suppliers, we further differentiate between internal and external company standards. In this aspect Blind and Großmann (2014) differentiate company standards based on their origin respective to the focal company where the company standard is implemented. Following their definition, company standards that are developed within the company and that are implemented in the company as well as cooperating companies and suppliers are called "internal company standards". Company standards, on the other hand, that are implemented within the focal company, but that have been developed by another company are so-called "external company standards". Hence, the internal company standards are those of the standard-developing company and the external company standards those of the standard-receiving company.

As the definitions have shown, formal and company standards can be differentiated by their scope of application: Company standards are only used in the company and within the cooperating companies, whereas formal standards can be applied by anyone. Furthermore, formal and company standards can differ for example by their user group, by individual, level and intention of use, scope and content as well as the development (both on the process level as well as in their developing group). The characteristics and differences are shown in table 1.

There are fundamental differences in the development process of formal and company standards. Formal standards are not developed on the company-level, but within external standardization committees that include actors of multiple organizations. The result is therefore a joint effort of interested members that partake in the standardization committee. The content of this final formal standard needs to be accepted by consensus voting within the committee. By being public goods, standards are accessible for the public and anybody can use them for a reasonable amount of money, with non-rivalry in consumption. This means that the implementation of these formal standards by one company do not affect the use by another company. As the companies participating in formal standardization activities are incurring costs of participation, we expect the benefit from participating to exceed the costs of the participation. These benefits can include for example cost savings or an increase in market share Researchers found

Table 1. Characteristics of company and formal standards (adapted from Feuchter, 1996, Swann, 2000 and DIN, 2000).

	Company Standards	Formal Standards
Body of experts	closed	open
Technological system	closed	open and transparent
Consensus	low	high
Development time	low	high
Public participation	no	yes
Bindingness	low	high
Development	independent	joint effort
Contributors	one	many
Type of good	private good	public good
Market form	monopoly	perfect competition
User group	company and partners	everybody who is willing to pay
Level of use	in the company and at partners	national, international or/and European level
Market positioning	value chain managing and optimization of internal processes	market entry preserving and reduction of trade barriers

that standards can lead to cost savings and an increase in market share (Swann, 2000; de Vries, 2007; Weissinger, 2013). Due to the accessibility of formal standardization committees, the content of formal standards is generally not specific to one particular company. Company standards, on the other hand, are primarily aimed at the economic benefit of one particular company. Furthermore, there are not only companies which influence the outcome of the standardization process at *standard development organizations* (abbr.: SDOs). but also e.g. environmental, political or other public organisations.

Company standards are developed by a company for the use of the company itself. Hence, they are specifically developed for the needs and interests of the company. However, they can also be developed and implemented when formal standards on a particular topic are not specific enough, or when no formal standards on a topic exist. Furthermore, company standards fuse and help to increase the product quality and are faster developed than formal standards (Großmann & von Gruben, 2014).

The general development process of company standard proceeds can be exemplary outlined as follows[1]: The respective department aiming to draft a new company standard will write a formal application addressed to the standardization-responsible department or the standardization responsible person. The standardization department correspondingly reviews this application based on the merit of the technical contents or the general need for the standard for the specific company. Hereby it is important that the company standard does not infringe on any intellectual property rights. If all the requirements are fulfilled, the respective department together with the standardization department will develop a company standard draft. After this, company-wide consent is sought with representatives of all relevant departments. Usually, the relevant departments are given a period of one month for a potential appeal to the company standard. If the draft version of the company standard is accepted, in will be widely distributed either only within the company or also with external partners, such as suppliers.

To emphasize the different uses of the standard types for the companies, we consider data from the German standardization panel (German: Deutsches Normungspanel; abbr.: DNP) compiled by the Technical University of Berlin. This large-scale survey was answered by more than 1,300 companies from different sectors[3]. As can be seen in figure 1, company standards are rated on a scale of -3 to +3 as an average related to the importance for the companies (the averages are 1.43 for internal and 0.88 for external company standards), however less important compared to formal standard (2.02) (DNP, 2014). When we consider the different economic effects of standards, however, as shown in section 3,

Figure 1. Importance of different types of standards (DNP, 2014)[2]:

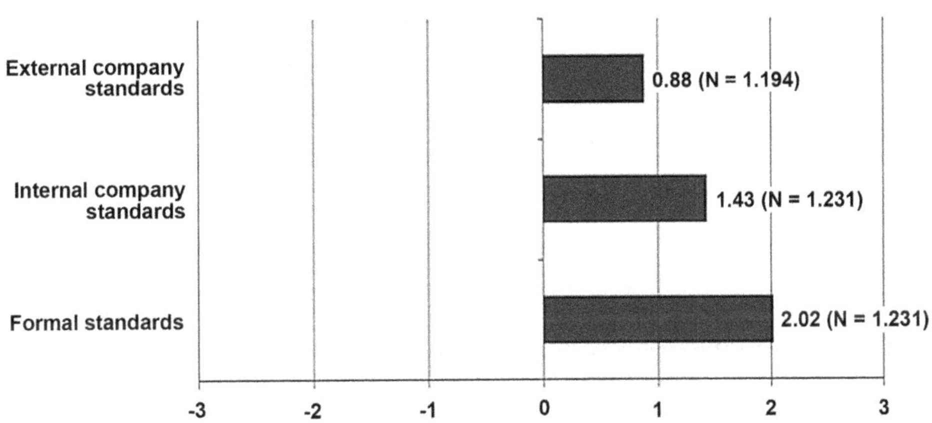

it will become apparent that compared to the overall importance of the standard types that this depends on the purpose that the standard aims to fulfil. In the following section we therefore consider different types of standards and compare these between company and formal standards to investigate whether differences exist.

Types of Company Standards

DIN EN 45020 defines eight types of formal standards: basic, terminology, testing, product, process, service, interface standards as well as standards on 'data to be provided'. In table 2 the existing formal standard types are compared with company standards that are used by a number of reviewed global German companies. The first company in the table is an engineering and manufacturing company of e.g. car parts, the second is an original equipment manufacturer and the third is an electronic component manufacturer.

In the following we provide a detailed description of the company standard types that are provided in table 2. We enhance the official definitions provided by the DIN with company-specific definitions in table 3 and include demonstrative examples.

These definitions of formal standards by the DIN EN 45020 however do not provide clarifications for *material standards*, although these type of standards exist in formal standardization, as shown in the table. They should therefore be explained separately, also because some companies have these as an additional type of company standards (see e.g. engineering and manufacturing company of e.g. car parts in Table 2).

Table 2. Comparison of the firms' equivalent company standards (of three global German companies) with formal standard types

Types of Standards	Engineering and Manufacturing Company	Original Equipment Manufacturer	Electronic Manufacturer
Basic standards		Basic standards	Basic standard
Terminology standards			Terminology standards
Testing standards	Testing standards	Testing standards, development testing standards	Testing standards i.e. test procedure standards
Product standards (Maßnorm)	Dimensional standards (Maßnorm)	Product standards	Product standards (Geometrical Product Specifications)
Process standards	Process standards	Process standards	Process standards
Service standards			
Interface standards			
Standards on data to be provided			
Material standards	Material standards	Material standards	Material standards
Quality standards	Quality standards	Quality standards	Quality standards
Delivery standards	Delivery standards	Delivery standards	Delivery and packaging standards
Construction standards		Construction standards	Construction standards

Table 3. Comparison of demonstrative examples for formal and company standard types

Standard Type	Description / DIN EN 4520	Formal Standard Example	Company Standard Example
Basic standard	A standard that has a wide-ranging coverage or contains general provisions for one particular field	"Basic standard for the evaluation of human exposure [...]" DIN EN 50505; VDE 0545-21:2009-03	Specifies the dimensions for the plug gauge
Terminology standard	A standard that is concerned with terms, usually accompanied by their definitions, and sometimes by explanatory notes, illustrations, examples, etc	"Preparation and layout of terminological specifications in standards" DIN 2344:2000-05	Numbering system or marking of characteristics in drawings
Testing standard	A standard that is concerned with test methods, sometimes supplemented with other provisions related to testing, such as sampling, use of statistical methods, sequence of tests.	"Testing of sealing compounds in buildings constructions [...]" DIN 52455-1:2003-05	The test procedure for the covering powder on a surface
Product standard	Is a standard that specifies requirements to be fulfilled by a product or a group of products, to establish its fitness for purpose [...]	"Product standard to demonstrate the compliance of fixed equipment for radio transmission [.. .]" DIN EN 50401:2012-06	Information on the hazardous substances in a vehicle
Process standard	A standards that specifies requirements to be fulfilled by a process, to establish its fitness for purpose	"Requirements for procedures and their application in job related proficiency assessment" DIN 33430: 2002-06	Assembly of non-metric threads about the surface treatment
Service standard	A standard that specifies requirements to be fulfilled by a service, to establish its fitness for purpose [...]	"Integrated Services Digital Network (ISDN); Connected Line Identification Presentation (COLP) supplementary service [...]" DIN EN 300097-1	Not used as particular named service standards in company standards
Interface standard	A standard that specifies requirements concerned with the compatibility of products or systems at their points of interconnection	"Postal services - Open standard interface between image controller and enrichment devices [...]" DIN CEN/TS 15448: 2007-01	Not used as particular named interface standards in company standards
Standard on data to be provided	A standard that contains a list of characteristics for which values or other data are to be stated for specifying the product, process or service [...]	"Sterilization of medical devices - Information to be provided by the manufacturer for the processing of resterilizable medical devices" DIN EN ISO 17664:2004-07	Not used as particular named as standard on data to be provided in company standards
Material standards	***NOT USED IN DIN EN 4520*** *Material standards* can be seen as a special form of product standards, but they describe more the texture and elementary composition rather than product features. Hence they are listed as a separate more specific category.	"Ground granulated blast furnace slag for use in concrete, mortar and grout [...]" DIN EN 15167-1:2006-12	Information of powder coating as surface protection

All these standards can be primarily differentiated by their purpose and theme. The above table 2 highlights the different types of company standards as they are implemented in representative companies. However, it is not always easy to make a clear distinction between these standard types. Most of the mentioned formal standard-types have their equivalent on the company scope in form of company standards. As seen in table 2, the exemplary companies also have company standards for the basic-, testing-, process-, product- and material standards and are used as company standard types.

It can further be noted that the company standard types do not differ much between the selected companies. However, within their company standard portfolios, the companies seem to differentiate more finely between additional types of standards. This aspect indicates that such additional company standard types provide the standard-developing firms with additional benefits. We therefore proceed to provide definitions for the additional types i.e. delivery, quality and construction standards, to delineate their scope from the already introduced standards. A *delivery standard* is a standard that is concerned with the requirements for purchased parts and their testing[4]. An example can be a delivery standard for fuels, hoses and lines. This type of standard is closely related to *product* and *testing standards*. Because this standard type is very specific to the firm, this is not a topic considered relevant in formal standardization. Another, more specific, example of a company delivery standard could be a standard that detects the conditions sets out for the material requirements of hose fuel system detections for example with specific information on the delivery terms (e.g. hose length for testing), nature, execution, identification, storage time of the delivered parts, etc.

A *quality standard* can also be very specific to the focal company similar to the delivery standards. These are mainly used by the internal production department of the standard-developing company and the suppliers to ensure a specific level of quality[5]. Examples are for instance the consistence of paint and the strength of adhesives or level of underbody protection. A further example is provided by quality requirements for topcoat[6]. The quality is a significant aspect for companies. This can be the driver to develop corresponding company standards because it does not only allow for internal quality assurance but these standards can also be required to manage a supplier's quality (de Vries & Slob, 2007).

A *construction standard* is a standard that is concerned with the description of the engineering of parts. An example could be the car design guide for light metal casting alloys which describes the recording points for measuring and straightening[7]. Similar to the delivery standards the construction standards are very firm specific and are usually not a topic for formal standardization.

These additional definitions show that formal and company standards can differ in their types, depending on the purpose of the standard as well as in their economic effects. As mentioned, company standards are generally used only within the company and cooperating companies whereas formal standards are used by a wider range of many different organizations. Consequently, companies need these additional types of company standards to complement their specific needs. Some formal standards that are primarily used in the context of formal standards are for example the terminology standards, because these are provided for a global base and are relevant across companies. It therefore seems natural to unify them for all interested circles and not provide company-specific standards on the terminology.

Standards like delivery standards or testing standards describe the relevant information for the supplier or external companies and can therefore be also passed on to supply chain or cooperation partners. However, company standards that are exclusive to internal processes exist. These are mainly for internal use or even have "confidential" information that the standard-setting company does not want to provide to other companies, and can be company standards like process or construction standards, developing or testing standards (used in the developing process).

Lazina (2014) found that the formal standards that are implemented within a global car manufacturer mainly concern materials and are therefore material standards. These types of standards can be used well in the format of the formal standards, because they have a very general content and are not company-specific. Standards with the content of vehicle specific information and vehicle parts data, manly describe product requirements. These figures are usually captured in product standards, which are mostly company specific and therefore formulated in company standards.

Independent of the type of standards, they all are important for the companies and other interest communities to diffuse knowledge. Standards help to codify the state of art and support the legal certainty to the user of a standard on a specific level or context, furthermore on these issues in following section. The use of formal standards is generally voluntary, but in the realm of company standards they can be a requirement from suppliers in order to get into trade relation.

As described in this chapter, company and formal standards differ. The next chapter will show that company standardization goes beyond the positioning achieved with external formal standardization.

PURPOSES OF COMPANY STANDARDS AND ACCOMPANYING EFFECTS

The "creation and diffusion of standards underlying new technologies is a driving element of contemporary globalization" as emphasized by network sociologists (Grewal, 2008).

As was introduced in the last section, company standards can be differentiated into internal and external company standards to get a decisive understanding of their purpose. Internal company standards are developed by the company which applies them. External company standards are developed by a business partner and transferred to the focal firm which then has to fulfil them (Blind & Großmann, 2014). In line with this explanation we distinguish between the purposes for developing company standards as well. First we describe the purposes with a company's internal origin e.g. internal optimization, quality assurance and performance improvement or differentiation opportunities. Then, in the following, we describe purposes for using company standards externally (by passing them to business partners) e.g. to manage the suppliers in terms of product quality or delivery terms.

Internal Purposes and Effects of Company Standards

The reason for companies to invest time and money to develop company standards concern issues which can be essential for day-to-day business such as testing or delivery terms and production processes. However, next to these more obvious factors, research has shown that also economic and legal issues are reasons to get involved in the development of company standards (Großmann & von Gruben, 2014). For those purposes especially product, material and quality standards appear to be of high relevance.

In figure 2 we present more results from the German Standardization Panel 2014 (DNP, 2014). The results show how the participated companies rate the importance of either formal standards, external company standards or internal company standards related to different business aspects (e.g. legal certainty, compliance with market access conditions, bargaining position, etc.). The rating is based on a Likert scale from -3 (very unimportant) to +3 (very important). It can be derived from the figures that company standards are on average rated especially important for *internal optimization factors* such as "quality improvement"[9]. Furthermore, the importance of company standards on *differentiation creating factors* such as "bargaining position"[10] and "competitiveness"[11] was rated high. Moreover, especially the importance of internal company standards in relation to the *performance improvement factor* "productivity"[12] was rated high by the participants.

On the basis of the findings from the standardization panel we will differentiate three types of factors to describe why firms develop company standards. These are 1) *internal optimization factors*, 2) *differentiation creating factors* and 3) general *performance improvement factors*.

Figure 2. Evaluation of the factors for the development of standards[8] (DNP, 2014)

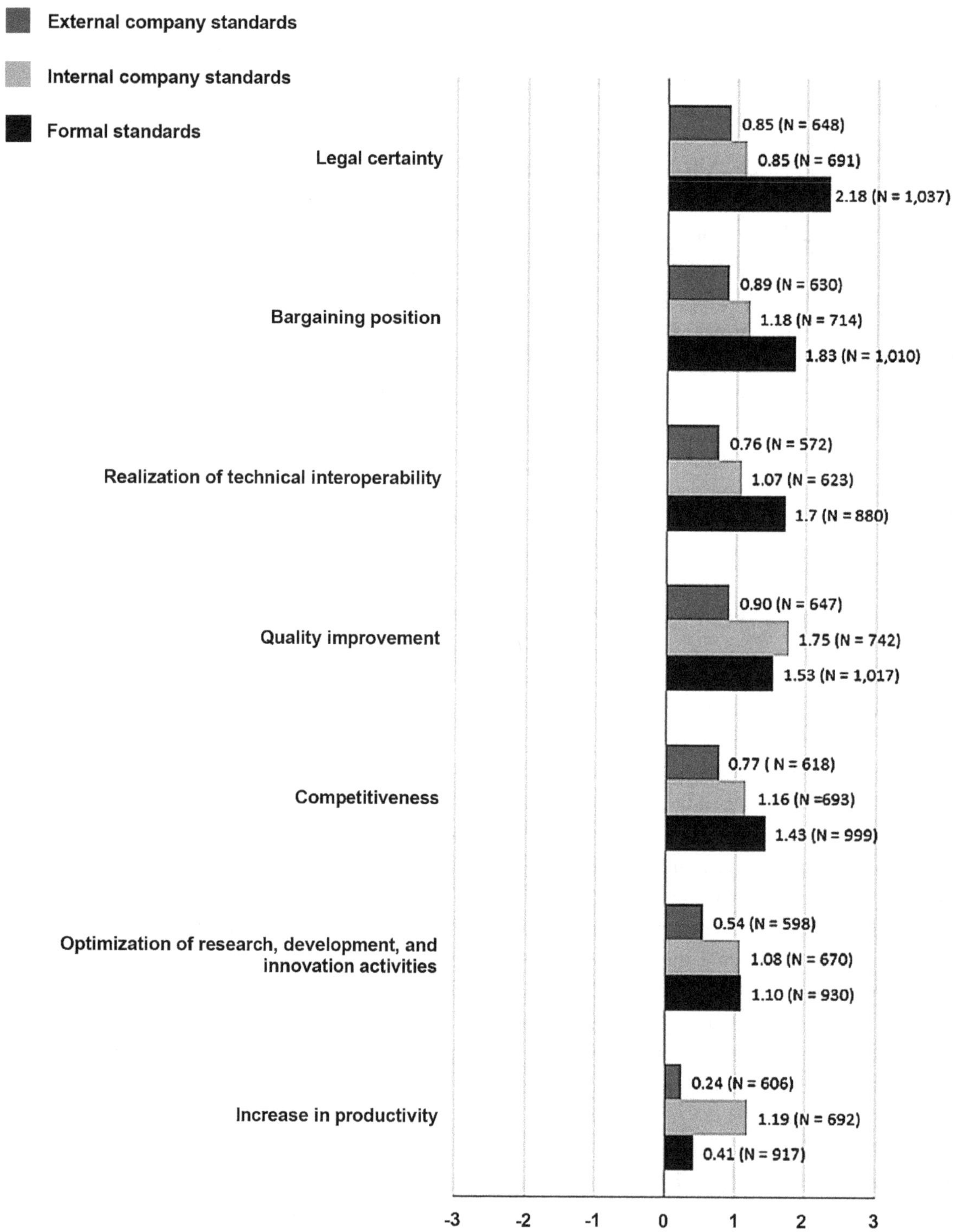

Internal Optimization Factors

Internal optimization factors include purposes for developing company standards which aim to improve products, processes and operations. These company standards are for example used as mechanisms, which helps to ensure *steady quality levels*. This means having the same quality levels throughout the whole company (see de Vries & Slob, 2007). Company standards for quality assurance can for example include minimum requirements, regarding tolerances of a product, pollution limits for a material or minimum life expectancies of a product. In addition, they can include procedure specifications to test for quality e.g. the number of testing cycles, the length of exposure to stress on a material or the time interval for the tests. Furthermore, company standards are used to document security and safety requirements as well as environmental requirements of a product.

In many cases company standards include specific knowhow of a company, for instance the requirements for materials which are used to achieve certain qualities. In this manner company standards may be used to document the knowledge accumulated by a firm. They help to describe the technological state of the art of the company (e.g. from the research and development department) and ensure the conformity with the latest technological state of the market (Bauer, 1980). Furthermore, as a company representa-

Figure 3. Overview on the purposes of company standards with internal origin (self-generated).

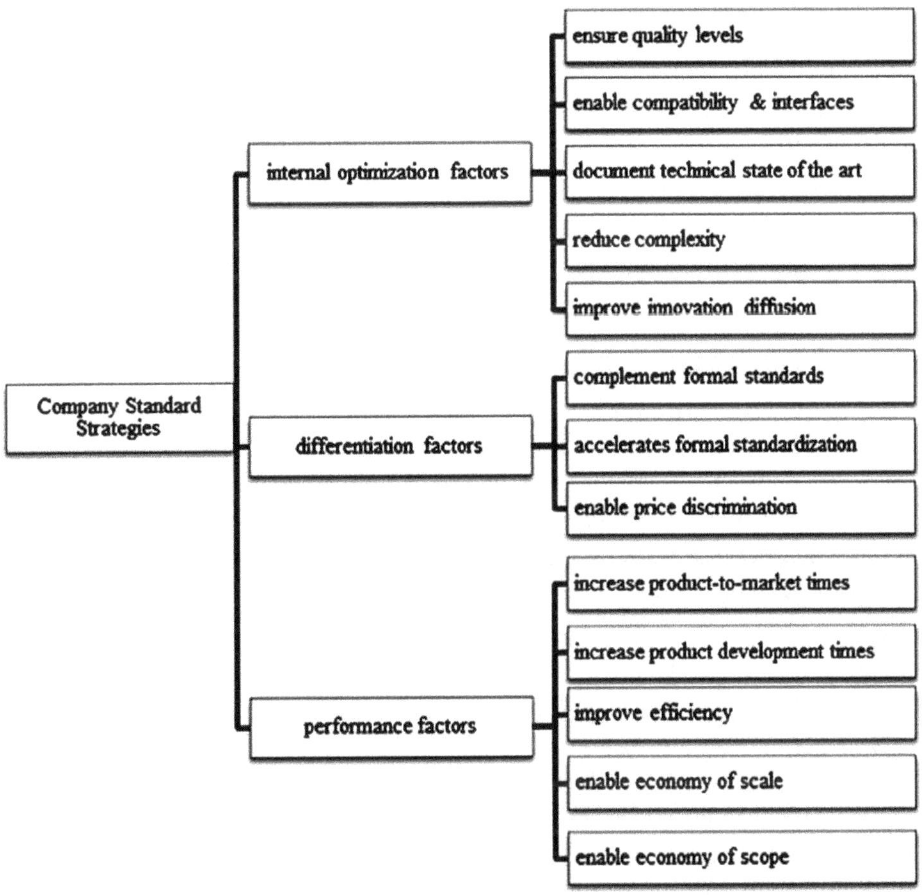

tive stated at a standardisation committee meeting[13], company standards are a medium to make certain knowledge accessible to the respective [legitimate] users within the company. For example when a specific issue is defined by the expert (e.g. from the R&D department) in form of a company standard, this can thereafter be applied by all business units (e.g. engineering or production division). Therefore, internal optimization can be fostered by the documentation and diffusion of knowledge via company standards.

By assisting to define the state of the art, company standards have the ability to align the different departments of a company on the same knowledge base, as described in the previous example. In addition to the already mentioned factors, this can help for faster diffusion and implementation of new technologies and innovations (Modi & Mabert, 2007; Iversen & Oversjoen, 2004) throughout the whole company. Creating a company internal standard on an innovative product or service and providing the standard to the equivalent departments eases the flow of that knowledge. Rapid diffusion of that knowledge thereby helps to improve the efficiency of a company. For example, if the R&D department improves the properties of a material (e.g. durability or solidness), incorporating this renewal into the company standard makes it instantly accessible to all other user of this material (Blind, 2004).

Further advantages result out of the collective use of the same information and knowledge codified in company standards. It ensures the compatibility of the different sub-systems of a company regarding their variety of components. This further helps to reduce complexity and simplifies the internal communication. Hence, less administration and documentation efforts are necessary (Blind & Bühring, 1996). Furthermore, company standards reduce the duplication of these efforts and consequently help to increase the productivity of the resources employed (Blind, 2004). For example during the product development process, company standards create a common ground for the different departments involved. Starting with the mechanical engineering, on to the electrical construction and even the information and communication technology developer can access a unified standard portfolio. This surely simplifies the consensual development and in the end supports the compatibility of the different components of a product. Furthermore, standardization helps to reduce the number of varieties which has an additional positive effect on the compatibility and reduces complexity.

Differentiation Creating Factors

This second section considers differentiation creating factors and provides examples for the use of company standards to distinguish own products from those available in the market. One way to look at it is to define formal standards as the common market knowledge; by using own standards with potentially proprietary knowledge instead, it creates opportunities for differentiation.

Three scenarios shall illustrate different ways how to use company standards compared to formal standards for differentiation purposes. In the first one, company standards are developed on the basis of a formal standard as a complementary element. Secondly, they are developed on issues which have not yet been standardized by SDOs. Third, company standards are developed instead of a formal standard to keep sensitive knowhow within the company; protecting it from spill-overs to competitors or the market (de Vries & Slob, 2007).

In scenario one, company standards can be seen as complementary element to formal standards[14]. They are used to reduce or focus the information codified in formal standards. Specifying the content and using a mutual language means that redundant and incomprehensible instructions can be excluded. Simplifying the formal standard furthermore increases the usability of that standard. Adapting the formal standard to the circumstances of the company or even to the respective department applying it reduces the complexity of formal standards in the form of a company standard. Additionally, company standards can adjust a formal standard to the terms of a company's specific requirements. For example regarding quality levels or environmental and safety regulations. Thus, the company standard can be identical to the formal standard regarding basic terms however they are more detailed on requirements where the company wants to differentiate themselves. This is especially applicable to product, material or testing standards.

The second scenario describes the case where due to the novelty of the matter, formal standards do not exist yet. Especially in fast changing markets and high-technology sectors, formal standards have not yet established or are too static; here company standards can be a substitute. Nevertheless, to exploit the full advantages of a newly developed product or service standardization in form of company standards is favourable. Creating a company standard will increase the diffusion of that innovation throughout the company, as mentioned before (Modi & Mabert, 2007; Iversen & Oversjoen, 2004). On the other hand, developing a company standard and pushing it to become a de facto standard in the industry or leading the way to formal standardisation can be a strategic advantage (Malisius & Weidner, 1998; Salop & Scheffman, 1983). More on the strategic use of company standards will be provided in the last section of this chapter.

The third scenario describes how company standards are employed when firms want to prevent knowledge spillovers (de Vries & Slob, 2007). Involvement in formal standardization activities holds the risk of leaking proprietary knowhow to competitors. Blind (2004) acknowledges that especially R&D intensive companies may be more reluctant to join standardization processes. These innovative firms prefer to market their products individually without a joint product design. Using company standards instead of formal standards moreover preserves the differentiation opportunities of a company. Although formal standards are beneficial for some characteristics of a product e.g. the interface or connection elements, yet for other qualities of a product formal standards would destroy the specific strength i.e. their *unique-selling-proposition* (*USP*). This however is the basis for their competitiveness.

Performance Improvement Factors

The following third part of this section describes the performance improvement factors for which company standards can be used. With the shortening of product life-cycles and the need for fast amortisation of (new) product designs; efficient product developments are required (see Tassey, 1995; Holler, 1996). Therefore, fast product-to-market times are increasingly important to generate sales from product developments earlier (e.g. Flaherty, 1986). Nevertheless, the speed of the standardization process on a national or even international level is rather slow (Farrell & Saloner, 1988). The standardization process of a committee of companies to strategically harmonize the contradictory effects of their current technological development can take up to three years (Kang & Motohashi, 2015). The opportunity costs of a delayed market entry of a product due to extensive negotiations on a common standard can be substantial (see Farrell & Saloner 1985; Katz & Shapiro 1985). Furthermore, product development and R&D are topics which often involve highly sensitive knowhow which companies prefer to keep within the boundaries of their company to avoid unfavourable spill-overs to their competitors. Hence, by using company standards, firms can improve the efficiency of their product development process on the one hand and retaining the knowhow private on the other. Revenue and costs are two exemplified performance improving effects achieved by using company standards. Firstly, earlier market rollout of a product due to shorter product developments increases total revenues and secondly they reduce the product development cost due to decreased excess duplication efforts for the market introduction of a new technology (Blind, 2004).

A further company performance improving factor is achieved through economies of scale. The already mentioned ability of company standards to diffuse knowhow across a company additionally enables the multiple uses of findings such as new or improved technologies from different areas of a company. This also enhances compatibility between products of different divisions of a company. A further exploitation of economies of scale comes from the variety reducing effect of standardization (see e.g. Dixit & Stiglitz, 1977; Blind, 2004). This is also true for the firm level. Reduction of varieties is seen as a chance for internal savings (see Farrell & Saloner, 1992; Hesser & Meyer, 1993; Meyer, 1995).

External Purposes and Effects of Company Standards

Reasons for Passing-On of Company Standards

Research on the use of company standards has shown that company standards are not only applied for internal purposes but also passed on to business partners and especially to suppliers (Großmann & von Gruben, 2014). The forwarded company standards can be used as a medium to manage the supplier base. They can be helpful in simplifying the correspondence and reduce the transaction costs. The passing on of company standards furthermore provide a contractual requirement and can therefore enhance legal certainty.

One effect of passing on company standards has on the buyer-supplier relationship is the reduction of transaction costs. This can be achieved by providing for example minimum qualities to the supplier, which do not have to be tested separately, and are therefore regarded as set (Blind, 2004). This can lower the transaction costs (see Tassey, 1982). They can also lower the uncertainty of the buyer and the risk for the supplier (of compensation or litigation). No independent and repeated testing is necessary. Since the standards are referenced in contracts and in requirement specifications they are thereby legally binding. The buyer can rely on the fulfilment of the requirements by the supplier (Dacin, Ventresca, & Beal, 1999; Hudson & Jones, 2001). Additionally, Kindleberger (1983) argues that standards do not only reduce transaction costs, but also with an increasing number of transactions enables further economies of scale.

Company standards can be used as a reinforcement of contractual relationships between buyers and suppliers or between cooperating companies. This can diminish problems of the usual relational agreements and contracts. Plus these contracts are generally aimed to achieve durable business relationships, which are often regarded as implicit and incomplete (Dölle, 2013). Company standards can enhance such relational contracts by providing additional specific information on aspects such as products, process, qualities and testing procedures. These standards can be a decisive element in the contractual relations and therefore should be considered when product and delivery specifications with supply chain partners are concerned.

Furthermore, next to formal standards, company standards can also sustain the fulfilment of specific requirement levels of suppliers especially regarding quality. On the downside this means that companies are forced to pass on some of the knowledge which is included in their company standards such as material composition, testing procedures, among others as described in section one (Großmann & von Gruben, 2014).

Apart from the above mentioned incentive to pass on their company standards, firms can furthermore use them to tie their supplier closer to them. If the buyers describe very specific requirements in their company standards for which the supplier needs to make product specific investments, which can be seen as potential switching costs. Increased switching costs can lead to a lock-ins which will be discussed in more detail in the next part. Furthermore, by the use of company standards, a firm is able to restrict a supplier to produce only for the specific customer. These component products are then only compatible with products of the focal firm limiting the exchange to that one buyer.

Dealing with Company Standards of External Business Partners

Analysing the buyer-supplier relationship from the perspective of the company standard-receiver, studies have shown that especially downstream companies (e.g. Original Equipment Manufacturer of the automotive industry such as Volkswagen AG or Daimler AG) develop a lot of company standards whereas upstream companies (e.g. suppliers) need to fulfil these external company standards (DNP, 2014; Großmann & von Gruben, 2014).

The dealing with externally-developed company standards describes the ability to operate independently in comparison to largely depending on the customer specifications. The ability to deal with the external company standards of different buyers can lead to difficulties which arise due to the heterogeneous requirements included in such company standards. Evidence from the automotive industry provides examples of three scenarios. In the first case, fulfilling the standard of a customer is mandatory to enter into the trade relation without room for further negotiation. In the second case, the supplying company which needs to fulfil the external company standard uses its bargaining strength to negotiate the received company standard in order to get to an agreed upon standard. In the third case the companies which are usually highly engaged in R&D and the development of their own company standards with high requirements already fulfil automatically the highest requirement of all of their customers which makes adjustments unnecessary (Großmann & von Gruben, 2014).

Some suppliers furthermore are able to bundle external company standards of different customers into their own internal company standard, for simplification reasons. In other cases companies need to split up external company standards so that they can be passed on separately to the respective sub-supplier.

Whereas passing on company standards to suppliers was found to decrease transaction costs on the side of the issuing company, it seems to increase transaction cost for the receiving company according to a study on the automotive industry (Großmann & von Gruben, 2014). The same could be observed regarding design and development times of new products. Self-developed company standards are perceived with a positive impact on internal processes but have a negative effect on the development times when received from external partners.

The fulfilment of external company standards was found to often increase the costs of a product. This means a supplier of a product has higher production costs due to specific requirements formulated in external company standards. Higher costs can arise due to extra testing procedures, higher quality levels than usual in the market, special customized solutions which are different from the standardized mass-product and thereby needs special efforts on the producer side.

RECOMMENDATIONS FOR THE STRATEGICAL EMPLOYMENT OF COMPANY STANDARDS AND THEIR EFFECS

At general management's core is strategy: defining a company's position, making trade-offs and forging fit among activities. (Porter, 1996)

A strategy is a firm's deliberate choice of its unique and valuable market position, where it becomes clear what a firm can (and cannot) do while it aims to integrate their numerous activities successfully. A strategy for company standardization activities will need to be in accordance with the overall managerial strategy of a firm. In addition, the standardization strategy can be defragmented into purely company internal standardization and industry-wide formal standardization. As outlined in section 3.1 these are often linked together. So, before a strategy with respect to company standards can be derived, the necessary first step is to decide whether a company wants to differentiate itself from the market. Otherwise existing measures such as formal standards could be used. Company internal standardization can be, for example, in terms of production techniques or quality levels. Or, as stated in previous sections, it provides standards interfaces that are specific to the firm.

Numerous companies proclaim to have the strategy to "set standards" – and thereby intend to differ from the market standard to win over customers. AirBerlin, for example, claims it is "*setting standards* in the aviation industry"[15]. Daimler AG wants to inspire their customers with "exciting premium automobiles that *set standards* in the areas of design, safety, comfort, perceived value, reliability and environmental compatibility"[16]. A comprehensive strategy how to develop a) own company standards, b) deal with other companies' standards or customize existing external (formal) standards to company needs and c) influence market standardization activities (e.g. formal or consortia standards) allows firms to achieve these noble goals of setting standards.

Making strategic decisions with regards to company standards requires the managers to consider activities both internal and external to the company. These internal and external activities are interlinked: Internal optimization, for example, can imply that variety-reducing standards or specific procedures are set and implemented according to internal company standards. Passing these company standards to development partners or suppliers can in turn enable them to achieve similar benefits through supply chain optimization. This moreover drives down costs, not only at the suppliers but also reduce the overall purchasing price of the product for potential buyers. Similarly, if a buyer imposes his company standards on the own supplier, this supplier needs to decide how to deal with such specific requirements – whether it makes sense to renegotiate such external company standards to align them with the own standards or whether to include these external company standards within the company's portfolio of standards, as they are.

As the "trade-offs that companies make are what distinguishes them strategically from other firms" (Collis & Rukstad, 2008) we need to consider the choices where a company decides what *not to do* (trade-offs). It is therefore important to point out whether a company aims its company standards portfolio primarily at *internal optimization*, *optimization of the own supply chain (i.e. supplier network)* or *strategic market positioning* for attaining a market or industry dominant standard. However, these goals are not mutually exclusive and companies can aim to find the most beneficial combination of them for their overall *strategy*. An effective *standardization strategy* depends on the objective as well as the company's environment.

The Development of a Strategy for Internal Company Standards

Internal Optimization, Variety Reduction and Differentiation

We assume three major *internal goals,* where company standards aim to influence internal activities. These are internal optimization, differentiation on the market, and internal learning. The latter enhances a subsequent conversion of former company standards into an industry or formal standard. This can amplify a first-mover advantage. Each of these goals requires a set of different types of standards.

Internal product and process as well as development, interface and compatibility standards on a company level can enhance *internal optimization* by communicating the importance of the firm-specific standard and to standardize processes. "Repetition" in products and production methods and therefore the development of routines can be supported by company standards (Anh, 2007). Internal company standards can enhance economies of scale and scope and ensure a constant quality of products and processes. *Variety reduction* occurs if a company clearly communicates that a certain product is the company's standard for internal applications and development, e.g. a certain type of screw used for multiple purposes. In this case, developers and engineers are compelled to apply the standard and not invent a new product or process for each application. This then induces economies of scale, where a higher volume of a smaller variety of inputs reduces costs.

Also platform strategies and complementary modularization especially benefit from matching and interface standards. These strategies are often employed by car developers as well as in other industries. Platforms are "standardized parts of a product" (Muffatto, 1999). These require technical compatibility of respective component products. The platform modularity can be enhanced by providing specifically developed interface standards between the platform and the add-on products. Apart from the aforementioned internal aspects, internal company standards can also aim at greater flexibility between individual plants of a company, increasing the use of plants and leading to improved performance as they reduce costs (Muffatto, 1999).

Another economic trend, modular production systems, is highly dependent on company specific standardization. The importance of compatibility of systems from different units of the same company or with cooperation partners supports the internal standardization detached from formal standards. The compatibility effects of a company standard furthermore enhance innovation. In the context of multiple proprietary components and product designs, standards provide the interface for coexistence, enabling innovation at the component level (Blind 2004).

By increasing own quality via internal quality standards, testing standards and setting higher product and process standards, a firm can *differentiate itself from their competitors in the market.* Companies, such as AirBerlin or Daimler AG, can proclaim to set new standards higher than existing (company) standard levels of their competitors. Such standards can provide signals to the consumer that the quality aspect is of such a specific concern to the company, that it has standardized this quality. Such quality standards suggest that technical requirements are made transparent and that a specific reputation is communicated. This differentiated product quality warrants a price premium above lower qualities in the market. If a company's standards have high recognition and acceptance, this can solve the problem of adverse selection (see Akerlof, 1970; Leland, 1979). If these standards can be credibly communicated and the reputation of the company can provide a "collateral" against the standard, then company standards provide a quality signal to consumers.

The aspect of differentiation, however, can also be targeted with a platform strategy by using standardized components to produce different products (Muffatto, 1999) which reduces costs for the company. However, a drawback of a "standard quality" can be the problem that consumers feel their individual needs neglected as they receive a standard product. To some extent, this can be avoided via a clever modularization strategy, where the end-consumer receives his preferred customized product made up of high-quality, standardized parts.

If the ultimate strategy of a company requires pushing own products, technologies or other aspect as the dominant standard into the market, an earlier internal development and implementation can enable internal learning and the "ripening" of a company standard that is subsequently introduced in the market or included at the beginning of the formal standardization process. If this standard is later successful in becoming a market or industry-wide standard, the firm can derive a *first-mover advantage*. Via their repeated application, aspects described in these company standards can furthermore enhance company-wide *organizational learning* (March, 1991).

An example provides the case of the Type-2-Plug for electric cars by Mennekes Electronics[17]. When the international standardisation body (ISO) decided to set a common international standard for enhancing the compatibility with charging stations, the company Mennekes Electronics had acquired lead time and presented an advanced solution for a plug for electric vehicles. The early experience and a detailed proposal for the formal standard on basis of their company standard had an influence on the successful conversion of Mennekes' company standard into the formal ISO/IEC standard (ISO/IEC 62196[18]). Company standards are developed through an internal learning process which allows the firm to present an applicable product as the standard proposal. Their first mover advantage and experience from the development also meant a lead in production after the Type-2-Plug was standardized. Additionally, Mennekes Electronics experienced brand recognition as pioneers in the sector.

Supply Chain Optimization

Additionally to internal goals, also *external goals* can be achieved with the development of company standards. By passing such standards on to supply chain partners for implementation allows a certain kind of governance. This governance can be structured between the extremes of market and the hierarchy, as a form of intermediate hybrid (see Ebers & Oerlemans, 2013) because they ultimately allow forms to implement a medium intensity of administrative controls. With regard to the intensifying supply chain competition this kind of governance enables business partners especially suppliers to be optimally equipped to maximize benefits for the supply chain. In times of globalisation, this kind of competition is increasingly concerned with entire value or supply chains (Gereffi & Lee, 2012). Hence, managerial strategies should take this aspect more and more into account.

Governing supply-chain partners via company standards can allow the standard developer (e.g. the buyer) to enforce and ensure a specific *quality level*, for example from suppliers. It can enhance the compatibility of sourced products with own products by enabling necessary *interfaces*. This however requires a specific level of power on part of the company standard developer. The power can either be manifested through the size, the position within the supply chain network with regards to both closeness to the end-customer and importance of the position itself. Furthermore, it depends on the unique capabilities a firm possesses – and whether it makes sense to potentially incur higher costs from fulfilling customer-specific requirements. Setting specific company standards that require suppliers to alter their production techniques or increase qualities can moreover increase the lock-in into the trade relationship. Such lock-ins result from relationship-specific investments that cannot be used beyond the specific business involvement. This effect arises on both sides; buyer as well as supplier can get locked-in.

Setting own company standards and passing these on to subsidiaries and other supply-chain partners located in different regions of the world facilitates development and engineering of products in compliance with the standard setting company. "Reductions in the numbers of platforms worldwide and the development of region dedicated platforms are valuable indications of the importance that is being attributed to product standardization" (Muffatto, 1999). The platform-owner can provide his compatibility or interface standards to supply chain partners and thereby enable their wider boundaries.

Strategic Market Positioning

We now consider two different ex ante scenarios that will determine strategies of how to set and implement internal company standards in the market: (1) the case where no company-specific standards exist for a product or technology and the company can be the first mover and (2) the case where competing standards exist and where the company is participating in the standard development process at a formal or informal standard-setting organization.

In the first scenario when a company is developing a new product, technology or process, standardizing certain elements of it can enable the differentiation via platform strategies. Furthermore, it can enable cost reductions, as mentioned above, but also accelerates *setting a completely new standard*. However, as all interested parties can participate in the formal standardization process, this provides any company with the possibility to influence formal standards by contributing content from own company standards. The possibly to contribute a company standard to the development process of a formal standard can be very valuable. A well-developed standard can allow the introduction and potential wider diffusion for implementation in the market. This could mean that a previous company standard gets widely accepted either as a de facto or a formal standard – not just on the side of the consumers but also on part of the suppliers and producers. If these standards include proprietary aspects, e.g. by including patented technology that has a potential for becoming an *essential patent* (see e.g. Lerner and Tirole, 2014), they can lead to a source of income for the standard setting company. If these, on the other hand, are free to be used within the whole market this can have a completely different effect. They could, for example, enable the sourcing of products according to company standards from a wider range of suppliers and at lower costs. An example of this latter case is screws openly sold via the large German screw retailer Würth to a company standard classification of Daimler AG[19].

Company standards also reduce the number of variations that are available for a product which is used or produced. The accompanying effect of this variety reduction is economies of scale. An example is provided when different business units or product classes are restricted to use the same standardized items defined in a company standard. The reduced number of variations allows mass sourcing of input materials (i.e. bulk purchasing) and therefore a reduction in costs. Furthermore, by restricting the use to certain products only, larger amounts of these particular products are needed instead of different variations. This therefore supports mass production. Both examples of economies of scale have a decreasing effect on the costs (Blind 2004).

Scenario two presents the case where specific characteristics of a company standard result in the existence of multiple company standards of different companies on a similar issue. This gives rise to an ex ante uncertainty about "which one will 'win' [and therefore dominate] the market"(de Vries, 2007).

"*Standard wars*" are the battles for market dominance between incompatible technologies (Shapiro & Varian, 1999). If a company is able to push its own company standard into the market by winning such a "standard war", it can benefit from a high market share. Examples of such standards wars are the battle between the DVD versus DIVX standard (Dranove & Gandal, 2003) or the case of Blue-Ray versus HD-DVD (Daidj, Grazia, & Hammoudi, 2010).

How much individual own content can become part of a formal or de jure (i.e. as imposed by the law) standard depends on the regulatory environment (Hesser, Czaya, & Riemer, 2007). Depending on the scope of state control and the possibility of private actors to operate independently, an individual company has the chance to include entire company standards or just parts of their company standards in formal standards. According to the DIN 820, some company standards manage to find their way into the formal standardization process (see e.g. Blind, 2004). An advantage of having an own company standards available within the standard-setting process is the readily formulated and internally proven technology documentation, which provides a solid groundwork in the negotiation process of the standardization committees.

The Development of a Strategy for Dealing with Externally Developed Company Standards

If a company receives company standards from an external organization, e.g. a buyer or cooperation partner, it needs to carefully assess whether it is able to implement these standards in accordance with its own strategy. Three strategic options are available in this case; in a first step: either accepting the external company standards as it is, or to try to renegotiate their content and requirements or else to decline them, running the risk of losing the trade relationship. The next step, after this decision is made, then includes whether to include the external company standards in the own portfolio of company standards.

A company can accept external company standards "as they are" if they can be implemented in the company without unreasonable costs or if this implementation provides an economic benefit for the firm. The first can be the case where, for example, quality or process requirements are such that they are already fulfilled by the firm anyway. The latter can occur if, for example, delivery specifications are required by external company standards that are not "so easy" implemented by the company and associated with higher costs. In this case it needs to be considered whether the compliance to the external company standard is indeed a necessity. In a qualitative study of the automotive industry, Großmann & von Gruben (2014) found that the compliance with external company standards is a necessary prerequisite to supply to the automobile manufacturer. However, if companies are in a stronger position within the supply chain network or have own technological developments that provide economically and technically advantageous solutions, they can try to renegotiate their buyer's company standards in favour of their own internal standards. Some suppliers even develop company standards together with their buyer. However, this can also induce the aforementioned problem of lock-ins into existing trade relationship. As a result only the particular supplier is able to produce to the required company standards and only the buyer is willing to pay the price premium for this non-market-standard solution.

If companies altogether neglect the implementation of specific external company standards this might induce that they completely lose the business relationship with the issuer of the company standard. However, if the company that is asked to implement the standard offers alternative, superior standards, such as its own internal company standards, a new solution could potentially be reached. Of course this depends on the economic viability of this alternative suggestion. It is therefore important to balance the implementation of own standards versus potentially available alternative company standards with regards to their economic benefits and strategically wider implications.

Finally, the handling of external company standards should include how they are incorporated in the own portfolio of standards, including own internal company standards. If they are of fundamental importance and relevant for long-term trade relationships, for example, inclusion of external company standards into the own portfolio of internal company standards – either in parts, as reference or as whole documents – then they can have considerable benefits. The communication, however, that an external standard becomes an internal standard, should carefully be analysed, as this can be a strong internal and external signal. In this light the potential length of the trade relationship should be considered.

FUTURE RESEARCH DIRECTIONS

This chapter has highlighted important aspects of company standards and their influences on firms. Due to the increasing globalization and the outsourcing of manufacturing, company standards become increasingly relevant (Gereffi & Lee, 2012). This is because company standards can spread company-specific know-how, not just within the individual firms, but also to supply chain partners and thereby optimize the entire supply chain. Company standards support this trend, for example via the provision of quality standards to suppliers.

This chapter provided a comprehensive perspective with reference to large-scale data and examples from companies in the manufacturing industries. Although the empirical data provides only a descriptive overview of the topic, it still enables the delineation of company standards from formal standards. This comparison could be enhanced by further analyses of synergies between formal and company standards. Even though the effects of both kinds of standards are known, the link between them is not yet been researched in depth. For example, the depth of topics covered in certain formal standards could be compared to the depth of these topics as they are covered in company standards. This understanding could enhance companies with an optimization of enhancing formal standards for internal use and converting them into company standards.

Extending the research on company standards to other sectors such as the service industry would allow delineating the strategies followed with company standards and the effects of these in such industries from the manufacturing perspective presented in this chapter. Company-specific standards in the service sector, for example, could include compliance standards or other process-related standards that need to be implemented. However, as a general rule we assume that the general outlook and strategic considerations of this chapter also hold for the service sector.

Furthermore, the research for this chapter is primarily based on findings from Germany for example from the German Standardization Panel (DNP) as well as insights from multinationals with headquarters in Germany. A wider perspective of different countries and also a comparison between countries would be valuable. This could be further enhanced with different time series data to analyse the implementation of both company and formal standards within companies over time and to get a better understanding of the dynamics.

Further research directions that arise from our excursion are the effects of company standards for standard wars and their outcomes. Although we looked at the influence of company standards on the development of formal standards by SDO committees, standard wars are rather won by de facto standards that result from networking effects. These effects are however increased by the use of standards (Blind, 2004).

CONCLUSION

This chapter provided definitions of different types of company standards as well as an in-depth understanding of the economic effects of company standards on companies. Comparing these company standards to the portfolio of formal standards that are defined by SDOs reveals that firms often use these two types of standards for different purposes. Firms rely on formal standards when it comes to general topics, for example described in basic standards and terminology standards or regarding interface standards. They do not individualize such terms and further have to ensure the compatibility with the interfaces on the market. Nevertheless, interface company standards are used when firms apply modular and platform strategies within their firms or partnerships. However, when it comes to products and processes, companies differentiate in more detail and create standard categories which are not defined by SDOs. Examples include quality standards, delivery standards and construction standards. The motivation to develop company standards can have a variety of reasons. The purposes of company standards were found to include internal optimization factors, differentiation creating factors and performance improving factors. These factors describe the benefits that company standards have on internal operations. However, firms can also use their company standards to help improve their supplier management for example to ensure steady quality of component deliveries. Governing the supply chain partners by passing on company standards as well as providing necessary knowledge helps improve the overall supply chain performance.

This chapter has further analysed the strategies that can be supported by the development and implementation of company standards. In their development process, company standards can be created on the basis of existing formal standards to adapt the requirements to the appropriate company. In other cases, company standards are developed and implemented to specifically differentiate from the common market standards. For example the setting of higher quality standards can be a strategic tool to compete in the market as they allow for a differentiated product. Furthermore, company standards deal with topics which have not yet been standardized by SDOs and can therefore enhance the standard-developing company with a lead time or first mover advantage. When formal standards are developed within the SDO committees, the company that has experienced internal learning effects by the use of internal company standards holds a strategic advantage when the design and technology has to be agreed upon within the standardization committee. Outcomes are thereby likely that favour the proven technologies or aspects that have already been developed and implemented and therefore tried by companies.

Introducing standard levels which are higher than the market standards was found to be used as a strategic instrument for example by Daimler AG. The signalling of their premium standards in the area of design, safety, comfort, etc. can be assumed to improve their brand recognition. Considering the development of a strategic company standard portfolio, as introduced in this chapter, is therefore important not only for standardization managers but also for general managers.

REFERENCES

Akerlof, G. A. (1970). Market for Lemons - Quality Uncertainty and Market Mechanism. *The Quarterly Journal of Economics*, *84*(3), 488–500. doi:10.2307/1879431

Anh, V. T. (2007). An Introduction - The History of Standardization. In W. Hesser, A. Feilzer, & H. de Vries et al. (Eds.), *Standardisation in Companies and Markets* (2nd ed.; pp. 45–76). Hamburg, Germany: Helmut Schmidt University.

Bauer, C.-O. (1980). Technische Normen – ihre Methoden und Wirkungen als Gegenstand von Forschung und Lehre. *DIN Mitteilung*, *59*, 329.

Blind, K. (2004). *The Economics of Standards: Theory, Evidence, Policy*. Cheltenham, UK, Northampton, MA: Edward Elgar Publishing Limited.

Blind, K., & Großmann, A.-M. (2014). An Exploratory Analysis of Company Standards –Evidence from the German Standardization Panel. In I. Mijatovic & K. Jakobs (Eds.), *EURAS Proceedings 2014: Co-operation between Standardisation Organisations and the Scientific and Academic Community: EURAS Annual Standardisation Conference* (Vol. 19). Aachen, Germany: Verlagshaus Mainz GmbH Aachen.

Blind, S., & Bühring, J. (1996). Die ökonomische Theorie der Standards und ihre Anwendung auf den Medienbereich. *Homo Oeconomicus*, (13), 515–560.

Collis, D. J., & Rukstad, M. G. (2008). Can you say what your strategy is? *Harvard Business Review*, *86*(4), 82–90. PMID:18435009

Dacin, M. T., Ventresca, M. J., & Beal, B. D. (1999). The embeddedness of organizations: Dialogue & directions. *Journal of Management*, *25*(3), 317–356. doi:10.1177/014920639902500304

Daidj, N., Grazia, C., & Hammoudi, A. (2010). Introduction to the Non-Cooperative Approach to Coalition Formation: The Case of the Blu-Ray/HD-DVD Standards' War. *Journal of Media Economics*, *23*(4), 192–215. doi:10.1080/08997764.2010.527206

de Vries, H. (2007). Fundamentals of Standards and Standardization. In W. Hesser, A. Feilzer, & H. de Vries et al. (Eds.), *Standardisation in Companies and Markets* (2nd ed.; pp. 1–44). Hamburg, Germany: Helmut Schmidt University.

de Vries, H., Slob, F., & Zuid-Holland, V. G. (2007). Best practice in company standardization. *International Journal of IT Standards and Standardization Research*, *4*(1), 62–84. doi:10.4018/jitsr.2006010104

Deutsches Institut für Normung e. V. (DIN). (Ed.). (2000). Gesamtwirtschaftlicher Nutzen der Normung: Volkswirtschaftlicher Nutzen. Der Zusammenhang zwischen Normung und technischem Wandel, ihr Einfluss auf den Außenhandel und die Gesamtwirtschaft. Berlin: Beuth Verlag GmbH.

DIN 820-1. (2014). Berlin: Beuth Verlag GmbH.

DIN EN 45020. (2007). Berlin: Beuth Verlag GmbH.

Dixit, A. K., & Stiglitz, J. E. (1977). Monopolistic Competition and Optimum Product Diversity. *The American Economic Review*, *67*(3), 297–308.

DNP. (2014). *Indikatorenbericht 2014: Deutsches Normungspanel.* Retrieved June 10th, 2015, from projects.inno.tu-berlin.de/DNP/Bericht_2014.pdf

Dölle, J. E. (2013). *Lieferantenmanagement in der Automobilindustrie. Struktur und Entwicklung der Lieferantenbeziehungen von Automobilherstellern. Schriften zur Unternehmensentwicklung.* Wiesbaden, Germany: Springer Gabler Verlag.

Dranove, D., & Gandal, N. (2003). The DVD-vs.-DIVX standard war: Empirical evidence of network effects and preannouncement effects. *Journal of Economics & Management Strategy*, *12*(3), 363–386. doi:10.1162/105864003322309518

Ebers, M., & Oerlemans, L. (2013). The variety of governance structures beyond market and hierarchy. *Journal of Management.* doi: .10.1177/0149206313506938

Farrell, J., & Saloner, G. (1985). Standardization, compatibility, and innovation. *The Rand Journal of Economics*, *16*(1), 70–83. doi:10.2307/2555589

Farrell, J., & Saloner, G. (1988). Coordination through Committees and Markets. *The Rand Journal of Economics*, *19*(2), 235–252. doi:10.2307/2555702

Farrell, J., & Saloner, G. (1992). Converters, Compatibility, and the Control of Interfaces. *The Journal of Industrial Economics*, *40*(1), 9–35. doi:10.2307/2950625

Farrell, J., & Shapiro, C. (1988). Dynamic Competition with Switching Costs. *The Rand Journal of Economics*, *19*(1), 123–137. doi:10.2307/2555402

Feuchter, H. (1996). *Werknormung. Normungskunde* (Vol. 37). Berlin: Beuth Verlag GmbH.

Flaherty, T. (1986). Coordinating International Manufacturing and Technology. In M. Porter (Ed.), *Competition in Global Industries* (pp. 83–110). Boston: Harvard Business School Press.

Gereffi, G., & Lee, J. (2012). Why the World Suddenly Cares About Global Supply Chains. *Journal of Supply Chain Management*, *48*(3), 24–32. doi:10.1111/j.1745-493X.2012.03271.x

Grewal, D. S. (2008). *Network power: The social dynamics of globalization.* New Haven, CT: Yale University Press.

Großmann, A.-M., & von Gruben, P.-V. (2014). The Role of Company Standards in Supply Chains - The Case of the German Automotive Industry. In T. Blecker (Ed.), Innovative Methods in Logistics and Supply Chain Management: Current Issues and Emerging Practices (pp. 99–121). Berlin: epubli GmbH.

Hesser, W., Czaya, A., & Riemer, N. (2007). Development of Standards. In W. Hesser, A. Feilzer, & H. de Vries et al. (Eds.), *Standardisation in Companies and Markets* (2nd ed.; pp. 123–170). Hamburg, Germany: Helmut Schmidt University.

Hesser, W., & Meyer, R. (1993). Parameter der Wirtschaftlichkeit von Normungsvorhaben – Das Wachstum der Typenvielfalt als überbetrieblich verwendbarer Parameter. *DIN Mitteilung*, (72), 349–354.

Holler, M. J. (1996). Die Rationalität strategischer Normung in Europa. In K.-E. Schenk, D. Schmidtchen, & M. E. Streit (Eds.), Jahrbuch für Neue Politische Ökonomie (pp. 137–153). Tübingen: Mohr-Siebeck.

Hudson, J., & Jones, P. (2001). Measuring the efficiency of stochastic signals of product quality. *Information Economics and Policy*, *13*(1), 35–49. doi:10.1016/S0167-6245(00)00029-9

ISO/IEC Guide 2. (2004). Geneva: ISO/IEC.

Iversen, E. J., Oversjoen, E., & Lie, H. T. (2004). Standardization, innovation and IPR. *Teletronikk*, *2*, 65–79.

Jakobs, K. (2000). Standardization Processes in IT: Impact: Problems and Benefits of User Participation. Wiesbaden, Germany: Vieweg+Teubner.

Kang, B., & Motohashi, K. (2015). Essential intellectual property rights and inventors' involvement in standardization. *Research Policy*, *44*(2), 483–492. doi:10.1016/j.respol.2014.10.012

Katz, M. L., & Shapiro, C. (1985). Network externalities, competition and compatibility. *The American Economic Review*, *75*, 424–440.

Kindleberger, C. P. (1983). Standards as public, collective and private goods. *Kyklos*, *36*(3), 377–396. doi:10.1111/j.1467-6435.1983.tb02705.x

Lazina, L. K. (2014). Standardization Landscape of Company Standards at Volkswagen AG: Short Version. In I. Mijatovic & K. Jakobs (Eds.), *EURAS Proceedings 2014: Cooperation between Standardisation Organisations and the Scientific and Academic Community: EURAS Annual Standardisation Conference* (Vol. 19). Aachen, Germany: Verlagshaus Mainz GMbH Aachen.

Leland, H. (1979). Quacks, lemons, and licensing: A theory of minimum quality standards. *Journal of Political Economy*, *87*(6), 1328–1346. doi:10.1086/260838

Lerner, J., & Tirole, J. (2014). A Better Route to Tech Standards. *Science*, *343*(6174), 972–973. doi:10.1126/science.1246439 PMID:24578566

Malisius, G., & Weidner, T. (1998). Die Normenabteilung – fit für die Zukunft. In *DIN Mitteilung*, (pp. 272-275). Academic Press.

March, J. G. (1991). Exploration and Exploiting in Organizational Learning. *Organization Science*, *2*(1), 71–87. doi:10.1287/orsc.2.1.71

Meyer, R. (1995). *Parameter der Wirksamkeit von typenreduzierenden Normungsvorhaben. DIN-Normungskunde* (Vol. 34). Berlin: Beuth Verlag.

Modi, S. B., & Mabert, V. A. (2007). Supplier development: Improving supplier performance through knowledge transfer. *Journal of Operations Management*, *25*(1), 42–64. doi:10.1016/j.jom.2006.02.001

Muffatto, M. (1999). Introducing a platform strategy in product development. *International Journal of Production Economics*, *60-1*, 145–153. doi:10.1016/S0925-5273(98)00173-X

Porter, M. E. (1996). What is strategy? *Harvard Business Review*, *74*(6), 61–78. PMID:10158474

Riillo, C. A. F. (2013). The engagement in stadarization activities: A firm level analyse of formal and company standardization. In K. Jakobs, H. J. De Vries, A. Genesh, A. Gulacsi, & I. Soetert (Eds.), *EURAS Proceedings 2013. Standards: Boosting European Competitiveness: EURAS Annual Standardisation Conference* (vol. 18, pp. 325–339). Aachen, Germany: Verlagshaus Mainz GMbH Aachen.

Salop, S. C., & Scheffman, D. T. (1983). Raising Rivals' Costs. *The American Economic Review*, (73), 267–271.

Shapiro, C., & Varian, H. R. (1999). The art of standards wars. *California Management Review*, *41*(2), 8–32. doi:10.2307/41165984

Shin, D.-H., Kim, H., & Hwang, J. (2015). Standardization revisited: A critical literature review on standards and innovation. *Computer Standards & Interfaces*, *38*, 152–157. doi:10.1016/j.csi.2014.09.002

Swann, G. P. (2000). The Economics of Standardization: Final Report for Standards and Technical Regulations Directorate Department of Trade and Industry. Manchester Business School, University of Manchester.

Tassey, G. (1982). The role of government in supporting measurement standards for high-technology industries'. *Research Policy*, *11*(5), 311–320. doi:10.1016/0048-7333(82)90053-1

Tassey, G. (1995). The Roles of Standards as Technology Infrastructure. In Standards, Innovation and Competitiveness (pp. 161–71). Camberley, UK: Edward Elgar.

Weissinger, R. (2013). Economic benefits of standards - Results and future perspectives of the "ISO Methodology". In K. Jakobs, H. J. de Vries, A. Genesh, A. Gulacsi, & I. Soetert (Eds.), *EURAS Proceedings 2013. Standards: Boosting European Competitiveness: EURAS Annual Standardisation Conference* (vol. 18, pp. 385–395). Aachen, Germany: Verlagshaus Mainz GmbH Aachen.

KEY TERMS AND DEFINITIONS

Company Standard: Documents with basic content, wide application and repeated use to set rules or characteristics for objects and operations. These are established by consensus within the company and accepted by applying or competent actors that provide, for common and repeated use, rules, guidelines or characteristics for activities or their results, aimed the maximum benefit for the company.

Differentiation Strategy: A company's definition to what degree it positions its product or services on the market different to its competitors' products.

Formal Standards: Documents, established by consensus and approved by a recognized body, that provide, for common and repeated use, rules, guidelines or characteristics for activities or their results, aimed at the achievement of the optimum degree of order in a given context.

Knowledge Diffusion: The spread and adoption of useful ideas among individuals or organizations.

Quality Management: The definition of how a company ensures that its services, products and processes perform consistently.

Standardization Management: The definition of how a company uses and monitors internal and external standards.

Standardization Strategy: The definition of how a company will position itself with regards to internal and external standardization activities, which includes the application of standards as well as the participation in standardization committees and the development of own standards.

Strategic Management: The definition of a company's unique position in the market with regards to its internal and external environment.

Supply Chain Management: The management of the flow of goods and services along the supply chain, which includes the development, control and monitoring of supply chain activities.

ENDNOTES

1 Please note that the development process differs between companies, because this is an internal process. We here present an exemplary approach used by some larger manufacturing firms.

2 The Effects are rated on a scale of -3 „completely disagree" to +3 „completely agree". The data and table originally comes from (DIN, 2014).

3 Sectors included were automotive industry, machine building industry, metal production industry, electrical industry, construction industry, energy, water and oil industry, information and communication industry, service industry, consumer goods industry, pharma and chemical industry.

4 This is according to internal documentation of a German automobile manufacturer.

5 This is according to internal documentation of a German automobile manufacturer.

6 This is according to internal documentation of a German automobile manufacturer.

7 This is according to internal documentation of a German automobile manufacturer.

8 The Effects are rated on a scale of -3 „completely disagree" to +3 „completely agree". The data and table originally comes from DIN (2014).

9 The rating of "Quality improvement" is 1.75 for internal Company Standards and 0.90 for external Company Standards.

10 The rating of "Bargaining position" is 1.18 for internal Company Standards and 0.89 for external Company Standards.

11 The rating of "Competitiveness" is 1.16 for internal Company Standards and 0.77 for external Company Standards.

12 The rating of "Productivity" is 1.19 for internal Company Standards and 0.24 for external Company Standards.

13 German Committee of Standards Users (ANP); Essen, Germany in December 2014

14 This scenario was described by the standardization manager of a global steel and technology company at the meeting of the German Committee of Standards Users (ANP); Essen, Germany in December 2014.

15 Emphasis added; {http://ir.airberlin.com/en/about-airberlin/strategy-and-business-model, last accessed: 13.01.2015}.

16 Emphasis added; {www.daimler.com/company/strategy; last accessed: 12.01.2015}.

17 MENNEKES Elektrotechnik GmbH & Co. KG, a manufacturer of electronic components and plugs.

18 {http://www.iec.ch/newslog/2011/nr1511.htm; last accessed on19.02.2015 and http://www.mennekes.de/uploads/media/The_solution_for_Europe_type_2_charging_sockets_with_or_with.pdf ; last accessed on 19.02.2015}.

19 MBN is the ICS of Daimler Ag. See { http://eshop.wuerth.de/Gesamtkatalog/MBN-10105-Stahl-10.9-Zink-Lamelle-Regel-u.-Fein-/14013511110501.cyid/1401.cgid/de/DE/EUR/, last accessed on 16.06.2015}.

Chapter 6
The Pressure Cooker Approach for Open Standards Development

Erwin Folmer
University of Twente, The Netherlands

Jasper Roes
TNO, The Netherlands

ABSTRACT

The new open world requires new approaches for IT development, and in particular for open standards. In the old world an average development time of 36 months (for European standards) is current practice, but not aligned with current and future needs. This chapter introduces a pressure cooker approach in which openness plays a crucial role in the development leading to an open standards with positive effects on adoption. Several case studies show that the average development time is reduced to an average of 25 weeks, that the approach is still improvable, and that users of the developed standards are to a certain extend satisfied with the quality of the standard. The pressure cooker concept fits within the development and management model of open standards (BOMOS) mainly for dealing with the operational development steps.

INTRODUCTION

Today´s businesses are in a dynamic and more and more connected environment. Organizations try to connect their business processes for improving productivity and higher efficiency (Harvey and Novicevic 2006). Organizations need interoperability to align their processes. Data standards support the achievement of interoperability by providing clear and straightforward definitions, layouts and processes for everyone who is willing to use it (Folmer & Verhoosel 2011). Data standards make sure that the different parties in the cooperation use the same names, measurements and way of doing business during their inter-organizational work. Standards set specifications for interfaces that make it possible to connect these systems to each other (Van Wegberg 2004). When the organizations 'speak the same language',

DOI: 10.4018/978-1-4666-9737-9.ch006

connecting their business processes is a much easier task and in the end reaching interoperability is easier. This is the reason that it is generally accepted that interoperability is supported by Data standards. The importance of the standards in the achievement of interoperability explains that the Standardization is essential for organizations. Companies lose a lot not having the advantage of interoperability. Almost $3.9 billion annual loss in the electronics industry and $5 billion annual loss are estimated in the automotive industry, just because of missing standards (Steinfield, Markus et al. 2011). Moreover, the lack of integration and data Standardization is making health care services inefficient and costly. Hospitals have $29 billion cost yearly because of errors. Venkatraman et al. (2008) claim that three out of four errors can be eliminated by better use of information technology. These losses could be eliminated with appropriate levels of interoperability. However, the IS Data standards Development processes are far from the desired maturity. One main concern is the average Development time of the Data standards. Van Wegberg (2004) claims that speeding up Standardization will be valuable if the benefits from the standard are time-dependent. Currently, the average time of standards Development is 36 months (EC 2011). The importance of the problem is shown by the European Commission too, who demands that before 2020 the average time is reduced by 50%, down to 18 months (EC 2011). However one could argue whether 18 months is acceptable or not, why 18 months, and how about quality? The need of fast solutions is especially important in the ICT field. Various actors of the same or different industries work together and their productivity can be seriously hindered by the lack of appropriate connectivity of their IT systems. To fully realize the benefits of e-business, common standards are required to define the syntax and semantics of web-based information sharing among firms (Zhao, Xia et al. 2005). To have an even more complicated situation, companies are waiting for the solutions in a fast changing technology environment.

In the ICT domain, industry consortia set the vast majority of important standards, in contrast to formal standards organizations (Rada and Ketchell 2000). Industry consortia are growing in number and importance; they cannot be neglected anymore in government policies (EC 2009; Kroes 2010; EC 2011). In the Data standards area there is often one dedicated consortium that maintains one specific data standard for a specific domain. However both large industry consortia and formal bodies are aiming for the inclusion of more of these Data standards. For instance OMG, The Open Group and W3C, are all industry consortia involved in Data standards for different domains, and currently W3C is offering a free online platform that can be used by Data standards initiatives. In contrast to popular belief, industry standards are often more open than formal standards: many of the industry consortia have open characteristics while the democratic process and free availability of formal standards is overestimated (Egyedi 2003; West 2007). The formal bodies bring their formal status into play, while the industry consortia offer their expertise and flexible processes (Folmer 2012).

The Openness of Data standards is of highest importance for the adoption of the standard within an industry domain, the adoption of closed standards is fairly impossible. In general approaches for Development and Management of Data standards, such as the BOMOS approach (Folmer and Punter 2011) endorsed by the Dutch government, implement the 10 requirements view (Krechmer 2009) that shows that Openness is more than the license, and that Openness is to a large extent related to the Development approach of the standard.

Speed and Openness are part of the broader concept quality of standards (Folmer 2012). Speed on itself is not that important however the related concept of the timing of standards is. Timing means that standards should be available when needed by the stakeholders. Since standards often arrive too late on the market, the concept of timing is often incorrectly simplified too faster speed, which might cause too much emphasize on speed only. With too much emphasize standards can also be released too early. Quality also includes the content of the standard, the product quality or intrinsic quality (Folmer 2012). Other quality aspects that causes standards to fail is the lack of stakeholder involvement and support through the Standardization process. In the end standards should solve an actual business need. Standards with adequate timing, content and stakeholder involvement have better chances to become adopted in corporate settings, and in the end become a successful standard.

The general perspective of this chapter is to introduce an innovative and validated Development approach for Open standards, focusing on speed/timing, and taking stakeholder involvement and other quality aspects into account.

RESEARCH APPROACH

In the research presented in this chapter we studied whether it is possible to reduce the Development time of open Data standards Development and still achieve an outcome with the desired quality for practical (pilot) implementations (as starting point for further Development). We studied this by developing an artifact, a method for developing standards, which is typical design science research (Hevner, March et al. 2004; Hevner & Chatterjee 2010).

Based on a state of the art study a first build of the method has been created, which is called the Pressure cooker approach. This first build was directly tested in the field in a first case study and loosely evaluated, this was repeated twice which resulted in 3 build-evaluation cycles, with 3 slightly different (improved) versions of the Pressure cooker and 3 practical implementations. After three iterations we changed the scope of our evaluation: not only aiming on finding improvement suggestions but more focussed on testing the original research questions to find out if the Pressure cooker approach leads to a shorter Development time, more satisfaction and to find out the effects on the quality and adoption of the data standard.

We conducted a survey on the participants of the case studies and combined this with expert interviews with the Standardization experts involved. Surveys are often appropriate for problem clarification and verification of problem relevance. They offer a way of getting more stakeholders involved and getting structured and comparable results in a time-efficient manner (Creswell 2009). In order to make the survey capable to gain valuable insight from the former Pressure cooker participants, we decided to create a model with variables. To do so, the survey drafting principles were followed from different sources (Maxwell 2005; Creswell 2009; Blumberg, Cooper et al. 2011). Survey questions addressing the variables have been formulated based on the model. Research on communication (Louhiala-Salminen and Kankaanranta 2012) showed the role of the language in enhancing knowledge sharing and in the Development of trust between communicators. It was found that the aforementioned factors can be addressed better by using native language instead of English in the response collection. Based on this finding, the survey prepared has been translated to Dutch, in order to eliminate the language barriers.

This chapter presents a short summary of the state of the art study, introduces the Pressure cooker approach, and presents some highlights of the survey results before answering the main research question in the final conclusion section.

Figure 1. Survey model

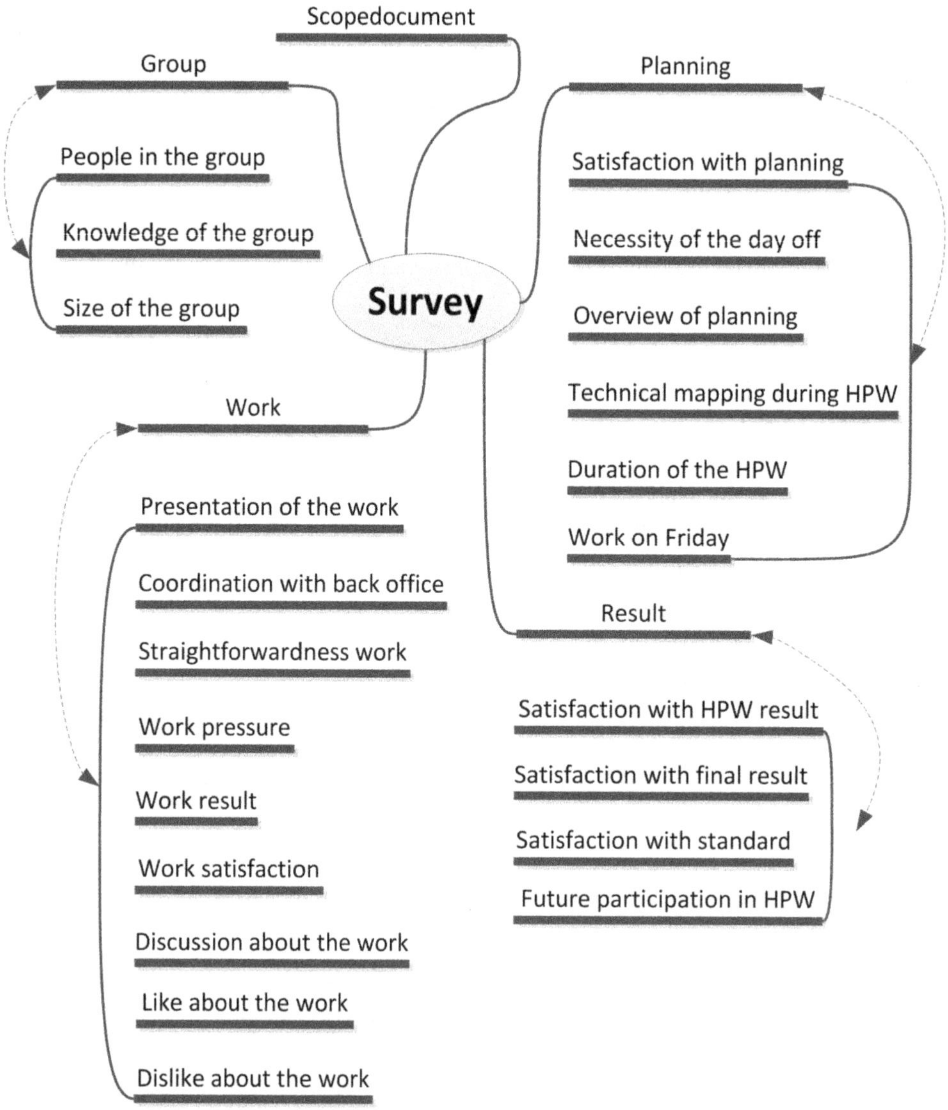

STATE OF THE ART

Although earlier work (Folmer, Berends et al. 2009), based on systematic literature review, already declared several topics within the area of Data standards as research gaps, still a lot can be learned from literature. The prior literature on the Development of Data standards is mainly concerned with the reasons for joining a standard Development organization (Folmer and Verhoosel 2011). One of the reasons to getting involved is to contribute and to orient the standard towards one`s own business practices (Zhao, Xia et al. 2007). Boh et al. (2007) presented the paradox of participation in standards Development. According to the paradox, the greater the number of stakeholders, the more difficult it is to achieve consensus. It will slow down the process. On the other hand, involved stakeholders will be early adopters.

Boh et al. (2007) also described strategies used in the Development of Rosettanet, a successful data standard in the electro technical industry. These are:

- Commitment of resources to the milestone program
- Clear roles and restrictions
- Validation beyond full implementation
- Informal norms and social networks

The lessons learnt from Rosettanet are:

- Only involve the organizations that are committed to solving the problem
- Focused, quick, problem solving approach to standard setting
- There is no one right approach for the standards Development process, not even a full open approach

In the MISMO (mortgages data standard) case, Markus et al. (2006) pointed out a certain success factor for standard Development. It is necessary to ensure participation of representative members of heterogeneous user groups, and avoid the natural tendency to splinter into rival homogeneous groups. Other studies (Jakobs and Kritzner 2009) specify this heterogeneous user group in direct users, mediators and indirect users (users of standards implementations), and define the key challenge to involve all in standards Development.

Zhao et al. (2007) claim that the better the standard and the faster it is developed, the greater is the direct benefit for the developers. By being involved in the Development of the standards, there is an increase in the understanding of the standard details which helps to reduce future implementations costs. Werle and Iversen (2006) made four suggestions for the standard Development process, based on observations:

- Openness to and direct representation (participation) of all actors interested in or potentially affected by a standard.
- Work in accordance to impartial and fair procedural rules.
- Decision-making should be based on consensus.
- All interests are considered in the Standardization process.

Spivak and Brenner (2001) emphasize that the standard developing organization should keep in mind that the process they select must serve the industry and not the other way round. Löwer (2005) found another problem: Although there are many existing standards, during a lot of standard Development, the standard developing organizations are reinventing the wheel and do not use each other`s results. This is in contradiction with the requirements of Open standards. (Krechmer 2009). The One World assumption means the same standard for the same function, worldwide. This assumes optimal re-use of existing Open standards, without developing competing standards. Other viewpoints on Openness include Open Meeting, Consensus Decision Making, Due Process, Open IPR, Open Documents, Open Change, Open Interfaces, Open Access and Ongoing Support (Krechmer 2009).

Another example of an open Development process is the process of ebXML that has been studied (Choi, Raghu et al. 2004) and has led to the following propositions (adapted from (Choi, Raghu et al. 2004)):

1. An Open Standardization process helps collaborators to create a functionally comprehensive standard.
2. An Open Standardization process promotes the convergence of technologies in the long run, paving the way to its domination over "proprietary" standards.
3. User participation is a moderating factor in an Open Standardization process for achieving a comprehensive and converged standard.
4. Interoperability, backward compatibility, feasibility and sponsor support (both SDO and technology providers) are critical factors that influence the creation of standards.

There are several standards on standards Development (ANSI/SES-1-2002, SES 2:2006, ISO/IEC Directives Part 2 and the British Standards Institution also developed a standard) but these initiatives provide only dictionaries and definitions of processes, they miss to offer more than a very general overview of the process of standards Development, and overlook the content characteristics of different kind of standards.

The Pressure Cooker Approach

This section presents a high-level overview of the Pressure cooker approach, a more detailed description of the Pressure cooker approach can be found in (Folmer, Zsolt Kulcsar et al. 2013).

On a high level the Pressure cooker approach consists of 13 steps that can be completed in a 15 weeks. These 13 steps lead from a need for a standard, by creating the standard in a working group, to the start of the adoption of the standard or a pilot with the standard. Next to the process of creating the standard, there is also a 14th step, that decides on the long term Management, governance, finance and adoption of the standard which is equally important as the process of creating the standard. In an ideal world the Pressure cooker concept can be completed in 15 weeks, due to practical issues like full agenda's and political reasons the three Pressure cooker implementations that were performed took a little bit more time, but all finished within 6 months.

The Pressure cooker approach starts with a need for a standard. As the Pressure cooker approach is a highly intense process, it is very important that this need is high and that the parties involved are fully committed to creating the standard. When the need for the standard is formulated, the process is followed by the step of setting up a steering committee. In the steering committee representatives of the companies that have a need for the standard and want to show commitment to the creating process take place and from that point on steer the Development of the standard. This steering committee is supported by one or two Standardization experts. The first job of the steering committee is to decide on the scope of the standard. This is an important step as a clear definition of the scope is essential to be able to complete the standard in a short period. Next to defining the scope the steering committee sets up the planning and selects the workgroup members in this step. After defining the scope and making sure everything is in place to create the standard the steering committee organizes an information meeting for which every party with an interest in the standard is invited. It is important to invite all parties with an interest, even a negative one, as Openness is important in creating the standards as became clear in the state of the art. Based on the information collected in the information meeting the scope of the standard is refined and the high pressure week is prepared by the Standardization experts.

The next step in the process is the high pressure week: this is exactly one week in which the working group and two Standardization experts spend 4 full working days on discussing the contents of the

Figure 2. Pressure cooker approach

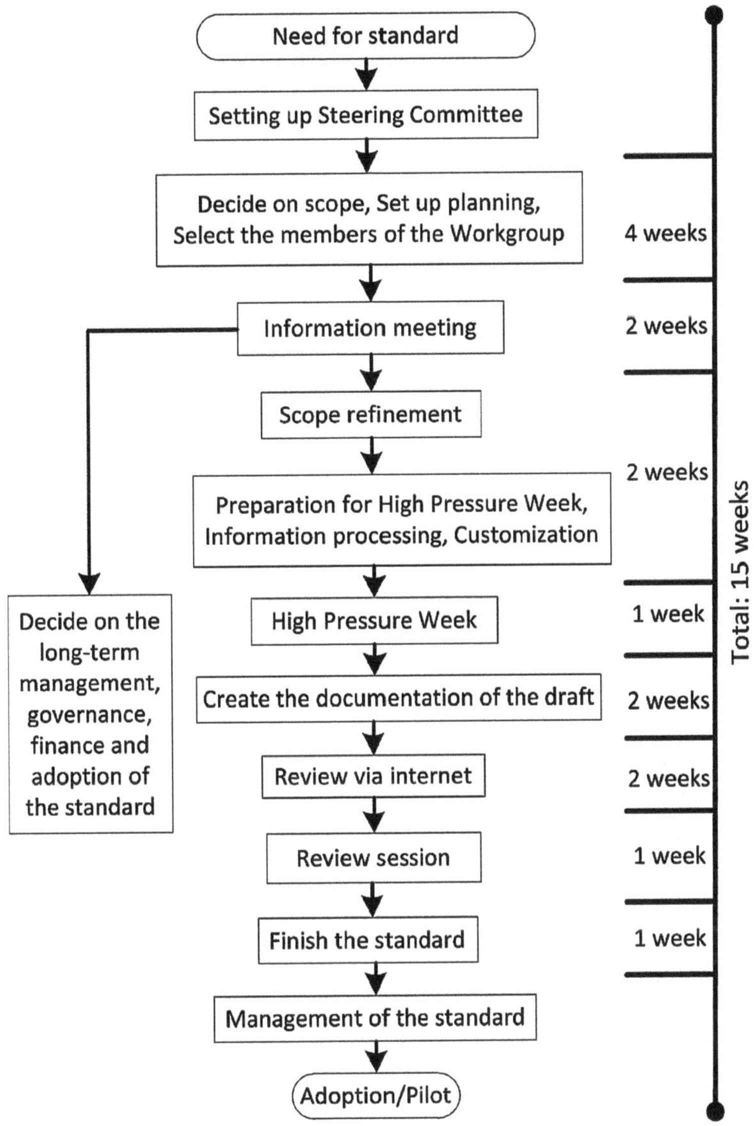

standard. They furthermore define all messages and information elements of the standard based on the predetermined scope by the steering committee. The necessity of a clear defined scope comes from this week: if the scope is not clear discussions during the week will be about the scope and not about the contents of the standard. The high pressure week is organized on one location, often in the centre of the country to limit travel time and expenses for the participants and to ensure that all companies (including Small and Medium Enterprises) have the possibility to join without high costs.

After the high pressure week the Standardization experts create a draft of the standard based on all information collected. In this work the Standardization experts also define the technical mapping of the standard, which enable the companies to actually exchange information with the standard. The technical mapping is often based on XML. The draft of the standard is send out to the working group members to review. The working group members send their comments to the Standardization experts who prepare a session to go through all the comments with the working group. After this session the Standardization experts finish the standard after which the standard is handed over to the steering committee that are responsible for the Management of the standard. After this step the adoption of the standard starts and a first pilot can be commenced.

It is important to realize that this Pressure cooker approach cannot lead to a 100% complete standard, the process is aimed at reaching a 80% standard that can be used for pilots and first implementation but which needs refinements after the first experiences. The difference between the 80% standard from the Pressure cooker and the 100% standard later on is that the 80% standard often contains more optional elements than the 100% version in which the optional elements might become mandatory, or have been removed based on the pilot implementations.

Evaluation Results

To evaluate and validate the Pressure cooker approach a survey was conducted under the participants of three Pressure cooker applications that were performed when conduction the research. Next to the survey the Standardization experts that were involved in the Pressure cooker applications where also interviewed to get there opinion on the Pressure cooker applications.

The population of the survey consisted of the workgroup members of the following three applications of the Pressure cooker approach:

- STOSAG (garbage collection)
- Digitale Rotonde (ground cables)
- EBA (waste transport)

The STOSAG standards are used for garbage collection. They describe the transponders for mini containers and the information that is exchanges about emptied mini containers. The STOSAG standards furthermore define the cards for accessing underground containers and provide the possibility to exchange information between underground containers and back-offices of the garbage collection companies. The STOSAG standards are currently widely applied in the garbage collection domain in the Netherlands.

The Digitale Rotonde standard is used in the process of installing underground cables and tubing for water, gas, electricity, telephone and fiber. To optimize the process of installing these cables and tubing a standard is define to manage the process between the grid operators and the constructors that actually lay the cables and tubing. The standards enable and facilitate the process of exchanging information between the mentioned parties and improve digital planning. Within the Digitale Rotonde project there was a mandatory requirement to use the VISI approach for data modelling, and expressing the Digitale Rotonde standard in VISI technology/syntax. VISI is a Dutch standard for the building & construction industry, without much usage, expertise and tooling available, but adopted by the Dutch government.

The EBA standard is a standard that defines a digital waste transport bill. Waste transport bills are required in the Netherlands when transporting waste, the standard enables the use of digital bills which do not require a visit to the office to get paper waste bills when transporting waste.

The questionnaire was designed to survey the workgroup members' impressions and experiences of the high pressure week mainly focusing on group setting, satisfaction, and work related questions. The questionnaire was administered online on the 1st of March in 2013 to a population of 43 persons, who were participating in one of three Pressure cooker projects. In the end, twenty-seven valid responses were received.

Survey Result

The invitation for the survey were sent out to the workgroup members via email and the responses were collected in an online survey system. The invited parties were allowed to fill in the survey between the 1st of March 2013 and the 21st of March 2013. The sample size was 43 of which 27 responded. This means 63% response rate overall. For the three cases the response rated is presented in Table 1.

The following paragraphs summarize the most important results from the survey based on the survey model.

With respect to the group setting 33% of the respondents claimed to be business-oriented, 22% of them claimed to be technical-oriented and 45% considered themselves to have knowledge in both sides as presented in Table 2. This indicates that a bit more business knowledge was represented in the workgroups. This is not a problem for the respondents as 74% of them agreed that the mix of knowledge was good and 70% agreed that the necessary knowledge was present in the workgroup.

On the group size the respondents did not agree, almost half of the people indicated that the size of the group was good (48%) and the other indicated that it was too big (52%). When filtering out the different cases we found that within the STOSAG group more than 50% of the people indicated that the group size was right, while the group size was not different than that of the other cases. The group size might therefore not be the real issue.

The scope document provided by the Steering Committee was found to be helpful or extremely helpful by the majority (67%) and roughly the rest (26%) considered it being all right.

The planning of the week is sufficient (78%) according to the respondents. Most (70%) found it easy to keep track of the actual status of the work during the week. The respondents that indicated that the planning was not sufficient gave the following explanations (translated from Dutch):

- "Too little attention for the preparation"
- "Especially during the last part decisions were taking to quick. This made the rest of the process less clear"
- "The high pressure week was planned on a too short timeframe and in a busy period of the year. Therefore I could no clear my agenda on all days of the high pressure week"
- "A lot of changes have been made to the scope during the process which have led to many improvements later"

Table 1. Response rate per case

Case	Sample Size	Responses	Response Rate
STOSAG	14	8	57%
Digitale Rotonde	17	12	70%
EBA	12	7	58%

Table 2. Viewpoint workgroup members

Viewpoint	Percentage	Responses
Business-oriented	33%	9
Technical-oriented	22%	6
Both	45%	12

Neglecting the technical mapping in the high pressure week proved to be supported by the workgroup members (63%). 33% of the business representatives agreed and another 33% of them strongly agreed that the technical mapping is not necessary in the high pressure week. The result is a bit different among technical representatives of the workgroup. 50% of them did not mind the absence of technical mapping at all, 16% partly accepted it, 33% missed it. The representatives of both the technical and business sides did not miss the technical mapping from the work. The majority (67%) absolutely and the rest (33%) partly agreed on not to include it.

The 4 work-day length of the high pressure week was generally considered as right by the majority (70%). The last day of the working week was considered sufficient and good. Those who call it productive, are referring to the reviewing of what had been done in the previous days. The majority of the respondents think that the free day was useful and good. The EBA case was left out for this result as they did not have a free day. Most common responses are outlining that the free day was used to consult with the back-office and it was good to give some time for the already achieved status to settle down and to be able to reflect on it. Four respondents were not absolutely happy with the free day. One of them reported that he got out of the flow, but two of them found it useful in another way. They used this day to be able to devote attention to other projects.

Considering all the responses, 85% indicated that they found it easy to work in the high pressure week, but 4 respondents (15%) not. 70% of the participants considered the information representation good during the work sessions. The big majority (85%) more or less agreed that there were enough opportunities to consult with the back-office next to the free day. 50% agreed or strongly agreed with this statement as presented in Table 3.

The time pressure was helpful in the Development of the standard according to 78%. The majority (70%) agrees that the objectives of the high pressure week were satisfied. 3 respondents disagreed and one strongly disagreed with the previous statement. Two of these participants were involved in the Digitale Rotonde project, which proved to be technically the most challenging according to the Standardization experts because the steering committee decided that the VISI technology needed to be used instead of XML. The two people that indicated that the time pressure was not helpful, also indicated that the objectives of the high pressure week were not satisfied.

Most of the respondents (78%) were satisfied with the result of the high pressure week. Four out of the six, who were not satisfied, are from the Digitale Rotonde project. In average, these six respondents had problems following the current status of the work and found the mix of knowledge not sufficient. Half of them indicated that the work in the high pressure week difficult. Most of those who filled in the survey (67%) think that the Chairman let the right amount of time for discussions. However significant amount (22%) indicated it as bit too little. These people were involved either in the STOSAG or in the Digitale Rotonde projects (3-3 members). These results do not seem to be a huge surprise since the

Table 3. Consult back-office

Question	Way to little	A bit to little	Exactly adequate	A bit to much	Way to much	Response Count
Did you have sufficient time to consult your collegues with one free day available?	3	6	17	1	0	27

STOSAG was the first test run of the Pressure cooker concept and the Digitale Rotonde project had the largest number of participants. Both aforementioned factors can explain the possibly less time provided by the Chairman.

The workgroup members emphasized three things they especially liked in the Pressure cooker. First, most of them reflected on the great collaborative work of the group. Second, the members liked the pressure too. Third, some participants from the Digitale Rotonde project explicitly mentioned the satisfaction with the location and environment of the high pressure week, which can be explained by the fact that this high pressure week was held at a little castle with tremendous views. Some of the reactions to the things people liked were (translated from Dutch):

- "Everybody was motivated and everybody want to have the results. A must for success"
- "Reaching a shared solution with all partners"
- "Working together"
- "Openness of het participants. Quick switches by Standardization experts from discussion to solutions. "
- "Participant got the time and space to discuss ideas"

Eight respondents (2-2 from the EBA and STOSAG projects and 4 from the Digitale Rotonde project) did not find any specific problem with the current method. However, a group of people (7) were complaining about the repeated discussions. Two workgroup members, one from the STOSAG and another from the Digitale Rotonde project, did not find the available time sufficient for the decision making. Some respondents emphasized that the group size was too large, but they did not suggest a better size. Two Digitale Rotonde workgroup members pointed out that they did not like the lack of a real application of the standard. They wanted more tangible and practical result after the high pressure week. Some of the reactions to the things people did not liked were (translated from Dutch):

- "Repeated discussions"
- "Too little time for optimizations"
- "The force to deliver a standard"
- "The group of people was too big and their background to different"
- "Not everybody was there during all days, or had themselves replace on some days"

Most respondents (78%) at least partly agreed that the high pressure week resulted in a good standard. Although, 11% (3 respondents) disagreed with this and another 11% strongly disagreed. In total, 22% of the workgroup members do not consider the standard as a good result. 4 out of the 6 respondents, who felt the result of the high pressure week was not good, were in the Digitale Rotonde project involved where the mandatory use of the VISI standard by the steering committee made things more difficult. The remaining two Workgroup members are distributed between STOSAG and EBA equally with 1-1 respondents. 56% of the survey respondents were satisfied and 44% were not satisfied with the final standard of the Standardization. Most (7 people) of those who were disappointed (12 people), were workgroup members in the Digitale Rotonde case. In the STOSAG, 3 and in the EBA project 2 participants were unsatisfied. The explanation of two of the three disappointed respondents from the STOSAG project reveal that they found the core standard appropriate as the result of this short Development time.

Chronologically reading the results, we can conclude that the workgroups perceived better results in the EBA project after the STOSAG, which had good evaluation too. The last Pressure cooker application, the Digitale Rotonde, was the least successful according to the experts and this is supported by the survey. 58% was not satisfied with the final standard. However it is a bit difficult to compare the Digitale Rotonde result to the other two project, because of its special condition. As it was mentioned before, the VISI standard was used instead of XML, which caused much more difficulties for the Standardization experts, who did not have a lot of practical experience with it. Still, the majority of this group (67%) thinks that the result of the high pressure week was good. Considering the survey responses from the three projects, there is obviously room for improvement, but it is visible that the workgroup is usually satisfied with the results.

DISCUSSION

It would be interesting to compare the above evaluation results with the evaluation results of traditional Development approaches used for Data standards. Unfortunately to our knowledge these results are not publicly available. So theoretically our evaluation results might be less satisfactory then when using traditional approaches. But it might also imply that even our most critical evaluation (Digitale Rotonde) still is more satisfactory compared to traditional approaches.

The Pressure cooker was intended for mainstream Data standards Development. Upfront it was already obvious that there would be situations where the Pressure cooker concept may not lead to desirable results, for instance when the number of stakeholders is too high, or the complexity of the standard is an issue. The evaluation of the Digitale Rotonde Pressure cooker seems to support this since this was by far the project with highest complexity (due to the VISI standard), and diversity of stakeholders. Therefore it would be interesting to define some situation characteristics in which the Pressure cooker concept will be appropriate. Based on the evaluation results of the Pressure cookers we constructed a first list of important characteristics to determine the fit of the Pressure cooker approach as depicted in Table 4.

These six characteristics could indicate the feasibility of the Pressure cooker concept. The number of stakeholders is important as it has a direct impact on the size of the steering committee of the Pressure cooker and the working group as we assume that all stakeholders are represented in the steering committee and have at least one representative in the working group. A big number of stakeholders will lead to a to large steering committee to efficiently make decisions and a too large working group to work efficiently.

Table 4. Possible characteristics

Characteristics
The number of the stakeholders
The number of specifications to develop
Knowledge and experience requirement of the Standardization experts
The target of the project
Level of commitment
Complexity of the business network

Due to the short Development time during the Pressure cooker, there is an obvious limitation of the number of specifications that can be created within the high pressure week. The number cannot be stated clearly, because the difficulty to create a specification greatly differs from another. Next to this the knowledge and the experience of the Standardization experts play a role. If they are unexperienced or do not know the domain, it will be difficult for them to lead the working group to optimal results in the short time available. The target of the project the Pressure cooker is used for is another important criteria as the Pressure cooker focussus on 80% standards that are ready for a first implementation but not fully done. If the target of a project is a 100% standard, the Pressure cooker approach is not wel suited. With respect ot the commitment, this can be split into two aspects. First, the more obvious one is to give high priority to the project and facilitate the planning of it in order to manage to keep the short deadlines of the Pressure cooker method .Second, the workgroup members have to be committed to the project as the Standardization experts cannot do it without the workgroup with their domain and background knowledge. Lastly the complexity of the business network that will use the standard can be of importance. The more complex the business network, the more stakeholders will be involved and this might lead to difficult discussion and difficulty in reaching consensus on the standard in the Pressure cooker.

The Pressure cooker is specific aimed at Data standards. It does not limit itself to the process (the steps), but also the content of the standard. E.g. there are templates for the Standardization document, the definition of data elements, etc. UML diagrams are being used, tooling for transformation to XML Schema standards. And based on the Digitale Rotonde results the conclusion might be drawn that it is better not to be flexible in changing the chosen technology/standards upon customer's request.

It can be argued if the goal of an 80% standard is appropriate. For sure it is not intended as end point, but only as starting point for further Development. In particular based on the believe that the first practical implementations of the standards will lead to required changes in the standard. This can be compared to Standardizations that have the use of "reference implementations" included in the Standardization process. However if 80%, or 90%, or 50% is most appropriate as starting point is hard to determine. Compared with software Development; 80-90% result can be compared to the more traditional waterfall approach, while currently agile approaches are more popular. Introducing concepts from agile software Development to standards Development might lead to a 20% ready standard, as first step.

Standards Development and Management

A standard that is not being managed is not a standard (Folmer & Punter, 2011). Unfortunately standards Management, and in particular the maintenance of a standard, receives far less attention both in literature and in practice. For in particular Data standards, the Management is of even more importance, since these standards change more frequently and therefore have the risk of being outdated.

While originating from a Dutch government setting, BOMOS has been adopted by the European Commission (ISA/JoinUp) to become an European approach for developing and maintaining Open standards (Folmer et al. 2014). BOMOS is intended to be used in Open Standardization projects and has a strong focus on semantic standards. In recent versions (Dutch Standardisation Forum, 2012) BOMOS has been applied for the Standardization of a broader range of information objects including IT-based boundary objects (such as services, processes, data (Eckartz, Folmer, 2015)). Figure 3 depicts the main asset of BOMOS: a normative framework of activities required for the Development and Management of an Open standard. It includes three core areas: strategy, tactics, and operational, which are flanked by two supporting areas of activities –communication and implementation support. The implementation of the

Figure 3. BOMOS activity model

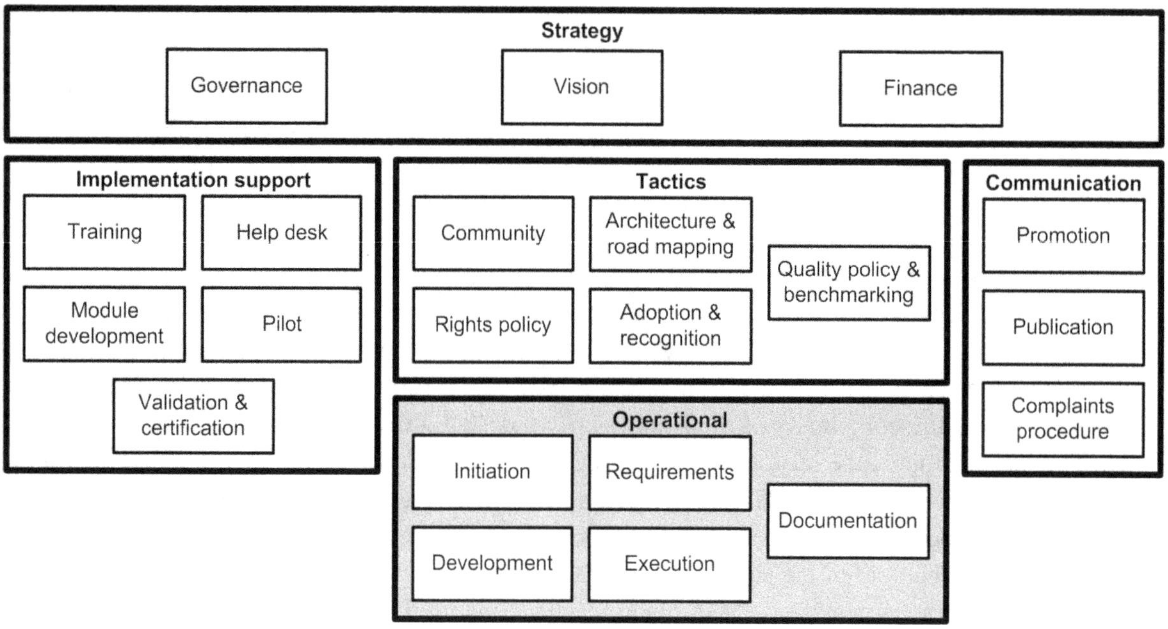

Development and Management activities is situation-dependent. This means that not all of the elements have to be implemented in each situation, or in similar ways. In essence, BOMOS is a collection of best practices for the activities. The core activity areas are briefly described below.

The *strategy* area summarizes activities related to the long-term Management of the standard. The *tactics* area is concerned with maintaining a community, an adoption strategy and a quality policy as well as setting property rights and defining architecture guidelines. The *operational* area is concerned with the executive activities that lead to new versions of standards. First, the initiation addresses the identification of new ideas (for example, for a new specification or new working group) and all activities associated with setting them up successfully (e.g. analysis of interests, business case, agenda). Second, the draft of the requirements of the specification needs to be developed and managed. Third, the Development follows. On a conceptual level, the intrinsic Development of solutions for the ideas, preferences and requirements set during previous phases need to be merged. These solutions are, separate from technology where possible, intended for further elaboration in the specification or a new version of it. Fourth, execution refers to implementing the actual amendments based on the conceptual solutions in the specification and any technical filling in. Finally, documentation provides a suitable reflection of the results of the primary Management process.

The Pressure Cooker within the Development and Management Model (BOMOS)

The Pressure cooker is an approach for initial standards Development and by that is focusing on the operational aspects. It covers the collection of requirements from the stakeholders, the Development of the solution and the execution and documentation blocks from the BOMOS Management model.

Other steps from the Pressure cooker approach relates to the strategy and tactics parts from BOMOS: E.g. The first step in the Pressure cooker, setting up the steering committee is tied to the Governance block in the Strategy part of BOMOS

However, is it important to realize that the Pressure cooker concept only covers parts of the Management model; e.g. setting up a financial model for the standard, or how to deal with implementation support or communication not covered in the Pressure cooker approach, while these are important aspects in standards Management.

CONCLUSION AND FURTHER RESEARCH

The research question started with was: Is it possible to reduce the Development time of open Data standards Development and still achieve outcome with desired quality for practical (pilot) implementations? This research has shown that we can answer this question positively; Development time can be significantly reduced by using the Pressure cooker approach and still achieve acceptable results. However the validation (survey and the experts interview) also pointed out that this is not some kind of holy grail. There are disadvantages and context situations in which the Pressure cooker will not result in optimal results.

We do however see some issues that need further clarification and research: e.g. there are some contradicting findings related to the group size. Probably group size itself is not the main issue, but group size together with for instance the chairmen's capabilities, and other factors, are interrelated and in combination determine the best way to organize the Pressure cooker sessions. Another interesting issue is the timing of the survey. It might be that when surveying is issued directly after finishing the Pressure cooker the results will be more positive than when the survey will be done months after the Pressure cooker. That might have to do with the fact that at the end of the high pressure week almost all participants are really proud and satisfied with the result they achieved in that week. That feeling might change over time.

Based on the Pressure cooker applications the Standardization experts confirm the statements of Löwer (2005) that re-inventing the wheel occurs where solutions/standards are already available. Although one of the main principles in the Pressure cooker approach is to be fully open, including making optimal re-use of existing standards, in practice this is fairly difficult. The reason is that the standards to be re-used are often only partial open (including many formal standards such as ISO), which when being re-used limits the Openness of the outcome of the Pressure cooker. To be fully open in the Pressure cooker, these partial Open standards cannot be re-used and then re-inventing the wheel will occur.

From a scientific point of view our validation is rather limited because of the low sample size of each case. Since the Pressure cooker groups are rather small this can only be covered by using other research techniques to gather more in depth responses, for instance interviews. In our current research we also include new Pressure cooker case studies. For this we decided to make the survey part of the standard Pressure cooker process.

As recent research has shown that the quality of standards is measurable (Folmer 2012), it would be interested to measure the quality of standards developed with the Pressure cooker approach and compare this with the quality of standards in traditional approaches.

Based on the survey results further iterations of the build & evaluate cycles are planned, leading to a more advanced Pressure cooker approach, but also leading to additional knowledge about when to use the Pressure cooker, and when it is better to use other approaches. One of these situation characteristics will be Openness: the Pressure cooker fits perfectly when an open transparent process is needed, with an independent chairmen, optimal re-use of open materials, and leading to an Open standard as the end result.

REFERENCES

Blumberg, B., & Cooper, D. R. (2011). *Business Research Methods*. London: Mc-Graw Hill.

Boh, W. F., Soh, C., & Yeo, S. (2007). Standards Development and diffusion: A case study of RosettaNet. *Communications of the ACM*, *50*(12), 57–62. doi:10.1145/1323688.1323695

Choi, B., Raghu, T. S., & Vinze, A. (2004). Addressing a standards creation process: A focus on ebXML. *International Journal of Human-Computer Studies*, *61*(5), 627–648. doi:10.1016/j.ijhcs.2004.04.002

Creswell, J. W. (2009). *Research Design - Qualitative, Quantitative, and Mixed Methods Approaches*. Los Angeles, CA: Sage.

Dutch Standardisation Forum. (2012). *BOMOS2i: The practical approach.* Author.

EC. (2009). *Modernising ICT Standardisation in the EU - The Way Forward*. Brussels: European Commission.

EC. (2011). *A strategic vision for European standards: Moving forward to enhance and accelerate the sustainable growth of the European economy by 2020*. Brussels: European Commission.

Eckartz, S., & Folmer, E. (2015). *BOMOD, Management and Development Model for Open Data*. TNO Publication.

Egyedi, T. M. (2003). Consortium problem redefined: Negotioting 'democracy' in the actor network on Standardization. *Journal of IT Standards and Standardization Research,* (2), 22-38.

Folmer, E. (2012). *Quality of Semantic Standards*. Enschede, The Netherlands: TNO, University of Twente, CTIT, NOiV.

Folmer, E., & Berends, W. (2009). Top IS research on quality of transaction standards, a structured literature review to identify a research gap. In *Proceedings of 6th International Conference on Standardization and Innovation in Information Technology (SIIT 2009)*. Verlag Mainz.

Folmer, E., Heddier, M., Matzner, M., Räckers, M., & Becker, J. (2014). A Method for Managing IT-Based Boundary Objects: Design and Application in the Public Sector. In *Proceedings of the European Conference on Information Systems (ECIS 2014)*. Tel Aviv, Israel: ECIS.

Folmer, E., & Punter, M. (2011). *Management and Development Model for Open standards (BOMOS) version 2*. The Hague, The Netherlands: Netherlands Open in Connection.

Folmer, E. & Verhoosel, J. (2011). *State of the Art on Semantic IS Standardization, Interoperability & Quality*. Enschede, The Netherlands: TNO, University of Twente, CTIT, NOiV.

Folmer, E., & Zsolt Kulcsar, I. (2013). Instant Standard Concept for Data standards Development. *Americas Conference on Information Systems (AMCIS)*, Chicago, IL.

Harvey, M. G., & Novicevic, M. M. (2006). The World is Flat: A Perfect Storm for Global Business? *Organizational Dynamics*, *35*(3), 207–219. doi:10.1016/j.orgdyn.2006.05.002

Hevner, A., & Chatterjee, S. (2010). *Design Research in Information Systems - Theory and Practice*. New York: Springer. doi:10.1007/978-1-4419-5653-8

Hevner, A. R., & March, S. T. (2004). Design science in information systems research. *MIS Quarterly: Management Information Systems*, (1), 75–105.

Jakobs, K. & Kritzner, J. (2009). How to Select the Best Platform for ICT Standards Development. In *Information Communication Technology Standardization for E-Business Sectors: Integrating Supply and Demand Factors*. Hershey, PA: Information Science Reference.

Krechmer, K. (2009). Open standards: A Call for Change. *IEEE Communications Magazine*, *47*(May), 88–94. doi:10.1109/MCOM.2009.4939282

Kroes, N. (2010). *How to get more Interoperability in Europe: Openness at the heart of the EU Digital Agenda*. Retrieved 06-17, 2010, from http://europa.eu/rapid/pressReleasesAction.do?reference=SPEECH/10/300&format=HTML

Louhiala-Salminen, L., & Kankaanranta, A. (2012). Language as an issue in international internal communication: English or local language? If English, what English? *Public Relations Review*, *38*(2), 262–269. doi:10.1016/j.pubrev.2011.12.021

Löwer, U. M. (2005). *Interorganisational Standards - Managing Web Services Specifications for Flexible Supply Chains*. Heidelberg, Germany: Physica-Verlag - A Springer Company.

Markus, M. L., & Steinfield, C. W. (2006). Industry-wide Information Systems Standardization as collective action: The case of U.S. residential mortgage industry. *Management Information Systems Quarterly*, 439–465.

Maxwell, J. A. (2005). *Qualitative Research Design - An Interactive Approach*. Thousand Oaks, CA: Sage.

Rada, R., & Ketchell, J. (2000). Sharing standards: Standardizing the European information society. *Communications of the ACM*, (3): 21–25.

Spivak, S. M., & Brenner, F. C. (2001). *Standardization essentials: principles and practice*. New York: Dekker.

Steinfield, C. W., & Markus, M. L. (2011). Cooperative Advantage and Vertical Information System Standards: An Automotive Supply Chain Case Study. *44th Hawaii International Conference on System Sciences (HICSS)*, Hawaii. doi:10.1109/HICSS.2011.130

Van Wegberg, M. (2004). Standardization and Competing Consortia: The Trade-off Between Speed and Compatibility. *International Journal of IT Standards and Standardization Research*, *2*(2), 18–33. doi:10.4018/jitsr.2004070102

Venkatraman, S., Bala, H., Venkatesh, V., & Bates, J. (2008). Six strategies for electronic medical records systems. *Communications of the ACM*, *51*(11), 140–144. doi:10.1145/1400214.1400243

Werle, R. and E. J. Iversen (2006, March). Promoting Legitimacy in Technical Standardization. *Science, Technology & Innovation Studies*, 19-39.

West, J. (2007). *The economic realities of Open standards: Black, white, and many shades of gray. In Standards and Public Policy* (pp. 87–122). Cambridge, UK: Cambridge University Press.

Zhao, K., Xia, M., & Shaw, M. (2005). Vertical e-business standards and standards developing organizations: A conceptual framework. *Electronic Markets*, *15*(4), 289–300. doi:10.1080/10196780500302690

Zhao, K., Xia, M., & Shaw, M. (2007). An integrated model of consortium-based e-business Standardization: Collaborative Development and adoption with network externalities. *Journal of Management Information Systems*, *23*(4), 247–271. doi:10.2753/MIS0742-1222230411

KEY TERMS AND DEFINITIONS

Data Standards: Standards that define the definitions and meaning of data.

Development: The process of creating a data standard.

Management: The process of keeping a data standard up to date and ensuring the standard is used and adopted.

Open Standard: A standard that is maintained by a not-for-profit organization, available free of charge or against a nominal fee, with not license fees and not limitations on the reuse.

Openness: The grade on how easy people and parties can participate in the development of a standard.

Pressure Cooker: Method for working on a data standard in which the available time is limited and work pressure is high.

Standardization: Determining with a group of companies that need to exchange data how the data is exchanged.

Chapter 7
An Innovative Approach to the Development of Project Management Processes for Small-Scale Projects in a Large Engineering Company

Claude Y. Laporte
École de Technologie Supérieure, Canada

Frédéric Chevalier
Tetra Tech, Canada

ABSTRACT

A 400-employee Canadian division of a large American engineering company has developed and implemented project management processes for their small-scale and medium-scale projects. The company was already using a robust project management process for their large-scale projects. The objectives of this project were to reduce cost overruns and project delays, standardize practices to facilitate the integration of new managers, increase the level of customer satisfaction and to reduce risk-related planning deviations. For this project, the engineering organization used the ISO/IEC 29110 standards developed specifically for very small entities, i.e. organizations, having up to 25 people. An analysis of the cost and the benefits of the implementation of small and medium scale project management processes was performed using the ISO economic benefits of standard methodology. The engineering enterprise estimated that, over a three-year timeframe, savings of about 780,000$ would be realized due to the implementation of project management processes using the ISO/IEC 29110 standard.

INTRODUCTION

Standards are sources of codified knowledge and studies have demonstrated the benefits of standards, such as product interoperability, increased productivity, market share gains, and improved interaction with stakeholders such as enterprises, government organizations and the public. Standards and associated

DOI: 10.4018/978-1-4666-9737-9.ch007

technical documents could be considered as a form of technology transfer and, if the right standards are selected and used correctly they should have an economical impact in an organization.

Many advantages or benefits as well as disadvantages or costs have been reported regarding the use of voluntary standards. Table 1 lists a few of the advantages and disadvantages reported.

The most recent study on the economic benefits of standardization (Miotti, 2009), performed by the French standardization organization AFNOR, showed that standardization made a significant contribution to growth of the French economy during the 1950-2007 period, i.e. 0.81% per year or almost 25% of GDP growth. The study was based on a survey of 1,790 French companies or organizations of all sizes and from all sectors of activity where 30% of respondents were from enterprises of less tan 20 employees, 47% from small and medium enterprises (i.e. 250 employees or less) and 23% from large companies (i.e. more than 250 employees). The contribution of standards to the French economy is in line with data illustrated in Table 2 for other countries, such as Germany and Australia. In addition to known benefits of standards, five major lessons emerge from the French study (Miotti, 2009):

- Company value enhancement: The knowledge capital contributed by corporate involvement in standardization work represents true value.
- Innovation: Standardization promotes the dissemination of innovation. It emphasizes a product's advantages and constitutes a product selection tool.
- Transparency and ethics: Standards contribute to better compliance with the rules of competition. By establishing the rules of the game, standards make it easier to eliminate players who fail to comply.
- International: By promoting the development of international exchanges, standardization provides companies with a genuine passport for exporting their products.
- Product and service quality: Standardization gives companies a great degree of control over safety-related problems and provides a genuine guarantee of quality.

Table 1. Advantages and disadvantages of voluntary standards reported (adapted from Miotti, 2009; Land, 1997)

Advantages or Benefits	Disadvantages or Costs
• Promote innovation • Improve efficiency of an organization • Increase competitiveness • Facilitate the access to a wider market • Clarify the rules of a market • Improve quality of products and services • Promote improvement of Processes • Facilitate partnerships • Improve the image, credibility of organizations • Promote a uniform terminology • Regularly updated • Facilitate the selection of suppliers and partners • Facilitate access to recognize knowledge • Facilitate access to investments and financing	• Difficult to understand • Cost of acquire standards • Cost of standard implementation • Cost of certification • Require outside expertise to implement them • Conflicting standards • High number of standards available • Describe only 'what to be done' not 'how to do it' • Insufficient guidance to select and apply them • Slow evolution of standard may impede innovation • Difficult and costly to apply in small organizations • Difficult to demonstrate 'savings' • Many producers of standards • Perception that standards add unnecessary bureaucracy to an organization

Table 2. Comparative contribution of standards to national economies (adapted from Miotti, 2009)

	Germany (DIN)	UK (DTI)	Standards Council of Canada	Australia Standards	France (AFNOR)
Period subject to analysis	1961-1990	1948-2001	1981-2004	1962-2004	1950-2007
Growth rate of GDP (%)	3.3	2.5	2.7	3.6	3.4
Contribution to growth of GDP (%)	27.3	11.0	9.0	21.8	23.8
Impact in % points on GDP growth	0.9	0.3	0.2	0.8	0.8

Any organization, especially a large organization, can benefit from the lessons described above by using consensus-based standards in the products, services or research activities they conduct. Smaller organizations can also benefits from the use of standards. Also, even large organizations, such as multinational enterprises or government agencies, have a wide range of project size to manage. Their multi-million dollar projects may involve hundreds of people while their smallest projects may involve many teams of 5 to 10 people.

Unfortunately, many standards, such as systems engineering and software engineering standards, have been developed by large organizations without having in mind smaller settings. Most small organizations do not have the expertise or the resources to participate to standard development. A large majority of enterprises worldwide are very small entities (VSEs). In Europe, for instance over 92% of enterprises, called micro-enterprises, have up to 9 employees and another 6.5% have between 10 and 49 employees (Moll, 2013).

In 2004, an ISO working group (ISO/IEC JTC1 SC7[1] Working Group 24) was mandated to develop a set of standards and guides to address the issues facing VSEs (Laporte et al., 2013a; Laporte & O'Connor, 2014a). Members of the working group had developed a list of hypotheses about VSEs (adapted from Laporte et al., 2008a):

- The VSE context requires light and well-focused life-cycle standards and guides.
- Particular business contexts require particular standards and guides.
- There are significant differences in terms of available resources and infrastructure between an enterprise employing up to 10 people and an IT department of the same size in a larger organization.
- VSEs are limited in both time and resources, which leads to a lack of understanding about how they can use standards for their benefit. VSEs need guidelines to help them apply standards.
- Benefits for VSEs might include recognition through an assessment or an audit by an accredited body.

A survey performed by the ISO working group showed that VSEs found it difficult to relate ISO standards to their business needs and to justify the application of the standards to their business practices. Also, the survey indicated that most of the VSEs were not able to afford the resources—in number of employees, cost, and time—or see a net benefit in establishing Processes (Laporte et al., 2008a). The results of the survey as well as the experience of the members of the working group have guided the development of a four-stage roadmap providing a progressive approach to satisfying the needs and expectations of a vast majority of VSEs. This roadmap has been published as a set of ISO/IEC 29110 standards and technical reports.

We describe in this chapter the main activities associated with the development and implementation of project management Processes for small and medium scale engineering projects in a Canadian division of an American engineering company (Chevalier, 2012). Even though this organization is considered a large enterprise, a large number of their projects are within the scope of the new ISO/IEC 29110 developed specifically for VSEs (i.e. enterprises, organizations, departments and projects having up to 25 people). We present the business objectives of the improvement project, the approach to select the frameworks used to document the Processes, a brief description of the ISO/IEC 29110 standards used to document the project management Processes, the methodology used to estimate the cost and benefits of the Process documentation activities and a description of the Processes and their support tools.

MAIN FOCUS OF THE CHAPTER

The project management Process improvement program was targeted at one Canadian division of a large American engineering company, which was created a decade ago and now boasts around 400 employees across 10 offices throughout Canada. This division, as a relatively new entity, has established a Quality Management System (QMS) and has been certified against ISO 90001:2008 standard. The division had no efficient tools or project management Processes suited to managing small-scale projects. The strong growth of the division in recent years made management aware of the need to improve its methods in order to remain competitive. For this reason, most of the projects managed by this division include project plans and cost-time estimates. In most cases, these projects involve updating or improving existing electric power infrastructures. Hence the challenge of handling multiple small-scale, fast-moving projects allowing little room for unwieldy management Processes, but still requiring an efficient and straightforward monitoring Process.

The improvement program's objective was to avoid cost overruns and project delays, to standardize practices to facilitate the integration of new managers, to increase the level of customer satisfaction and to reduce risk-related planning deviations.

MANAGING PROJECTS OF VARYING SCALE

Projects in one division, of the engineering enterprise, were classified into three categories according to duration, size, number of disciplines involved and engineering fees. It was decided to classify the engineering projects into three categories: small-, medium- and large-scale projects. As illustrated in Table 3, over 95% of the projects fall in the small- and medium-scale categories.

The goal-problem approach, developed by Potter and Sakry (2002), was used to set the improvement program's priorities and to ensure that the goals set by the program addressed tangible problems that the company wished to solve. This approach includes the following steps:

- Determine the business goals (see Table 4) and the problems that the company wishes to solve
- Group goals and problems
- Prioritize problems
- Develop and implement an action plan

Table 3. Classification of the division's projects (in Canadian Dollar)

	Small Project	Medium Project	Large Project
Duration of project	Less than 2 months	Between 2 and 8 months	More than 8 months
Size of team	Up to 4 people	Between 4 and 8 people	More than 8 people
Number of engineering specialties involved	One	More than one	Many
Engineering fees	Between $5,000 and $70,000	Between $50,000 and $350,000	Over $350,000
Percentage of projects	70%	25%	5%

Table 4. Division's business goals

Objective ID	Description
O-1	Facilitate the integration of new project managers.
O-2	Achieve a global customer satisfaction level of 80%.
O-3	Meet the deadlines and costs planned for the projects, within a margin of 5%.
O-4	Reduce resource overload by 10%.
O-5	Reduce time delays to one week and cost overruns to 5% of the initial budget.
O-6	Reduce corrective work during the quality control phase by 10%.
O-7	Reduce non-chargeable time for resources by 10%.

Table 5 presents a list of 16 problems, ranging from estimation problems to lack of defined project management practices and tools, that slow down the achievement of the 7 business objectives listed in Table 4.

Managers of the engineering division grouped the problems relative to the 7 goals. Finally, they estimated the expected cost and benefit of each objective in order to prioritize them and group them in different implementation phases. Table 6 shows the estimated cost, benefits and the prioritization of the business objectives.

To verify the achievement of the objectives, a set of questions and indicators have been defined. Table 7 shows the set of questions and indicators used to assess the progress of the improvement program against established objectives.

Since not all objectives could be addressed at the same time, it was decided to address objectives 1, 2, 3 and 5. Table 8 shows an example of an objective and a task to achieve it.

As the last activity of this objective setting phase, a risk management plan was developed in order to prevent – i.e., reduce the probability and minimize the impact of – certain events on the project Process.

SELECTION AND EVALUATION OF IMPROVEMENT FRAMEWORKS

There are several frameworks which describe recognized project management practices, such as the Guide to the Project Management Body of Knowledge (PMBOK® Guide) published by the Project Management Institute (PMI, 2013); maturity models such as the Capability Maturity Model Integration (CMMI®) for

Table 5. Problems that slow down the achievement of business objectives

Problem ID	Description
P-1	Difficulty in integrating new project managers
P-2	Lack of knowledge of existing tools
P-3	Difficulty faced by new project managers to understand the ways of doing business of the division
P-4	Projects in difficulty due to poor time management
P-5	Projects in difficulty due to poor management of resources
P-6	Project implementation cost overruns
P-7	Difficulty in managing change requests
P-8	Difficulty in communicating change requests
P-9	Difficulty in estimating workload and resources
P-10	Difficulty in monitoring project progress due to a lack of project standardized project management practices
P-11	Incorrect estimation of resources to successfully carry out projects
P-12	Lack of experience or knowledge of some project managers
P-13	Lack of defined Processes and appropriate tools
P-14	Customers dissatisfied with schedule slippage
P-15	Lack of risk monitoring
P-16	Slippage of project schedules

Table 6. Prioritization of the objectives (from 1 (low) to 10 (high))

Objective Identification	Objective Description	Estimated Benefits [1-10]	Estimated Cost [1-10]	Priority (Benefits/Cost)
O-1	Facilitate the integration of new project managers.	5	10	0.50
O-2	Achieve a global customer satisfaction level of 80%.	7	4	1.75
O-3	Meet the deadlines and costs planned for the projects, within a margin of 5%.	10	10	1.0
O-4	Reduce resource overload by 10%.	6	6	1.0
O-5	Reduce time delays to one week and cost overruns to 5% of the initial budget.	10	7	1.43
O-6	Reduce corrective work during the quality control phase by 10%.	8	8	1.0
O-7	Reduce non-chargeable time for resources by 10%.	5	8	.63

Development of the Software Engineering Institute (SEI, 2010); and standards such as the new ISO/IEC 29110 series for VSEs (ISO, 2011b). Table 9 presents a few well-known project management frameworks.

A meeting with the improvement program sponsors helped define a set of criteria to determine the most suitable project management framework for the engineering division. The following criteria were selected:

Table 7. Questions and indicators to assess the achievement of objectives

Objective ID	Questions and Indicators
O-1	- Do the new project managers follow the project management Process? - Do they complete the required forms of the Process? - Are they able to classify project documents at the right place on the repository?
O-2	- What is the level of customer satisfaction about the management of his project?
O-3	- Does the project manager meet the deliverable dates as agreed with the customer? - Does the project meet its budget?
O-4	- Do the project resources have to work overtime?
O-5	- Has the project manager performed a risk analysis? - Has the project manager performed risk management? - Has the project manager addressed the risks?
O-6	- Does the quality control team have to rework engineering products when verifying deliverables?
O-7	- What is the amount allocated for non-billable resources to a customer?

Table 8. Example of intermediate objectives for one business objective

Objective ID	Main Objective/ Intermediate objective	Goal of an Objective	Task
Business Objective O-1	**Facilitate the integration of new project managers.**	Reduce the cost of integrating new employees Raise the skill level (or effectiveness) of managers.	
	Define and standardize project management practices.		Develop Processes and tools to support project execution.

- The framework is suitable for the management of small-scale projects (small team and limited means),
- The company's management knows the framework,
- The framework is recognized by the company's customers,
- Tools are available to facilitate the use of the framework,
- The framework may easily be used and integrated into the existing Processes,
- A recognition mechanism through accreditation for the company is available,
- The framework is readily available.

Before analyzing the selected frameworks, each criterion was weighted by its importance according to the project sponsors. Table 10 describes the 7 criterion used to evaluate the frameworks as well as the justification for the weight assigned to each criterion.

The sponsors evaluated the 5 frameworks using the weighted selection criteria. Figure 1 illustrates the evaluation of the frameworks selected. The final score for each framework is obtained by multiplying each individual score by the weight of each criterion and then adding the individual scores. As illustrated in Figure 1, the PMBOK® Guide and the ISO/IEC 29110 standard obtained the highest overall score.

Table 9. Well-known project management frameworks[2]

Framework	Description
Guide to Project Management Body of Knowledge (PMBOK® Guide)	Developed by the Project Management Institute, it provides project managers with the fundamental practices needed to achieve organizational results and excellence in the practice of project management (PMI, 2013a).
Organizational Project Management Maturity Model (OPM3®)	Developed by the Project Management Institute, it provides the tools organizations need to measure their maturity against a comprehensive set of organizational best practices (PMI, 2013b).
PRINCE2®	A structured project management method based on experience drawn from thousands of projects and from the contributions of countless project sponsors, Project Managers, project teams, academics, trainers and consultants (AXELOS, 2009).
Capability Maturity Model® Integration (CMMI®)	Developed originally by the Software Engineering Institute and now under the management of the CMMI Institute, the set of CMMI models are collections of best practices that help organizations to improve their Processes (SEI, 2010).
ISO/IEC/IEEE 16326 - Systems and software engineering — Lifecycle Processes — Project management	This standard provides normative content specifications for project management plans covering software projects, and software-intensive system projects. This International Standard also provides detailed discussion and advice on applying a set of project Processes that are common to both the software and system life cycle as covered by ISO/IEC/IEEE 12207:2008, Systems and software engineering – Software life cycle Processes, and ISO/IEC/IEEE 15288:2008, Systems and software engineering – System life cycle Processes, respectively. The discussion and advice are intended to aid in the preparation of the normative content of project management plans (ISO, 2009).
ISO/IEC/IEEE 16085 - Systems and software engineering — Lifecycle Processes — Risk management	This standard defines a Process for the management of risk in the life cycle. It can be added to the existing set of system and software life cycle Processes defined by ISO/IEC/IEEE 15288 and ISO/IEC/IEEE 12207, or it can be used independently. ISO/IEC/IEEE 16085 can be applied equally to systems and software (ISO, 2006).
ISO/IEC/IEEE 12207 – Software lifecycle Processes	This standard establishes a common framework for software life cycle Processes, with well-defined terminology, that can be referenced by the software industry. It contains Processes, activities, and tasks that are to be applied during the acquisition of a software product or service and during the supply, development, operation, maintenance and disposal of software products. Software includes the software portion of firmware (ISO, 2008a).
ISO/IEC/IEEE 15288 – System lifecycle Processes	This standard establishes a common framework for describing the life cycle of systems created by humans. It defines a set of Processes and associated terminology. These Processes can be applied at any level in the hierarchy of a system's structure. Selected sets of these Processes can be applied throughout the life cycle for managing and performing the stages of a system's life cycle. This is accomplished through the involvement of all interested parties, with the ultimate goal of achieving customer satisfaction (ISO, 2008b).

Table 11 describes the frameworks evaluated, the cumulative score of each framework evaluated as well as the justifications for each score and the decisions (e.g. select, discard) about each framework.

The ISO/IEC 29110 was the framework selected for the improvement project. Even if the company's division comprises more than 400 employees a significant number of small- and medium-scale projects are carried out by small teams. Since the ISO/IEC 29110 standard applies to Very Small Entities (VSEs), i.e. enterprises, organizations, departments and projects having of up to 25 people, this standard was very suitable for the engineering division of the company.

OVERVIEW OF ISO/IEC 29110 STANDARDS AND TECHNICAL REPORTS

This section describes the approach used by WG24 to develop a set of standards and technical reports to address the needs of VSEs. This section also describes the management and engineering guide used by the engineering division to develop their small-scale and medium-scale project management Processes.

Table 10. Weights of the criteria used to evaluate the five frameworks

Description of the criteria	Weight assigned to each criterion (1 = Low importance to 3 = High importance)	Justification
Adapted for the management of small projects	3	- The majority of projects are small projects. - A complex method will need to be adapted to small and medium scale projects to be effective. - Using standards already adapted to small projects could reduce the effort required for the development of small and medium scale project Processes.
Known to the management of the organization	2	- Using a known framework could promote the commitment of management to solutions that will be developed.
Recognized by the company's customers	2	- Some customers have project management practices based on standards. - The use of similar frameworks could facilitate communication and the monitoring of projects with customers.
Tools to facilitate the use of standards are available	2	- Using standards supported with tools could reduce the effort for the development of Processes.
Ease of integration with existing organizational Processes	3	- It is very important that the standards can be used in the current business context. - Existing Processes are not under the responsibility of the Division and therefore cannot be changed easily. - Changes made in the Division should not force changing the practices employed by other divisions.
Accreditation/ Certification available	1	- The company does not currently show interest for accreditation related to project management practices.
Ease of access to frameworks	1	- The company has the monetary means to acquire the frameworks. - The impact of this criterion is small.

There is a wide spectrum of development approaches for organizations developing software. Figure 2 illustrates the spectrum of approaches. On the vertical axes are illustrated the approaches based on the level of risk. The upper part of the vertical axis illustrates a low risk linear approach using a waterfall approach while the lower part of the axis illustrates a risk-driven project using an iterative approach. On the horizontal axis, the level of ceremony varies from a low ceremony approach with little or no documentation (e.g. agile approach) to a high ceremony approach with a comprehensive documentation and the use of standards (e.g. ISO/IEC/IEEE 16085 risk management standard) and frameworks (e.g. CMMI®). As we will illustrate in the next paragraphs, the new standard developed specifically to address the needs of very small entities, i.e. the family of ISO/IEC 29110 International Standards (ISs) and Technical Reports (TRs), is located at about the centre of both axes. VSEs will be able to use ISO/IEC 29110 in a wide spectrum of approaches.

Many international standards have been developed to capture such proven management and engineering practices. However, these standards were not written for very small organizations and were consequently difficult to apply in such settings. An ISO Working Group (Working Group 24, WG24) has been established to address these difficulties (ISO, 2005).

WG24 used the 6-stage model of the innovation-development Process model, developed by Rogers (2003) to develop the set of ISs and TRs. Rogers defines the innovation-development Process as follows:

Figure 1. Scores of the five frameworks selected

	Evaluation criteria							
	Adapted for the management of small projects		Known to the management of the organization		Recognized by the company's customers		Tools to facilitate the use of standards are available	
Framework	Weight assigned to each criteria	Evaluation of framework using criteria	Weight assigned to each criteria	Evaluation of framework using criteria	Weight assigned to each criteria	Evaluation of framework using criteria	Weight assigned to each criteria	Evaluation of framework using criteria
CMMI-DEV, version 1.3	3	1	2	1	2	0	2	2
CMMI-SVC, version 1.3	3	1	2	0	2	0	2	2
PMBOK Guide (PMI)	3	1	2	3	2	3	2	2
PRINCE2 2009 Edition	3	1	2	0	2	0	2	2
ISO/IEC 29110	3	3	2	0	2	0	2	3

	Evaluation criteria						
	Ease of integration with existing organizational processes		Accreditation/Certification available		Ease of access to documents		
Framework	Weight assigned to each criteria	Evaluation of framework using criteria	Weight assigned to each criteria	Evaluation of framework using criteria	Weight assigned to each criteria	Evaluation of framework using criteria	Total Score
CMMI-DEV, version 1.3	3	1	1	3	1	3	**18**
CMMI-SVC, version 1.3	3	1	1	3	1	3	**16**
PMBOK Guide (PMI)	3	3	1	0	1	3	**31**
PRINCE2 2009 Edition	3	1	1	2	1	1	**13**
ISO/IEC 29110	3	3	1	2	1	3	**29**

Table 11. Analysis and justification of the five frameworks

Frameworks Evaluated	Cumulative Result	Decision about a framework	Justification
CMMI® for Development, V1.3	18	Not selected	- This document is not well known by the management of the company and its customers. - Sponsors of the improvement project showed little interest in this document.
CMMI® for Services, V1.3	16	Not selected	- This document is not well known by the management of the company and its customers. - Sponsors of the improvement project showed little interest in this document.
PMBOK® Guide 4th edition	31	Selected for the large-scale projects	- It is a known document of the management of the company and its customers. - Sponsors of the improvement project have expressed interest in this document. - This framework is not specifically designed for small projects. It will be used only to complete the documentation for the large-scale projects Process.
PRINCE2®, 2009 edition	13	Not selected	- It is a repository little used in Canada. - Sponsors of the improvement project have little interest in this repository.
ISO/IEC 29110	29	Selected for documenting the Process of small projects and medium-scale projects	- It is an unknown repository from management of the company and its customers. - This framework is specially designed for small projects. - However, the concepts of project management of this repository are in line with those described in the PMBOK Guide. - This framework will be used for the development of project management Processes for small and medium scale projects.

Figure 2. Positioning of the ISO/IEC 29110 (adapted from Kroll and Kruchten 2003)

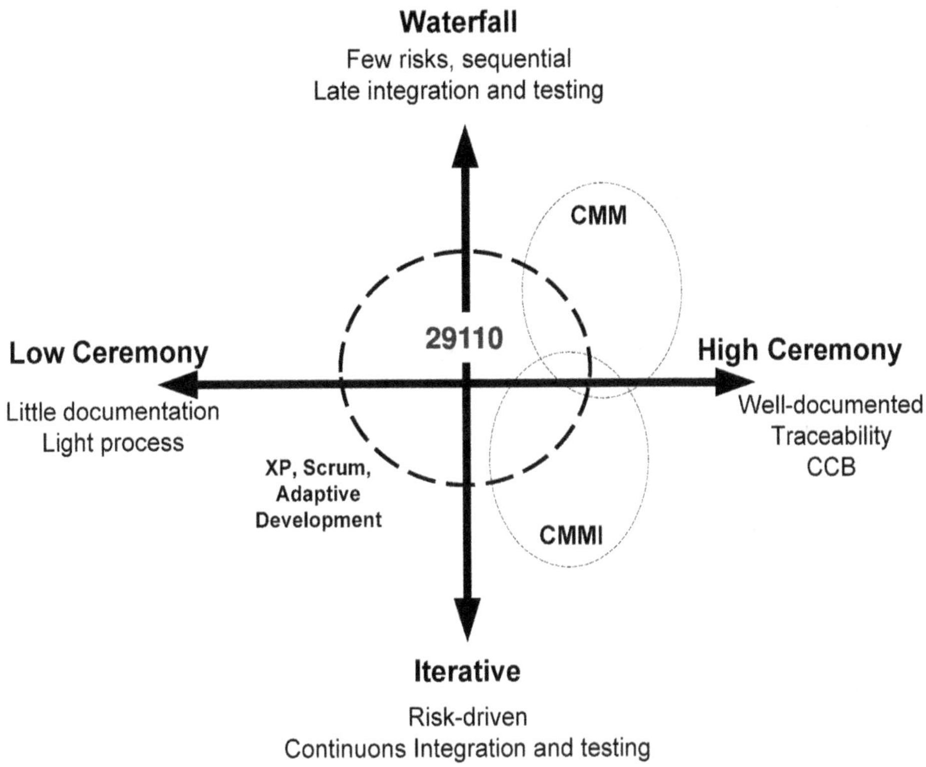

all the decisions, activities, and their impacts that occur from recognition of a need or problem, through research, development, and commercialization of an innovation through diffusion and adoption of the innovation by users, to its consequences.

We briefly outline the stages of Rogers model and the actions done by WG24 to design and develop the new standard (Laporte, 2009):

- *Stage 1 The Recognition of Needs and Problems:* Is primarily concerned with the awareness of a problem. In this stage, WG24 documented the lack of awareness and acceptance of existing standards by small and very small software companies and justified the need for a new approach to the development of Process standards aimed specifically at very small companies.
- In *Stage 2 Basic and Applied Research:* WG24 designed and conducted an international survey of very small software companies (described below), leading to an enhanced understanding of the needs of very small software companies from a Process standard.
- In *Stage 3 Development:* WG24 used an innovative approach for the development of the new standard as well as a novel approach to support the implementation, i.e. the development of a set of Deployment Packages.
- In *Stage 4 Commercialization:* ISO is the organization that is publishing and selling ISs and TRs. ISO/IEC 29110 ISs and TRs and made them available on its web site. WG24 obtained from ISO that the set of TRs would be made available at no cost on the ISO web site.

- In *Stage 5 Diffusion and Adoption:* WG24 developed a set of activities to accelerate the dissemination aimed at those VSEs wishing to put in place a minimal set of software development and management practices. As an example, a series of mini-research pilot projects with very small software companies were conducted as a means to accelerate the adoption and utilization of ISO/IEC 29110.
- In *Stage 6 Consequences of an Innovation:* A list to present the potential negative consequences of the publication of standards was developed (Laporte, 2009) and the evaluation of the benefits of ISO/IEC 29110 were estimated using the ISO methodology presented in this chapter. For VSEs, the ISO methodology is too demanding. Two examples of the benefits, using the cost of software quality classification developed by Krasner (1998) have been published recently (Garcia et al., 2015; Laporte et al., 2014c). In these two examples, the cost of rework *(that is, wasted effort)* for projects of about 900 hours was between 13% and 18% of the total project effort. Organisation with a high level of maturity, i.e. having a defined organizational Process, typically have about 15% to 25% of rework (Krasner, 1998).

At stage 2, WG24 conducted a survey, in 9 languages (English, French, German, Korean, Portuguese, Thai, Turkish, Russian and Spanish), of small and very small organizations to question them about their use of standards as well as to collect data to identify problems and potential solutions to help them apply standards. The survey was made up of 20 questions structured in 5 parts: General information, Information about standards utilization in VSEs, Information about implementation and assessment problems in VSEs, Information about VSE needs and Information about justification for compliance to standard(s). Over 400 responses were collected from 32 countries. The main findings were documented in (Laporte et al., 2008c), however some salient points are discussed here.

WG24 anticipated the weak use of standards by VSEs by asking questions designed to provide a better understanding of the reasons for this. There are three main ones, as shown in Figure 3. The first

Figure 3. Why don't VSEs use standards? (Laporte et al., 2008a)

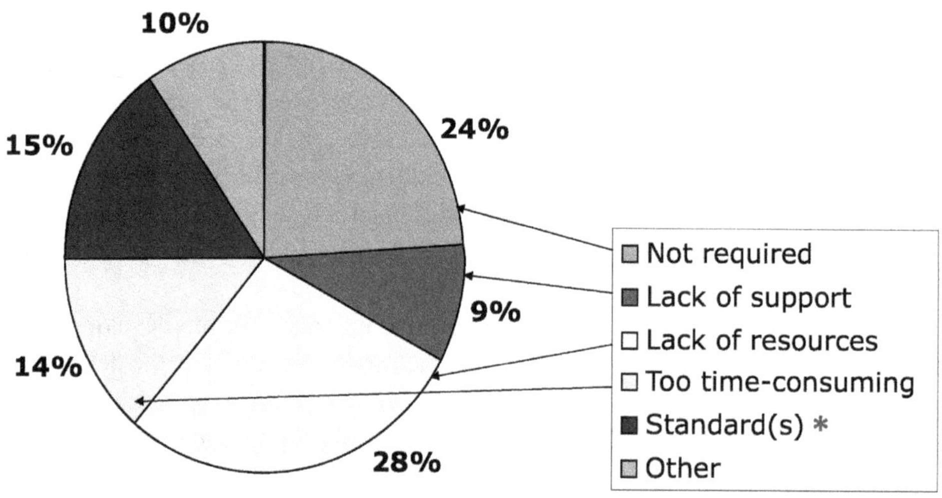

* Difficult, Bureaucratic, not enough guidance.

is a lack of resources (28%); the second is that standards are not required (24%); and the third derives from the nature of the standards themselves: 15% of the respondents consider that the standards are difficult and bureaucratic, and do not provide adequate guidance for use in a small business environment.

Over 62% of the responses, received from 32 countries, indicated they would like more guidance with examples, and 55% were asking for lightweight and easy-to-understand standards, complete with templates. Finally, the respondents indicated that it had to be possible to implement standards with minimum cost, time, and resources (Laporte et al., 2008a).

The main objective of the new ISO/IEC 29110 standard is to assist and encourage VSEs to improve their systems engineering and/or software engineering Processes. The approach used by the ISO working group mandated to develop ISO/IEC 29110 standards and guides was to develop a 4–stage roadmap by selecting a minimal set of Process elements and documentation elements from existing international standards such as the software life cycle standard ISO/IEC/IEEE 12207 (ISO, 2008a), the systems engineering life cycle standard ISO/IEC/IEEE 15288 (ISO, 2008b) and the documentation standard ISO/IEC/IEEE 15289 (ISO, 2011a).

The ISO working group decided to develop a set of 4 profiles, i.e. a four-stage roadmap (Entry, Basic, Intermediate, Advanced) providing a progressive approach to satisfying the needs and expectations of a vast majority of VSEs. The first set of profiles, as illustrated in Table 12, called Generic profiles, is targeted at VSEs not involved in the development of critical systems or critical software. VSEs targeted by the Entry Profile are VSEs working on small projects (e.g. at most six person-months effort) and for start-up VSEs. The Basic Profile describes development practices of a single application by a single project team of a VSE. The Intermediate Profile is targeted at VSEs developing multiple projects within the organizational context taking advantage of it. The Advanced Profile is targeted to VSEs that want to sustain and grow as an independent competitive development business.

The set of ISO/IEC 29110 documents has been developed to address the needs of various audiences. As an example, the Management and engineering guide is mainly targeted at VSEs and their customers. At the request of WG24, all ISO/IEC 29110 technical reports (i.e. Part 1, 3 and 5) are available at no cost from the ISO web site. The management and Engineering Guides, the most valuable documents for VSEs, are available in English, French, Japanese, Portuguese and Spanish.

For illustration purposes, we will briefly describe one profile used by the engineering division, i.e. the Basic profile, to develop the medium-scale project management Process of the engineering organization. At the time that the project management improvement program was launched, only the software engineering Entry and Basic profiles had been published by ISO. The ISO/IEC 29110 systems engineering profiles were published after this improvement project had been completed. Even though, as illustrated

Table 12. 4-Stage ISO/IEC 29110 Roadmap (O'Connor et al., 2010)

Entry	Basic	Intermediate	Advanced

in Figure 4, the Entry and Basic profiles are composed of a project management (PM) Process and a software implementation (SI) Process, only the project management Processes were used to develop the small-scale and medium-scale project management Processes of the engineering organization.

As illustrated in Figure 4, the customer's Statement of Work (SOW) is used to initiate the PM Process. The project plan will be used to guide the development of the software. The PM Process closure activity will deliver a Software Configuration (i.e. product and documentation) and will obtain the customer's acceptance to formalize the end of the project.

The purpose of the PM Process is to establish and carry out in a systematic way the tasks of the project in order to meet the objectives in the expected quality, time and costs. The seven objectives of PM Process of the Basic profile are (ISO, 2011b):

- PM.01 - The *Project Plan* for the execution of the project is developed according to the *Statement of Work* and reviewed and accepted by the Customer. The *Tasks* and *Resources* necessary to complete the work are sized and estimated.
- PM.02 - Progress of the project is monitored against the *Project Plan* and recorded in the *Progress Status Record.* Corrections to remediate problems and deviations from the plan are taken when project targets are not achieved. Closure of the project is performed to get the Customer acceptance documented in the *Acceptance Record.*
- PM.03 - The *Change Requests* are addressed through their reception and analysis. Changes to software requirements are evaluated for cost, schedule and technical impact.
- PM.04 - Review meetings with the Work Team and the Customer are held. Agreements are registered and tracked.
- PM.05 - *Risks* are identified as they develop and during the conduct of the project.

Figure 4. Processes and activities of the software engineering Basic and Entry profiles (Laporte & O'Connor, 2014b)

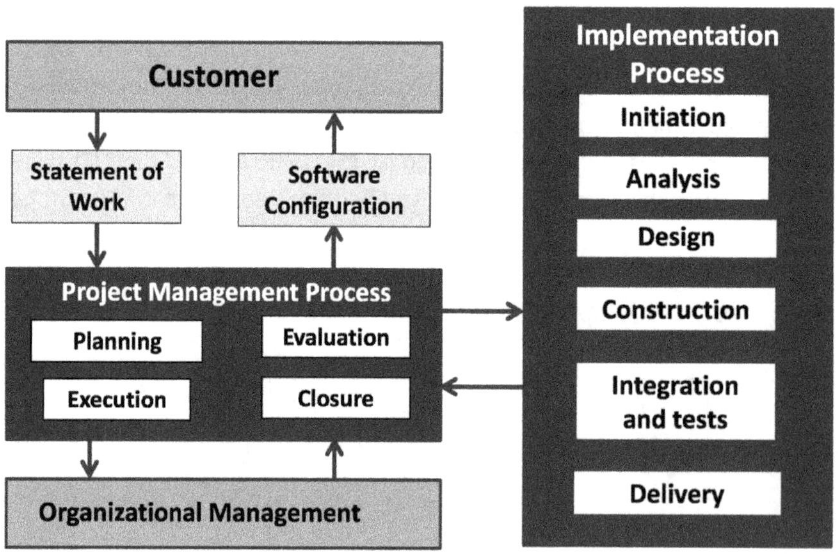

- PM.06 – A software *Version Control Strategy* is developed. Items of *Software Configuration* are identified, defined and baselined. Modifications and releases of the items are controlled and made available to the Customer and Work Team. The storage, handling and delivery of the items are controlled.
- PM.07 - Software Quality Assurance is performed to provide assurance that work products and Processes comply with the *Project Plan* and *Requirements Specification.*

During the project planning activity, a project plan is developed. Then, the assessment and control tasks are used to assess the project's progress against the project plan. Action is then taken, if needed, to eliminate deviations from the project plan or to incorporate changes to the plan. The project closure activity groups together the deliverables produced by the implementation Process, such as the software or the user manual, and gets the customer's written acceptance to finalize the project. A repository is established to save the work products and to control their versions during the project.

Figure 5 illustrates the 4 activities of the PM Process as well as their input and output products. Each activity is composed of a set of tasks. The task description doesn't impose any technique or method to perform it. Even though Figure 5 describes the PM Process in a waterfall or sequential approach, ISO/ IEC 29110 is not intended to preclude the use of different life cycles such as iterative, incremental, evolutionary or agile.

To show the links between ISO/IEC/IEEE 12207 and the objectives of the PM Process, Figure 6 shows the outcomes of the ISO 12207 project planning Process and measurement Process used to develop Objective PM.01 of the PM Process of ISO/IEC 29110.

Note that only a subset of the ISO/IEC/IEEE 12207 project planning and measurement Processes has been selected for the Basic profile.

For illustration purposes, two tasks of the Project Planning activity are described in Table 13. On the left side of the table are listed the roles involved in a task: the project Manager (PM) and the Customer (CUS). Each task statement starts with an active verb to make sure that there is no ambiguity (e.g. versus 'Project Planning'), the artefacts needed to perform each task are listed as well as the output of each task. The status of input and output artefacts is also indicated.

The customer is also involved, during the execution of the project, when he submits change requests, during project review meetings, for the validation and approval of the requirements specifications and for the acceptance of the deliverables.

Table 13. Example of 2 tasks of the project planning activity (ISO, 2011b)

Role	Task list	Inputs	Outputs
PM CUS	PM.1.2 Define with the Customer the *Delivery Instructions* of each one of the *Deliverables* specified in the *Statement of Work.*	*Statement of Work [reviewed]*	*Project Plan* *- Delivery Instructions*
PM CUS	PM.1.14 Review and accept the *Project Plan.* Customer reviews and accepts the *Project Plan,* making sure that the *Project Plan* elements match with the *Statement of Work.*	*Project Plan [verified]*	*Meeting Record* *Project Plan [accepted]*

Figure 5. Project management Process diagram for software (ISO, 2011b)

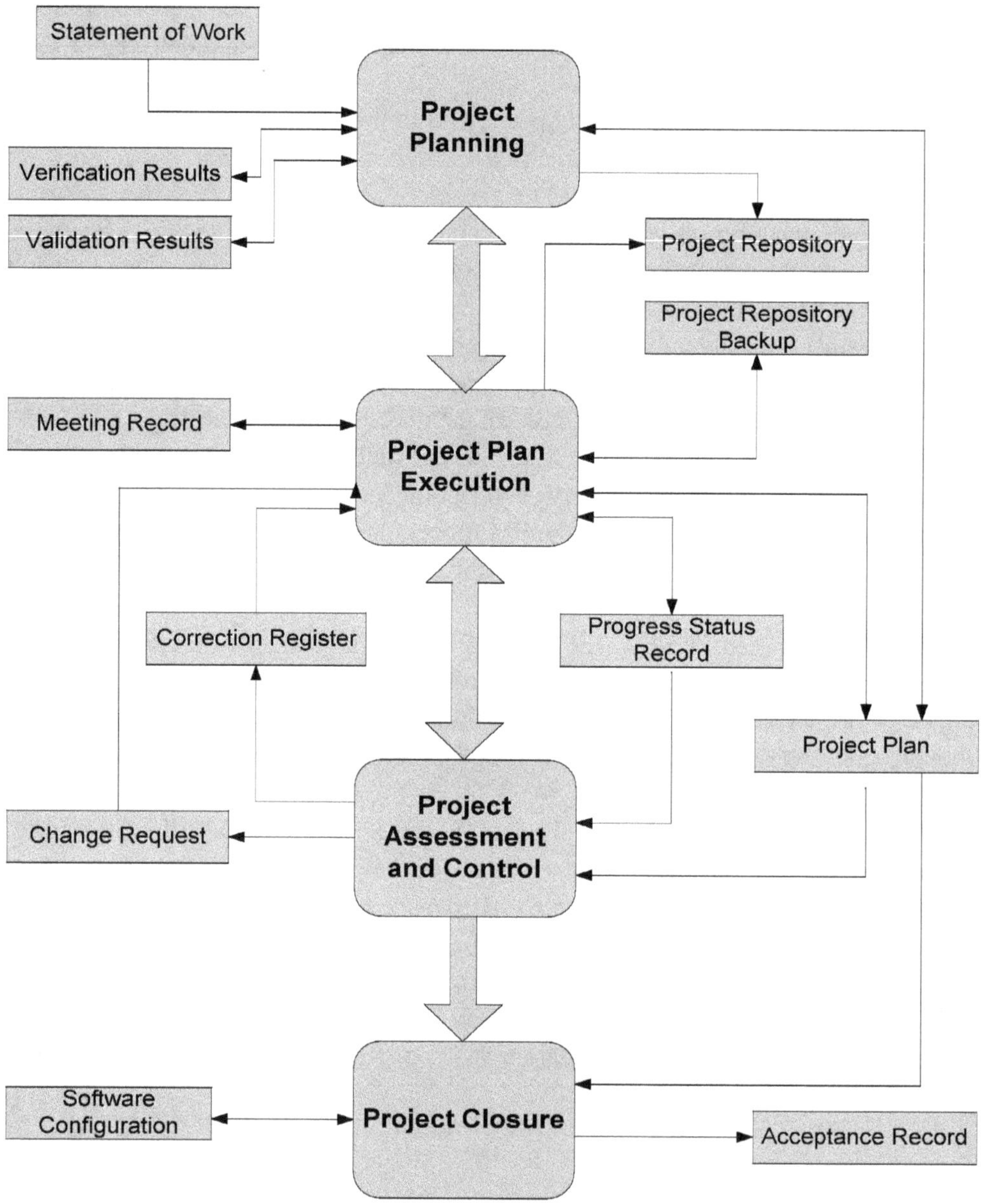

Management Process of the Engineering Organization

The project management practices used by the engineering division's project managers were assessed against the Entry and Basic profiles of ISO/IEC 29110 by interviewing project managers. Figure 7 shows the overall results of the assessment of the activities of the PM Process of the Entry profile. The figure displays the percentage of the tasks performed for each of the following activities of the ISO/IEC 29110 management and engineering guide (ISO, 2012):

Figure 6. Links between ISO/IEC/IEEE 12207 outcomes for Objective 1 of the PM Process (ISO, 2011b)

6.3.1 Project Planning Process
a) the scope of the work for the project is defined;
c) the tasks and resources necessary to complete the work are sized and estimated;
e) plans for the execution of the project are developed; and
f) plans for the execution of the project are activated.

6.3.7 Measurement Process
a) the information needs of technical and management processes are identified.

[ISO/IEC/IEEE 12207:2008, 6.3.1, 6.3.7]

- Project planning (11 tasks)
- Project plan execution (2 tasks)
- Project assessment and control (3 tasks)
- Project closure (2 tasks)

Table 14 illustrates the assessment of 2 of the 15 tasks of the Project Planning Activity of the Basic profile. As illustrated in this table, not only each task was evaluated, but a Low/Medium/High implementation priority was assigned as well as an estimation of the implementation effort and impact.

Figure 8 shows the overall results of the assessment of the activities of the PM Process of the Basic profile. The figure displays the percentage of the tasks performed for each of the following activities of the ISO/IEC 29110 management and engineering guide (ISO, 2011b):

Figure 7. Assessment of project management Process against the ISO/IEC 29110 Entry profile

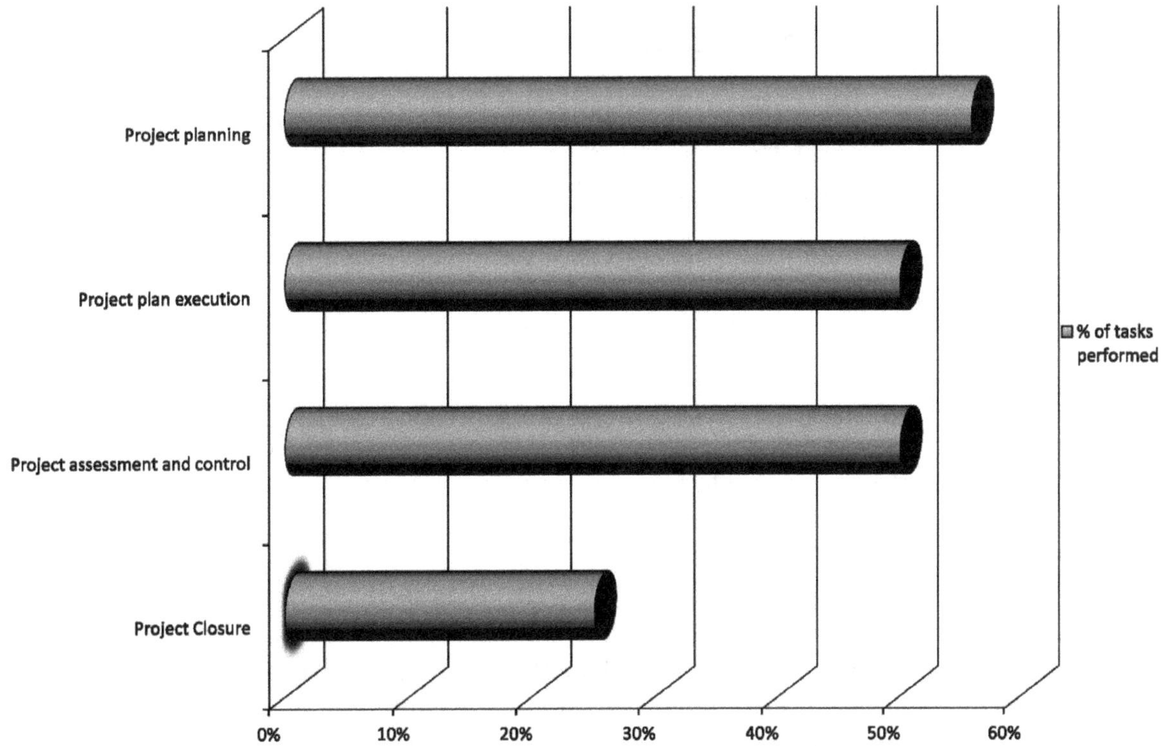

Table 14. Example of assessment of 2 of tasks the project planning activity against the Basic profile

ID	Roles	Tasks	Input Products	Output Products	Implemented	Implementation Priority	Implementation Effort	Implementation Impact
PM.1.1	PM TL	Review the Statement of Work	Statement of Work	Statement of Work [reviewed]	Partially	Medium	Low	Medium
PM.1.2	PM CUS	Define with the Customer the Delivery Instructions of each one of the Deliverables specified in the Statement of Work.	Statement of Work [reviewed]	Project Plan - Delivery Instructions	Partially	Medium	Low	Medium

Figure 8. Performance assessment against the Basic profile

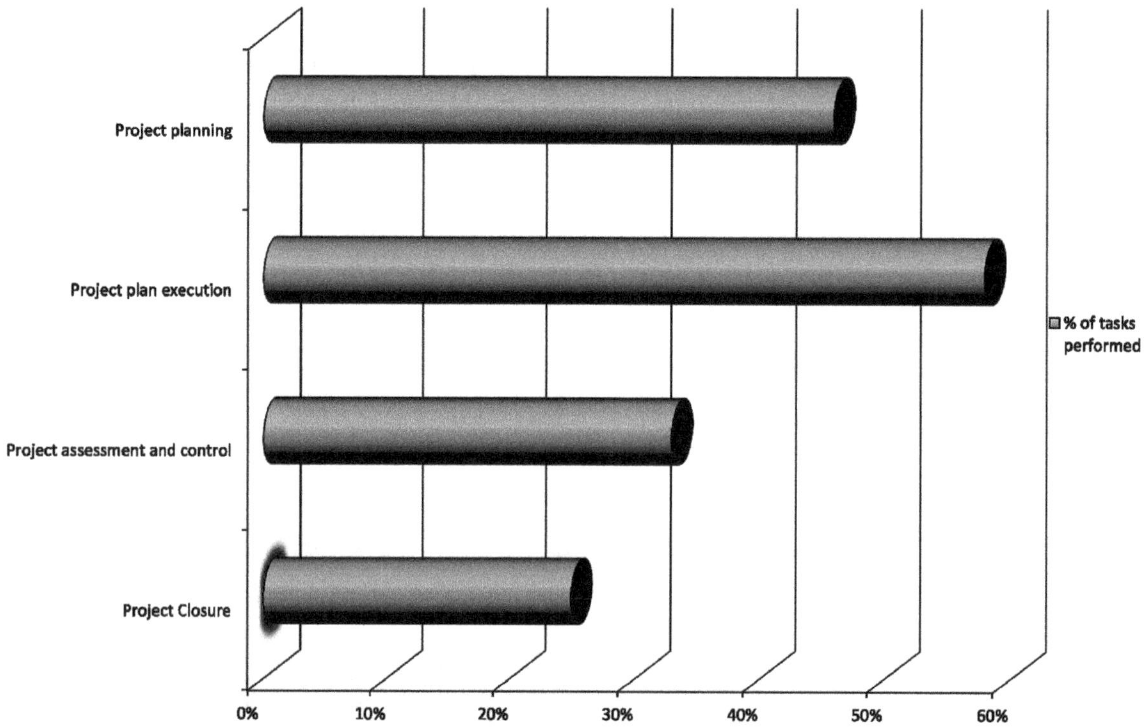

- Project planning (15 tasks)
- Project plan execution (6 tasks)
- Project assessment and control (3 tasks)
- Project closure (2 tasks)

We note that a low level of implementation of ISO/IEC 29110 activities was achieved, for medium scale projects, within the engineering company at the beginning of the improvement program. Also, during the interview with managers, it was noted the PM tasks were not performed systematically. In addition, the assessment revealed that PM practices varied from project manager to project manager and that no guideline had been defined for a few tasks.

We note that the level of implementation of the project management tasks of the Basic profile is slightly lower than that of the Entry profile. This is explained by a higher number of tasks in the Basic profile. However, a significant gap remains between the profile and the current situation analyzed.

DEVELOPMENT OF THE PROJECT MANAGEMENT PROCESSES

The development of Processes and tools, such as checklists, templates and forms, was the central element of the solution to the problems identified. They have been developed taking into account the 3 categories of company projects.

Project Management Processes Developed

A Process is defined as a set of related activities that transforms inputs into output elements. The central element of the proposed solution is a project management Process guide. Three groups of project management Processes have been developed:

- Project management Process of small projects
- Project management Process of mid-sized projects
- Project management Process of large projects

Each project management Process is composed of 4 stages as illustrated in Figure 9.

A significant portion of projects undertaken by the company, about 70%, are small-scale projects with duration of less than four months with a team of less than five people. These projects have the following constraints:

- The completion dates are often too short to set-up a complex project management structure,
- The number of projects executed in parallel has a significant impact on the human resources assigned to them,
- Some project managers do not necessarily have all the required project management knowledge. They are very technically competent individuals who are assigned management tasks. They perform technical and management tasks.

Figure 9. Four stages of all project management Processes

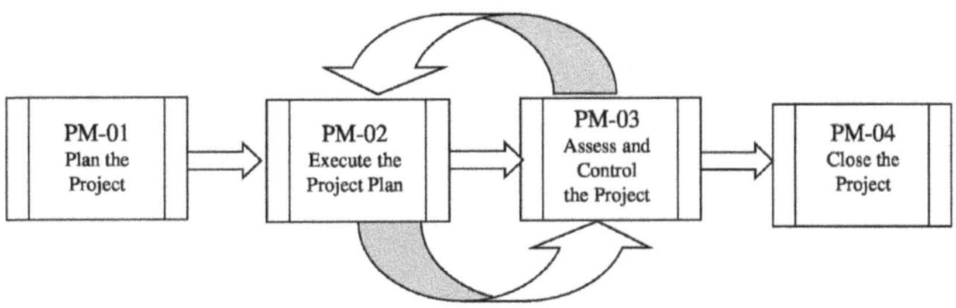

The small-project project management Process was based on the Entry profile of the ISO/IEC 29110 standard. The medium-project project management Process was based on the Basic Profile. The large-project project management Process was improved and supplemented using the PMBOK® Guide of the Project Management Institute.

Notation Used to Document the Small- and Medium-Scale Processes

The ETVX notation (i.e. Entry-Task-Verification-eXit) was developed in the 80s by IBM (Radice & Roth, 1985). Given its simplicity to use, it has been adopted by many organizations such as NASA. Figure 10 describes an ETVX notation that was enriched by adding entry and exit criteria and measures.

This notation includes the following elements:

- Inputs
    - Elements received from outside the Process that are required for its execution. It is perhaps not mandatory that all inputs exist before starting activities in the Process. It is sometimes possible, for a first iteration, to use a subset of inputs. When the other inputs are available, further iterations may be performed.
- Tasks
    - Activities to be carried out to complete the Process and create the required artefacts. Usually, a task statement starts with an active verb and is followed with an object (e.g. Test code elements).
- Entry criteria
    - Criteria to ensure that the artefacts required for the tasks to be carried out meet quality criteria (e.g. Project budget has been approved)
- Exit criteria
    - Criteria to ensure that the tasks were carried out as required and that artefacts meet the quality criteria (e.g. Software Requirements have been reviewed and corrected)
- Outputs
    - Artefacts following the execution of the tasks that could be used in a next step of a Process.
- Measures
    - Measures collected during the execution of the tasks such as effort measured in hours, number of defects detected during a review.

Figure 10. Enriched ETVX Notation

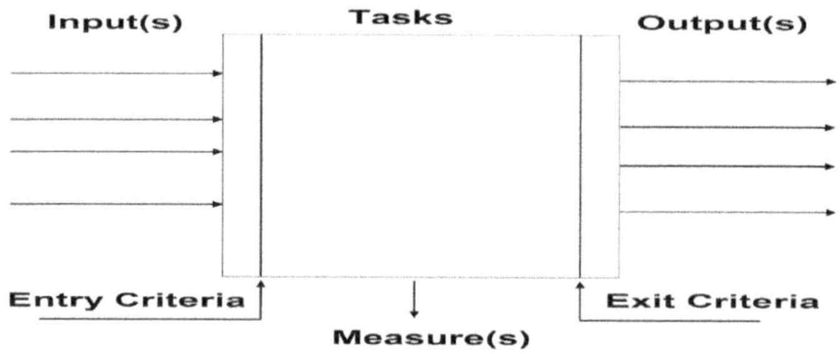

Figure 11 illustrates one step of the small-scale project management Process called plan the project. The identification of the Process is 'SPP' for Small-Scale Project; the title of the step of the Process is 'Plan the Project'. This step is composed of 6 tasks listed in the ETVX diagram.

It is also possible to use an enriched ETVX textual notation to describe a Process or a procedure. Figure 12 shows the template of this textual notation. This notation, as well as that of the figure above, could be used by an expert user.

A new user of a Process would require more information. As an example, each task would be described in more details such that it could be executed by new users and produce the same results. As an example, the task titled 'Estimate resources, effort and duration' could be enhanced by adding a detailed procedure showing, step by step, what has to be done.

Checklists Developed

Discussions with project managers of the organization revealed that they were often burdened with technical tasks in addition to the management of their project. This situation often impacted their ability to perform management tasks despite their level of expertise in project. It was therefore decided that a few checklists might provide a useful tool for project managers for the following reasons:

- They are a good way to explain or briefly summarize the tasks to be performed by the project manager
- They help identify quickly the forms and templates available to perform the project management tasks
- They provide quick links to additional references

Figure 11. Enriched ETVX description of a step of the project management Process

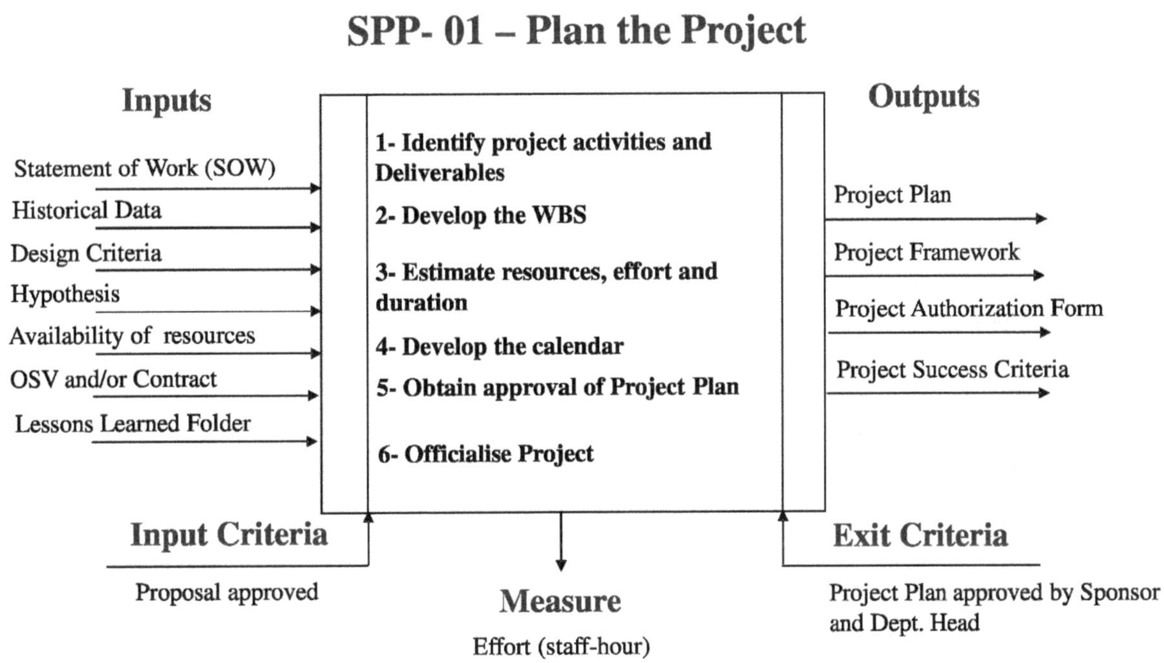

Figure 12. Textual representation of the enriched ETVX notation

Procedure: <Name of Process or Procedure>	**Phase:** <Name of phase where the procedure is used>
Process/Procedure Owner: <Owner of this process/procedure>	
Description: a brief description, background and purpose of the process/procedure	
Entry Criteria: • <entry criteria>	**Exit Criteria:** • <exit criteria>
Inputs: • <work products as input>	**Outputs**: • <work products as output>
Roles: • <list of all the actors and their responsibilities>	
Reference(s) • <Document required to use this procedure>	
Assets: • <Tools; methodologies; references; guidelines; checklists; other procedures>	
Tasks: • <Itemized list of tasks (summarized) which need to be accomplished to satisfy this process/procedure (using an active verb and a noun)>	
Measures: • <Measures captured during execution of process/procedure>	

- The provide guidance to the project manager for storing the project management documents
- They provide an easy means of assessing the implementation of Processes

As part of this project to improve management practices projects, five checklists were developed:

- Project management Process of small projects
- Project management Process of medium projects
- Project management Process of large projects
- Preparation of service offerings
- Preparation of detailed project planning

Project Management Forms and Templates Developed

Forms and templates were developed to guide Project Managers in the execution of management tasks and enable a consistency of results. They also serve to guide managers unfamiliar with some project management practices.

Testing Solutions Developed

To test the solutions developed, pilot projects have been performed. Checking solutions in the context of real project validated that the proposed solutions were consistent, feasible and complete. The small and medium Processes and tools have been tested. The pilot projects consisted of running three different projects where project managers implemented the Process and the associated tools.

Managers then evaluated the proposed Processes, identified problems and potential improvements. The lessons learned sessions conducted at the end of the pilot projects have identified minor adjustments to the Processes and tools. It was also noted, when executing the pilot projects that project managers would appreciate examples of how to implement the tools.

Developing a Deployment Strategy

Once the final adjustments were made to the project management Process and tools a strategy was developed for the deployment of these solutions to all project managers in the division, i.e. about thirty persons. The strategy consisted of the following three components: communication, training and diffusion of the Processes and their supporting documents.

The first phase was to inform project managers. In order to reduce questioning and to mitigate the negative impacts that 'unknown' can generate project changes, various means have been implemented to inform all managers. Here are some of them:

- Informative emails were sent;
- Articles were published in the monthly newsletter of the company;
- Informational messages have been added to the company intranet.

In addition, one-day training sessions have been prepared for project managers. The training allowed the project managers to learn the new Processes and tools developed.

The second phase was used to distribute documents to all managers of the division located in ten offices of the company in Canada. The most appropriate medium was the company's intranet. A section of the intranet, dedicated to project management, was created and served as a main access to project management documents such as project management Process guides, checklists, forms and templates. This section of the intranet also contains information relevant to project management as links to websites, the identification of project management standards and other information such as projects management books.

Table 15 lists a sample of the first projects that are or have been carried out by 4 project managers using the Processes and tools developed during the improvement project.

DAY-TO-DAY APPLICATION OF THE PROCESSES

After an agreement or a contract with a customer has been concluded, the selection of the project management Process must be made, by the project manager, using a spreadsheet tool containing the criteria listed in Figure 13. Once the appropriate project management Process has been selected, the sponsor of the project as well as by a Director or a Vice-President approve this selection.

Once the selected project management Process has been approved, it is applied in conjunction with the engineering Processes of the division. Figure 14 illustrates the day-to-day use of the project management Process and the relation with the knowledge management cycle. An organization can continually improve over time by learning from its success and its failures, and enhance the organizational performance in project management. Effective project management Process involves the identification, creation, acquisition, dissemination, and reuse of knowledge asset, i.e. lessons learned and best practices.

Table 15. First projects conducted using the Processes and tools developed

Project Budget (CAD$)	Process Used	Project Manager ID
120 000$	Medium-scale project	PM-1
27 000$	Small-scale project	PM-1
200 000$	Medium-scale project	PM-1
400 000$	Medium-scale project	PM-2
65 000$	Medium-scale project	PM-2
130 000$	Medium-scale project	PM-2
250 000$	Medium-scale project	PM-2
6 000$	Small-scale project	PM-1
40 000$	Small-scale project	PM-4
38 000$	Small-scale project	PM-5

Figure 13. Tool developed to select the appropriate PM Process

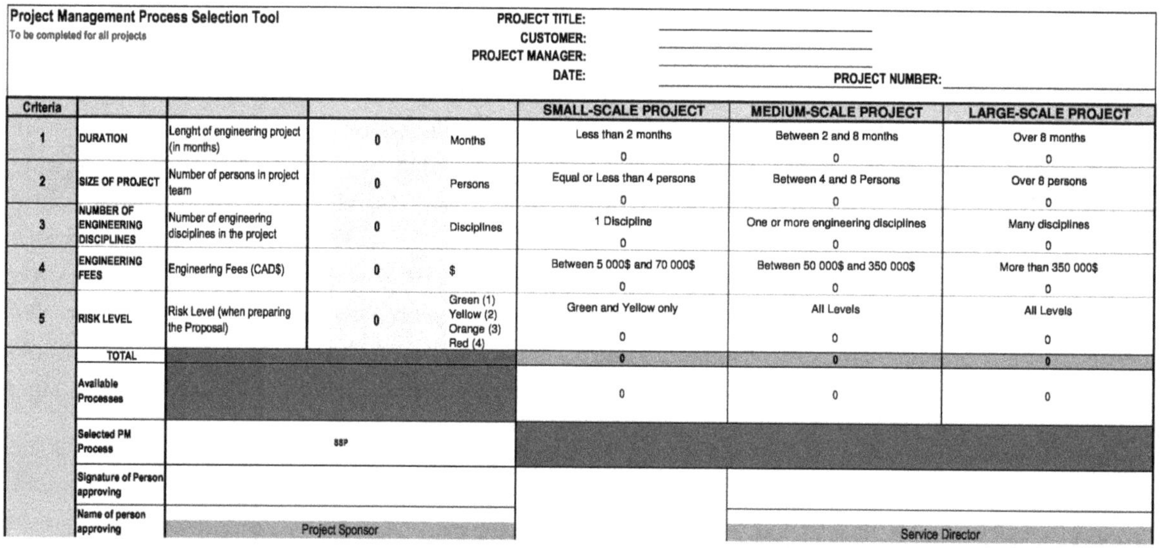

Project Management Process Selection Tool
To be completed for all projects

PROJECT TITLE:
CUSTOMER:
PROJECT MANAGER:
DATE:
PROJECT NUMBER:

Criteria					SMALL-SCALE PROJECT	MEDIUM-SCALE PROJECT	LARGE-SCALE PROJECT
1	DURATION	Lenght of engineering project (in months)	0	Months	Less than 2 months 0	Between 2 and 8 months 0	Over 8 months 0
2	SIZE OF PROJECT	Number of persons in project team	0	Persons	Equal or Less than 4 persons 0	Between 4 and 8 Persons 0	Over 8 persons 0
3	NUMBER OF ENGINEERING DISCIPLINES	Number of engineering disciplines in the project	0	Disciplines	1 Discipline 0	One or more engineering disciplines 0	Many disciplines 0
4	ENGINEERING FEES	Engineering Fees (CAD$)	0	$	Between 5 000$ and 70 000$ 0	Between 50 000$ and 350 000$ 0	More than 350 000$ 0
5	RISK LEVEL	Risk Level (when preparing the Proposal)	0	Green (1) Yellow (2) Orange (3) Red (4)	Green and Yellow only 0	All Levels 0	All Levels 0
	TOTAL				0	0	0
	Available Processes				0	0	0
	Selected PM Process	SSP					
	Signature of Person approving						
	Name of person approving	Project Sponsor				Service Director	

COST AND BENEFITS ANALYSIS OF THE USE OF ISO/IEC 29110 IN THE PROJECT MANAGEMENT PROCESSES

As discussed above, since a standard is a document describing knowledge and recognized practices about a specific domain, it should be possible to systematically assess the cost of its implementation and, more importantly, the economic benefits of its implementation. ISO has sponsored the development of a methodology to quantitatively assess the contribution of standards, consensus-based standards as well as company standards, to corporate value creation. The ISO methodology does not consider the negative impacts of the implementation of standards in an organization. The ISO methodology's main objectives are to provide (Gerundino & Hilb, 2010):

Figure 14. Day-to-day application of the PM Processes

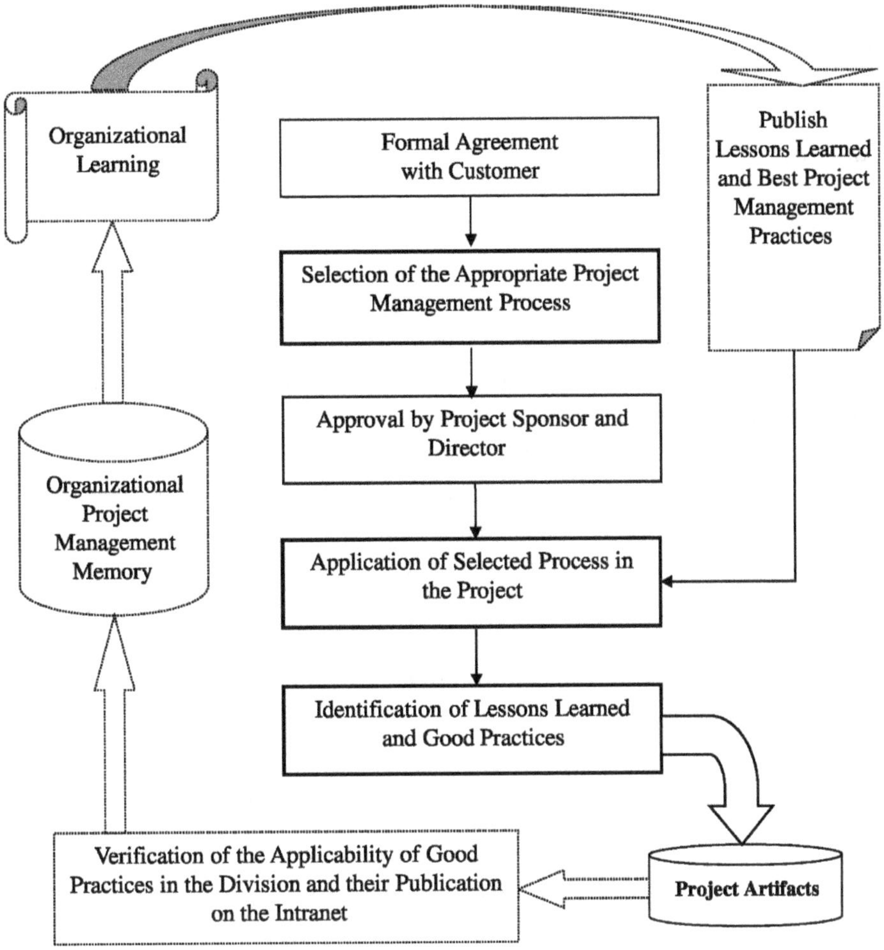

- A set of methods to measure the impact of standards on organizational value creation (with an emphasis on businesses)
- Decision-makers with clear and manageable criteria to assess the value associated with using standards
- Guidance on developing studies to assess the benefits of standards within a particular industry sector.

The methodology addresses the following key questions (Gerundino & Hilb, 2010):

- What is the contribution of standards to corporate value creation?
- How do industry and company specifics impact corporate value creation arising from standards?
- How can companies maximize the value generated by standards?

The approach used by the engineering division, to estimate the cost and benefits, comprised four steps:

- Understanding the company's value chain
- Analyzing the value drivers
- Determining the impacts of standards
- Assessing and consolidating results

The four steps of the ISO methodology are described below.

Step 1: Understanding the Company's Value Chain

The "value chain" is a concept described by Michael Porter of the Harvard Business School (Porter, 2008). Porter describes the value chain as a tool to understand the competitive advantage that a company can have in the actions it undertakes. The value chain is a representation of the different steps for an organization to create value in the form of goods or services to customers.

The performance of an activity can have an impact on cost and create a differentiation from competitors. Hence the advantage of using this tool to determine the impact of the project management improvement project to improve project management practices of the engineering division. Figure 15 illustrates the value chain of the company according to Porter's model. The vertical activities produce the main outputs of an organization while support activities displayed horizontally, such as procurement, influence and assist the primary activities.

In this model, the competence domains of an engineering division of the company are:

Figure 15. Value chain of the engineering division (adapted from ISO, 2010)

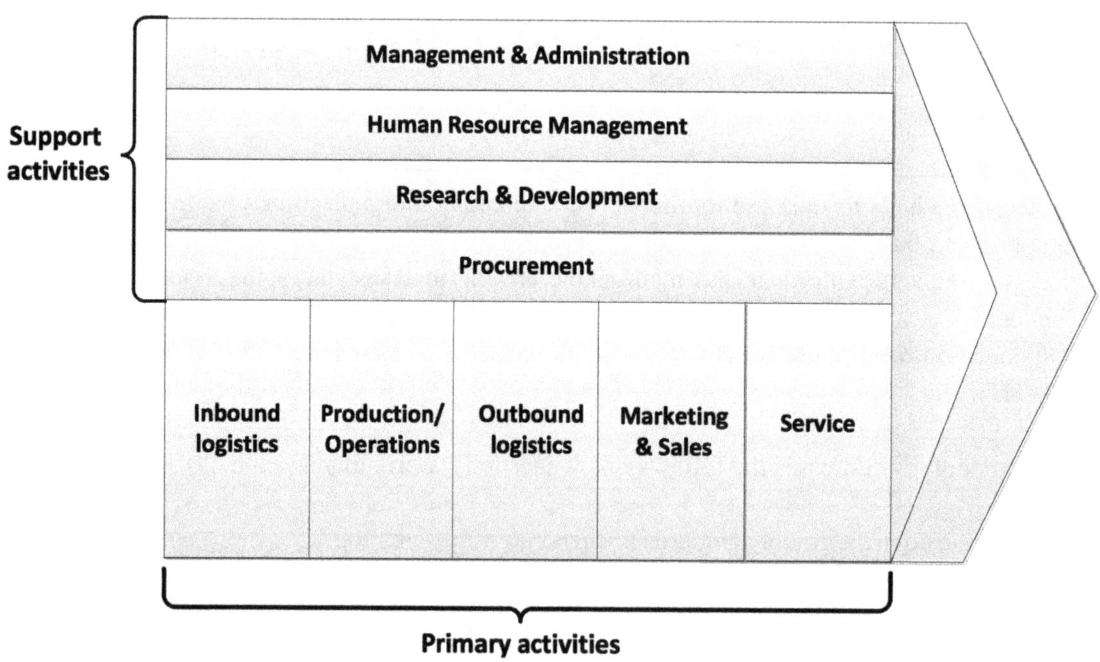

- Operations;
  - detailed engineering including the design of plans and specifications;
  - engineering linked to the achievement of specialized studies.
- Marketing and sales;
  - activities related to business development;
  - the development of contractual agreements;
  - assessment of services.
- Service
  - activities related to procurement of construction and installation phases of a project;
  - monitoring of construction and implementation activities;
  - activities related to the management of strategic assets.

Step 2: Analyzing the Value Drivers

After discussing with the company's governance board, the elements shown in Table 16 were identified as the main value drivers for the engineering consulting firm. The importance (i.e. important (3), largely important (2), very important (1)) of each driver was also determined.

Step 3: Determining the Impacts of Standards

The objective of this step was to identify the significant impacts that the improvement project will have on the company. Impacts were selected from the "Standards Impact Map" of the ISO methodology. Table 17 illustrates these impacts for the production category, similar tables were also developed for marketing and service categories. The links between the impacts of standards and the performance indicators identified in the previous step are also presented.

Table 16. Table of value drivers

Value Driver	Description	Performance Indicators	Importance
Quality of the design Process	Quality in terms of execution time, costs and quality of deliverables	Time spent on corrective engineering work. Cost overruns related to quality control.	Very important (company viability)
Efficiency versus costs	Ability to complete the work at minimum cost	Meeting budgets allocated to each sub-project. Meeting overall project budget.	Very important (company viability)
Project management capacity	Capacity to manage projects according to plans	Cost Performance Index (CPI)	Very important (completing projects is the company's core activity)
Technical expertise	Ability to solve complex problems	Schedule Performance Index (SPI)	Important
Geographic positioning	Geographic proximity of customers	Resource usage time (e.g. overtime)	Average importance
Partnership	Capacity to initiate partnerships with other companies	Number of partnerships Recurring customers	Average importance
Flexibility	Capacity to adapt to different customer needs	Number of services provided Type of service compared with competitors	Important

Table 17. Impacts of standards

Category	Activity	Impact ID	Impact Title	Description	Priority [1-high, 3-low]	Performance Indicator
Operation (P)	All activities	P-1	Improvement of internal information transfer	The use of standardized documents and specifications allows a more efficient transfer of information internally	2	- Meeting budget allocated to each sub-project. - Meeting overall project budget - Cost performance index (CPI)
		P-2	Better training of staff	Staff can be better trained due to the standardization of Processes	3	-Meeting budget allocated to each sub-project. - Meeting overall project budget -Cost performance index (CPI).
		P-3	Additional cost of staff	Increased costs due to the implementation of Processes	1	-Cost of the Process improvement project
	Execution	P-4	Better quality of deliverables	Better quality of deliverables based on standardization of Processes which reduce errors and rework costs	1	-Time spent on reworking engineering products -Cost overruns required for quality control.
	Quality Assurance	P-5	Better management of quality	Management of quality can be implemented more efficiently due to the standard.	1	-Time spent on reworking engineering products. -Cost overruns required for quality control.
	Knowledge	P-6	More effective internal standardization	It is more efficient to implement standards which are defined on the basis of recognized standards compared to proceed with the development of own internal standards.	1	-Cost of the Process improvement project

Step 4: Assessing and Consolidating Results

During this final step, the identified impacts were analyzed and impacts were determined separately by two persons of the enterprise: the engineer responsible for the improvement project and his supervisor. Tables 18 to 20 illustrate these results of the estimation for the production category. Table 18 illustrates the financial impacts due to the use of standardized documents and specifications on the internal transfer of information.

Table 19 illustrates the Staff training impact on projects and cost of training.

Table 20 illustrates the financial impact due to the cost of the development of project management Processes and of the cost of training the project managers.

The expenses for the improvement of the project management practices were estimated at 59,600$ and the expected net gain before tax between were estimated, for the first year, to be between 250,900$ and 260,100$. Table 21 presents the expected savings over a period of one year estimated by the 2 evaluators.

Table 22 presents the cost estimates for the development and implementation of project management Process for the first year.

Implementation and maintenance costs were also estimated for year 2 and 3. Table 23 presents the costs over the first 3 years.

Table 18. Assessment of the financial impacts about the improvement of internal information transfer

Impact ID	Description of Impacts		Evaluator # 1	Evaluator # 2
Operation -1	**Improvement of Internal Information Transfer**			
	What is the overall cost overrun of projects?	Estimation based on the Profitability Report	555,500 $	555,500 $
	What percentage of the project was in trouble due to a problem of information transfer?		25%	20%
	How much does the project management Process can improve the transfer of information?		80%	75%
		Financial impact	**111,100 $**	**83,325 $**

Table 19. Staff training impact on projects and cost of training

Operation -2	Better Training of Staff		Evaluator # 1	Evaluator # 2
	What is the overall cost overrun of projects?		555,500 $	555,500 $
	What percentage of the projects had difficulties due to staff training?		20%	15%
	By how much (in percent) could the Process improvement project can improve staff training?		50%	50%
		Financial impact	**55,550 $**	**41,663 $**

Table 20. Financial impact due to the development of Processes and staff training

Operation -3	Additional Cost of Staff		Evaluator # 1	Evaluator # 2
	Costs of developing Processes and related documents	see Table 23 for details		
	Costs of staff training	see Table 23 for details		
		Financial impact	**- 59,600$**	**- 59,600$**

Table 21. Summary of the economic benefits for the first year of utilization

Earnings Before Interest and Taxes Evaluation	Evaluator # 1	Evaluator # 2
Operation	201 485 $	162 600 $
Marketing	49 438 $	97 500 $
Service	0 $	0 $
Total	**250 923 $**	**260 100 $**

Table 22. First year estimation of implementation costs

		Year 1	
Category	**Description**	**Schedule Estimation (hour)**	**Cost Estimation ($ CAD)**
Communication			
	Company Newsletter (2)	40	4000
	Official communication to use the Processes (3)	40	4000
	Official communications of updated Processes (2/year)	0	0
Pilot Projects			
	Secretarial support	40	3000
Employee training			
	Trainer	80	9600
	Participants (30 persons x 8h) – Initiation to Processes	240	28800
	Participants (30 persons x 4h) - Update	0	0
	Secretarial support	40	3000
Reserve			
	Reserve of 15% (100$/hour)	72	7200
	Total	**552 hours**	**59,600 $**

Table 23. Costs over the first 3 years of implementation

		Year 1	Year 2	Year 3
Cost Category	**Description**	**Cost Estimation**	**Cost Estimation**	**Cost Estimation**
Communication				
	Company Newsletter (2)	4,000	4,000	4,000
	Official communication to use the Processes (3)	4,000	0	0
	Official communications of updated Processes (2/year)	0	2,000	2,000
Pilot Projects				
	Secretarial support	3,000	1,500	1,500
Employee Training				
	Trainer	9,600	4,800	4,800
	Participants (30 persons x 8h) – Initiation to Processes	28800	14,400	14,400
	Participants (30 persons x 4h) - Update	0		
	Secretarial support	3,000	3,000	3,000
Reserve				
	Reserve of 15% (100$/hour)	7,200	6,000	6,000
	Total	**59,600 $**	**50,100$**	**50,100$**

Assessing and Consolidating Results

The most significant impacts of the improvement program on the company were then selected based on the Standards Impact Map of the ISO Methodology. Table 24 provides an example of such impacts. The link between the impacts of standards used and the performance indicators identified during the previous step are also shown in this table.

An estimate of anticipated costs and benefits over a period of three years was done by the improvement program project sponsors. Table 25 shows the results for the first three years.

The 400-person Canadian division has evaluated five potential project management frameworks and has selected the most appropriate one for its business context as well as a function of existing assets. The effort invested in the selection, documentation and deployment of the small and medium-scale project management Processes and support tools was costly but, without an appropriate selection of the framework, the development of a set of tools and their wide deployment on the intranet, the division would not have been able to achieve the business goals listed in Table 4 as well as the benefits listed in Table 25.

NEXT STEPS

Since the utilization of ISO/IEC 29110 was very successful in the development project management Processes, the recently published systems engineering ISO/IEC 29110, described below, will be used by the engineering division to redefine and improve its existing engineering Process. This Process will address the activities required from engineering requirements identification to final product delivery.

The systems engineering Basic profile, as illustrated in Figure 16, like the software engineering Basic profile, is composed of two Processes: a Project Management (PM) Process and a System definition and

Table 24. Example of impacts of ISO/IEC 29110

Functions	Activities	Impact number	Impact	Description	Prioritization [1-high, 3-low]	Performance indicator
Production	All activities	P-1	Better internal information transfer	The use of standardized documents and specifications enables a more efficient internal information transfer.	2	Meeting budgets allocated to each sub-project. Meeting overall project budget. Cost performance indicators (CPI)

Table 25. Anticipated costs and benefits from the improvement program ($ CAD)

	Year 1	Year 2	Year 3	Total
Implementation and maintenance costs	59,600$	50,100$	50,100$	159,800$
Net benefits	255,500$	265,000$	265,000$	785,500$

Figure 16. Processes and activities of the systems engineering Basic profile (Laporte et al., 2014 c)

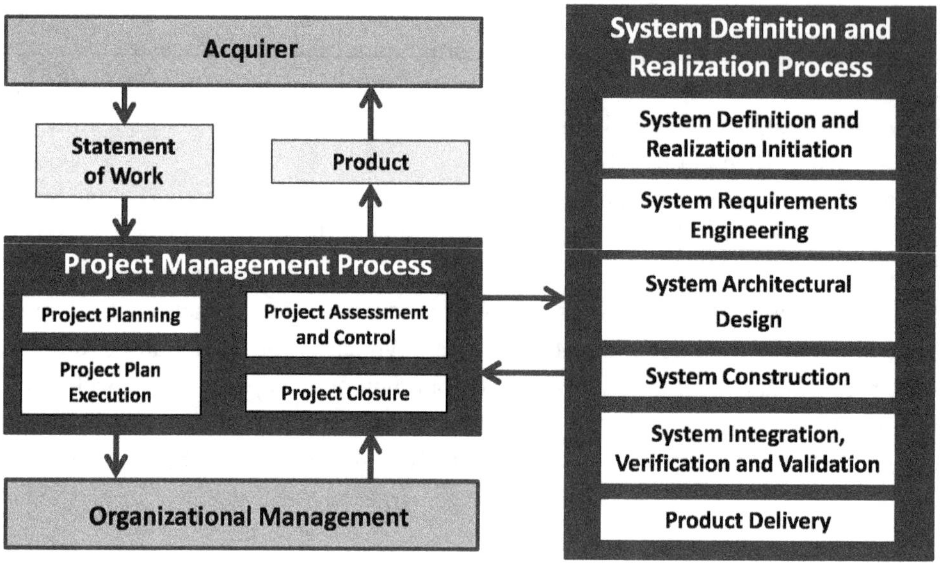

Realization (SR) Process. As defined in ISO/IEC 29110, the purpose of the Project Management (PM) Process is to establish and carry out in a systematic way the tasks of the system development, which allows complying with the project's objectives in the expected quality, time and cost.

The purpose of the System Definition and Realization (SR) Process is the systematic performance of the analysis, design, construction, integration, verification, and validation activities for new or modified system according to the specified requirements.

So far, one Canadian organization has implemented the Basic profile of the systems engineering ISO/IEC 29110: a 10-person start-up VSE in public transportation (Tremblay, 2012). The company, which began operations in 2011, specializes in the integration of interactive systems, communication and safety in the field of public transportation for trains, subways and buses as well as stations and stops. ISO/IEC 29110 standard for systems engineering has been used as the main reference for the development of its Processes.

ISO/IEC 29110 Certification of VSEs

VSEs requested in the WG24 survey described above, a mean to be recognized internationally. For VSEs, international certifications can enhance credibility, competitiveness and access to national and international markets. For VSEs, such a certification should not be too expensive and short. The ISO/IEC 29110 certification scheme, as illustrated in Figure 17, it is a four-stage certification Process.

Bit Perfect, a Peruvian VSE of 4 people created in 2012, was successfully audited in 2014 by a Brazilian auditor. The VSE provides software development services and automation of business Processes with information systems solutions. During its first two years of existence, the VSE has been involved in over 80 projects of which most have been lasted less than two months. The VSE used agile practices to

Figure 17. ISO/IEC 29110 Four-Stage Certification Process (Laporte 2014b)

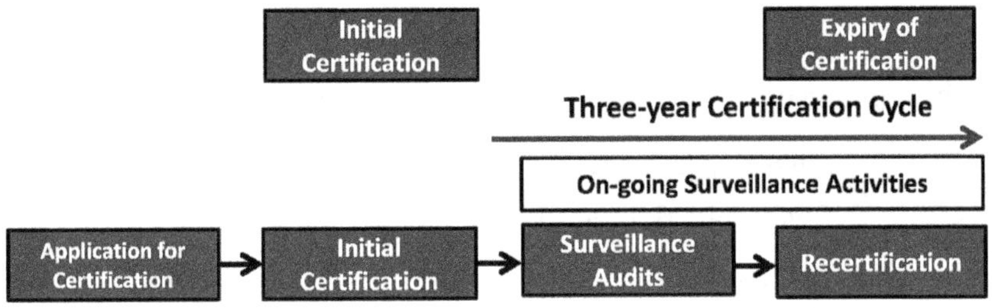

implement software solutions. The VSE of 4 people implemented the Project Management Process and the Software Implementation Process of the Basic profile of ISO/IEC 29110 (Garcia et al., 2015). The VSE spent about 22 hours for Phase 1 of the certification Process. The VSE spent 63 hours for Phase 2 of the certification Process. The total cost of the audit was about $1,500. For a VSE, the effort and cost are considerably lower than the cost of conducting an assessment using a framework such as the CMMI®.

CONCLUSION

The new ISO/IEC 29110 standard enabled an engineering consulting firm to develop project management Processes for their small-scale and medium-scale projects that offered a structured approach to its project managers. The actions required by such Processes are restricted to the most essential ones, in order to limit the management effort per project.

The tools developed to support the project management Processes proved very useful and helped the project managers rapidly integrate the knowledge required to execute the Processes.

For the first time, the company has documented management Processes for small-scale projects. Besides, some project managers have joined forces to promote project management practices within this engineering firm's division.

The improvement program was so successful that managers of the company's other divisions have shown an interest in learning this approach in order to implement it within their respective divisions.

Since the utilization of ISO/IEC 29110 was very successful in the development project management Processes, the systems engineering ISO/IEC 29110, published in 2014, will be used to redefine and improve the existing engineering Process of the Canadian division of the American engineering company. This Process will address the activities required from engineering requirements identification to final product delivery.

The management of the 400-person division has used an engineering approach to this project management Process development. It has treated the project as any other industrial project performed by their staff in assessing the context, selecting the framework and developing and deploying the Process and its associated support tools. As indicated by Garcia « When an organization selects a standard that fits its context well, and plans the adoption thoughtfully, it's most likely to achieve the standard's advertised benefits » (Garcia, 2005).

REFERENCES

AXELOS. (2009). *Managing Successful Projects with PRINCE2, 2009 Edition*. Available from: https://www.axelos.com/best-practice-solutions/prince2/what-is-prince2

Chevalier, F. (2012). *Programme d'amélioration des Processus de gestion de projets adaptés à l'envergure des projets*. École de technologie supérieure. (in French)

Garcia, L., Laporte, C. Y., Arteaga, J., & Bruggmann, M. (2015). Implementation and Certification of ISO/IEC 29110 in an IT Startup in Peru. Software Quality Professional Journal, 17(2).

Garcia, S. (2005). How standards enable adoption of project management practice. *IEEE Software, 22*(5), 22–29. doi:10.1109/MS.2005.122

Garcia, S., & Turner, R. (2006). *CMMI Survival Guide: Just Enough Process Improvement*. Addison Wesley Professional.

Gerundino, D., & Hilb, M. (2010, June). *The ISO Methodology - Assessing the economic benefits of standards*. ISO Focus, International Organization for Standardization.

ISO. (2005, May). *ISO/IEC JTC1/SC7 N3288, New Work Item Proposal – Software Life Cycles for Very Small Enterprises*. Retrieved from http://www.jtc1-sc7.org/

ISO. (2006). *ISO/IEC/IEEE 16085:2006 - Systems and software engineering — Lifecycle Processes — Risk management*. Geneva, Switzerland: International Organization for Standardization/International Electrotechnical Commission.

ISO. (2008a). *ISO/IEC/IEEE 12207:2008, Systems and software engineering– Software life cycle Processes*. Geneva, Switzerland: International Organization for Standardization/International Electrotechnical Commission.

ISO. (2008b). *ISO/IEC/IEEE 15288:2008, Systems and software engineering– System life cycle Processes*. Geneva, Switzerland: International Organization for Standardization/International Electrotechnical Commission.

ISO. (2009). *ISO/IEC/IEEE 16326:2009 - Systems and software engineering — Lifecycle Processes — Project management*. Geneva, Switzerland: International Organization for Standardization/International Electrotechnical Commission.

ISO. (2010). *Economic Benefits of Standards: Implementation Guide*. International Organization for Standardization. Retrieved from http://www.standardsinfo.net/info/benefits/benefits_s1.html

ISO. (2011a). *ISO/IEC/IEEE 15289:2011, Systems and software engineering - Content of systems and software life cycle Process information products (Documentation)*. Geneva, Switzerland: International Organization for Standardization/International Electrotechnical Commission.

ISO. (2011b). *ISO/IEC TR 29110-5-1-2:2011, Software Engineering - Lifecycle Profiles for Very Small Entities (VSEs) - Part 5-1-2: Management and engineering guide - Generic profile group: Basic profile.* International Organization for Standardization/International Electrotechnical Commission. Available at no cost from ISO at: http://standards.iso.org/ittf/PubliclyAvailableStandards/c051153_ISO_IEC_TR_29110-5-1_2011.zip

ISO. (2012). *ISO/IEC TR 29110-5-1-1:2012, Software Engineering - Lifecycle Profiles for Very Small Entities (VSEs) - Part 5-1-1: Management and engineering guide - Generic profile group: Entry profile.* International Organization for Standardization/International Electrotechnical Commission. Available at no cost from ISO at: http://standards.iso.org/ittf/PubliclyAvailableStandards/

ISO. (2014). *ISO/IEC TR 29110-5-6-2:2014 - Systems Engineering – Lifecycle Profiles for Very Small Entities (VSEs) - Management and engineering guide: Generic profile group: Basic profile.* International Organization for Standardization/International Electrotechnical Commission. Available at no cost from ISO at: http://standards.iso.org/ittf/PubliclyAvailableStandards/index.html

Krasner, H. (1998, November). Using the cost of quality approach for software. *Crosstalk, 11*(November), 6–11.

Kroll, P., & Kruchten, P. (2003). *The Rational unified Process made easy: A practitioner's guide to the RUP*. Boston: Addison-Wesley Longman Publishing Co.

Land, S. K. (1997). *Results of the IEEE Survey of Software Engineering Standards Users*. Software Engineering Standards Symposium and Forum, 1997. Emerging International Standards, Walnut Creek, CA. doi:10.1109/SESS.1997.595987

Laporte, C. Y. (2009). *Contributions to software engineering and the development and deployment of international software engineering standards for very small entities*. (Doctoral dissertation). Université de Bretagne Occidentale, Brest.

Laporte, C. Y., Alexandre, S., & O'Connor, R. (2008b). A Software Engineering Lifecycle Standard for Very Small Enterprises. In R.V. O'Connor et al. (Eds.), EuroSPI 2008. Dublin, Ireland: CCIS 16.

Laporte, C. Y., Alexandre, S., & Renault, A. (2008c, June). The Application of International Software Engineering Standards in Very Small Enterprises. *Software Quality Professional Journal, ASQ, 10*(3), 4–11.

Laporte, C. Y., Chevalier, F., & Maurice, J.-C. (2013c, February). *Improving Project Management for Small Projects*. ISO Focus+, International Organization for Standardization. Retrieved from http://www.iso.org/iso/home/news_index/iso_magazines/isofocusplus_index/isofocusplus_2013/isofocusplus_2013-02.htm

Laporte, C. Y., Hébert, C., & Mineau, C. (2014b). Development of a Social Network Website Using the New ISO/IEC 29110 Standard Developed Specifically for Very Small Entities. *Software Quality Professional Journal, 16*(4), 4-25. Available at not cost at: http://asq.org/pub/sqp/past/vol16_issue4/index.html

Laporte, C. Y., Houde, R., & Marvin, J. (2014c). *Systems Engineering International Standards and Support Tools for Very Small Enterprises*. Paper to be presented at the 24th Annual International Symposium of INCOSE (International Council on Systems Engineering), Las Vegas, NV. doi:10.1002/j.2334-5837.2014.tb03167.x

Laporte, C. Y., & O'Connor, R. (2014a). Systems and Software Engineering Standards for Very Small Entities - Implementation and Initial Results. QUATIC'2014. *9th International Conference on the Quality of Information and Communications Technology*, Guimarães, Portugal.

Laporte, C. Y., O'Connor, R., & Fanmuy, G. (2013b, May/June). International Systems and Software Engineering Standards for Very Small Entities. *Crosstalk*, *26*(3), 28–33.

Laporte, C. Y., & O'Connor, R. V. (2014b). A Systems Process Lifecycle Standard for Very Small Entities: Development and Pilot Trials. In B. Barafort, R. V. O'Connor, & R. Messnarz (Eds.), *Systems, Software and Services Process Improvement, CCIS* (Vol. 425). Springer-Verlag. doi:10.1007/978-3-662-43896-1_2

Laporte, C. Y., & Palza Vargas, E. (2012). The Development of International Standards to facilitate Process Improvements for Very Small Enterprises. In *Software Process Improvement and Management: Approaches and Tools for Practical Development* (pp. 34–61). Hershey, PA: IGI Global. doi:10.4018/978-1-61350-141-2.ch003

Laporte, C. Y., Renault, A., & Alexandre, S. (2008a). Applying ISO/IEC Software Engineering Standards in Very Small Enterprises. In *Software Process Improvement for Small and Medium Enterprises: Techniques and Case Studies* (pp. 42–70). Hershey, PA: Idea Group Inc. doi:10.4018/978-1-59904-906-9.ch002

Laporte, C. Y., Séguin, N., & Villas Boas, G. (2013a, February). *Seizing the benefits of software and systems engineering standards*. ISO Focus, International Organization for Standardization. Retrieved from http://www.iso.org/iso/home/news_index/iso_magazines/isofocusplus_index/isofocusplus_2013/isofocusplus_2013-02.htm

Miotti, H. (2009, June). The economic impact of standardization technological change. In *Standards growth in France*. AFNOR.

Moll, R. (2013, February). *Being prepared – A bird's eye view of SMEs and risk management*. ISO Focus +. Geneva, Switzerland: International Organization for Standardization.

O'Connor, R. V., & Laporte, C. Y. (2010). Towards the Provision of Assistance for Very Small Entities in Deploying Software Lifecycle Standards. In *Product Focused Software Development and Process Improvement* (PROFES 2010). PROFES.

PMI. (2013a). *A Guide to the Project Management Body of Knowledge (PMBOK® Guide)* (5th ed.). Newtown Square, PA: Project Management Institute.

PMI. (2013b). *Organizational Project Management Maturity Model (OPM3®)* (3rd ed.). Newtown Square, PA: Project Management Institute.

Porter, M. E. (2008). *On Competition, Updated and expanded*. Boston: Harvard Business School Publication.

Potter, N., & Sakry, M. (2002). *Making Process Improvement Work*. Addison-Wesley – Pearson Education.

Radice, R. A., Roth, N. K., O'Hara, A. C., & Ciarfella, W. A. (1985). A Programming Process Architecture. *IBM Systems Journal*, *24*(2), 79–90. doi:10.1147/sj.242.0079

Rogers, E. M. (2003). *Diffusion of Innovations* (5th ed.). New York: Free Press.

SEI. (2010). *Software Engineering Institute. (2010). CMMI for Development, Version 1.3*. Pittsburgh, PA: Carnegie Mellon University. CMU/SEI.

Tremblay, N. (2012). *Implantation de la norme d'ingénierie système ISO/IEC 29110 chez CSinTrans (Rapport de maitrise)*. École de Technologie Supérieure. (in French)

KEY TERMS AND DEFINITIONS

Basic Profile: A profile that describes development practices of a single application by a single project team of a very small entity (ISO/IEC TR 29110-1:2011 Software engineering-Lifecycle profiles for Very Small Entities (VSEs)-Part 1: Overview).

Certification: A third-party attestation related to products, Processes, systems or persons (ISO/IEC 17000).

Entry Profile: A profile targeted at very small entities working on small projects (e.g. at most six person-months effort) and for start-ups (ISO/IEC TR 29110-1:2011 Software engineering-Lifecycle profiles for Very Small Entities (VSEs)-Part 1: Overview).

Micro-Enterprise: An enterprise having up to 9 employees.

Generic Profile Group: A profile group applicable to VSEs (very small entities) that do not develop critical systems or software products and have typical situational factors (ISO/IEC TR 29110-1:2011 Software engineering-Lifecycle profiles for Very Small Entities (VSEs)-Part 1: Overview).

Process: System of activities, which use resources to transform inputs into outputs (ISO/IEC 25000:2014 Systems and software Engineering--Systems and software product Quality Requirements and Evaluation (SQuaRE) - Guide to SQuaRE, 4.41).

Process Notation: Means of concrete representation for a particular type of a model, expressed as a grammar and suitable glyphs for its terminal symbols (ISO/IEC 10746-2:2009 Information technology -- Open Distributed Processing - Reference Model: Foundations, 7.5).

Profile: A set of one or more base standards and/or profiles, and where applicable, the identification of chosen classes, conforming subsets, option and parameters of those base standard, or standardized profiles necessary to accomplish a particular function. (ISO/IEC TR 29110-1:2011 Software engineering-Lifecycle profiles for Very Small Entities (VSEs)-Part 1: Overview).

Project Management: The application of knowledge, skills, tools, and techniques to project activities to meet the project requirements (A Guide to the Project Management Body of Knowledge (PMBOK® Guide) - Fifth Edition, 2013).

Software Engineering: The application of a systematic, disciplined, quantifiable approach to the development, operation, and maintenance of software; that is, the application of engineering to software (ISO/IEC/IEEE 24765:2010 Systems and software engineering -Vocabulary). On-line vocabulary: http://pascal.computer.org/sev_display/index.action.

Standard: Document, established by consensus and approved by a recognized body, that provides, for common and repeated use, rules, guidelines or characteristics for activities or their results, aimed at the achievement of the optimum degree of order in a given context (ISO/IEC TR 29110-1:2011 Software engineering-Lifecycle profiles for Very Small Entities (VSEs)-Part 1: Overview).

Systems Engineering: An interdisciplinary approach governing the total technical and managerial effort required to transform a set of customer needs, expectations, and constraints into a solution and to support that solution throughout its life (ISO/IEC 15940:2013 Systems and software engineering-Software Engineering Environment Services).

Very Small Entity: An enterprise, an organization, a department or a project having up to 25 people (ISO/IEC TR 29110-1:2011 Software engineering-Lifecycle profiles for Very Small Entities (VSEs)-Part 1: Overview).

ENDNOTES

1. International Organization for Standardization/ International Electrotechnical Commission Joint Technical Committee 1/ Sub Committee 7.
2. The table describes information extracted and adapted from web sites of organizations which have developed the frameworks (e.g. Project Management Institute, Software Engineering Institute and International Organization for Standardization).

Section 4
Dealing with Resistance

Chapter 8
Standardizing Knowledge Work:
How "Homegrown" Standards Overcome Resistance among Professional Employees

Kurt W. Sandholtz
Brigham Young University, USA

ABSTRACT

This chapter analyzes how standards were implemented among design engineers in two divisions of the same corporation. Although both divisions achieved ISO 9000 certification, engineers in one of the divisions resisted the standard work practices and a culture of cynicism developed. In the other division, the standardized work practices were internalized by engineers and are still voluntarily, and often enthusiastically, followed. Analysis of the contrasting cases results in three recommendations for managers who wish to increase the acceptance of external work standards among professional workers.

INTRODUCTION

The management of standardization has its roots in the late 1800s, when the first management theorists – predominantly engineers – sought to codify methods for controlling far-flung railroad operations or enforcing manufacturing efficiencies (Chandler, 1977; Taylor, 1993 [1903]). Now, in the second decade of the 21st century, standardization is increasingly applied in non-traditional settings such as product design, software engineering, and other types of knowledge work (Sandholtz, 2012; Timmermans & Epstein, 2010). This migration of standard work from the assembly line to professionals' cubicles raises a number of new questions: First, is it possible to standardize an abstract activity such as knowledge work? If so, how do the targets of the effort – well-educated and highly autonomous knowledge workers – feel about having their specialized, creative work standardized? And given their feelings, how can managers best proceed when attempting to introduce work standards in professional contexts?

This chapter will address these questions by examining a unique pair of case studies. Between 1997 and 2002, two engineering divisions of the same large manufacturer achieved ISO 9000 certification. In one of the divisions, the ISO 9000 initiative met stiff resistance from the engineers, who interpreted the standard as a threat to their professional autonomy and complied only symbolically. In the other

DOI: 10.4018/978-1-4666-9737-9.ch008

division, engineers created their own version of standardized engineering that exceeded ISO requirements, and the division's employees now follow the standard work not only willingly, but in many cases, enthusiastically. Analysis of the two cases yields a set of practical guidelines for managers tasked with implementing work standards among professional employees. As will be shown below, this chapter suggests that when knowledge workers can internalize standard practices *before* being "ordered" to do so, the substantive adoption of standards is a *fait accompli*.

The chapter is organized as follows. First, I touch briefly on two concepts from organization theory that have bearing on the management of standardization. Next, I describe the empirical setting and the fieldwork that documented the case studies. Third, I reconstruct the implementation of ISO 9000 in the two engineering groups and chronicle how the standards were rejected in one of the groups but accepted in the other. Finally, I present practical tips for managers who face the challenge of implementing professional work standards.

BACKGROUND

Standards as Unique Forms of Social Control

A fundamental characteristic of organizations is that they seek to control and coordinate the behavior of their members. Social scientists theorize three different ways that organizations can exert this control. (See Brunsson & Jacobsson, 2000a for a complete discussion.) First, the organization can issue directives, then impose rewards for compliance and sanctions for non-compliance. A directive is typically an order or command, sometimes presented as a request or an action plan, made by a manager who possesses formal authority. Second, organizations can foster (or import) informal norms that guide behavior. In this context, a norm is a social rule that rarely appears in written form, yet is accepted as natural in the organization. As Brunsson and Jacobsson (2000a: 12) put it, 'Norms are internalized rules that we can follow without having to reflect on them. . . . We appear to be following them because we want to, and not because we are exposed to any outside pressure to do so.' Third, organizations can voluntarily adopt and implement standards. Standards are unique in that they exhibit properties of both formal directives and informal norms (see Table 1). Like directives, standards are characterized by formal authorship; an individual or group somewhere articulated and recorded the standard. Like norms, however, standards are voluntarily adopted. Both of these aspects – authorship and volition – hold important lessons for managers wishing to implement standards in professional settings.

Table 1. Comparing Different Forms of Social Regulation

	Norms	Directives	Standards
Voluntarily Adopted?	Yes	No	Yes
Explicitly Stated?	No	Yes	Yes
Identifiable Author?	No	Yes	Yes

Source: Adapted from Brunsson and Jacobsson 2000a

Who Sets the Standard?

Typically, scientific and technical experts working within "standards developing organizations" (SDOs) preside over the creation of standards. Indeed, the largest global SDO is the International Organization for Standardization (ISO), made up of some 260 Technical Committees, each with its subcommittees and working groups (Tamm Hallström, 2004). In other words, a group of technocrats are often the ones charged with identifying, codifying, and disseminating – but not implementing – standards. Implementation takes place within organizations that are physically and culturally distant from the SDO, among employees who are themselves highly trained technical experts.

To these "local" experts, it matters greatly *whose expertise is distilled in the standard.* Professionals seek control over their own knowledge base and the types of problems to which it is applied. Abbott (1988: 3) asserts that 'jurisdiction is the defining relation in professional life,' and Scott and Meyer (1991: 130) note that 'as individual practitioners, professionals jealously guard their discretion over instrumental decisions.' Thus, numerous studies have shown that standards are likely to be resisted by individuals who consider themselves 'highly autonomous, innovative, and different' (Brunsson & Jacobsson, 2000b:134).

For Whom is the Standard Voluntary?

Timmermans and Epstein (2010: 81) observe that 'the trick in standardization appears to be to find a balance between flexibility and rigidity and to trust users with the right amount of agency to keep a standard sufficiently uniform for the task at hand.' This statement runs counter to one of the defining characteristics of standards: their voluntariness. To argue that users are entrusted with varying amounts of agency suggests that following the standard is not entirely voluntary. Likewise, scholars of transnational governance often refer to international standards as 'soft law,' where 'soft' emphasizes the voluntary aspect of standards, while 'law' implies that non-compliance carries consequences (Djelic & Sahlin-Andersson, 2006).

This paradoxical language points to a second unique aspect of standards: their seeming inseparability from traditional hierarchical control. According to Brunsson and Jacobsson (2000a:13), 'a standard in society at large becomes a directive for the members of an organization whose management has decided that the organization is to follow the standard.' In other words, senior managers exercise agency in adopting a standard; the rank and file exercise obedience. This is not to imply that managers have the ability to enforce perfect compliance with the standard. On the contrary, professionals often resist being told what to do (Abernethy & Stoelwinder, 1995; Brunsson & Sahlin-Andersson, 2000). When "ordered" to follow a standard that they do not believe in, professionals can preserve their autonomy by making a symbolic show of compliance while continuing to do the work in their own preferred (non-standard) way. Social scientists refer to this process as 'decoupling,' and we will examine this phenomenon next.

Decoupling: A Way for Professionals to Resist Control

A core tenet of organizational behavior is that organizations strive to buffer their internal work processes from external interference (Thompson, 1967). No manager wants outsiders telling him or her how to run his business. Yet organizations also crave the regulatory approval and social legitimacy that come from acceding to external demands (Meyer & Rowan, 1977). To resolve this dilemma, managers of organizations often resort to a practice known as decoupling: They make a symbolic show of complying with the requests of external stakeholders, while striving to change the organization's production activities as little as possible (Boxenbaum & Jonsson, 2008). That is, the supposed adoption of certain procedures is deliberately 'decoupled' from their implementation. For example, a manufacturing firm may create a Quality Control department and staff it with Six Sigma certified employees, yet fail to change its assembly line in conformance with Six Sigma principles. Such actions may seem hypocritical, but numerous scholars have argued that organizations faced with complex and contradictory demands from various external stakeholders have little choice but to engage in strategic decoupling (Oliver, 1991; Ruef & Scott, 1998).

Studies of the implementation of ISO 9000 and other quality management systems show rampant decoupling among those expected to implement the standard (Zbaracki, 1998; Walgenbach, 2001; Vallas, 2003; Storz, 2007). Too often, however, decoupling is described in binary terms; it is either present or absent (Westphal & Zajac, 2001; Fiss & Zajac, 2006; Stevens et al., 2005). This oversimplification ignores qualitative differences in processes of decoupling. The cases presented in this chapter illustrate two different types of decoupling: a malignant variant that resulted from professional opposition to an imposed standard, and a benign variant that reflected a consensual implementation of a standard.

Furthermore, studies of decoupling have paid little attention to 'the unintended effects of decoupling, such as whether it affects morale and fosters cynicism within the organization' (Boxenbaum & Jonsson, 2008: 91). An exception is the work of Boiral (2003), who studied 47 managers and workers at companies that had achieved ISO certification. Deep cynicism was expressed by 21% of the respondents; 43% were supportive and positive toward the ISO certification even though they openly acknowledged that it was decoupled from technical work; the remaining 36% – more than half of whom were internal quality consultants – saw no evidence of decoupling and were characterized as 'quality enthusiasts.'

In one of the two cases presented in this chapter, engineers decoupled their daily work activities from the requirements of ISO 9000 certification. In large part, this decoupling was a way for the engineers to preserve professional autonomy: by symbolically complying with ISO standards, the engineers retained the ability to conduct their work as they saw fit. In the other case, however, a robust system of self-monitoring evolved into a feedback loop ensuring that a homegrown system of standards remained closely linked to actual practice. We might call this arrangement 'tight coupling,' a phenomenon that deserves further explication. Decoupling has achieved such presumptive status in the social sciences that scholars have coined a new term – recoupling – to describe cases in which organizations attempt to overcome decoupling's destructive consequences (Espeland, 1998; Hallett, 2010). The study of tight coupling is thus a vital counterpoint to the assumptions prevalent in organization theory. More importantly, such close alignment between standards and work practices is presumably the goal of any manager who embarks on a standardization project. Thus, understanding how to achieve tight coupling is of practical interest as well.

In summary, the theoretical aim of this chapter is to expand our understanding of the dynamics that come into play when a standard is implemented by an organization. We will focus on the connections between standards, directives, and norms and examine how an organization implemented a common external standard in two of its divisions. The chapter highlights a previously opaque but nontrivial factor that influences the path of standardization: the professional norms and practices of those expected to standardize their work. This focus on the implications of standardization at the level of concrete work practices contributes to our understanding of decoupling, a key factor that managers must take into account in their attempts to implement standards among professional workers.

RESEARCH SETTING AND METHODS

The research reported in this chapter was conducted within Global Amalgamated Manufacturing, or GAM (a pseudonym). GAM is a century-old global manufacturing company headquartered in the eastern United States; it designs, manufactures, and sells a range of consumer durable goods, employing more than 200,000 people worldwide. At the time of the study reported in this chapter, GAM was in the process of migrating standards from its manufacturing plants to its engineering and design centers.

Guided by principles of theoretical sampling (Eisenhardt, 1989), I identified two engineering divisions that were in the midst of on-going efforts to standardize engineering work. Division A (Div-A) was tasked with designing the structural frame within which all subsystems and components were integrated. Division B (Div-B) was responsible for the design of a mechanically complex subsystem for GAM's key product line. The two groups did not interface directly with one another. Div-A had 936 employees and Div-B had 628. The workforce composition in both divisions was nearly identical: employees were predominantly engineers with similar educational backgrounds, career histories, and demographic makeup.

The nature of the day-to-day engineering work in both divisions was similar: mechanical engineers used computer-aided engineering software to mathematically define three-dimensional components. Both divisions were subject to the corporation's high-level product development process, which was diagrammed as a flowchart and included key development milestones or "gates" associated with specific deadlines. Both divisions' standardization efforts included the successful adoption of ISO 9000 standards.[1] The divisions' approaches to implementation, however, differed on a number of practical dimensions.

These differences became the primary object of study. My empirical approach follows the dictum that science and technology must be studied "in the making" (Latour, 1987). Accordingly, three types of empirical data were collected: "live" observation of the engineers' work practices, semi-structured interviews with the engineers and managers, and archival information from the company's on-line database. On-site observations and interviews were conducted between April 2008 and September 2008.

Interviews and observations were conducted among employees at all levels of the hierarchy in both divisions, representing the full range of design work conducted in the divisions. Interviews with 43 members of Div-B and 20 members of Div-A lasted for 60 minutes on average; all were recorded and transcribed, producing 850 double-spaced pages of text (see Table 2). These interviews constituted the primary data source, supplemented by observations of engineers at work. Supplementary archival materials totaled more than 800 pages of documents, including presentations (the favored medium of internal communication at GAM), training manuals, screen shots from intranet sites, and copies of emails.

Table 2. Summary of Division A and Division B Informants

	Div-A	Div-B	TOTAL
Executive	3	5	8
Manager	6	6	12
Engineer	9	25	34
Support Staff	2	6	8
External	0	1	1
Total	**20**	**43**	**63**

Using these data sources, chronological case studies of each division's experience with standardization were constructed. To minimize retrospective bias, interview data were compared with archival evidence, such as company training manuals, documents posted on the company intranet, or presentations provided by informants (Miles & Huberman, 1984). When analysis revealed gaps in the narrative, informants provided the missing information. Detailed reconstructions of events were reviewed by a cross-section of engineers, managers, and executives within each division, and corrected accordingly. In the end, these panels of informants attested to the accuracy of the case summaries.

With these summaries in place, analysis took place in three phases (Eisenhardt & Graebner, 2007). The first phase required a return to the raw data, searching for the key themes in each case summary. The second analytical phase involved more detailed examination of the key theme of decoupled vs. coupled implementation of the standard. The third phase of analysis consisted of comparison across the case summaries, iterating between the data and the literature on standardization and decoupling. Gradually, the case histories were rearranged not just chronologically, but by levels of analysis, noting how the ISO 9000 standard was transformed during implementation in the two divisions. This process produced the abbreviated accounts presented in the section that follows.

FINDINGS

Division A: How Implementation Led to Cynical Symbolic Compliance with the Standard

Division A's experience with ISO 9000 is consistent with findings from previous research: certification-related activities were mainly symbolic and kept separate from day-to-day technical work (Boiral, 2003; Walgenbach, 2007). However, this chapter sheds light on areas overlooked in prior studies: the processes of adoption, implementation, and decoupling, with particular attention to engineers' rationales for rejecting standardized practices.

The Pre-Adoption Context

The history of Div-A is relevant to its experience with the implementation of the standard. Div-A had its roots in an acquisition made early in GAM's history (mid-1920s). Div-A had been a supplier to GAM, and once acquired, Div-A became a center of expertise for all structural engineering. Because GAM had purchased Div-A to prevent competitors from accessing its expertise, the company was not allowed to provide its services to external customers. In addition, Div-A's role in the production process was highly integrative: its structural frames had to interface with at least six other design and engineering divisions, creating a complex system of interdependencies.

Following a wave of decentralization in the 1940s, all engineering activities were distributed to GAM's semi-autonomous product divisions. Over time, engineering divisions developed their own standard processes and work routines. In the 1980s, GAM began again to centralize its structural engineering work. A series of reorganizations culminated in 2004 with the consolidation of all structural engineering in the corporate research center. Div-A thus inherited work standards from a number of different internal sources, resulting in a 'tribal' knowledge base, as one informant put it.

Div-A's history, then, suggests two factors that affected its implementation of ISO 9000. First, the division had no direct contact with GAM's external customers. Second, the recent reintegration of disparate structural engineering divisions into a corporate center of expertise meant that the engineers' preferred work practices and analytical approaches were divisional inheritances that were far from standardized.

Adopting an External Quality Standard

Attempts to unify engineering processes around a common standard had begun in 1994, prior to corporate centralization. Senior managers in Div-A launched a series of workshops with the goal of 'synchronizing' the multitude of engineering processes within the division. There followed a sequence of short-lived quality initiatives, each involving slightly different rhetoric and exerting little impact on engineering processes (see Table 3). This cycle of starts, stops, and restarts had lasting effect on the engineers' attitudes. One Div-A engineer said that 'for a GAM person to be cynical about a new process tool is a normal state' (*Informant 53).* Another engineer, from a different part of Div-A, pointed out that:

We're famous for programs of the day. . . . You just kind of go through it, you take whatever training they want you to because they tell you to, and you do what you need to, but it does nothing for your job. Zero. Zilch. As a matter of fact, it just eats up more time. (Informant 49)

Among the Div-A informants, the most salient of these false starts was the executive-led initiative to achieve ISO 9000 certification, launched in 2000. The motivations to adopt the standard were threefold:

Table 3. The Succession of Standardization Initiatives in Division A

Initiative	Description
Synchronous workshops (1994)	Following the lead of the manufacturing organization, Div-A attempts to synchronize its product development processes to enable parallel engineering.
Engineering Factory (1995)	An attempt to re-energize the manufacturing-inspired synchronization effort. This is the first time engineering is formally referred to as a 'factory.'
'Lean Engineering Factory' Workshops (1996)	Further entrenchment of the 'engineering factory' concept. The initiative is workshop based; serious process redesigns are intended to follow.
'Lean Acceleration' (1997)	First effort to identify and implement common product development processes across all engineering groups. The groups remain structurally separated, however, and continue with separate identities and engineering approaches.
Product Development Process (PDP) (1998)	The PDP – a master project-schedule-coordinating mechanism – is designed and launched. This is the first time 'standard work' terminology is used in engineering.
Constraints Management Workshops (1999)	Workshops and consulting based on the ideas of Eliyahu Goldratt (*Theory of Constraints*) sweep through Div-A. The goal is to optimize the performance of the engineering factory.
'FastChange' Workshops (2000)	'FastChange,' a different method for holding continuous improvement meetings, is taught within the product development organization (including Div-A).
Work Instructions (2001)	For the first time, specific 'how to do your job' instructions are created by a central 'process expert' group and introduced to Div-A engineers. The principal objective is to prepare for ISO 9000 registration.
ISO 9000 Registration (2002)	External auditors grant Div-A the ISO seal of approval.
ISO Registration lapses (2004)	VP of Engineering announces that Div-A will no longer pursue external audits of its quality processes, but that engineers should continue to follow the guidelines.

First, a key competitor announced it had recently achieved certification for its entire U.S. engineering function. GAM executives thus felt pressure to 'keep up.' Second, GAM's manufacturing plants had been ISO certified within the past two years and the labor union was beginning to push for 'equal treatment' of headquarters' staff organizations. Third, Div-A's work was suffering chronic quality problems. In a highly complex, interdependent system, identifying and correcting root causes posed a significant technical challenge. An explanatory email from an internal quality expert who was directly involved in the ISO 9000 initiative is illuminating:

It was rumored that [Div-A] might be 'drunk and disorderly' around the notion of a quality system; executives would make off-the-cuff remarks to that point. There was a belief that [ISO] registering the product development activity of [GAM] would help to bring efficiencies to the organization and eliminate waste. The other product engineering centers around the globe all had ISO registrations. North America was the exception. Having North America achieve an ISO registration would be good for global relations. (Informant 1c)

These multifaceted motivations shared a common characteristic: the absence of a solid business rationale for adopting the ISO standard. Instead, the motivations were a mixture of 'copy-cat' tendencies (copying both an external competitor and other GAM engineering centers), corporate politics (appeasing the union), and the 'flavor of the month' tendency to try the next new thing (see Brunsson & Olsen, 1993). In sum, pursuit of ISO certification represented a readily available, externally legitimate and internally conciliatory gesture, the successful execution of which offered at least some hope of substantive impact. Perhaps more tellingly, the adoption of the ISO standard was a management decision, not a grass-roots movement among engineers.

Implementation: From External Standard to Decoupled Directive

Once GAM decided to achieve ISO certification, the standard was transformed into a directive. Managers took the lead in communicating the purpose of the ISO drive to their employees, but the certification effort itself was delegated to an internal staff department. Significantly, this staff group was located in a distant part of the building, on a separate floor from the product design engineers. The ISO specialists were tasked with gathering documentation and preparing detailed 'work instructions' – generalized descriptions of how each engineering job contributed to overall quality. An internal quality staffer who participated in the ISO certification effort noted:

Every individual had to have work instructions. Because if they didn't, we couldn't audit them – we couldn't be [ISO] registered. Some of it was driven by this fear of audit. We didn't want an auditor wandering up to somebody and saying, 'What are you doing?' and they go, 'I don't know.'. . . We kind of went overboard when we used this phrase 'work instructions,' and we wrote thousands of them, thousands, down to the minutia level. (Informant 1a)

Engineers were expected to memorize their work instructions so that they could recite them for ISO auditors, with punishment for poor performance. Consider the following recollection from an Div-A informant:

We went through a very expensive, non-value-added ISO process here, where we had to memorize what we did in our jobs, and when the ISO man came around, we had to regurgitate what we memorized and if we screwed it up, we were flogged by the highest portions of management: 'How could you fail the ISO? You failed ISO!' (Informant 51)

Despite being disconnected from its technical work, GAM's ISO preparations paid off with successful certification in 2002. Clearly, the ISO audit process was capable of being 'gamed' (Walgenbach, 2001; Boiral, 2003).

Abandoning the Standard

ISO registration requires recertification every three years. In 2004, prior to Div-A's first recertification audit, the executive over Div-A announced that the division would no longer seek ISO approval. Instead, Div-A engineers were urged to continue to apply the standards of their own accord. An excerpt from the executive's formal announcement is reproduced below:

. . . a recent decision has been made that will allow Headquarters' ISO 9000 registration to lapse by March 2005. . . . The primary rationale behind this decision is that the main benefit of ISO 9000 registration is the internal benefit, which we believe can be sustained without external, third-party audits. . . . Our employees have been thoroughly trained and are proficient at performing their assigned responsibilities. We are certain that this expertise will be used to make further improvements to excellent standard work practices that are already in place. In addition, much of the infrastructure that was in place to support ISO 9000 will be maintained to ensure that standardized work implementation continues to thrive within the Headquarters organization. This transition should yield few, if any, substantive changes to the way we perform our daily responsibilities.

When asked about this executive's statement, five of the engineers expressed the view that the claimed 'internal benefit' of the ISO process was non-existent and that the remaining ISO 'infrastructure' – a fully-staffed internal consulting department that produced quarterly updates to work instructions – was out of touch with the day-to-day realities of engineering work.

Irrelevant work instructions may have been symptomatic of decoupling, but they were not its cause. When asked what derailed continued use of the standards developed for the ISO 9000 certification drive, 16 of the 20 Div-A informants indicated that standardized approaches were antithetical to the work of design engineers. They expressed two variations on this theme. First, standardization was framed as the opposite of creativity and, thus, a threat to their autonomy. Engineers used pejorative labels such as 'cookbook engineering' (or following a 'recipe'), 'catalog engineering,' or becoming 'knowledge robots.' The following statements are illustrative:

I've never, in years, heard my manager say, 'Have you referred to the standardized work for this issue?' It never comes up. It's just like, 'Let's get the job done. You're an engineer, you know what you need to do.'. . . And there is probably some feeling that standardized work equals cookbook engineering, which equals 'don't use your brain.' (Informant 44)

Engineering in my mind cannot be that automated and that rigid. Sure enough, it needs to be guided by certain absolutes and by certain rules, but then there's a creative element of it that really makes engineering fun. And engineers tend to want to have that freedom. They tend to not want to be constrained. They tend to not want to be treated as, you know, virtually sort of knowledge robots that crank out solutions with very rigid sets of formulas. (Informant 55)

Second, some engineers objected to standardized work on more practical grounds, insisting that their work was unpredictable and therefore not amenable to routinization. Instead, they felt that they were always in 'fire-fighting mode' on the job. The following comments are representative of this theme:

I look at standard work . . . as a standard set of procedures that management has laid out for us that they would like us to follow on how to do a given task. All these tasks do not necessarily have an order to them. There is no set of standard work orders that encompasses everything that we do. A lot of what we do is trial and error, talking to your peers, figuring it out, finding a way to do it. (Informant 45)

Standardization doesn't work – every program is different. At the high level, you can come up with some sort of a standard flow, standard deliverables. But when you get down to the nuts and bolts of it, that's where it's always going to be a little different. (Informant 49)

It is interesting that the most common reasons for resisting standardization had to do with the professional self-image of the engineers, and the way they described the nature of engineering work. The historical context is noteworthy. A sequence of external quality standards had been brought into Div-A, only to be abandoned and replaced by the next 'new' thing. ISO 9000 was then introduced and compliance was mandated. Implementation, however, was carried out by a staff function that was administratively, spatially, and professionally decoupled from the design engineers. As a result, the ISO legacy – a voluminous catalog of work instructions – is essentially ignored by practicing engineers.

Division B: How Implementation Led to Substantive Compliance with the Standard

In contrast to Div-A, Div-B achieved ISO certification in 1997 and remains certified. Yet among Div-B engineers, the ISO requirements were almost never mentioned, having been superseded by a rigorous, internally-developed work protocol known by the acronym POE (Process of Engineering). Div-B's engineers executed their technical work in a way that was consistent with ISO requirements *but without explicit reference to the standard*. Indeed, in 43 interviews with Div-B engineers, ISO certification was mentioned exactly twice. This is a striking example of the type of substantive compliance referred to as 'tight coupling' earlier in this chapter.

To pinpoint the factors that contributed to this substantive adoption of the ISO standard, Div-B's history will be presented, with particular attention to its motivation for adopting the ISO standard and the subsequent evolution of standardization within the division. The early sequence of events bears strong resemblance to the experience of Div-A; the story's ending, however, is markedly different.

The Pre-Adoption Context

Div-B designs a component that was pioneered by GAM in the 1930s and incorporated into GAM's products in the 1940s. The technology was highly innovative and gave GAM a significant advantage that it has never relinquished. Beginning in the 1950s, Div-B began selling finished components and licensing the component's technology directly to other manufacturers. Also during this time period, a fire at a GAM facility required the relocation of most of Div-B's staff to a new engineering center attached to a manufacturing plant. In the early 1980s, all of Div-B's engineers and technicians were moved to this location, a site 60 miles from headquarters. The Div-B workforce thus developed strong cohesion over 20 years of working under the same roof.

Adopting an External Quality Standard

In 1994, Div-B's senior leadership announced to its suppliers that they would need to be ISO certified by 1995. Many of Div-B's customers responded with a reciprocal request that Div-B become ISO certified as well. Unless it wanted to lose these customers' business, Div-B had little choice but to adopt the standard. Thus Div-B's motivation for adoption was primarily commercial, in contrast to Div-A's mixture of 'copy-cat' and political motivations. Furthermore, Div-B had not experimented with a sequence of seven 'flavor of the month' quality management approaches before deciding to pursue ISO registration in 2000 (see Table 3).

Implementation: From External Standard to Decoupled Directive

Given such a business imperative, Div-B responded with an ISO implementation plan that, unlike Div-A, did not require engineers to memorize generic work instructions. Instead, engineers were to describe their actual work, but use designated ISO terminology. Certification was achieved in 1997 without significant changes to existing work practices, although the engineers bore a heavier burden of documentation. A seasoned engineer recalled the effort to document work practices in preparation for the ISO audit:

It was difficult to go to somebody's desk, file cabinet, or computer to prove that you meet their elements [i.e., ISO requirements]. This could be very, very harmful [to the ISO audit] because you wouldn't be able to show [compliance] even though it's in the back of somebody's mind. If you're trying to demonstrate compliance, it would be awful tough. So, again, I think it became a side benefit [of ISO certification] to prove that we have the integrity that we think we do. (Informant 20)

Of note is the engineer's framing of ISO documentation as a 'side benefit' that ensured the 'integrity' of the division's engineering work. Div-B's implementation of ISO standards did not insult the engineers' professional dignity, in contrast to the punitive audit practices that had engendered cynicism among Div-A's engineers.

Thus, Div-B's initial implementation of ISO 9000 was decoupled from core technical work practices but did not occasion the resentment registered by Div-A engineers. Div-B also departed from Div-A in a more crucial way: the standard was not abandoned. Instead, the engineers themselves came up with a standardization protocol that not only satisfied ISO requirements, but did so in a way that was consistent with the engineers' professional values and tightly coupled with their work practices. (See Table 4 for a chronology of Div-B's standardization initiatives.)

Transcending the Standard

In 2005, a subgroup of engineers in Div-B's new products group began to develop their own highly regimented engineering process. Prompted by inconsistencies between document formats and the difficulty of obtaining information on the latest design tolerances, simulation methods, and test evaluation criteria, the manager of the new products group told the subgroup of engineers that they would do 'little or no work on the [component] for the next two weeks':

We piled everybody into this room, and around-the-clock, around-the-shifts, we wrote out our process. (Informant 10a)

At the same time, a new executive was assigned to lead Div-B. He emphasized quality standards and, building on the subgroup's initial efforts, selected a team of the most respected engineers in the division and gave them the responsibility to extend the new product group's standardization effort across all of Div-B. In contrast to the structural isolation of the Div-A quality staff, Div-B's standardization team reported directly to the division's senior executive; its team leader attended the executive's staff meetings and gave regular updates on progress. The product of this team's efforts turned into the 'Process of Engineering' (POE); it became Div-B's core quality management tool, superseding prior programs.

Table 4. The succession of standardization initiatives in Div-B

Initiative	Description
ISO 9000 Preparations (1995)	At the request of external customers, Div-B begins preparations for ISO 9000 certification. The effort is designed to disrupt engineering work as little as possible.
ISO 9000 Registration (1997)	External auditors grant Div-B the ISO seal of approval. Since this time, registration has never been allowed to lapse.
'Engineering Standard Work' Training Program (2001)	Div-B taskforce develops a one-day workshop to help 'all people in the organization who produce similar deliverables for product program execution do it the same way every time' (GAM Training Manual). Attendance is mandatory for engineering team leaders and above, and optional for engineers.
New Product Development Team Creates Standards (2004)	To eliminate duplication and confusion, engineering team leader takes his engineers off the job for a week to create a structured sequence of input requirements, output requirements, and document formats for all engineering tasks.
New Executive Joins Div-B (2004)	New Div-B Top Executive supports the extension of the New Product Development Team's work to all of Div-B. The resulting set of work standards is labeled 'Process of Engineering' (POE).
Process of Engineering (POE) Rolled Out in Div-B (2005)	POE becomes a web-based tool that brings all work standards and supporting documentation into a single virtual location.
Internal Audits of POE Compliance (2006)	Monthly audits ensure engineers are using POE tools and uploading documentation to the POE website. Auditors also collect improvement suggestions from engineers and feed these back to a 'change review board' for inclusion in POE.
POE Spreads to Div-B Offices in Asia and Europe (2008)	POE becomes a global system for ensuring standardized work.

Many engineers in Div-B were wary of POE at first; 11 of the 43 Div-B informants specifically mentioned their own and others' initial resistance. The following comments are typical:

Engineers don't like to be saddled with systems, so we were not the first ones in line to say: 'Okay, we'll use it.' (Informant 20)

Initially I thought a very regimented management process is not conducive to an inventive type of process, and there's some invention in our work, in the sense that we are going to create a unique [component] almost every time. But [POE] works very well. (Informant 39)

Unlike Div-A's ceremonial (and cynical) ISO compliance protocols, POE was integrated into the way the engineers conducted their day-to-day work. It had been created by a group of Div-B's most respected design engineers, not by a separate group of quality experts. In addition, its continued use was mandated by the division's top executive. The combination of 'bottom up' endorsement with 'top down' executive sponsorship proved potent enough to overcome the engineers' reticence.

POE's appeal derived primarily from what it was *not:* It did not tell engineers how to do their jobs. As described by its users, POE standardized everything but the creative engineering tasks, a fact which the data make evident: 39 of the 43 Div-B informants asserted that POE did not limit their freedom to do creative engineering work. Only 2 informants intimated that POE hampered their creativity. (In contrast, recall that 12 of the 20 Div-A informants indicated that standardized work was anti-creative.) The following comments reflect the majority view in Div-B:

When you work with non-manufacturing organizations, especially engineering, the impression is that what they do is very creative and therefore it resists standardization. What we try to do is say: 'Okay, we'll give you that. What we will do is work on standardizing or making your job easier by taking the non-creative parts, which is generally administrative – document flow, information flow, how teams work together, how people interact – we'll take that portion that everybody agrees is non-creative and non-productive, and we'll make that work better.' (Informant 19)

POE isn't prescriptive. It doesn't tell you, 'Here's how you fix a specific problem.' It just says, 'At this point you've got to have all of this stuff done.' Unfortunately, we haven't made all of the mistakes that we could possibly make yet, so we still invent some new ones and we've still got to figure out ways to fix those new ones. I think we're still rich with creativity and creative solutions. There's still enough of that to satisfy engineers. (Informant 6)

POE's web-based interface defines what completion looks like for some 270 steps in the design process; it also requires engineers to keep track of project milestones. Informants reported that they consult the website as often as five times a day, and no less than once every two weeks. Usage statistics are readily available for the POE website. The total monthly number of distinct users increased steadily over POE's first 30 months (see Figure 1), suggesting a wide acceptance of the system. The nomenclature associated with POE was noteworthy as well. The website avoided the terms 'standardization' and 'engineering factory,' and Div-B informants never used the phrase 'cookbook engineering' in describing POE.

Figure 1. Total Number of Distinct Users per Month, 'Process of Engineering' (POE) Website
Source: GAM internal documentation

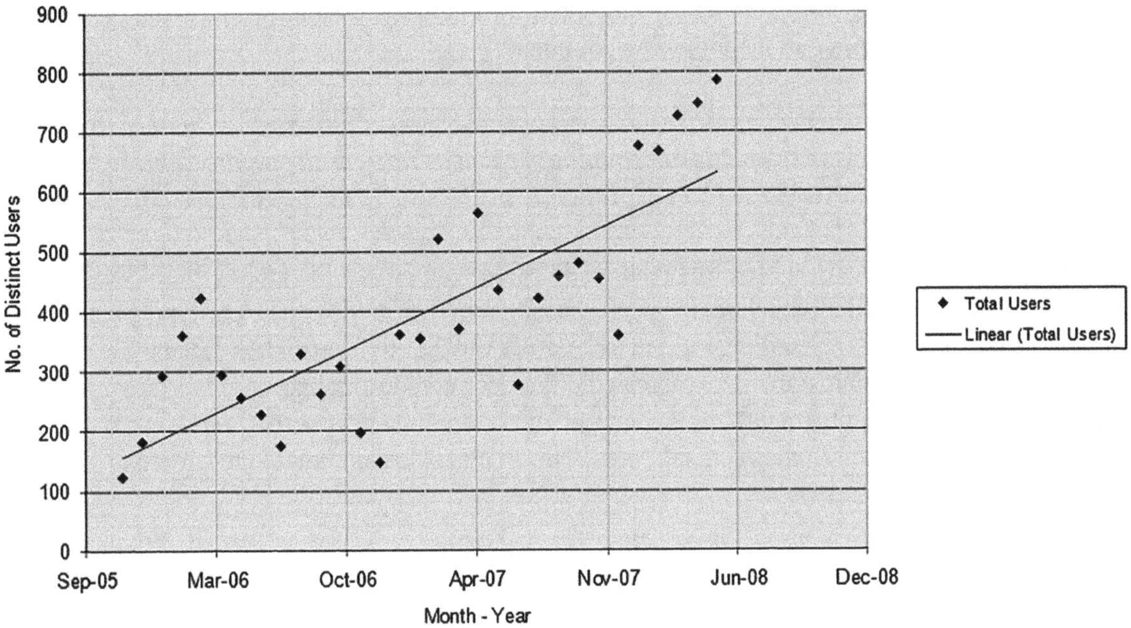

Finally, POE included internal audits, but they were used more for continuous improvement than punishment. Often during audits, engineers would explain that their failure to follow POE standards was because of a superior technique that they had developed. An auditor described one such exchange, in which the standard documentation was not uploaded to the POE website as required:

[The engineer said,] 'Well, we used to use this form. But now we use this – it's kind of expanded. So now, this [standard] really needs to be more in-depth and say this, this and this.' And we're like, 'We'll copy and paste it [the new process] and throw it in a Word document and . . . we'll get it to [the change review board].' So, we also play the go-between. (Informant 26)

In effect, the auditors' go-between role ensures that POE standards remain evergreen, constantly reflecting the engineers' improvements. The same sentiment was echoed by an engineer in describing the advantages of POE:

I think POE retains knowledge. Previously whenever you had to know what to do next, you had to go to people who were there who had worked in the company for a long time, who knew what the next step was. (Informant 8)

The POE website thus documented and made available to all engineers much of the core technical knowledge that had previously resided only in the heads of experienced engineers.

Not everyone was an enthusiastic follower of POE, however. There was a particular engineering team that had the reputation of being POE detractors. Accordingly, all nine members of this team were interviewed, and their work patterns were observed over the course of a week. Three of the team's engineers expressed annoyance at the burden of documentation that POE required. (This issue was echoed by four Div-B informants outside this team). Yet in each case, the engineers continued to use the POE system. Their sentiments are captured in the following comment:

The paperwork and the documentation are just not a typical engineer's strength or interest. And so one of the biggest problems is getting the [documentation] posted or referenced properly. I mean, again, it's not to badmouth the system, because its intent is noble, it's just the execution's tough. (Informant 20)

A second area of concern within this team of POE detractors was the feeling that POE had been over-engineered: the detailed, standardized work processes were geared toward developing new components, and were excessive for handling routine tasks. But even the engineers who voiced these criticisms complied with POE standards; their complaints did not occasion decoupling.

The following statement from a Div-B manager illustrates how deeply the POE standards had become embedded in Div-B's engineering practices. This manager had authorized his group to violate the standards in response to a crisis:

We had a program recently where we were asked to do everything in one-third the time that we normally do it, and we got the okay to not follow POE. All of my folks had tremendous anxiety. They all felt it was going to take them longer to not follow the process. . . . So to say, 'Hey, don't take the time to do all the calculations. Just do the ones you think are important.' 'Well, the ones I think are important are the ones that I always do.' 'Well, just do half the things you think are so important.' 'Well, does that mean it only has to be half-right?' That's probably the best example I can give you of how people believe in this. (Informant 10a)

To summarize, both Div-A and Div-B initiated and successfully executed ISO 9000 certification initiatives. However, the antecedents and effects of their efforts have little in common. Within Div-A, a sequence of failed top-down attempts at standardization engendered cynicism and fatigue among engineers, who expressed a distinct distaste for the hollow ceremony associated with ISO certification. In Div-B, adoption of ISO 9000 was a business necessity and was initially implemented as such, with minimal disruption to existing patterns of work. Both of these implementations were decoupled from core engineering processes. In Div-B, however, a robust standardization tool later evolved from the bottom up and was tightly coupled to the engineers' daily work. It was almost universally followed, making Div-B's continued ISO certification a mere formality.

LESSONS LEARNED

This chapter began with a puzzle: What factors explain how two engineering divisions within the same corporation achieved such divergent outcomes in the pursuit of standardization? The detailed presentation of these cases now serves as the basis for a general discussion of how managers can skillfully handle the adoption and implementation of an external standard among knowledge workers.

Lesson 1: Let Local Professionals Create and Own the Standard

The decision to adopt and implement a standard is typically made at the executive level in organizations. Once adopted, the standard can be implemented substantively or symbolically.[2] Either type of implementation necessarily involves groups of *knowledge workers* – the engineers, accountants, or other professionals who will need to change their daily work in accordance with the standard. It is at this level – the myriad technical and functional departments, work groups, and other subunits in complex organizations – that a standard's fortunes are ultimately determined.

Studies of standardization have neglected the crucial role of technical experts on the receiving end of standards. Indeed, professional expertise and standardization typically are portrayed as complementary: As groups of experts coalesce around shared knowledge and interests, they tend to codify their expertise and promulgate rules to govern its application and lay claim to occupational jurisdiction (Abbott, 1988). Some of these rules take shape as professional norms, while others are instantiated in standards (Jacobsson, 2000). But when such standards are to be implemented by professionals who did *not* create them, professionals become the objects rather than the agents of standardization. This role reversal represents an inflection point in the standardization process. To the degree that abstract external standards are enforced by managerial mandate, they become decoupled from technical work. In contrast, if experts codify their own tools and practices, the resulting standards reflect local realities and enable tight coupling between daily technical work and the abstract external standard.

For managers attempting to implement a standard, this is a crucial distinction. Once they adopt the standard, managers in effect transform it into a directive. Furthermore, managers often create additional bureaucracy around the standard: they locate the implementation effort in its own department with its own budget, and provide this department with the authority to issue incentives for compliance (and sanctions for non-compliance). Such bureaucratic trappings create an unintentional barrier to successful implementation; they isolate the standard from those doing the day-to-day work (Barley & Kunda, 2001).

In other words, the transformation from standard to directive is readily accomplished; the nature of managerial work is to transform abstract goals into actionable programs. The transformation from directive to norm, however, is much more difficult because it cannot be mandated. Knowledge workers must find the standard useful and therefore be willing to incorporate it into their work routines.

Div-B's successful implementation of the POE standards shows that when standards are homegrown, the standards assimilate the group's processes, rather than the group assimilating standard processes. In such cases, adoption – defined as 'taking for one's own use something originated by another' (Merriam-Webster, 2011) – never occurs. Rather, if the homegrown standard is of sufficient rigor, external certification becomes a simple word game: 'Existing practice is continued but now described in accordance with the [external] standard' (Brunsson & Jacobsson, 2000b: 128).

This raises a difficult question for managers: If adoption strips a standard of its unique voluntariness, how else might an organization engage a standard? In brief, Div-B's successful implementation of POE resulted from the senior executive's efforts to *create the conditions under which the professional group could generate its own set of standards.* The executive chartered a group of internal experts to standardize their existing work processes in a manner consistent with ISO 9000 demands. The eventual result was a comprehensive set of protocols that not only met ISO 9000 requirements, but also became so much a part of the engineers' culture that, when given the option to abandon the standards, the engineers chose to follow them anyway.

Lesson 2: Be Transparent about Managerial Objectives

Initially, both Div-A and Div-B failed in their effort to have ISO 9000 standards substantively embraced by their engineers. Div-A's implementation was in many ways doomed from the start. Past history led the engineers to doubt the division's commitment to quality; Div-A had tried one quality system after another, only to abandon it and try the next managerial fashion. This led to a reaction that the engineers described with the acronym BOHICA, which stands for, 'Bend over, here it comes again.' Little wonder that engineers struggled to take the ISO 9000 standards seriously. Their doubts were validated when, after three years and before the first ISO 9000 certification, Div-A's senior executive cancelled the ISO initiative, but inexplicably instructed engineers to continue to follow the standard work instructions (none of which they had been following).

Div-A's cynical type of decoupling has been observed in numerous other empirical studies. For example, Boiral's (2003) analysis of attitudes toward ISO 9000 among Canadian implementers identified a large group of 'dissident' workers who minimally (and cynically) complied with the quality management system; and Walgenbach (2001) found similar emotions among German ISO 9000 adopters. Common to these accounts is an element of hypocrisy: managers espoused substantive change but enacted only symbolic compliance (Brunsson, 2000).

In contrast, Div-B initially adopted the ISO 9000 standard as a commercial necessity. Div-B's leaders were straightforward regarding the symbolic nature of compliance, and thus made no serious effort to change technical work arrangements. Decoupling ensued, but without hypocrisy.

That these contrasting effects have been observed repeatedly prompts a natural question for managers attempting to implement standards: Under what conditions do we see which type of decoupling? The answer implicates senior management. When Div-A's senior executive was interviewed about the current implementation of work standards within his division, he displayed no awareness of the rampant decoupling taking place in the division. His involvement in the day-to-day use of the standard work instructions was minimal. He had delegated this activity to a group of quality specialists and therefore focused little time and attention on it. In contrast, the executive over Div-B at the time of the first ISO 9000 certification set a tone of openness and transparency. He did not pretend that the ISO standards should fundamentally change the engineers' work processes. Instead, he acknowledged that ISO certification was a commercial necessity, and supported the workforce in their attempts to meet the ISO standards with minimal disruption to their *modus operandi*.

Lesson 3: Conduct Audits that Emphasize Learning, Not Punishment

Internal audits played a crucial role in embedding Div-B's POE in daily practice. Existing research highlights the importance of *external* audits (e.g., Seidl, 2007), but the POE case indicates that *internal* audits can offer similar encouragement. As described above, the POE audits functioned as a feedback loop, ensuring that the standards continuously reflected current practice. In contrast, recall that ISO audits were primarily punitive within Div-A, where failure to follow the ISO 9000 program was an invitation to be 'flogged' by upper management. Such managerial sanctions run counter to the principle of professional autonomy enshrined in the knowledge-worker ethos (Bailyn, 1985; Petroski, 1992; Kunda, 1992; Alvesson, 2001).

Auditing performance is an indispensable role of management. However, audits can be designed to harvest continuous improvement suggestions from the knowledge workers being audited. This emphasis on development rather than discipline resonates with prior research on participatory processes. For example, a study of the implementation of quality standards in paper mills found that three of the initiatives failed due to conflict between corporate management's 'logic of *standardization*' and the 'logic of *participation*' necessary for worker-driven improvement efforts (Vallas, 2003: 238, italics in original). Commenting on the spread of evidence-based standards in medicine, Thévenot (2009: 794) observed that 'participation in the process of form-giving can be a means to prevent a standard from becoming external to one's own concerns.' When audits become a vehicle for such participation, standardization faces a much higher probability of successful implementation.

These three lessons for managers – with corresponding implications for knowledge workers – are summarized in Table 5 below.

CONCLUSION

This chapter has examined how two similar divisions of the same organization achieved disparate outcomes in their efforts to standardize their engineering work. The two cases show that, in the process of adoption, external standards become organizational directives. When such directives threaten the professional norms and practices of those expected to implement them, a malignant form of decoupled compliance is observed. If instead the directives accommodate the norms and practices of the targeted professional group, a benign form of decoupled compliance develops. If, however, the standards originate within the group and incorporate local professional norms and practices, then the resulting directive will encounter less resistance. Tightly-coupled compliance with the external standard is thus rendered possible.

This view of standards raises a cautionary question for managers. Quite possibly the most salient feature of standards is their tendency to beget bureaucracy. It is all too easy for managers to create organizational units to house their standardization efforts. Such structures may be difficult to dismantle. For example, Div-A's dalliance with ISO 9000 has resulted in a staff department that is not only decoupled, but vestigial. Assuming Div-A is not an isolated case, one wonders how much organizational clutter

Table 5. Actions that Aid or Undermine Successful Implementation of Standards Among Knowledge Workers

	Managerial Actions		**Knowledge-Worker Actions**	
	Constructive	***Destructive***	***Constructive***	***Destructive***
Standards	Empower workers to "own" the creation and implementation of the standard	Adopt the standard without clarifying objectives or involving workers	Contribute to the creation and interpretation of the standard	Resist the standard as foreign and irrelevant
Directives	Celebrate and mandate use of the homegrown standard to ensure consistency	Mandate inflexible adoption of the external standard	(Not applicable)	(Not applicable)
Norms	Sponsor audit processes that capture current professional practice; modify standard accordingly	Enforce audit processes that reward strict compliance and punish deviation from external standard	Suggest improvements so that the standard reflects up-to-date professional beliefs and practices	Attack the standard as contrary to professional beliefs and practices

gets left in the wake of attempted standardization. At what point might an organization crumble under the accumulated detritus of past standards-inspired directives? Such questions about the unintended consequences of standardization merit further research.

ACKNOWLEDGMENT

This research was made possible by National Science Foundation grant ITR 0427173.

REFERENCES

Abbott, A. (1988). *The system of profession: an essay on the division of expert labor*. University of Chicago Press.

Abernethy, M. A., & Stoelwinder, J. U. (1995). The role of professional control in the management of complex organizations. *Accounting, Organizations and Society*, *20*(1), 1–17. doi:10.1016/0361-3682(94)E0017-O

Adoption. (2011). In *Merriam-Webster*. Retrieved from http://www.merriam-webster.com/thesaurus/adopt

Alvesson, M. (2001). Knowledge work: Ambiguity, image and identity. *Human Relations*, *54*(7), 863–886. doi:10.1177/0018726701547004

Bailyn, L. (1985). Autonomy in the R&D Lab. *Human Resource Management*, *24*(2), 129–146. doi:10.1002/hrm.3930240204

Barley, S. R., & Kunda, G. (2001). Bringing work back in. *Organization Science*, *12*(1), 76–95. doi:10.1287/orsc.12.1.76.10122

Boiral, O. (2003). ISO 9000: Outside the iron cage. *Organization Science*, *14*(6), 720–737. doi:10.1287/orsc.14.6.720.24873

Boxenbaum, E., & Jonsson, S. (2008). Isomorphism, diffusion and decoupling. In The Sage handbook of organizational institutionalism. Sage Publications.

Brunsson, N. (2000). Standardization and fashion trends. In A world of standards. Oxford Univ. Press.

Brunsson, N., & Jacobsson, B. (2000a). The contemporary expansion of standardization. In A world of standards. Oxford Univ. Press.

Brunsson, N., & Jacobsson, B. (2000b). Following standards. In A world of standards. Oxford Univ. Press.

Brunsson, N., & Olsen, J. P. (1993). *The reforming organization*. New York: Routledge.

Brunsson, N., & Sahlin-Andersson, K. (2000). Constructing organizations: The example of public sector reform. *Organization Studies*, *21*(4), 721–746. doi:10.1177/0170840600214003

Chandler, A. D. Jr. (1977). *The Visible Hand: The Managerial Revolution in American Business*. Belknap Press.

Djelic, M., & Sahlin-Andersson, K. (2006). *Transnational governance: Institutional dynamics of regulation*. Cambridge University Press. doi:10.1017/CBO9780511488665

Eisenhardt, K. M. (1989). Building theories from case study research. *Academy of Management Review*, *14*, 488–511.

Eisenhardt, K. M., & Graebner, M. E. (2007). Theory building from cases: Opportunities and challenges. *Academy of Management Journal*, *50*(1), 25–32. doi:10.5465/AMJ.2007.24160888

Espeland, W. N. (1998). *The struggle for water*. University of Chicago Press.

Fiss, P. C., & Zajac, E. (2006). The symbolic management of strategic change: Sensegiving via framing and decoupling. *Academy of Management Journal*, *49*(6), 1173–1193. doi:10.5465/AMJ.2006.23478255

Hallett, T. (2010). The myth incarnate: Recoupling processes, turmoil, and inhabited institutions in an urban elementary school. *American Sociological Review*, *75*(1), 52–74. doi:10.1177/0003122409357044

Jacobsson, B. (2000). Standardization and expert knowledge. In A World of Standards. Oxford Univ. Press.

Kunda, G. (1992). *Engineering culture: culture and control in a high-tech organization*. Philadelphia: Temple Univ. Press.

Latour, B. (1987). *Science in Action*. Cambridge, MA: Harvard University Press.

Meyer, J. W., & Rowan, B. (1977). Institutionalized Organizations: Formal Structure as Myth and Ceremony. *American Journal of Sociology*, *83*(2), 340. doi:10.1086/226550

Miles, M. B., & Huberman, A. M. (1984). *Qualitative data analysis: A sourcebook of new methods*. Sage Publications.

Oliver, C. (1991). Strategic Responses to Institutional Processes. *Academy of Management Review*, *16*(1), 145–179.

Petroski, H. (1992). *To engineer is human: The role of failure in successful design*. Vintage.

Rombach, B. (1993). Success at the expense of control. In The Reforming Organization. Routledge.

Ruef, M., & Scott, W. R. (1998). A Multidimensional Model of Organizational Legitimacy: Hospital Survival in Changing Institutional Environments. *Administrative Science Quarterly*, *43*(4), 877–904. doi:10.2307/2393619

Sandholtz, K. W. (2012). Making Standards Stick: A Theory of Coupled vs. Decoupled Compliance. *Organization Studies*, *33*(5-6), 655–679. doi:10.1177/0170840612443623

Sauder, M., & Espeland, W. N. (2009). The discipline of rankings: Tight coupling and organizational change. *American Sociological Review*, *74*(1), 63–82. doi:10.1177/000312240907400104

Scott, W. R., & Meyer, J. W. (1991). The organization of societal sectors: Propositions and early evidence. In The new institutionalism in organizational analysis. Univ. of Chicago Press.

Seidl, D. (2007). Standard setting and following in corporate governance: An observation-theoretical study of the effectiveness of governance codes. *Organization*, *14*(5), 705–728. doi:10.1177/1350508407080316

Stevens, J. M., Steensma, H. K., Harrison, D. A., & Cochran, P. L. (2005). Symbolic or substantive document? The influence of ethics codes on financial executives' decisions. *Strategic Management Journal, 26*(2), 181–195. doi:10.1002/smj.440

Storz, C. (2007). Compliance with international standards: The EDIFACT and ISO 9000 standards in Japan. *Social Science Japan Journal, 10*(2), 217–241. doi:10.1093/ssjj/jym048

Tamm Hallström, K. (2000). Organizing the process of standardization. In A World of Standards. Oxford Univ. Press.

Tamm Hallström, K. (2004). *Organizing international standardization: ISO and the IASC in quest for authority*. Northampton, MA: Edward Elgar.

Taylor, F. W. (1993). *The Principles of Scientific Management, and Shop Management*. London: Routledge/Thoemmes Press. (Original work published 1903)

Thévenot, L. (2009). Governing life by standards: A view from engagements. *Social Studies of Science, 39*(5), 793–813. doi:10.1177/0306312709338767

Thompson, J. (1967). *Organizations in Action: Social Science Bases of Administrative Theory*. McGraw-Hill.

Timmermans, S., & Epstein, S. (2010). A world of standards but not a standard world: Toward a sociology of standards and standardization. *Annual Review of Sociology, 36*(1), 69–89. doi:10.1146/annurev.soc.012809.102629

Vallas, S. P. (2003). Why teamwork fails: Obstacles to workplace change in four manufacturing plants. *American Sociological Review, 68*(2), 223–250. doi:10.2307/1519767

Walgenbach, P. (2001). The production of distrust by means of producing trust. *Organization Studies, 22*(4), 693–714. doi:10.1177/0170840601224006

Walgenbach, P. (2007). Façade and means of control: The use of ISO 9000 standards. In L. Theuvsen, A. Spiller, M. Peupert, & G. Jahn (Eds.), *Quality Management in Food Chains* (pp. 29–43). Wageningen: Wageningen Academic Publishers.

Westphal, J. D., & Zajac, E. J. (2001). Decoupling policy from practice: The case of stock repurchase programs. *Administrative Science Quarterly, 46*(2), 202–228. doi:10.2307/2667086

Zbaracki, M. (1998). The rhetoric and reality of Total Quality Management. *Administrative Science Quarterly, 43*(3), 602–636. doi:10.2307/2393677

KEY TERMS AND DEFINITIONS

Adoption: The act of formally bringing a new practice, policy, or technique into the organization. Adoption is typically mandated by senior managers.

Decoupling: A process by which organizations make a symbolic show of complying with a standard or guideline, but in reality they make little substantive change to their internal processes (i.e., symbolic compliance with standards).

Norms: Informal, socially constructed rules for behavior.

Soft-Law: Informal social rules of behavior that are not the result of legislative action and therefore carry no legal authority for enforcement. However, because of the widespread adoption of such rules, failure to comply can bring negative consequences in the form of bad publicity, loss of goodwill, consumer boycotts, etc.

Standards Developing Organization (SDO): Typically a non-governmental organization (NGO) that creates and disseminates standards for other organizations or professionals to follow. The International Organization for Standards (ISO) is perhaps the prototypical example.

Tight Coupling: Congruence between the core technical work processes of the organization and externally mandated work process standards (i.e., substantive compliance with standards).

ENDNOTES

1 ISO 9000 is the most widely adopted management standard in the world. In 1987, the International Organization for Standards (ISO) first published guidelines defining the fundamental elements of a production system that would result in high-quality products. Since then, the standards have evolved to become general management guidelines for the consistency and reliability of any organizational process (Tamm Hallström, 2000).

2 See Brunsson and Jacobsson (2000b) for a helpful discussion of the distinction between practicing standards and standardizing practice.

Chapter 9
Managing In-Company Standardisation while Avoiding Resistance:
A Philosophical-Empirical Approach

Ries Haverkamp
Interface European Manufacturing, The Netherlands

Henk J. de Vries
Erasmus University, The Netherlands

ABSTRACT

Standards may be an advantage for a company but employees often resist them because they feel they are forced to behave in a certain way. This chapter uses a philosophical approach to study why staff tend to resist company standardisation initiatives. Foucault and Habermas provide insights into the reasons for this resistance but do not solve the tension between freedom and control. Dooyeweerd's philosophy seems to be more promising. This chapter uses a company standardisation project of an automotive supplier to examine three philosophical approaches to understand resistance to standards and to investigate how this resistance can be avoided by managing in-company standardisation in a more holistic way.

INTRODUCTION

'Conforming to standards means following the advice of others, relinquishing a certain degree of one's freedom of choice and self-control to others, and often becoming more similar to many others as well – none of which is very consistent with the concept of an actor. We may therefore expect a certain reluctance to follow standards – in particular those that are well known and followed by many – from individuals and organisations seeking to be highly autonomous, innovative, and different' (Brunsson and Jacobsson, 2000, p. 134). Actors may view standards negatively because these can limit responsibility, prevent them from doing or saying what they want, and incur high costs during conversion from current practice to the practice prescribed in the standards. Moreover, standards may lead to uniformity (Brunsson and

DOI: 10.4018/978-1-4666-9737-9.ch009

Jacobsson, 2000, p. 136). 'Many of the objections to standards are similar to the objections to rules and regulation in general. standardisation is often seen as an unwelcome, unnecessary, and harmful intrusion into a world of free, distinct individuals and organisations that are wise enough to decide for themselves, or into the world of civil society or free markets. standardisation, it is felt, will mean regulation from outside, whereby actors, things, and conditions are now to be shaped in a uniform manner' (Brunsson, 2000, p. 171). Standardisation is the activity of establishing and recording a limited set of solutions to actual or potential matching problems, directed at benefits for the party or parties involved, balancing their needs, and intending and expecting that these solutions will be used repeatedly or continuously, during a certain period, by a substantial number of the parties for whom they are meant (De Vries, 1997). In the case of in-company standardisation, the activity is carried out fully or mainly within the company, the standard may be used outside the company as well.

The issue of resistance to standards has received little attention in the literature on managing standardisation within a company. Even the studies that provide a broad managerial approach ignore it (Adolphi, 1997; AFNOR, 1967; Boh & Yellin, 2007; Wenström et at., 2000; Van Wessel, 2010). These objections may apply to all standards but probably in particular to management system standards because these affect human behaviour in a more direct way than technical standards. Therefore, this chapter starts by discussing management standards and examining the paradox between benefits of implementation on the one hand, and resistance to this implementation on the other hand. Inspired by Lelong and Mallard (1995) who referred to Foucault in their introduction to a special issue of *Réseaux* on Standardisation, and by an earlier study by Van Veldhuisen (1996), we analyse resistance to such standards using Foucault's (1977) analysis of anonymous power systems. This analysis provides an explanation for resistance but not for benefits. Habermas' (1987) concept of communicative action provides additional insights but does not solve the problem of the tension between control and freedom. Dooyeweerd's (1955, 1957) philosophy overcomes this tension and provides the basis for a holistic management approach. We apply this approach in the case of a standardisation project in a company. This project was related to the company's quality management system which was based on an international quality management standard. We evaluate the company standardisation project and the resistance to it during a six-year period, using quality management literature and philosophical studies by Foucault, Habermas and Dooyeweerd.

THEORETICAL BACKGROUND

The Paradox: Benefits of Standards and Resistance to Them

The paradox of benefits of standards and resistance to them may be related to the difference between common and individual benefits of standards mentioned by Arthur (1989). Psychologists emphasise human resistance to control – freedom versus order (Brehm and Brehm, 1981). The same terms are used by David (1995, p. 18): 'The kernel of the problem posed for private and public decision-making with regard to the setting of technology standards may be construed to be nothing more and nothing less than the fundamental issue with which all social organisations are confronted: where to position themselves on the terrain between the poles of 'order' and 'freedom'.'

In this chapter we focus on management system standards. More than one million organisations have obtained a certificate showing that their quality management systems meet the requirements in the international standard ISO 9001 (ISO, 2008; ISO, 2011). Several researchers report positive impacts of implementation of ISO 9001 (Cadadesús & Karapetrovic, 2005; Corbett et al., 2005; Lo & Chang, 2007; Nicolau & Sellers, 2002; Padma et al., 2008), others find hardly any or no positive impact (Dick et al., 2008; Martínez-Costa et al., 2009; Terziovski et al., 1997; Tzelepis et al., 2006), and some report mixed positive/neutral effects (Benner & Veloso, 2008; Boiral & Roy, 2007; Nair & Prajogo, 2009; Naveh et al., 2004; Yeung et al., 2003). The best explanation for these differences seems to be the motivation to implement the standard. Is the company internally motivated to implement the standard because it is a tool to improve performance, or is it just implementing it because it needs the certificate for their position in the market? (Nair & Prajogo, 2009; Terziovski et al., 2003; Yeung et al., 2003). Boiral and Roy (2007) find a positive effect of internal motivation on performance improvements related to ISO 9001 implementation. However, in case of low internal motivation, external pressure does not necessarily lead to more human resource problems and a sort of iron cage of bureaucracy (Weber, 1922; Boiral, 2003), and external pressure can encourage a better integration of ISO 9001 (Boiral & Roy, 2007, p. 240). In these studies, 'motivation' refers primarily to the motivation of the organisation's management but few studies take employee motivation into account. Some address the internalisation of practices related to ISO 9001 (Nair & Prajogo, 2009; more references in their paper) but they focus on the measures prescribed by the standard rather than on the attitude of the employees. Lazaric and Denis (2005) report that implementation of ISO 9001 may lead to stress because of a change in routines and consequent broadening of tasks, and an aversion to writing and recording the majority of tasks. Surprisingly, literature about organisational routines (Becker, 2008) pays little attention to resistance. However, Connor (1997) noted the importance of employees in achieving quality management objectives.

Van Kemenade et al. (2011) focuses on the employees themselves (his case is not about ISO 9001 implementation but about a more or less similar situation of accreditation of universities of applied sciences in Flanders (Belgium) and the Netherlands). He reports that professionals in these organisations are willing to contribute to the process of obtaining certification if this process has internal added value, has a motivating object, has procedure and subject, has simple rules, and does not have control as its dominant value. He also states that willingness to contribute also depends on employee characteristics (is loyal to the organisation, sees self-interest, and relates to the organisation and does not delegate to others) and organisation characteristics (does not have quality experts to do all the work, gives enough management support, and has its internal quality management system in place). In other words, the basic attitude is positive but many conditions have to be fulfilled. If this is not the case, professionals may react with 'dramaturgical compliance' (just pretending to comply, fooling the auditor). Involvement of employees is found to contribute to the success of ISO 9001 implementation (Brown et al., 1998; Mahadevappa & Kotreshwar, 2004).

Some studies report a positive impact of ISO 9001 on employees. In Egyptian manufacturing firms, ISO 9001 implementation improved employee motivation (Magd, 2008). Almost 50% of registered Spanish firms reported an increase in employee satisfaction (Casadesús & Karapetrovic, 2005). However, the information in these studies comes from respondents at management level, not from employees. Brown et al. (1998) report that ISO 9001 certification did not bring any significant improvements in staff motivation and staff retention within Australian SMEs. Employee commitment was the most often (50%) mentioned problem in these companies. In a Swedish study, several quality management approaches were compared. Users of ISO 9001 reported fewer improvements in the organisation than non-users (Lagrosen & Lagrosen, 2005).

ISO 9001 views the organisation as a network of manageable processes designed to satisfy customers and includes measurable objectives, management control, and documentation (Furusten, 2000). ISO 9001 has codified the layman's conception of a successful organisation. It sees organisations as manageable and controllable units which can be used by management as instruments for fulfilling predetermined goals, strategies, and visions. It ignores the scholars that find fault with rational models for control and management of organisations or that cast doubt on the whole conception of being modern (Furusten, 2000, pp. 83-84). In a study on the implementation of the 1994 predecessor version of ISO 9001:2008, Naveh and Erez (2004) found that the implementation of the standard focuses workers on documentation, Standardisation and conformity to rules and procedures and not on a culture of innovation.

In the 2000 and 2008 version of ISO 9001, the number of compulsory procedures was reduced from 19 to six. Although the standards do not prescribe a lot of paper work, in practice, this is often different. Manders (2015) shows that this practice may differ per employee. The attitude towards actual use of control measures depends on awareness, perceived usefulness and ease of use, and the willingness to contribute to continuous improvement depends on the extent they feel responsible for the functioning of the ISO 9001 system. If these conditions do not apply, employees may ignore the system or show resistance. The process approach prescribed in the 2000 and 2008 version of ISO 9001 and probably continued in ISO 9001:2015, 'forces' people to behave in a way prescribed by the system; in fact, they are part of the system. Its approach resembles the 'cybernetic' view as described by Hofstede (1978). ISO 9001:2008 refers to employees as 'human resources,' next to other resources such as equipment and money and this is exactly the consequence mentioned by Hofstede (1978, p. 460). It is not surprising that such a system may cause concern and unrest among employees. Different theoretical lenses may be used to study this: theory of planned behaviour (Ajzen, 1991; Manders, 2015), innovation diffusion theory (Rogers 2003), and the technology acceptance model (Davis, 1989; Davis et al., 1989). Theories on change management (e.g. Bevan, 2011; Carnall, 2007) provide approaches to involve employees in change projects. In this chapter, we intend to focus on resistance to standards in a broad and fundamental way by taking a philosophical approach.

Foucault and ISO 9001

Philosophers that studied man in relation to power systems include Michel Foucault and Jürgen Habermas. Both Foucault and Habermas criticise the historical transformations of modern forms of rationality (Kelly, 1994, p. 372). Foucault did not develop a complete theory but rather studied several domains that have been shaped by power relations (Foucault, 1976, p. 109). His objective was to create a history of different modes by which human beings are made subjects (Foucault, 1982, p. 208). Foucault (1977) studied the domain of torture, punishment, and prisons which can serve as an example for other systems of anonymous power. The emergence of the prison marks the institutionalisation of the power to punish. Traditionally, the focus was on the body, with the ultimate punishment of public torture and execution, to be seen by all almost as a triumph (Foucault, 1977, p. 34). This changed in punishing the soul instead, following a penal judgement (Foucault, 1977, p. 19). This soul is shaped by methods of punishment, supervision and constraints (Foucault, 1977, p. 29). The object of control is 'the economy, the efficiency of movements, their internal organisation. (…) Finally, there is the modality: it implies an uninterrupted, constant coercion, supervising the processes of the activity rather than its result and it is exercised according to a codification that partitions time, space, movement as closely as possible. These methods allowed the meticulous control of the operations of the body, which assured the constant subjection of its forces and imposed upon them a relation of docility-utility, might be called 'disciplines'' (Foucault, 1977, p. 137). The success of disciplinary power derives from the instruments of hierarchical observation (eyes that must see without being seen), normalizing judgement (a combination of comparing, differentiating, hierarchising, homogenizing and excluding), and examination (Foucault, 1977, pp. 170-187). Examination transforms the economy of visibility into the exercise of power. (...) It is the fact of being constantly seen, of always being able to be seen, that maintains the disciplined individual in his subjection (Foucault, 1977, p. 187). The examination that places individuals in a field of surveillance also situates them in a network of writing; it engages them in a mass of documents that captures and fixes them (Foucault, 1977, p. 189). The examination opens up the possibility of the constitution of a comparative system of measurements, description and registrations (Foucault, 1977, p. 190). It makes each individual a 'case'. The individual may be described, judged, measured and compared to others, in his very individuality; and it is also the individual who has to be trained or corrected, classified, normalised and excluded, etc. (Foucault, 1977, p. 191). Finally, the examination is at the centre of the procedures that constitute the individual as effect and object of power and as effect and object of knowledge (Foucault, 1977, p. 192). Public execution was replaced by penal detention as a calculated technique for altering individual behaviour (Foucault, 1977, p. 264).

Foucault refers to Bentham's (1838) Panopticon: a prison in the form of a circle, with a tower in the middle. The prisoners live in cells in this circle. These cells have barred windows in the outside of the building and in the inside of the circle so that the supervisors in the tower can observe them. The prisoners, however, cannot see whether they are observed or not. 'The surveillance is permanent in its effects even if it is discontinuous in its action. (...) The power relation is independent of the person who exercises it' (Foucault, 1977, p. 201). 'The side walls prevent each individual from coming into contact with his companions. He is seen, but he does not see; he is the object of information, never a subject in communication' (Foucault, 1977, p. 200). The Panopticon was also a laboratory; it could be used as a machine to carry out experiments, to alter behaviour, to train or correct individuals (Foucault, 1977, p. 203). 'Whenever one is dealing with a multiplicity of individuals on whom a task or a particular form of behaviour must be imposed, the panoptic schema may be used' (Foucault, 1977, p. 205). Moreover, the level of use may be extended from individuals to organisations: 'The seeing machine was once a sort of dark room into which individuals spied; it has become a transparent building in which the exercise of power may be supervised by society as a whole' (Foucault, 1977, p. 207).

The prison system has not been successful. 'Prisons do not diminish the crime rate, detention causes recidivism, the prison produces delinquents, encourages the organisation of a milieu of delinquents' (Foucault, 1977, pp. 265-268). As a result, initiatives to reform of the prison system have been taken at the national level. Illegal activities remain but society uses them or even needs them to show the limits of legality and to make illegality less dangerous. 'The carceral 'naturalizes' the legal power to punish, as it 'legalizes' the technical power to discipline' (Foucault, 1977, p. 303). The prison is 'linked to a series of 'carceral' mechanisms which seem distinct enough – since they are intended to alleviate pain, to cure and to comfort but which all tend to exercise a power of normalisation' (Foucault, 1977, p. 308).

Foucault is the co-founder of the Groupe de l'Information de les Prisons, a prison information group that provides a channel for prisoners to voice their concerns, and such protest is, in fact, Foucault's "solution" for those who suffer from the exercise of power. He "walked his talk" by joining students in occupying administration buildings and fighting with the police.

An ISO 9001 quality management system is a system of control to assure that the organisation's products (which may include services) meet the expressed or unexpressed requirements of the customer. This control is achieved through process management with a set of measures including policy development and deployment, written procedures, registrations, measuring of product characteristics against requirements, measuring customer satisfaction, corrective and preventive actions, internal and external audits, and management reviews. The 'human resources' are managed by providing them education. Employees have to behave in the way prescribed in the written instructions and procedures of the quality management system. Manders (2015, chapter 4) investigated the involvement of individual employees in their company's ISO 9001 system. She found a diversity of attitudes towards the system, from very negative to very positive, with some employees showing the kind of resistance described by Foucault. Others had a positive attitude to the system, acknowledging its usefulness and perceived ease of use. Only those who had a specific role within the system, for example, internal auditors, were also involved in improving the system, provided they felt responsible to do so. The manager or management team is not the 'sovereign king' of the organisation but, instead, part of the system. Management observes (management review) but is observed as well through internal and external audits.

We can see the similarities with Foucault systems. Once established, the quality management system exercises normalising power. It is an instrument of discipline. This power is exercised rather than possessed, it comes from the bottom up rather than from the top to the bottom, and it is not repressive (backed by sanctions) but productive (Kelly, 1994, p. 374). The prescriptions of behaviour related to described processes and the subsequent measurements, evaluations and audits make employee behaviour transparent. This does not mean that employees cannot take any initiative. On the contrary, but such initiatives are prescribed by the system as well because it defines their responsibilities, authorities and tasks which include corrective and preventive actions and even improvements of the quality management system. In forcing them to further strengthen the system, ISO 9001 even goes a step further than the old prison systems. Anyhow, the employee as well as his manager is captured by the system. It is not surprising that they may have feelings of resistance. At the end of the day, Foucault would have advised them to escape. This is indeed what Gouldner (1954) already observed: a discrepancy between official procedures and actual behaviour. This discrepancy is problematic for organisations striving for quality or being more or less forced by customers or government to implement ISO 9001. So Foucault helps us to understand resistance to standards but does not offer an acceptable solution.

Habermas and ISO 9001

Habermas (1987) distinguishes between 'System' and 'Lifeworld'. The lifeworld is 'that province of reality which the wide-awake and normal adult simply takes for granted in the attitude of common sense. By this taken-for-grantedness, we designate everything which we experience as unquestionable; every state of affairs is for us unproblematic until further notice' (Habermas, 1987, p. 130). 'The lifeworld is the intuitively present, in this sense familiar and transparent, and at the same time a vast and incalculable web of presuppositions that have to be satisfied if an actual utterance is to be at all meaningful, that is, valid *or* invalid' (Habermas, 1987, p. 131). Systems, on the other hand, are characterised by economy (resources), polity (goals), pattern-maintaining systems (values) and integrative subsystems (norms) (Habermas, 1987, p. 244). Society is in crisis because of the "colonisation" of the lifeworld by systems (Habermas, 1987, p. 394). Systems are fully rationalised and are related to material reproduction, rationality of money and power, and achievement of results. Principles of rationalisation are efficiency, calculability, predictability and control.

Conflicts are about defending and restoring endangered ways of life. (…) The new problems are related to quality of life, equal rights, individual self-realisation, participation, and human rights (Habermas, 1987, p. 392). The lifeworld consists of 'communicative action' and therefore Habermas proposes such communicative action that 'decodes the deformations of the lifeworld' (Habermas, 1987, p. 403). Communication has three functions: (1) reaching understanding (communicative action serves to transmit culturally stored knowledge), (2) coordinating action (it serves the fulfilment of norms in the form of social integration and establishment of solidarity), and (3) socialising actors (it serves the formation of personality structures) (Habermas, 1987, pp. 63, 137). The concept of consensus is important: 'participants in communication reach an agreement concerning the validity of an utterance; agreement is the validity claim the speaker raises for it' (Habermas, 1987, p. 120).

Implementing a quality management system based on ISO 9001 can be regarded as the introduction of a 'system' in the lifeworld of an organisation, and consequently, it is logical that feelings of resistance may surface. Organisations are more than just units for material reproduction; they are important for the identity and wellbeing of individuals. Habermas would protest against ISO 9001 implementation and in particular against the reduction of human beings to 'human resources'. If such a system is inevitable, Habermas would probably advise the lifeworld to penetrate into the system. This should be done through communicative action. Employees should discuss and negotiate all elements of the quality management system and the necessity of the system itself, taking into account both the external (market) situation and the lifeworld within the organisation. If they reach consensus, the system will better satisfy their needs or the system is no longer needed because common understanding provides an alternative way of coordination.

If Foucault had been confronted with this fictitious advice, he would have argued that the necessity to communicate and arrive at consensus becomes an element of the system. 'The problem is not of trying to dissolve them [relations of power] in the utopia of a perfectly transparent communication, but to give oneself the rules of law, the techniques of management, and also the ethics, the ethos, the practices of the self, which would allow these games of power to be played with a minimum of domination' (Foucault, cited by Kelly, 1994, p. 391). Habermas focuses on moral-practical relations between modern, autonomous subjects but that regime exists alongside a second regime analysed by Foucault determined by practices, institutions, and knowledge (Kelly, 1994, p. 376). Therefore, their approaches are complementary in their application to ISO 9001, and in both cases, there is a conflict between freedom and control. The organisation's management, or even the entire organisation, uses its freedom to opt for a (quality management) system, but, once established, this system restricts individual freedom, including that of the management.

Addressing the Conflict between Freedom and Control

Freedom and control are essential values in Western culture. Goudzwaard (1979) argues that freedom and control are the two main ideals of Humanism which originates from the Renaissance. Exercising individual freedom includes controlling one's environment including the creation of systems of control but this may affect the freedom of other people. How can this problem be solved?

Erasmus, one of the founding fathers of Humanism, talks about individual freedom in a way uncommon in our days: 'Nobody obeys more (…) than he who has drunk in the Spirit of Christ and starts getting free' (Allen, 1913, p. 374). He links freedom to Christianity. The Christian source document, the Bible, addresses the tension between freedom and control but uses different terms. The basic assumption in Renaissance and Humanism is man's autonomy, whereas man, according to the Bible, was created and, therefore, is in a subordinate position to his Creator. Man has been given freedom, but this freedom gets a purpose. 'Everything is permissible, but not everything is beneficial or constructive' (1 Corinthians 10:23). Man has been set free to love his neighbours and to look after creation and develop culture. His freedom includes the option to choose to go against his Creator. The latter is called 'sin' which means 'miss the goal'. Sin is related to autonomy: deciding for yourself what to do instead of being dependent on your Creator and respecting the structures of creation and God's commandments. So in the positive sense, man's freedom gets a purpose of serving, which includes the creation of artefacts, organisations and systems, and thus measures and systems of control. Such control, embedded in the attitude of serving neighbours, is claimed to be of benefit to other people and to creation. However, sin affects behaviour and man, seeking his autonomy, may use his freedom to control in a way that affects his environment. Such behaviour may lead to a system of control in which man is both offender and victim. This may include systems at societal level. The first biblical example is the tower of Babel and with numerous similarities with the systems described by Foucault (Ellul, 1970). Foucault provides a thorough analysis of such a system and its inherent reasons for resistance. However, his view is too negative. Reality shows a mixture of positive and negative developments, of emancipation and of oppression.

Standards imply control, which may result in organisational benefits but can cause employee resistance as well. We will now explore the feasibility of a Christian-philosophical approach to the control that is inherent to standards. Again, we take the case of ISO 9001 implementation and explore if such an alternative way of control may avoid resistance. This topic can be addressed using ideas by Dooyeweerd.

Dooyeweerd and the Freedom vs. Control Issue

The Christian philosopher Herman Dooyeweerd (1955; 1957; see for an introduction Basden, 2010 and Kalsbeek, 1975) addresses the issue of freedom versus control more indirectly (Basden, 2004). Dooyeweerd understands 'freedom' in another way than common in our culture. Freedom should be 'meaningful', otherwise it is no real freedom. For instance, we are free to design a car with square wheels but it makes little sense to do so. Meaningful freedom constitutes response to norms. Such norms apply to entities, such as organisations, and to aspects. We will focus on the diversity of aspects. Freedom is 'free to' rather than 'free from': free to do something meaningful, which our Creator granted us. For instance, if we had not been given the ability to communicate in common symbols, we would have had no freedom in that area (Basden, 2004).

Aspects Related to ISO 9001

We can distinguish several aspects of phenomena in reality. A table, for instance, may have four legs – a numerical aspect. It may have a certain economic value – an economic aspect. It may be beautifully designed – an aesthetical aspect. Dooyeweerd lists 15 aspects which are distinct ways in which reality can be meaningful and good, and can exist and occur. These are: the numerical, spatial, kinematic, physical, biotic (organic life), psychic (sensitive), analytical (logical), historical, lingual, social, economic, aesthetic, juridical, ethical and pistic (trust, faith) aspect. There are laws related to each aspect that should be honoured. For the first six aspects, these are natural laws which are generally obeyed without opposition. For instance, mathematical laws of adding, subtracting, multiplying, etc., apply to the numerical aspect. For the other nine aspects, these 'laws' are normative principles that man, in his freedom, can obey or violate, unintentionally (not being aware of certain normative principles) or intentionally. Much of the dissatisfaction with ISO 9001 quality systems may be explained by violating the 'human' side of quality management. This will not come as a surprise. Therefore, we examine two less evident aspects: the lingual aspect and the pistic aspect.

According to Dooyeweerd, the lingual aspect concerns symbolic signification. It is related to more than just language. For example, reserving the best parking place next to the company's main entrance for the general manager is a form of non-verbal communication that reveals something about the organisation's internal relationships. According to Dooyeweerd, each aspect has a normative component. Clarity is the normative component for the lingual aspect. In quality management practice, the lingual aspect is related to communication about establishing and using the system, and the system's description in the form of quality handbooks, procedures, and instructions. Violating the lingual aspect is a source of unnecessary dissatisfaction. Examples include:

- The organisation's management that does not clearly communicate their quality management policy or show their commitment. Perhaps there is a well-written and verbally expressed policy, but the absence of top management at meetings of the steering committee for quality, signals a lack of commitment.
- Quality Management systems that are designed by an 'expert' such as the organisation's quality management officer or an external consultant, instead of resulting from a joint effort of all employees. The latter has proven to be better (Brown et al.,1998; Mahadevappa and Kotreshwar, 2004). Involvement creates enthusiasm for quality management and lack of involvement is a source of dissatisfaction. 'Listening to employees' can be improved, for instance, by using interview techniques.
- The quality system's documentation that is often an impressive folio in the cupboard instead of a practical guide for the employees working on the shop floor. Sometimes, integration of a procedure into, for instance, software applications may replace a written instruction. In other cases, the process descriptions can be improved using clear text, or flow charts.
- One-way communication. Using feedback systems can eliminate dissatisfaction and improve customer service.

The pistic aspect is not related to a religion but to the possibility to believe. Reliability is the normative principle for this aspect. The organisation's mission statement is its 'confession' and its quality policy should be in line with this, otherwise both are untrustworthy. To be reliable, the organisation's actions should support its policy. In order to achieve this, quality management is needed. It is characterised as: 'say what you do and do what you say.' (Grimes, 2003, p. 11), sometimes with the addition 'and (be in a position to) prove it' (Dale, 2007, p. 293). Quality assurance should ensure that the organisation is able to promise only what it is able to fulfil, and, once promised, to deliver what was promised. Reliability can be honoured by a certification audit. However, is the certificate itself reliable? The above findings suggest that an ISO 9001 system can have little impact, no impact or a positive impact, and the main reason for this difference is the internal motivation. However, this implies that it is possible to achieve a certificate for a quality management system that has little or no added value. This questions the trustworthiness of the certification. Even ISO's own journal reports about lack of trust. An Italian survey revealed that 45% of the quality management systems of suppliers of an energy company did not sufficiently conform to the standard (Nicoloso, 2007). This problem does not only affect the energy company. Employees of the unreliable suppliers may feel uncomfortable with the window-dressing in their organisation and may prefer to have reasons to be proud of their employer.

The examples of these two aspects show that these normative 'laws' do not reflect an esoteric activity but common wisdom and issues addressed in professional and scientific literature. However, it is difficult and time consuming to examine all 15 aspects of an entire system. Therefore, the empirical part of our study will focus on one process only instead of studying the entire implementation of a quality management system. We will address the management decision to standardise one of the processes to enhance efficiency, and examine employee resistance, if any, to this standardisation.

RESEARCH APPROACH

During 2003 and 2004, we developed an approach based on the philosophy of Dooyeweerd to analyse a standardisation project in the company Voestalpine Polynorm. We offered solutions and recommendations. Six years later, we evaluated the results of this approach, and focused on how resistance to standardisation had developed during the years. We make a case reconstruction and then analyse the case using quality management literature and the ideas of Foucault, Habermas and Dooyeweerd (see Figure 1).

During 2003 and 2004, we applied Dooyeweerd's philosophy in the real-life case of a standardisation project in a company. Scientific research tends to focus on one or a few aspects, running the risk of remaining unaware of other aspects and not seeing reality as it is. Dooyeweerd emphasises the importance of what he calls 'naïve experience': an unprejudiced experience of the situation, similar with what Habermas (1987, p. 130) names the taken-for-grantedness typical for the 'lifeworld'. Therefore, we started our research with a 'lifeworld' approach which is also common in social-constructivist research. We evaluated the situation by being on the shop floor, talking informally to people involved in the process and conducting unstructured interviews. We further analysed the situation by examining the organisational embedding of the standardisation project, and the various aspects described by Dooyeweerd. For this purpose, we studied internal documents, observed behaviour and held interviews. This fieldwork, carried out in 2003, resulted in a report, which describes and analyses the situation and provides recommendations for the company (Haverkamp, 2004, summarised in Haverkamp, 2006). In 2009 and 2010, we conducted ten new interviews, mainly with people also interviewed in 2003, and studied new internal documents in order to see what had changed since then. As part of the interview protocol, we developed a measurement protocol for identifying, selecting, and accessing sources of evidence, and for generating a valid and reliable score for each of the variables, which were measured through the interviews. We measured the following concepts: efficiency of work in terms of time, work standardisation in terms of implemented standard working methods, and satisfaction. In particular, we addressed the concept of satisfaction related to the standardisation initiatives. The degree of satisfaction was defined as the extent to which workers perceived their daily work related to the project as satisfying, and refers to a value attributed to work by the workers. It is a characteristic of people's opinions, and the value of this variable ranges from 'not satisfying at all' to 'very satisfying'. The interview evidence was recorded and transcribed, and we analysed and evaluated the contents of these transcriptions. We were able to graphically visualise the (relative) degree of resistance to the standardisation activities in time (see Figure 4). The interviewees reviewed the case reconstruction and visualisation of resistance in time and agreed with our description and analysis.

Figure 1. Research approach

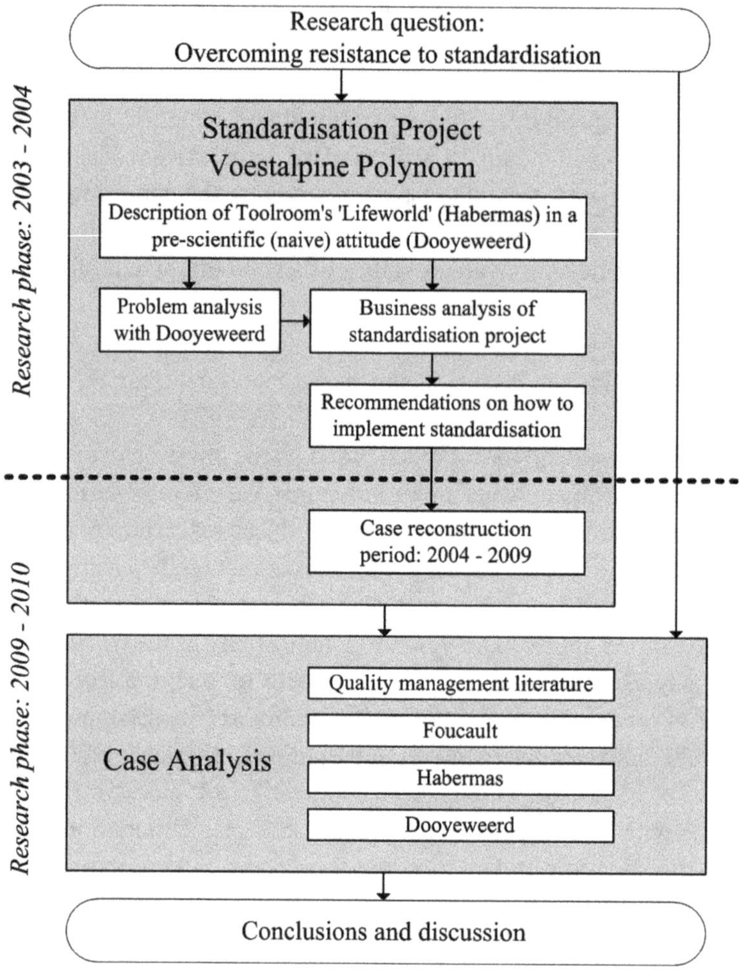

Below, we first describe the company, the department, the problem and the reasons for standardisation. Then, we give a chronological description including the extent to which our recommendations were implemented, and we relate resistance to the phases of the project. Next, we analyse the case and discuss whether Dooyeweerd's approach was feasible to tackle the standardisation problem.

STANDARDISATION PROJECT VOESTALPINE POLYNORM

Case Company and Department

The case company Voestalpine Polynorm Automotive in Bunschoten, the Netherlands, develops and produces automobile body parts, such as doors, bonnets and roofs for the automotive industry. The company is a preferred supplier of major automobile manufacturers in Europe. The production requires specific tools that determine the shape of the parts. These tools are made, maintained and, if necessary, repaired or adapted by the company's Toolroom. Professional technical skills are a prerequisite for accurate 'formability', the degree in which a sheet-metal is formed in the required shape.

The parts are made by a process called deep drawing, in which a punch forces a flat sheet-metal into a die cavity. A sheet-metal blank is placed over a die opening, and is held in place with a blank holder. The punch travels downward and forces the blank into the die cavity. The important variables in deep drawing are the properties of the sheet metal, the ratio of blank diameter to punch diameter, the clearance between punch and die, the punch radius and die-corner radius, the blank holder force, friction, and lubrication (Kalpakjian, 1995, p. 472). Especially deep drawing of non-symmetric parts can present significant difficulties in practice (Kalpakjian, 1995, p. 476). For example, various regions of a part undergo different types of deformation during drawing. In order to avoid sheet metal during forming, it is important to consider and control all relevant factors. In our research, we examined the Toolroom's activities regarding the control of these relevant factors of formability in the deep drawing process.

Voestalpine Polynorm has a certified quality management system based on the international standard ISO/TS 16949, which is an extension of ISO 9001 with additional requirements for the automotive supply chain. One of the additional prescriptive requirements of this standard is that management shall review the product realisation processes and the support processes to assure their effectiveness and efficiency (ISO, 2009a, p. 7).

The Problem

The Toolroom had already successfully implemented a standardised process for formability, as part of the quality management system. As a result, the lead times in getting tools ready for production had decreased and scrap percentages had reduced significantly. However, the process description (procedure) allowed different working methods and, indeed, the teams differed in the method they used. Management thought that they could achieve more efficiency by standardising working methods. They would modify the standard as soon as further improvements to the working methods were found. It would reflect 'best practices', see Figure 2.

Interestingly, this project can be seen as a Standardisation project in two ways. It is part of the company implementation of an international standard, but it also leads to the development and implementation of a new in-company standard. A researcher (first author of this chapter) was asked to investigate how the working methods in the Toolroom could be standardised.

Figure 2. Standardisation project in Toolroom of Voestalpine Polynorm

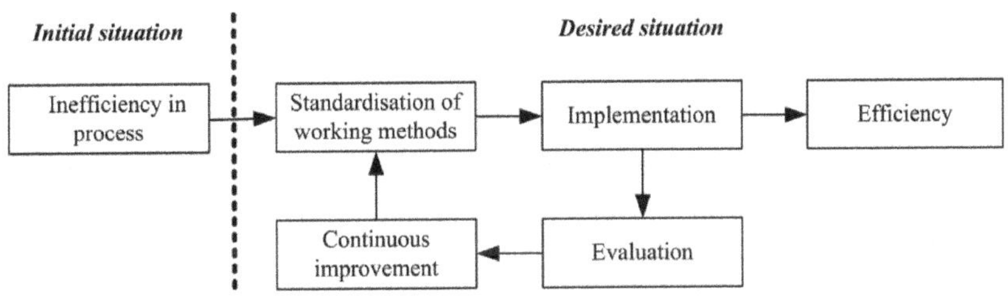

First Impression: Resistance to Intended Standardisation Project

The researcher joined the Toolroom for 10 months and was accepted by the workers immediately, partly because of his technical background. The first impressions and rather informal interviews with 12 employees (Toolroom manager, shift leader, team leaders (4), toolmakers (3), welder, project leader and formability specialist), revealed that the older employees expressed more resistance to the standardisation initiative than the younger generation. The following quote of a very experienced toolmaker is exemplary for the older generations 'Craftsmen are not clonable' (Haverkamp, 2004, p. 26). The younger generation, both toolmakers and team leaders, were more positive about standardisation. 'Our methods can and need to be standard; we could benefit considerably. However, it is important that everything remains understandable for everybody. Everything needs to be explainable.' (Haverkamp, 2004, p. 27).

Aspectual Analysis

Next, the researcher analysed the organisational embedding of the process and the aspects more systematically. The aspectual analysis was mainly based on many interviews on the shop floor, logbooks, internal documents, and observations (during meetings and on the shop floor). The functioning of the Toolroom was analysed using each of Dooyeweerd's 15 aspects. For example, the process of tool reparation relates to the numerical aspect in the form of several different working methods. The tools are located in the Toolroom (spatial aspect) and consist of a punch which travels down (kinematic aspect) to transform a blank (physical aspect) into a die cavity, in accordance with ergonomic standards (biotic aspect). However, standardisation of this process may lead to resistance (psychic aspect) and requires an investigation of practices and the choice of a best practice (analytical aspect). The current standard was developed and implemented in the past (historical aspect). However, it was expected that a written (lingual aspect) standard for working methods, implemented in an acceptable way (social aspect), would lead to more efficiency (economic aspect). The description has a certain format (aesthetical aspect). It should meet customer specifications (juridical aspect). The Toolroom's mission (pistic aspect) is to serve customers (ethical aspect) in a flexible way. Next, the researcher determined to which extent the functioning was in line with the specific (normative) 'laws' for each aspect.

A similar analysis was made for the change process and for the desired (standardised) situation. For each aspect, the researcher predicted the positive and negative consequences of process standardisation by investigating all possible changes in functioning in relation to the corresponding 'laws' of all aspects. For example, standardising existing working methods (numerical aspect) might lead to a more efficient way of working, which is in line with the norm of 'efficiency' in the economic aspect.

The aspectual analysis of the initial situation, change process and desired situation resulted in an overview of relevant aspects related to the research question. Although all aspects are relevant, some, for example, the psychic aspect were found to be more prominent, whereas others, such as the aesthetic aspect, were less apparent. The applicable normative requirements for each of these aspects should be found and honoured. From Dooyeweerd's point of view, this is a necessary condition for the success of the project. For example, the normative component of the analytical aspect is the ability to distinguish one thing from another and to abstract something from the concrete reality. Applied to the Toolroom, knowledge is required to be able to analyse a tool. Insufficient knowledge could (unconsciously) lead to a wrong analysis, which is an example of not functioning in line with the norms for the analytical aspect. Standardising a 'best practice' assumes that everybody has the same skills and knowledge to be able to 'analyse' according to this 'best practice'. The empirical data showed that this was not the case, and that more attention should focus on the analytical aspect. In a similar way, the researcher determined whether special attention was required for each aspect during the standardisation process. The following aspects needed special attention:

- Psychic aspect, in which the *motivation* of employees is crucial.
- Social aspect, in which the condition of true *respect* for each other is important.
- Lingual aspect, in which verbal and non-verbal *communication* is important.
- Pistic aspect, in which *trust and commitment* are important.
- Analytical aspect, in which homogeneous *knowledge* and *skills* are required.
- Historical aspect, in which *events and managerial behaviour in the past* partly explain current resistance to standardisation, and therefore need to be addressed.

Following Dooyeweerd, it is especially important to consider whether everything is in LINE with the normative requirements for these aspects in order to avoid resistance to the standardisation project.

Pragmatic business administration insights can help businesses to function in compliance with the normative requirements because these reflect experiences in what 'works' and what does not. The 'norms' in the psychic, social, lingual and pistic aspects can be understood and managed by insights about motivation. The 'norms' in the analytical aspect can be understood and managed by applying knowledge management concepts. Normative requirements in the historical aspect can be honoured by notions of willingness to change, see Figure 3.

Business Analysis of the Standardisation Project

We investigated the relation between the numerical (number of methods) and the economic aspect (efficiency) with insights related to standardisation. Notions such as 'standardisation', 'standards' and 'continuous improvement', as described in the problem field, were applied to the Toolroom. We used the ideas of Nakamura (2000), who sees continuous improvement as one of the key concepts of in-company standardisation. We applied a model of Slob and De Vries (2001) to understand how such standards can be developed. Furthermore, we determined how these standards can be developed and maintained, based on Simons and De Vries (2002), Kondo (2000) and Imai (1986).

In order to function in line with the norms in the analytical aspect in the standardised situation, knowledge management is required; employees first need to learn from each other. Different types of knowledge apply: implicit and explicit (Nonaka, 1994), and subject, method, social and acting knowledge (Sprenger & Ten Have, 1996). We used Buckler's model for learning to understand learning as a continuous process – a prerequisite for the Toolroom. Buckler (1996) sees learning as a process which requires three ingredients: focus, a stimulating environment and the right techniques. Because Buckler does not distinguish the different types of knowledge that need to be transferred, we used the notions of knowledge conversion (Nonaka, 1994) to get insight into the 'how' of learning. Nonaka distinguishes four types of knowledge conversion: socialising (from tacit knowledge to tacit knowledge), internalising (from explicit to implicit), externalising (from implicit to explicit) and combining (from explicit to explicit). Mom et al. (2002) conclude that there are three phases in knowledge conversion processes:

Figure 3. Problem analysis based on Dooyeweerd's philosophy

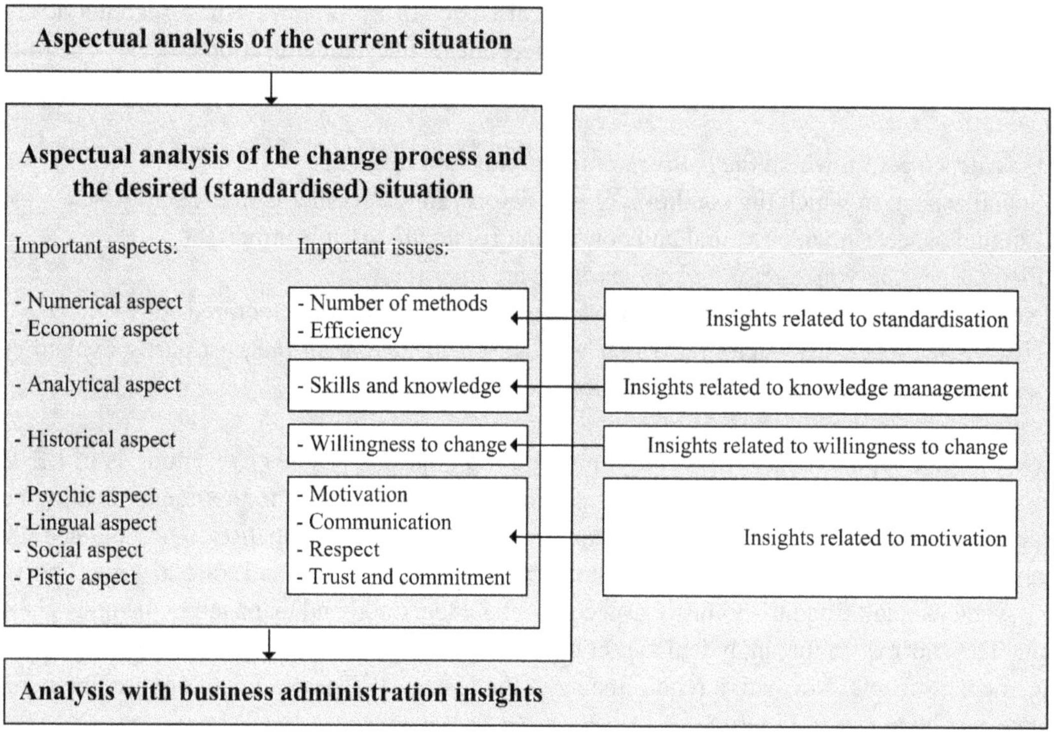

decision, transfer, and integration. Also according to Buckler (1996), learning requires a transition through different phases or stadiums (ignorance, awareness, understanding, commitment, enactment and reflection). We applied these insights in our recommendations for the Toolroom.

The psychic aspect relates to understanding and managing the feelings of employees during the standardisation process. In order to function in line with the psychic norms, people should remain (or become) motivated. Motivation originates in the introduction and unfolding of human aspects in daily work, such as creativity (the pleasure of thinking) and sociality (the pleasure of sharing things with colleagues) (Kondo, 1996). We applied this vision to the Toolroom.

The historical aspect relates to willingness to change. This willingness stems from employee motivation, whereas resistance to change originates from incomprehension regarding what, why and how management wants to do something (Wissema et al., 1991). Thus, in combination with Kondo's findings, the success of standardisation in the Toolroom depends on the willingness to change, which in turn depends on the amount of attention on the human aspects in this process.

This conclusion is in line with Dooyeweerd – all norms in the 'human Aspects' need to be honoured to function successfully. In addition to the findings of Wissema et al. (1991) and Kondo (1996), Dooyeweerd's philosophy helped us to describe exactly *which* human aspects needed special attention.

Recommendations for Implementation

We concluded that the Toolroom could benefit from standardising their working methods, but that the organisation was not fully prepared for the proposed standardisation project. First, the above-mentioned insights (described in detail in the report (Haverkamp, 2004, pp. 91-134)) were needed to help the Toolroom to function in line with the norms of the indicated aspects. Therefore, the researcher made the following recommendations to Toolroom's management:

- Establish a common goal and vision. The purpose of the standardisation needs to be crystal clear and should be communicated regularly (lingual aspect). This is important for an effective learning process (analytical aspect), motivation (psychic aspect) and the willingness to change (historical aspect).
- Invest in mitigating individual differences in relevant knowledge and skills (analytical aspect), because this allows all employees to meet agreed-upon standards.
- Strengthen social knowledge through which employees can learn from each other (Sprenger & Ten Have, 1996). Social knowledge gives insight in the communicative processes and is required to support and coach one another (analytical aspect). It gives access to the knowledge others obtain. Awareness and understanding of the need to learn are fundamental for this (Buckler, 1996). A decentralised and informal structure stimulates the development of social knowledge.
- Stimulate employees to express (verbally or non-verbally) their ideas, concerns or feelings about the standardisation project and give them feedback (lingual aspect). This increases understanding (analytical aspect) and, as a result, commitment (pistic aspect) and action.
- Pay attention to commitment ("Yes, we can") (psychic aspect).
- Give employees time to get used to the idea that change is needed (historical aspect).
- Be flexible – employees need freedom to influence changes in their work and want to be taken seriously, so adjustments during the process should be possible. This increases involvement (social aspect) and commitment (pistic aspect).
- Adopt a decentralised approach with a clear project structure (historical aspect).

Case Reconstruction 2004-2009

During the last quarter of 2009 and the beginning of 2010, we investigated if and how our recommendations had been implemented, and whether there had been any resistance among staff. We interviewed the Toolroom manager who described the changes and activities since 2004. The manager started the change project just after the researcher had left the company. The recommendations were implemented implicitly rather than explicitly and systematically. The Toolroom manager hired a consultant who applied a human-centred approach, founded in social constructivism. She did not receive the researcher's report with analyses and recommendations, and started from scratch and interviewed people again to gather data. She focused on knowledge management rather than on standardisation. However, part of this knowledge management was to make implicit knowledge explicit (Nonaka and Takeuchi, 1995) with the intention to use this explicit knowledge repeatedly, which is standardisation as well (De Vries & Van Delden, 2006). Important elements of her approach were: how to learn from each other, how to make implicit knowledge explicit, how to write down practices, and how to standardise and implement a best practice.

The start-up phase of the project involved examining the working methods in the Toolroom. The consultant arranged workshops and investigated how people perceived their job, which other departments needed to be involved and what was important to the employees. Central questions were 'who' 'what', 'where', 'why' and 'how'. Other questions included 'what are we good at?', 'what do you think is important?', 'when is standardisation successful?' and 'how can you contribute to the project?' The employees expressed their expectations and points of views, after which a common view was created by and for the Toolroom.

Next, an abstract level content of the standard was defined through mind mapping techniques. However, the project became much more extensive and complex than initially foreseen due to the company's ambition to further increase performance and lower cost because of fierce competition in the automotive market. Instead of a standardisation project, it became a project to transform the Toolroom in a streamlined and mature organisation, in which working, learning and innovating were to be key elements. The more complex project continued with an emphasis on knowledge management. Knowledge 'pilot production teams' were formed and 'experts' per specific area were nominated to produce teaching content to educate their colleagues. The teaching materials could be seen as standards. However, it was not successful because even the motivated leaders of the pilot teams did not understand the structure and complexity of the project. The rules and guidelines of the project were too complex and vague and, in their perception, unnecessary and therefore ineffective. The external consultant failed to clearly explain the objectives of the project, partly because the manager wanted to restrict her contact with Toolroom employees. As a result, the employees lost confidence and enthusiasm and no longer supported the project. Shortly afterwards, the external consultant was replaced by an internal project leader. This new project leader, a professional and experienced teacher, had other (didactic) ideas regarding the project. He discontinued the trial with the pilot production teams and tried to implement his own views and insights. However, Toolroom management did not support his initiatives, and consequently, the project made very little progress.

However, Toolroom employees acknowledged the need for more efficiency because of increased international competition. Together with one of production teams, the new project leader started bottom-up activities in developing the standard in a way he thought was right. According to the production team, the manager was not aware of this and they pretended to act in line with his original ideas in case he suddenly attended one of their meetings. This 'dramaturgical compliance' (Van Kemenade et al., 2011) continued until the manager left in April 2009. From then on, Toolroom employees openly continued their standardisation activity. They decided to keep it simple and defined three levels of craftsmanship. They used pictures rather than text in their descriptions. Training was provided group by group instead of the whole department at once. As a result, the lead time for writing these knowledge standards decreased considerably.

Resistance during the Standardisation Project

In this chapter, we analysed resistance to standards by applying the philosophical ideas of Foucault, Habermas and Dooyeweerd. In the empirical part of the chapter, we focused on the standardisation process at the case company Voestalpine Polynorm. We used insights from our philosophical analysis in our recommendations to the company. Six years later, we evaluated the results of this intervention by analysing internal reports related to the project and by conducting six in-depth interviews. These interviews had an 'open' character and were aimed at retrieving real feelings and opinions from the respondents. Based on a systematic questionnaire, we measured whether (1) the initial philosophical diagnosis of the problem field and the practical conclusions and recommendations were effective and adequate and (2) whether the Dooyeweerd approach of implementing company standards was effective in overcoming resistance to standardisation. The amount of resistance was determined by measuring the degree of satisfaction for each stage in the standardisation process. The objects of measurement were the evaluation of the project as formulated in the project management reports and the judgements, opinions and recollections of individual persons (face-to face in-depth interviews). Several ways of recording the interview evidence were applied. We used voice recording and made notes of verbal and non-verbal communication. The interview evidence (comments, experiences and suggestions) was recorded and transcribed (reviewed and approved by the respondents). The content of these transcriptions and the project reports were analysed, evaluated and used for a reconstruction. Based on the evaluation and reconstruction, we could visualise the relative amount of resistance during each project phase (Figure 4). This tentative visualisation was approved and confirmed by the respondents.

1. Initial resistance to the Toolroom management's plan to standardise.
2. Resistance decreases during the initial project phase.
3. Resistance increases during the project.
4. Resistance peaks during the phase of implementation of the standard.
5. Resistance decreases in the final stage of the project.

We now explain these five changes in resistance.

1. In 2003, particularly older employees were unhappy with the management decision to standardise. The subsequent involvement of the researcher reduced this resistance slightly.
2. The project that started in 2004 had a multi-sided approach. Implicitly, most of the researcher's recommendations were applied. A common goal and vision were established in the workshops, the purpose was clearly communicated and knowledge management became an (important) element

Figure 4. Resistance to standardisation 2003-2009

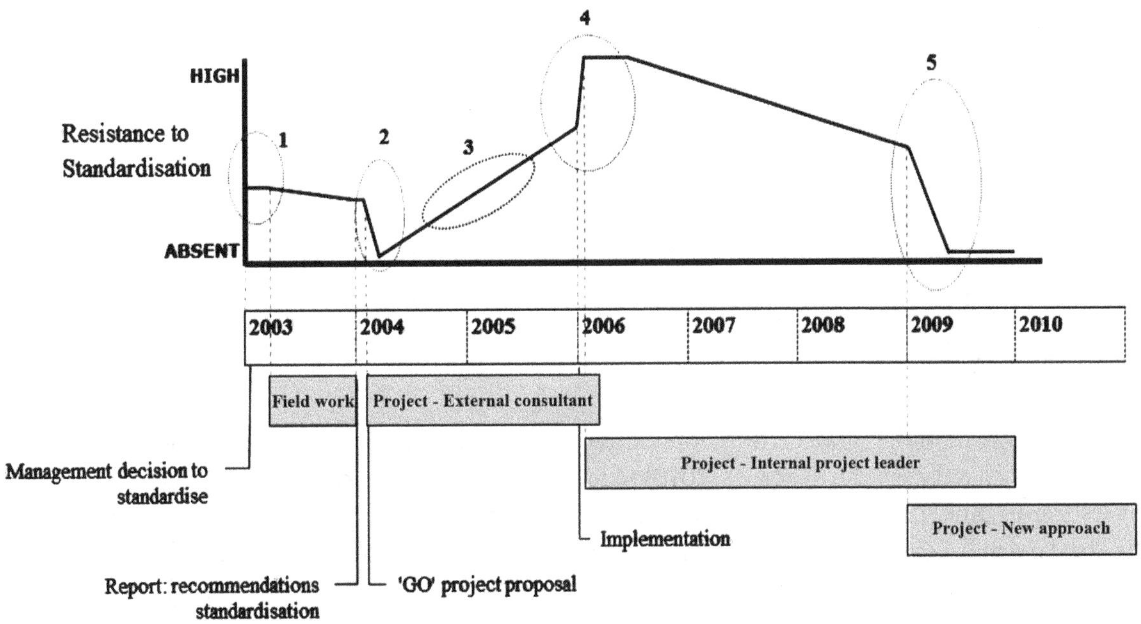

of the project. Everyone understood the need to change, the project was managed on commitment, and a gradual implementation strategy was developed in which the role of the employees seemed to be important. During this start-up phase, employees were enthusiastic, satisfied and, as a consequence, cooperative.

3. However, the last recommendation was not implemented. The project was not decentralised or clearly structured. The project team changed its bottom-up approach into a top-down approach in which several of the other recommendations (unintentionally) were ignored. It developed a project structure which was only clear to the team itself. The common goal and vision became unclear and the standard became too complex. The possibilities for employees to propose or make adjustments seemed very limited, and feedback and communication were lacking. As a consequence, key employees lost confidence and enthusiasm, their commitment decreased, and resistance to the Standardisation initiative grew.
4. Resistance reached a peak during the implementation phase. Employees no longer supported the project and the standard was not implemented.
5. However, team leaders were convinced that the Toolroom could benefit from standardisation. After a while (in Figure 4: between 4 and 5), a small team started to convert implicit practical knowledge into explicit knowledge in the form of a standard. By providing employees with standard packages of knowledge, it was assumed that the tool maintenance process could be made more efficient. The implementation of this new standard was successful. The standard was easy to understand and pictures rather than text were used. Communicating about the relevance of the standard and verifying whether it was understood became important elements. Moreover, the standard was now seen as a new way to approach a certain problem, rather than a prescribed way to control the behaviour of the toolmakers. As a result of standardising knowledge and skills, toolmakers automatically adopted the most efficient working method for similar problems. They were free to adopt other methods and were not forced to use the standard. As a result, in their perception the standard was not a limitation of their freedom, but a necessary tool to become mature craftsmen. The initial goal was achieved: efficiency in terms of quicker repair times and shorter lead times to get tools ready for series production.

CASE ANALYSIS

In this section, we analyse the case and in particular the five events that mark changes in resistance by using standardisation and quality management literature and by applying the insights of the philosophers Foucault, Habermas and Dooyeweerd.

Standardisation and Quality Management Literature

As mentioned, most literature on the adoption and use of standards in a company ignores possible staff resistance to standards. Van Wessel's (2010) main message is that proper governance and management of standardisation projects in the phases of standard selection, implementation and use lead to success. The best practice for company standardisation provided by De Vries (2006) addresses policies, processes, facility management and funding, but does not include individual employees. In this sense, some studies related to management system standards (Boiral, 2003; Manders, 2015) are an exception. They notice differences between employees, whereas in our case, all employees share a certain amount of resistance, only age seems to make a small difference. Boiral (2003) and Manders (2015) interviewed employees, and show patterns in employee perceptions but do not provide an in-depth diagnosis of their resistance, and as a consequence, they cannot provide fundamental solutions. This in-depth diagnosis might be a research area for psychologists (Brehm & Brehm, 1981) but then the relation to standards management is missing.

In our case company, Toolroom management was highly motivated to start the standardisation project for economic reasons: they were convinced it could improve their competitiveness. Such motivation is a good starting point for achieving results. However, their commitment did not avoid resistance. Resistance grew, in particular, due to lack of communication. In line with Brown et al. (1998) and Mahadevappa and Kotreshwar (2004), the initial enthusiasm and satisfaction disappeared because the normative requirements of the 'human' aspects, especially the lingual aspect, were violated.

Van Kemenade et al.'s (2011) findings are helpful in sharpening our diagnosis of resistance. They developed criteria for professionals' willingness to contribute to quality assurance. (1) A main reason for the initial resistance to standardisation was the limitation of the toolmakers' freedom. However, as we have seen in the aspectual analysis, there were more reasons for this initial resistance. It slowly decreased due to better communication. The absence of resistance in the start-up and final phase of the project (2 and 5) can be understood using Van Kemenade et al.'s prerequisites as well. The criteria were met both times, but in different ways. The growing resistance after the start-up phase (3 and 4) was because two of Van Kemenade et al.'s criteria were not satisfied: the rules and templates were too complex and needed to be followed strictly. This rigidity caused control, rather than freedom as the dominant factor during implementation. Moreover, there was a decrease in ownership by Toolroom professionals since the work was done by a small development team directed by an external expert. Since management was part of this development team, support for management decreased as well. Although Van Kemenade et al.'s prerequisites were met in phases 2 and 5, this was not sufficient to avoid resistance to the standard. An explanation is that Van Kemenade et al.'s prerequisites do not cover all of Dooyeweerd's (human) aspects. For example, Van Kemenade et al. do not mention the lingual aspect, which proved to be important in this case.

Foucault

Foucault's findings on power systems can be used to analyse quality management systems such as the case company's ISO/TS 16949 system (Van Veldhuisen, 1996) but for the employees in the Toolroom, it is not the system as such which is of interest but its impact on their department. Our case description shows that in line with the ISO/TS 16949 requirements, the Toolroom management intended to standardise processes on the shop floor. Employees were expected to behave according to predefined rules and guidelines. The plan to implement this (1) caused resistance. The more concrete the plan became (3 and 4), the stronger the resistance. The attempt to force production teams to use the standard (4) resulted in the strongest resistance. Foucault helps us to understand this resistance. Interestingly, the revolt against the standard is also in line with Foucault. The employees revolted against the proposed way of working and, finally, escaped from the 'system' by no longer supporting it. However, their voluntary decision to develop another standard (5) and the lack of resistance in phase 2 are not in line with Foucault.

Habermas

Resistance to this standardisation project can also be understood with Habermas' insights. (1) The initial resistance in 2003 decreased due to communicative action. (2) The absence of resistance in the start-up phase can partly be understood by this communicative action as well. However, as mentioned earlier, this successful project start was not only dependent on honouring the normative requirements regarding communication. (3) Next, besides the Toolroom's 'lifeworld', a 'system' was being developed which the workers perceived as questionable, it was not taken for granted in the attitude of common sense. (4) This system had to be implemented into the lifeworld of the Toolroom which resulted in strong resistance because it was seen as too complex and too theoretical. Apart from the lingual aspect, Habermas does not mention any of Dooyeweerd's other human aspects and these, as we have seen, in this case, the economic, analytical, historical, psychic, social and pistic aspects are important for success as well. On the other hand, it seems that violating the norms of only the lingual aspect, the one Habermas mentions, could be enough for the project to fail.

Dooyeweerd

According to Dooyeweerd, all of the sides, or aspects, of the project are important and all corresponding normative requirements should be honoured. Our problem analysis using Dooyeweerd's philosophy showed which aspects required additional attention. This philosophical analysis was fundamental for the selected combination of business-administrational approaches and, as a result, the recommendations for standardisation. As we have seen in the systematic inventory of the Toolroom, not all the normative requirements of the (human) aspects were honoured, which explains the initial resistance to the standardisation plan (1). In the start-up phase (2), most of our recommendations were implicitly implemented. Attention focused on the 'human' side of standardisation. As expected, there was no resistance to the proposal and even the older employees were cooperative. However, soon these normative requirements were violated. Due to inadequate communication (lingual aspect) and ignoring the comments of the employees (social aspect), the motivation (psychic aspect) and trust (pistic aspect) in standardisation eroded. Also, commitment (pistic aspect) and the willingness to change to the unintelligible standard (analytical aspect) decreased dramatically (3 and 4). In the new project approach (5), all recommendations were implicitly implemented again. Instead of resistance, a form of standardisation was achieved in which the tension between control and freedom seemed to have been eliminated.

Common Case Analysis

Foucault and Habermas help us to understand staff resistance to the standardisation project. However, they fail to explain the initial lack of resistance and do not offer an acceptable solution to overcome resistance to the project. Habermas' call for communicative action and Foucault's advice to escape do not solve the conflict between control (of the development team) and freedom (of the production team).

The analysis based on Van Kemenade et al. and Dooyeweerd considers the economic necessity for standardisation which is ignored by Foucault and Habermas. According to Dooyeweerd, the economic aspect is the leading aspect for a company. 'Frugality in managing scarce goods' is fundamental for this aspect (Kalsbeek, 1975, p. 100, based on Dooyeweerd, 1955, p. 66). By planning and executing a standardisation project, Toolroom management acknowledged this. Toolroom employees also seemed to recognise the leading role of this aspect. Toolmakers perceived the quality management system as a 'necessary evil' which ensured that customers received the right products. Moreover, without ISO/TS 16949 certificate, Voestalpine Polynorm would be out of business. This explains both the absence of resistance and the strongest resistance in Figure 4 (2, 4 and 5). The strongest resistance occurred in phase 4 when Toolroom employees perceived the standard to be unnecessarily complex, ineffective, and therefore not in line with 'frugality in managing scarce goods'. The economic aspect also shows an additional explanation for the increasing resistance in phase 3. The company hired an external consultant who was unsuccessful in implementing the project. The employees felt they could have achieved more for the company without these extra costs.

We can conclude that the philosophy of Dooyeweerd provides a better explanation for (1) staff resistance to Toolroom management's plan to standardise, (2) the decreasing resistance when the recommendations based on Dooyeweerd's philosophy were implicitly implemented, (3 and 4) the increasing resistance when the 'laws' related to the aspects were increasingly violated and (5) the decreasing resistance in the final stage.

CONCLUSION AND DISCUSSION

Standards can generate benefits but may cause resistance as well because they may impact human behaviour directly. This applies in particular to management system standards. The implementation of such standards in an organisation leads to process standardisation in the form of written procedures and work instructions. Resistance to this form of standardisation can be partly explained using standardisation and quality management literature. In particular, the prerequisites as formulated by Van Kemenade et al. (2011) appear to be useful.

The paradox of the benefits of standards and the resistance to them may be related to the fundamental tension between the poles of 'order' and 'freedom' in a social organisation. To get a deeper understanding of this issue, we applied some concepts of philosophers that study man in relation to power systems. We found that Foucault's analysis of systems and Habermas' notions of communicative action give us a deeper understanding of resistance, but their approaches are not of help in understanding the benefits of standards. Moreover, neither Foucault nor Habermas provides a solution for the tension between freedom and control, two essential ideals of Humanism. Exercising individual freedom includes controlling one's environment, and this may involve the creation of systems of control which may affect the freedom of other people. We explored how this problem can be solved by following Erasmus, one of the founding fathers of Humanism, in linking the notion of freedom to Christianity. In Dooyeweerd's Christian-philosophical approach, freedom is not absolute, but it should be 'meaningful', otherwise it is not real. Meaningful freedom constitutes response to norms. These norms can be related to (organisational and other) structures and to the 15 aspects in which 'things' (e.g., processes) function.

In this study, we focused on these aspects and applied them in a company standardisation project. We found that the economic goals of standardisation can be achieved without resistance from employees. Our analysis shows that social needs are not necessarily detrimental to economic goals. On the contrary, they can support these and have a beneficial influence. However, too much focus on human aspects can undermine the leading economic aspect of an organisation. By addressing all relevant aspects and associated norms, Dooyeweerd's philosophy has helped us to determine the interrelated set of causes for resistance to standardisation, and to develop an approach to overcome this resistance. We found that

Dooyeweerd's theory of modal aspects was especially useful for analysing the problem, for systematically selecting relevant business theories for the project, and for giving a holistic analysis as a basis for recommendations. This was done by linking insights from the field of business administration to those aspects that were most prominent in the case. These aspects relate to the original situation, the intended new situation, and the transition from the existing to the desired situation. Our findings suggest that such an approach can be applied in other standardisation cases as well, but we expect that modifications will be required due to situational differences. Moreover, it is a holistic approach and not just a management tool, and rigid application would create rather than solve problems because it would mean forcing employees to accept a system. Therefore, the solution is not to prescribe but rather to familiarise managers with the approach. Then they will learn that business issues can be seen from different angles and that by doing so rigid solutions that may harm rather than benefit the organisation can be avoided. By becoming sensitive to the multitude of aspects and getting a feeling which are the most relevant ones, they may also look for – case-specific – managerial and quality management approaches and tools to support them in finding sustainable solutions. It illustrates that standardisation is a multidisciplinary field of research and that contributions from many disciplines are needed for a solid understanding, and that a business approach fed by philosophy may help to interrelate these contributions (De Vries, 2015).

Our approach differs from 'best practice' approaches for in-company standardisation (De Vries, 2006; Van Wessel, 2010) or for ISO 9001 implementation (Briscoe et al., 2005; Lin & Jang, 2008). First, this is a single case study and replication studies might show whether certain patterns apply that distinguish good from worse practices. However, assuming that indeed a best practice model can be developed, then this model as such might serve as a system which would be implemented in other organisation. This as such may cause resistance (Foucault) and affect the 'life world' (Habermas). Following Dooyeweerd, a best practice may be of help because it provides insights into what helps in similar situations, and the more aspects covered in this best practice the more informative the model, but each situation is unique as well and may deserve other emphasis, also depending on the people involved. So our approach would help researchers as well as managers to not forget aspects that might be important but we think it does not make sense to see the 'model' developed for our case company as a general model to be applied in other companies as well.

We encourage researchers to replicate our study. A longitudinal multiple case study would be best, but requires a substantial amount of time and research capacity. Action-research in which our approach is applied to one or more cases would be a feasible alternative. It seems that our findings reach beyond the area of in-company standardisation and can be beneficial to other areas as well. Our approach may be used to combine insights from mono-disciplinary studies of organisations into a balanced multidisciplinary approach which is relevant for both research and practice.

REFERENCES

Adolphi, H. (1997). *Strategische Konzepte zur Organisation der betrieblichen Standardisierung. In DIN Normungskunde* (Vol. 38). Berlin: Beuth Verlag.

AFNOR. (1967). *La normalisation dans l'entreprise*. Paris: AFNOR.

Ajzen, I. (1991). The theory of planned behavior. *Organizational Behavior and Human Decision Processes*, *50*(2), 179–211. doi:10.1016/0749-5978(91)90020-T

Allen, P. S. (Ed.). (1913). *Opus Epistolarum Des Erasmi Roterodami* (vol. 2). Oxford, UK: Oxford University Press.

Arthur, W. B. (1989). Competing Technologies, Increasing Returns, and Lock-in by Historical Events. *The Economic Journal, 394*(99), 116–131. doi:10.2307/2234208

Basden, A. (2004). Dooyeweerd and Freedom. In *The Dooyeweerd Pages*. Retrieved January 7, 2010, from http://www.dooy.salford.ac.uk/freedom.html

Basden, A. (2010). *The Dooyeweerd Pages*. Retrieved January 7, 2010, from http://www.dooy.salford.ac.uk/index.html

Becker, M. C. (Ed.). (2008). *Handbook of organisational routines*. Cheltenham, UK: Edward Elgar. doi:10.4337/9781848442702

Benner, M. J., & Veloso, F. M. (2008). ISO 9000 practices and financial performance: A technology coherence perspective. *Journal of Operations Management, 26*(5), 611–629. doi:10.1016/j.jom.2007.10.005

Bentham, J. (1838). Plan of the Panopticon. In J. Bowring (Ed.), *The Works of Jeremy Bentham* (vol. 4, pp. 172–173). Edinburgh, UK: William Tait.

Bevan, R. (2011). *Changemaking: Tactics and Resources for Managing Organizational Change*. Seattle, WA: ChangeStart Press.

Boh, W. F., & Yellin, D. (2007). Using Enterprise Architecture Standards in Managing Information Technology. *Journal of Management Information Systems, 23*(3), 163–207. doi:10.2753/MIS0742-1222230307

Boiral, O. (2003). ISO 9000: Outside the iron case. *Management Science, 14*(6), 720–737.

Boiral, O., & Roy, M. (2007). ISO 9000: Integration rationales and organisational impacts. *International Journal of Operations & Production Management, 27*(2), 226–247. doi:10.1108/01443570710720630

Brehm, S., & Brehm, J. W. (1981). *Psychological Reactance – a Theory of Freedom and Control*. New York: Academic Press.

Briscoe, J., Fawcett, S., & Tood, R. (2005). The Implementation and Impact of ISO 9000 among Small Manufacturing Enterprises. *Journal of Small Business Management, 43*(3), 309–330. doi:10.1111/j.1540-627X.2005.00139.x

Brown, A., Van Der Wiele, T., & Loughton, K. (1998). 'Smaller enterprises' experiences with ISO 9000. *International Journal of Quality & Reliability Management, 15*(3), 273–285. doi:10.1108/02656719810198935

Brunsson, N. (2000). The Pros and Cons of Standardisation – An Epilogue. In *A World of Standards* (pp. 169–173). Oxford, UK: Oxford University Press.

Brunsson, N., & Jacobsson, B. (2000). Following Standards. In *A World of Standards* (pp. 125–137). Oxford, UK: Oxford University Press.

Buckler, B. (1996). A learning process mode to achieve continuous improvement and innovation. *The Learning Organization, 3*(3), 31–39. doi:10.1108/09696479610119660

Carnall, C., & Todnem, R. (2007). *Managing Change in Organizations* (6th ed.). Harlow, UK: Pearson Education.

Casadesús, M., & Karapetrovic, S. (2005). Has ISO 9000 lost some of its lustre? *International Journal of Operations & Production Management*, *25*(6), 580–596. doi:10.1108/01443570510599737

Connor, P. E. (1997). Total Quality Management: A selective commentary on its human dimensions, with special reference to its downside. *Public Administration Review*, *57*(6), 501–509. doi:10.2307/976961

Corbett, C. J., Montes-Sancho, M. J., & Kitsch, D. A. (2005). The Financial impact of ISO 9000 Certification in the United States: An Empirical Analysis. *Management Science*, *51*(7), 1046–1059. doi:10.1287/mnsc.1040.0358

Dale, B. G. (2007). Quality Management Systems. In B. G. Dale, T. Van der Wiele, & J. Van Iwaarden (Eds.), *Managing Quality* (pp. 279–309). Malden, MA: Blackwell Publishing.

David, P. A. (1995). Standardisation policies for network technologies: the flux between freedom and order revisited. In Standards, Innovation and Competitiveness – The Politics and Economics of Standards in Natural and Technical Environments (pp. 15-35). Aldershot, UK: Edward Elgar.

Davis, F. D. (1989). Perceived usefulness, perceive ideas of use, and user acceptance of information technology. *Management Information Systems Quarterly*, *13*(3), 319–340. doi:10.2307/249008

Davis, F. D., Bagozzi, R. P., & Warsaw, P. R. (1989). User acceptance of computer technology: A comparison of two theoretical models. *Management Science*, *35*(8), 982–1003. doi:10.1287/mnsc.35.8.982

De Vries, H.J. (1997) Standardization – What's in a name? *Terminology – International Journal of Theoretical and Applied Issues in Specialized Communication*, *4*(1), 55-83.

De Vries, H. J. (2015). Standardisation – A Developing Field of Research. In P. Delimatsis (Ed.), *The Law, Economics and Politics of International Standardsation*, (pp.15-41). Cambridge, UK: Cambridge University Press.

De Vries, H. J., & Slob, F. J. C. (2006). Best Practice in Company Standardization. *International Journal of IT Standards and Standardization Research*, *4*(1), 62–85. doi:10.4018/jitsr.2006010104

De Vries, H. J., & Van Delden, M. (2006). How to integrate standardisation in knowledge management. *ISO Focus*, *3*(3), 37–40.

Dick, G. P. M., Heras, I., & Casedesús, M. (2008). Shedding light on causation between ISO 9001 and improved business performance. *International Journal of Operations & Production Management*, *28*(9), 687–708. doi:10.1108/01443570810881811

Dooyeweerd, H. (1955). A new critique of theoretical thought. In The general theory of the modal spheres (vol. 2). Philadelphia: The Presbyterian and Reformed Publishing Company.

Dooyeweerd, H. (1957). A new critique of theoretical thought. In The structures of individuality of temporal reality (vol. 3). Philadelphia: The Presbyterian and Reformed Publishing Company.

Ellul, J. (1970). *The meaning of the city*. Grand Rapids, MI: Eerdmans.

Foucault, M. (1976). *Histoire de la sexualité I – La volonté de savoir*. Paris: Éditions Gallimard.

Foucault, M. (1977). *Discipline and Punish; Birth of the Prison* (A. Sheridan, Trans.). London: Penguin. (Original work published 1975)

Foucault, M. (1982). Afterword. The subject and power. In H. L. Dreyfus & P. Rabinov (Eds.), *Michel Foucault. Beyond structuralism and hermeneutics* (pp. 208–226). Brighton, UK: University of Chicago Press.

Furusten, S. (2000). The Knowledge Base of Standards. In *A World of Standards* (pp. 71–84). Oxford, UK: Oxford University Press.

Goudzwaard, B. (1979). *Capitalism and Progress: A Diagnosis of Western Society*. Toronto: Wedge.

Gouldner, A. W. (1954). *Patterns of industrial Bureaucracy – A case study of modern factory administration*. New York: The Free Press.

Grimes, K. R. (2003). *ISO 9001:2000: a practical quality manual explained*. Milwaukee, WI: ASQ Quality Press.

Habermas, J. (1987). The Theory of Communicative Action: Lifeworld and System: A Critique of Functionalist Reason (vol. 2; T. McCarth, Trans.). Boston: Beacon. (Original work published 1981)

Haverkamp, A. (2004). *Bezint eer gij begint – Standaardisatie van werkmethoden bij Polynorm Automotive – Onderzoek op basis van een christelijk-filosofische analyse*. Rotterdam: Erasmus Universiteit, Faculteit Bedrijfskunde.

Haverkamp, A. (2006). Een veelzijdig kwaliteitsverbeteringsproject. *Synaps*, *4*(20), 21–26.

Hesser, W. (2010). Standardisation within a Company – a Strategic Perspective. In W. Hesser (Ed.), *Standardisation in Companies and Markets* (3rd ed.; pp. 215–258). Hamburg.

Hofstede, G. (1978). The Poverty of Management Control Philosophy. *Academy of Management Review*, *3*(3), 450–461.

Imai, M. (1986). *Kaizen – the key to Japan's competitive success*. New York: Random House.

ISO. (2008). *ISO 9001:2008 Quality Management systems – Requirements*. Geneva: International Organisation for Standardisation.

ISO. (2009). *ISO/TS 16949:2009 Quality Management systems – Particular requirements for the application of ISO 9001:2008 for automotive production and relevant service part organisations*. Geneva: International Organisation for Standardisation.

ISO. (2011). *The ISO Survey of Management Standard Certifications 2011*. Geneva: ISO Central Secretariat.

Kalpakjian, S. (1995). *Manufacturing Engineering and Technology* (3rd ed.). Reading, MA: Addison-Wesley.

Kalsbeek, L. (1975). *Contours of a Christian Philosophy: An Introduction to Herman Dooyeweerd's Thought*. Toronto: Wedge.

Kelly, M. (1994). Foucault, Habermas and the Self-Referentiality of Critique. In M. Kelly (Ed.), *Critique and Power – Recasting the Foucault/Habermas Debate*. Cambridge, MA: The MIT Press.

Kondo, Y. (1996). Are creative ability and work standardisation in contradictory relationship? *Training for Quality*, *4*(3), 35–39. doi:10.1108/09684879610125734

Kondo, Y. (2000). Innovation versus Standardisation. *The TQM Magazine*, *12*(1), 6–10. doi:10.1108/09544780010287177

Lagrosen, Y., & Lagrosen, S. (2005). The effects of Quality Management – A survey of Swedish quality professionals. *International Journal of Operations & Production Management*, *25*(10), 940–952. doi:10.1108/01443570510619464

Lazaric, N., & Denis, B. (1995). Routinization and memorization of tasks in a workshop: The case of the introduction of ISO norms. *Industrial and Corporate Change*, *15*(5), 873–896.

Lelong, B., & Mallard, A. (1995). Presentation. *Reseaux*, *18*(102), 9–34.

Lin, Ch.-I., & Jang, W.-Y. (2008). Successful ISO 9000 implementation in Taiwan: How can we achieve it, and what does it mean? *International Journal of Productivity and Performance Management*, *57*(8), 600–622. doi:10.1108/17410400810916044

Lo, L. K., & Chang, D. S. (2007). The difference in the perceived benefits between firms that maintain ISO certification and those that do not. *International Journal of Production Research*, *48*(5), 1881–1897. doi:10.1080/00207540600733709

Magd, H. A. E. (2008). ISO 9001:2000 in the Egyptian manufacturing sector: perceptions and perspectives. *International Journal of Quality & Reliability Management*, *25*(2), 173–200. doi:10.1108/02656710810846934

Mahadevappa, B., & Kotreshwar, G. (2004). Quality Management Practices in Indian ISO 9000 Certified Companies: An Empirical Evaluation. *Total Quality Management*, *15*(3), 295–305. doi:10.1080/1478336042000183389

Manders, B. (2015). *Implementation and Impact of ISO 9001. ERIM PhD Series Research in Management*. Rotterdam: Erasmus Research Institute of Management – ERIM.

Manders, B. (2015). *Implementation and Impact of ISO 9001. ERIM PhD Series in Research in Management 337*. Rotterdam: Erasmus Research Institute of Management – ERIM.

Martínez-Costa, M., Choi, T. Y., Martinez, J. A., & Martínez-Lorente, A. R. (2009). ISO 9000/1994, ISO 9001/2000 and TQM: The performance debate revisited. *Journal of Operations Management*, *27*(6), 495–511. doi:10.1016/j.jom.2009.04.002

Mom, T. J. M., Van den Bosch, F. A. J., & Volberda, H. W. (2002). Exploratie én exploitatie van kennis - Het managen van de determinanten van horizontale kennisuitwisseling. *Management en Organisatie*, *56*(5), 23–43.

Nair, A., & Prajogo, D. (2009). Internalisation of ISO 9000 standards: The antecedents role of functionalist and institutionalist drivers and performance implications. *International Journal of Production Research, 47*(16), 4545–4568. doi:10.1080/00207540701871069

Nakamura, S. (2000). *The new Standardisation: Keystone of continuous improvement in manufacturing*. Portland, OR: Productivity Press.

Naveh, E., & Erez, M. (2004). Innovation and Attention to Detail in the Quality Improvement Paradigm. *Management Science, 50*(11), 1576–1586. doi:10.1287/mnsc.1040.0272

Naveh, E., Marcus, A., & Moon, H. K. (2004). Implementing ISO 9000: Performance improvement by first or second movers. *International Journal of Production Research, 42*(9), 1843–1863. doi:10.1080/00207540410001662912

Nicolau, J. L., & Sellers, R. (2002). The stock market's reaction to quality certification: Empirical evidence from Spain. *European Journal of Operational Research, 142*(3), 632–641. doi:10.1016/S0377-2217(01)00312-5

Nicoloso, E. (2007). Bridging the gap between Quality Management systems and product quality. *ISO Management Systems, 7*(1), 7–11.

Nonaka, I. (1994). A Dynamic Theory of Organisational Knowledge Creation. *Organization Science, 5*(1), 14–37. doi:10.1287/orsc.5.1.14

Nonaka, I., & Takeuchi, H. (1995). *The knowledge creating company: how Japanese companies create the dynamics of innovation*. New York: Oxford University Press.

Padma, P., Ganesh, L. S., & Rajendran, C. (2008). A study on the critical factors of ISO 9001:2000 and organisational performance of Indian manufacturing firms. *International Journal of Production Research, 46*(18), 4981–5011. doi:10.1080/00207540600793299

Rogers, E. M. (2003). *Diffusion of innovations* (5th ed.). New York: Simon & Schuster.

Simons, C. A. J., & De Vries, H. J. (2002). *Standaard of maatwerk – Bedrijfskeuzes tussen uniformiteit en verscheidenheid*. Schoonhoven, The Netherlands: Academic Service.

Slob, F. J. C., & De Vries, H. J. (2001). Wie die (petro)chemische Industrie von Normen profitiert. *DIN-Miteilungen, 80*(6), 440–442.

Sprenger, C. C., & Ten Have, S. (1996). Kennismanagement als motor van de lerende organisatie. *Holland Management Review, 51*, 69–76.

Tamm-Halström, K. (2004). *Organizing International Standardisation – ISO and the IASC in Quest of Authority*. Cheltenham, UK: Edward Elgar.

Terziovski, M., Power, D., & Sohal, A. S. (2003). The longitudinal effects of the ISO 9000 certification process on business performance. *European Journal of Operational Research, 146*(3), 580–595. doi:10.1016/S0377-2217(02)00252-7

Terziovski, M., Samson, D., & Dow, D. (1997). The business value of Quality Management systems certification – Evidence from Australia and New Zealand. *Journal of Operations Management*, *15*(1), 1–18. doi:10.1016/S0272-6963(96)00103-9

Tzelepis, D., Tsekjouras, K., Skuras, D., & Dimara, E. (2006). The effects of ISO 9001 on firms' productive efficiency. *International Journal of Operations & Production Management*, *26*(10), 1146–1165. doi:10.1108/01443570610691111

Van Kemenade, E. A., Hardjono, T. W., & De Vries, H. J. (2011). The willingness of professionals to contribute to their organisation's certification. *International Journal of Quality & Reliability Management*, *28*(1), 27–42. doi:10.1108/02656711111097535

Van Veldhuisen, A. (1996). *De Macht van het Systeem – Een studie naar kwaliteitssystemen gebaseerd op het werk van Michael Foucault*. Rotterdam: Erasmus Universiteit, Faculteit Bedrijfskunde.

Van Wessel, R. (2010). *Toward Corporate IT Standardization Management: Frameworks and Solutions*. Hershey, PA: IGI Global. doi:10.4018/978-1-61520-759-6

Weber, M. (1922). *Wirtschaft und Gesellschaft*. Tübingen: J.C.B. Mohr.

Wenström, H., Ollner, J., & Wenström, J. (2010). *Focus on Industry and Standards – An Integrated Approach*. Stockholm: SIS.

Wissema, J. G., Messer, H. M., & Wijers, G. J. (1991). *Angst voor veranderen? Een mythe! – Of: Hoe u veranderingsbereidheid op de werkvloer vergroot*. Assen: Van Gorcum.

Yeung, A. C. L., Lee, T. S., & Chan, L. Y. (2003). Senior management perspectives and ISO 9000 effectiveness: An empirical research. *International Journal of Production Research*, *41*(3), 545–569. doi:10.1080/0020754021000033896

KEY TERMS AND DEFINITIONS

Aspects: Irreducible dimensions of reality in which all 'things' function in a meaningful way.

Aspectual Laws: The enablers of meaningful being or functioning in each aspect.

ISO 9001: An international standard for organisational quality management systems to increase customer satisfaction through the deliverance of quality products and services.

ISO/TS 16949: An international quality management standard specifically intended for organisations in the automotive industry that produce and deliver products, parts, materials or services to OEM manufacturers. The standard is based on ISO 9001 with additional (sector-specific) requirements.

Panopticon: A design of a building (or system) to allow a 'watchman' to observe others without the others being able to tell whether or not they are being observed. It is an inspection mechanism of constant overbearing surveillance and a mechanism of power over mind to normalize or modify behavior.

Philosophy: The discipline of disciplines that enables to recognise, respect and respond to the complexity of coherence and diversity of the reality.

Standardisation: The activity of establishing and recording a limited set of solutions to actual or potential matching problems, directed at benefits for the party or parties involved, balancing their needs, and intending and expecting that these solutions will be repeatedly or continuously used, during a certain period, by a substantial number of the parties for whom they are meant.

Quality Management: Maintaining and improving purposeful value adding processes.

Section 5
Intellectual Property Rights

Chapter 10
A Primer on Intellectual Property Policies of Standards Bodies

Jorge L. Contreras
University of Utah, USA

Andrew Updegrove
Gesmer Updegrove, USA

ABSTRACT

This chapter offers business managers an introduction to the intellectual property rights (IPR) issues that should be taken into account when an organization is considering joining or participating in a standards development organization. Copyright, trademark and patent issues, particularly patent disclosure and licensing, are addressed with an emphasis on the potential benefits and risks offered by the most common policy approaches today. Managers are encouraged to review the IPR policies of standards organizations with a view toward their own organizations' preferred IPR strategies and approaches, whether these tend toward product manufacture and distribution without significant attention to IPR monetization, research and development with a focus on IPR generation, or the generation of substantial revenue through IPR licensing.

INTRODUCTION

One of the most important and complex factors that the corporate manager must face when determining whether his or her organization should join and participate in a Standards Development Organization (SDO) is the degree to which the SDO's policies governing patents, copyrights, trademarks and other intellectual property rights (IPR) is consistent with the organization's goals, strategy and policies. In some cases an organization may be presented with multiple SDOs working in similar areas, and a comparison of their IPR policies will be warranted in evaluating which, if any, to join. If an SDO's IPR policies are inconsistent with the organization's approach to IPR, then the manager will face a number of choices. They include making an exception from normal practice, due to the importance of the standardization

DOI: 10.4018/978-1-4666-9737-9.ch010

effort to the organization's business; providing special instructions or supervision to the organization's participating representatives; seeking changes to the SDO's IPR policy (occasionally, but not often feasible); seeking alternative venues for standards development; or even proposing the formation of a new SDO having a more acceptable IPR policy. For these reasons, it is critical that corporate managers involved in standardization understand and be prepared to evaluate the implications of SDO IPR policies and how they operate in practice.

SDO IPR policies began to assume their current forms in the late 1990s, after Dell Computer entered into a consent decree with the U.S. Federal Trade Commission (FTC) relating to alleged misconduct within the Video Electronics Standards Association (VESA).[1] A second wave of policy adoption and updating occurred in the mid-2000s, following widespread litigation involving semiconductor design firm Rambus, Inc. and its alleged manipulation of the IPR rules of the Joint Electron Device Engineering Council (JEDEC).[2] The *Rambus* litigation affected firms in the U.S., Europe and Asia, and led to a general overhaul of IPR policies by SDOs around the world. Most notably, many such policies were revamped to add, or ensure the effectiveness of existing, obligations of SDO members to disclose patent claims essential to standards approved by the SDO, and to license those claims to producers of standardized products on terms that are "fair, reasonable and nondiscriminatory" (FRAND). Most recently, standards-essential patent litigation has proliferated (notably in the mobile platform industry), resulting in further evolution in IPR policies.

This chapter summarizes the key IPR issues that are likely to appear in contemporary SDO IPR policies. Managers should develop a solid understanding of these issues in order to make informed decisions regarding SDO participation for their organizations. Managers should also determine whether their organization requires additional levels of internal review prior to committing the enterprise to the IPR policy of an SDO. Frequently an organization will require that membership in a given SDO must be approved by the organization's legal department, and also that certain communications from an SDO (e.g., notices relating to the review of draft specifications prior to an adoption vote) be sent directly to the legal department.

BACKGROUND

There is a large literature relating to standards, standardization and related IPR policies. This chapter is intended to offer the business manager a concise and non-technical introduction to the field. Those desiring a more detailed treatment are referred to the sources listed at the end of this chapter.

Biddle et al (2012), Ernst (2012), EC (2014) and Contreras (2015a) describe the institutions and organizational structures that have evolved to address the needs of contemporary standards-development communities. These include large, multi-national bodies simultaneously addressing a broad range of standardization projects, industry-specific groups that develop solutions for specific industry sectors, and smaller consortia focusing on single products or standards. Within this range of organizations a variety of IPR policies have emerged. These are categorized and analyzed by Lemley (2002) and Bekkers & Updegrove (2012). ABA (2007) offers a detailed clause-by-clause analysis of typical SDO IPR Policies. Cannady (2013) and Herman (2010) offer practical advice regarding the negotiation of standards-related patent license agreements, and Lemley and Shapiro (2013) and Contreras and Newman (2014) discuss alternate dispute resolution mechanisms for resolving standards-related conflicts.

Much of the current theoretical literature concerning standards and patents supports or responds to claims by Shapiro and Varian (1999), Farrell et al (2007) and Lemley and Shapiro (2007) regarding threats of "hold-up" and "stacking" (defined below) in standards-development environments. This literature is surveyed in NRC (2013). Miller (2007), Lichtman (2010) and Blind et al. (2011) assess the impact that patents may have on standardization practices and organizations.

A significant amount of attention has been paid by various authors to the interpretation of FRAND commitments, including Swanson and Baumol (2005) (proposing an auction model for establishing FRAND royalty rates); Layne-Farrar, Padilla and Schmalensee (2007) (offering an economic analysis of different FRAND royalty calculation methodologies); Brooks & Geradin (2011) (advocating bilateral negotiation among licensors and licensees); Layne-Farrar (2010) and Gilbert (2011) (analyzing the non-discrimination prong of the FRAND commitment); Kesan and Hayes (2013) (focusing on the binding nature of FRAND commitments following transfers of underlying patents); Sidak (2013) (outlining methodologies for calculating FRAND royalties based on recent case law); and Contreras (2015b) (analyzing legal theories supporting enforcement of FRAND and related commitments). Another strand of the literature, including Michel (2011), Cotter (2014) and Sidak (2015), addresses the availability of injunctive relief to patent holders that have made FRAND commitments, including recent judicial decisions on this subject.

Much of the legal analysis concerning patents on standards is based on antitrust and competition law. These themes are addressed broadly in DOJ-FTC (2007), Carrier (2009), ABA (2011), Cary (2011) and Lundqvist (2014). Skitol (2005) and Contreras (2013) propose the relaxation of antitrust scrutiny to permit collective negotiation of FRAND license rates, a proposal that is challenged by Layne-Farrar, Llobet and Padilla (2014) among others. Contreras (2015c) offers a historical perspective on the evolution of FRAND commitments, and Cotter (2014) offers a comparative assessment of the legal doctrines applied to standardization in various jurisdictions.

Copyright pertaining to standards has attracted less scholarly attention than patents, though some areas have been explored in detail. Samuelson (2007) questions the applicability of copyright law to standards, Strauss (2013) challenges the use of copyright to prevent standards incorporated into U.S. law from being made publicly available, Contreras and Hernacki (2014) explore the implications of copyright license termination on standards in the marketplace, and Vetter (2007) and Updegrove (2009) discuss implications of open source code licensing for standardization. In the area of trademarks, Chon (2009) and Contreras, Lewis and Roth (2011) offer critiques of current certification marks and programs.

IPR POLICIES AND ISSUES

Copyright

Copyright law protects works of authorship, including written documents, software programs and visual works such as drawings and computer screen displays. As such, copyright generally applies to most aspects of technical standards. In most of the (currently 168) countries that adhere to the Berne Convention for the Protection of Literary and Artistic Works,[3] copyright arises automatically when a work is created and recorded in a tangible or electronic medium, without the need to register with a governmental agency. Works that are protected by copyright cannot be reproduced, displayed, or modified without permission of the copyright owner.

Copyright in Contributions

Because most standards documents are generally protected by copyright, it is important to understand who owns and controls the copyright in these documents. The great majority of SDOs claim ownership of the copyright in their published standards and compilations, and many designate them with the copyright symbol © and the year of publication in order to avoid any ambiguity on that subject.

However, there is more variability when it comes to ownership of copyright in the underlying technical contributions that have been made with respect to a standard. Technical standards are typically the result of collaboration among multiple organizations. These organizations come together under the aegis of the SDO, but most SDOs are themselves "thin" organizations with a few, mostly administrative employees, but little technical expertise. The bulk of the technical work that goes into a standard thus comes from participant organizations.

Most standards begin with a written technical proposal submitted by one or more SDO participants. As written documents, such proposals are themselves protected by copyright. Under the "work made for hire" doctrine in many countries, an employer owns the copyright in works that are created by its employees within the scope of their employment. Thus, technical proposals submitted to an SDO are typically owned by the organizations employing the individuals who participate in the relevant SDO working groups when these individuals do so at the behest of their employers.

However, as noted above, the owner of a copyrighted work generally controls the right to reproduce, publish and modify it. Thus, in order for a technical contribution to be distributed to other SDO participants, modified during the course of technical discussions, and eventually published as a standard, the copyright owner must grant permission. This permission generally comes in one of two forms: (1) the SDO may require that participant organizations assign or transfer copyright in their technical contributions to the SDO, making the SDO the owner of the copyright in such contributions, or (2) the SDO may require that participant organizations grant the SDO and other participants a *license* to reproduce, modify and publish its technical contributions within the scope of SDO activities.

The first option (transfer of copyright to the SDO) gives the SDO greater flexibility regarding the use of technical contributions and eliminates any doubt regarding the SDO's ability to control ultimate publication (and sale) of its standards. However, unless the SDO's policy explicitly provides a license back to the contributor, this option limits the participant's ability to utilize its own technical contributions. In SDOs that only obtain a license to use technical contributions, participants typically retain an unrestricted right to utilize those original contributions (excluding any modifications made by the SDO and other participants) independently of the SDO in any way they wish.

These considerations are important for managers to consider, particularly when technical contributions include software code. If an SDO obtains ownership of the copyright in a technical contribution, absent an appropriate license back from the SDO, the contributing organization may lose the right to utilize it internally and to submit it to other SDOs. This later point is particularly important if the SDO in question is small, untested or one of several different groups working on the standardization of similar technologies. Organizations sometimes find that their technology can be utilized most effectively in a different setting than the one to which they originally submitted it, and retaining this flexibility may be in the best interest of the organization

To this end, a manager evaluating an SDO should determine whether the license or assignment of rights in a technical contribution to an SDO can be terminated or revoked by the organization making the contribution. The right to revoke may be important if the organization decides that the SDO is no longer operating in a manner consistent with its goals and objectives, and that it wishes either to contribute its technology to a different SDO or to pursue it as a proprietary competitive technology. Because the right to revoke a license in a contribution is usually not permitted once it has been submitted to the SDO (assuming that the contribution becomes embodied in a standard), managers should also investigate whether a given SDO requires that a detailed charter must be approved for each working group. Where this is the case, a manager can make a contribution with less concern that the enterprise may regret the decision at a later date.

Availability of Standards

If an organization is planning on participating in an SDO, it may have an interest in how widely the resulting standards will be made available. Standards that are made available only to SDO members or only for a high fee may be less widely adopted than standards that are more generally available.[4] Thus, if the organization's goal is to achieve the broadest possible dissemination of a particular technology or platform, the manager may wish to consider whether the SDO's dissemination policies are consistent with this goal.

Software in Standards

Many technical standards describe how products must be designed in order to interoperate. Such standards are typically used in the product design and manufacturing process, but will not themselves be distributed as part of products. When software code is included in a standard document or specification, however, the situation is somewhat different. Some software code included in standards is sample code intended to be emulated but not utilized verbatim. But in other instances, a standard may permit or require that a product implement the exact code contained in the specification or standard. As noted above, software code is typically protected by copyright, and a copyrighted work can be reproduced and distributed only with the permission of the copyright owner or licensee. As a result, these cases pose several important questions.

First, does the IPR policy of the SDO under consideration address the subject of software code at all? As of this writing, there are SDOs that are either unaware of the fact that software code is sometimes being included in standards by working group members, or have not yet embarked upon updating their IPR policies to address this fact. In such a case, those that wish to implement the standard in question may at best have an implied right to include the code.

Assuming that the IPR policy or related documents of the SDO do address this issue, what are the conditions under which software code appearing in standards documents can be used in products? Can such code be reproduced and distributed freely, or are there limitations? Can the software code be modified, or must it be used in the exact form specified in the standards document? Are there any royalty or other payment obligations associated with the reproduction and distribution of software in products?

Some SDOs, such as the Internet Engineering Task Force (IETF), license code on different terms than other text appearing standards documents. Specifically, code that is intended to be reproduced and distributed in products implementing a standard is licensed under the open source BSD License.[5] The BSD License is extremely permissive, and allows a user to reproduce and modify software code without any obligation or restriction, save that it must attribute the code to the copyright owner. Importantly, the terms of the BSD License are more permissive than the license that is granted with respect to other text appearing in IETF standards. Thus, the manager considering participation in an SDO must determine what use the organization wishes to make of the code contained in SDO standards, if any, and whether such use will be permitted under the SDO's rules and policies.

Standards and Open Source Software

Over the last fifteen years, IPR policies have also been impacted by the increasing popularity of "open source software," a term which relates both to the form in which the software in question is distributed (in both human-readable source code as well as machine-readable object code) as well as to the license terms to which it may subject. These terms are found in a wide variety of licenses that range from minimalistic and permissive to lengthy and restrictive. The implementation of standards subject to popular permissive licenses such as the BSD and MIT licenses is usually not problematic. However, the same may not be true with respect to software subject to many of the more restrictive licenses, such as the popular GNU General Public License (GPL) versions 2 and 3.

The reason is that a number of licensing terms that are impliedly or expressly consistent with a FRAND licensing obligation may not be consistent with the terms of many such open source licenses. For example, the owner of a patent claim essential to the implementation of a standard would usually be permitted to provide in a license that the rights received are non-assignable, with the result that an integrator or distributor of a compliant product would be required to secure a license from the patent owner as well. Under the GPL License, code cannot be included if, as a result, a downstream party would be required to go back to the original source to obtain a license. Similarly, while IPR policies frequently permit the owner of an essential patent claim to require an implementer to cross-license any of its own patent claims that are essential under the same standard. This would not be permissible with respect to technology incorporated into GPL-licensed software. Nor could the owner of an essential patent claim require that the terms of its license remain confidential.

As a result, some standards organizations, such as the World Wide Web Consortium (W3C), have amended their IPR policies to narrow the range of permissible patent licensing terms to those that are consistent with open source licensing. Others, such as OASIS, have bifurcated their policies to permit working groups to be chartered under IPR terms that have been similarly narrowed. The negotiation (among dozens of member legal counsel) of the changes to the W3C Patent Policy was contentious, spanning a three year period. While terms and norms for reconciling IPR policies with open source realities have evolved and achieved some degree of consensus, they continue to evolve and remain subject to particular attention whenever they are proposed or adopted.

The difficulty of reconciling IPR rules for open source software and open standards is compounded by the fact that standards gain their value in large part because they can only be changed with the permission of the standards developer, while the appeal of open source software derives in significant part from the freedom of the licensee to make whatever changes it wishes to the code in question. This inherent inconsistency has sometimes led to confusion as well as friction between the open source and open standards communities. Convergence at the technical level continues to challenge those seeking to reconcile and streamline IPR rules relating to both disciplines. For example, standards today may include actual computer code instead of merely descriptive text, meaning that license terms must be clarified for the code as well as for the specification. Again, major IT vendors today are launching open source projects to enable dramatic new business models, such as cloud computing and the Internet of Things. In so doing, they lower their own development costs and encourage customer uptake, since the danger of lock in to a single vendor is ameliorated. At the same time, other vendors (and sometimes even the same vendors) are launching new standards projects to create interoperability standards supporting the same business models. In the former case, the open source project may later create interface standards to increase interoperability, and in the latter, the standards organization may develop an open source reference implementation of its own standards to lower costs and increase uptake. The result is a blurring of disciplinary lines and a more complex IPR landscape.

The lesson for business managers is that the proliferation of open source software has added an extra legal dimension to participation in a given SDO. In order to remain competitive, IT companies and their legal counsel need to acquire a solid understanding of both the legalities as well as the practicalities of open source licensing. They must also be careful to evaluate the IPR policy of a given SDO in the context of its scope of work and their own licensing strategies and tolerances.[6]

Trademarks

A trademark is a word, symbol or other designation that is used to identify the source of a particular product or service. Though SDOs generally do not offer products or services, per se, creating and protecting trademarks is an extremely important tool for SDOs, in part because it provides them with the legal rights necessary to control the authoritative version of a standard, as well as to determine and assert what does and does not comply with that standard. As a result, trademark law can become relevant to participation in an SDO in several ways, and SDOs will frequently register one or more marks in multiple, and sometimes scores, of countries.

First, SDO participants must be careful when making submissions to technical working groups to avoid granting rights in their trademarks. That is, using a company trademark in the title or text of a technical specification could grant the SDO and other participants a license to use the mark freely in connection with the resulting standard or on their own products. For example, if an SDO participant has developed a proprietary technology that it refers to as "Gizmo" and then submits to an SDO a technical proposal entitled "Gizmo Interface Specification", it is possible that the SDO could gain the right to refer to the resulting standard by the name "Gizmo", even if changes are made during the standardization process and the resulting standard is no longer identical to, or even compatible with, the participant's original Gizmo technology. Potentially, this could result in the participant losing the right to enforce its trademark by reason of its having become generic. To avoid such conflicts, the manager should (a) review SDO technical submissions to ensure that they are not designated by the organization's proprietary product or technology names, and (b) ensure that the names of the organization's proprietary technologies are clearly designated as the organization's trademarks (e.g., "this specification ensures interoperability with Company X's Gizmo® technology platform").

By the same token, an SDO participant should not assume that it can utilize the name of a standard in its own product names without the permission of the SDO and, possibly, another SDO participant. In some cases, using the name of a standard requires certification that a particular product complies with the standard. In the U.S. there is a special category of registered marks called "certification marks". These marks can only be used as indications that a particular product or service complies with a particular set of criteria for certification, and cannot be used by the mark's owner on its own products. For example, the IEEE 802.11 suite of wireless networking standards is commonly referred to as "Wi-Fi". However, the term "Wi-Fi" is a registered trademark of the Wi-Fi Alliance, an independent certification organization that separately licenses use of the Wi-Fi mark for 802.11-compliant products. Under the licensing terms of Wi-Fi Alliance, a company cannot use the term Wi-Fi in a product name, and may only use this term in specified configurations (e.g., "Wi-Fi Compatible"). This being said, under the "nominal fair use" principle that exists under the trademark laws of many countries, it is permissible to refer to other parties' trademarks in a factual sense to indicate the participant's assertion that it has produced the product using the associated standard, so long as the trademark itself is not used in the name of the compatible product. For example, a product's documentation could explain that "the XYZ Device is compatible with the ABC-123 standard," without an express license from the owner of the mark.

Patents

Background

Standards are written descriptions of particular attributes of specified products and processes. Assuming that statutory requirements of utility, novelty, and non-obviousness are met, patents may be obtained on products and processes that conform to particular standards. The scope of protection afforded by a patent is described in a series of written "claims", each of which is a discrete element entitled to protection. When it is impossible to implement a standard as intended without infringing upon a patent claim, that claim is referred to in a typical SDO IPR policy as being either "essential" or "necessary." Patents covering standards are ordinarily not obtained by the SDO in which the standard was developed, as SDOs seldom produce complete products and almost never seek to patent their work. Rather, such patents are typically obtained by participants in the SDO or outsiders, and sometimes both.

The Threat of Patent Hold-Up and Stacking

Two general patent-related risks arise in the context of technology standardization: "patent stacking" and "patent hold-up." Patent or royalty "stacking" occurs when multiple entities each hold patents claiming aspects of a single standardized technology. In order for a producer to implement the standardized technology, it must obtain licenses from multiple parties, each acting independently and each seeking to maximize its gains. The risk of stacking is that the sum of individual royalty demands may be excessive in relation to the overall value of the product, making the standardized product uncompetitive in comparison to products that do not conform to the standard.

The second major patent issue that can arise in the standards context is patent "hold-up". Patent hold-up can occur when a patent holder seeks to assert a previously undisclosed patent or to collect royalties from implementers of a standard after the standard has been adopted. If such an "ambush" occurs after the industry has devoted significant resources to production, marketing, and training with respect to standardized products (in economic terms, after the marketplace has become "locked-in"), unexpected royalty demands from patent holders can have a disruptive effect on the market. This phenomenon can result in an increase in the cost of standardized products, sometimes to levels that are inefficient and uncompetitive with alternative technologies.

Patent hold-up has been alleged in several cases within the ITC industry. The first of these cases to gain significant attention involved Dell Computer, which allegedly failed to disclose patents covering the voluntary VL-bus standard that it helped to develop at the Video Electronics Standards Association (VESA).[7] Following approval of the standard, Dell sought to enforce its patents against other computer manufacturers. The Federal Trade Commission (FTC) sought to enjoin Dell's enforcement of its patents on the ground that Dell had violated a VESA rule that required disclosure of an essential claim if a participant wished to reserve the right to later charge for its use. The FTC action resulted in the entry of a 1996 consent decree that, among other sanctions, prohibited Dell from enforcing the essential claims under its VL-bus patents against any third party product implementing the VESA standard. The Dell case caught the attention of participants for two reasons: not only could they become subject to a patent ambush if members did not abide by the terms of an IPR policy, but they could also lose the right to assert an essential claim if they did not disclose it in a timely fashion (a concern that continues today).

Perhaps the best-known case of alleged disclosure failure within an SDO involved Rambus, Inc., a developer of semiconductor memory technology. Much has been written about the decade-long legal battles in which Rambus sought to assert its patents against implementers of dynamic random access memory (DRAM) technology standardized by the Joint Electron Device Engineering Council (JEDEC), a voluntary SDO in which Rambus participated in the early 1990s. In the cases (which involved U.S. and EU enforcement agencies, as well as multiple semiconductor companies), Rambus escaped liability largely because some (but not all) of the triers of fact determined that that the JEDEC patent policy was too vague to prohibit the conduct that Rambus was alleged to have committed. Or, as concluded by one judge, it suffered from "a staggering lack of defining details" that left SDO participants with "vaguely defined expectations as to what they believe the policy requires."[8]

The *Dell* and *Rambus* cases each provided a wake-up call to the standards-development community, alerting it both to the possibility of patent hold-up, and to the deficiencies of many existing SDO policies. In response, many SDOs adopted or enhanced their formal policies to address ambush issues more clearly. They did so by imposing one or both of the following obligations on SDO participants: (1) an obligation to disclose patents essential to implementation of a standard, and/or (2) an obligation to license patents essential to implementation of a standard, either on a royalty-bearing or a royalty-free basis, but in either case on FRAND terms.

These related obligations are intended to ensure that standards developers have sufficient information to assess the relative costs and risks of technologies under consideration for standardization. That is, disclosure obligations (when included in a policy) seek to ensure that standards developers know whether and which patents owned or controlled by their members cover technologies under consideration, giving them the opportunity to "design around" such patents if they so wish. Likewise, licensing obligations are intended to ensure that such patents will be licensed to implementers of the standard on terms that are bounded, at least to some degree. Below we discuss the principal issue that an enterprise should consider when evaluating an SDO's policies in this area.

Disclosure

Antitrust and Competition Law Concerns. Under ordinary circumstances, significant antitrust and competition law risks arise when groups of competitors share proprietary competitive information. However, due to the widely recognized pro-competitive effects of standards development and adoption, SDOs are able to host the disclosure of certain information under appropriately controlled circumstances. Historically, this has been limited solely to disclosure of essential patent claims and to intentions to license (or not) on FRAND terms, as well as whether associated, non-quantified fees will or will not be charged. Within the past ten years, a few SDOs have adopted rules requiring (VITA), or permitting (IEEE-SA and ETSI) disclosure of additional licensing terms, including price, under strictly controlled processes.

Benefits and Burdens of Disclosure. The incorporation of disclosure terms into an IPR policy represents a classic example of weighing the burdens of an obligation against the benefit anticipated from it. The benefit, as noted above, is the ability for standards developers to choose between alternative technologies on a more informed basis before the marketplace becomes locked in to a standard. The burden is the time and cost for members to determine whether they have potentially essential claims to disclose. For a small company that participates in only one or a few SDOs and owns only a few patents, the burden may be small. But for companies that own thousands of patents and actively participate in multiple working groups within scores, and even hundreds, of SDOs, the burden can be very great. For such companies, conducting a conscientious patent review in connection with every draft standard their representatives helped develop could consume the full time of a very large team of attorneys.

For this reason, IPR policies today almost universally state that a complete "patent search" is not required by a member, although they do not define explicitly what this phrase is intended to mean. The statement is nevertheless significant for two reasons: first, the member has latitude to decide how thorough an examination of its patent portfolio it will undertake. And second, the possibility of an innocent failure to disclose is implicitly acknowledged. And indeed, regardless of the level of review, the question of whether a given patent claim is indeed "essential" often consumes substantial legal fees when the question is litigated. And even then, one court may reach a different conclusion than another.

SDOs have developed a variety of approaches to disclosure. The strictness of the approach adopted by a given SDO will generally reflect one (or several) of the following factors: prevalence of patented technology in its standards; tolerance of the industry for patent royalties; number of standards adopted by the SDO per year; prevailing disclosure practices among peer SDOs; and whether it is an established SDO with a difficult to change, legacy IPR policy or a recently launched organization.

Thus, as an alternative to an absolute requirement that SDO members disclose every patent essential to a standard under development, most SDOs qualify the Disclosure Obligation by the "actual knowledge" of some group of individuals, which may or may not be defined. This may be the knowledge of the engineers participating in the working group developing the standard, all of the company's engineers who participate in the SDO, or the members of the company's legal/patent department.

For a manager evaluating whether or not to recommend membership in a given SDO, a balancing of potential burdens and benefits should be conducted. Burden factors that an organization will find relevant include: the overall size of its patent portfolio and internal legal department; whether it is likely that it will own essential claims covering the SDO's standards; whether it would be tolerant or intolerant of payment obligations associated with standards; whether it generally charges royalties for its own essential claims; and how great the anticipated internal costs of disclosure would be. On the benefit side of the equation, an organization will want to assess how important the work of a given SDO is to its strategic mission; whether it hopes to monetize any of its patents through inclusion of its technology in a final standard; and whether it hopes to secure indirect benefits by affecting the final design of the standard or having early access to the standard during its development. Because an organization's answers to questions such as these will often vary widely in a given case, it may be willing to join one SDO and not another, even when both organizations have substantially identical IPR policies.

Scope and Process of Disclosure. The least burdensome approach to disclosure in an IPR policy is, of course, to omit mention of disclosure at all. While this may be common in SDOs that are not active in patent-heavy industries, the majority of SDOs in the ICT arena will at least require participants to identify essential claims that they are *not* willing to license on a FRAND basis. A few SDOs (primarily those developing standards for software and the Worldwide Web) do not require disclosure because their policies do not permit participants to charge royalties for their essential claims, thereby dramatically decreasing the value of knowing in advance whether patent claims owned by a participant will be essential (i.e., there is less motivation to "work around" a patent if it will be made available for free). Many of these SDOs still include an "escape hatch" mechanism whereby a participating member can avoid such licensing commitments by disclosing an essential claim it wishes to withhold. Usually, there is a disincentive to doing so, however. Most commonly, this will be a requirement that the member leave the working group in question, and sometimes resign its membership in the SDO itself (without a refund of its dues). Other SDOs permit their members to make "blanket" commitments to license their patents on FRAND terms, and may thus treat disclosures as optional, on the theory that access will be granted to the patents anyway. SDO licensing commitments are discussed in greater detail in the next section.

Some SDOs, and particularly those that have been in existence the longest, may reference disclosure of essential claims as something that is permitted or encouraged at any time during the development process. The same SDOs may also require disclosure only of the patents involved, and not the specific patent claims that might be infringed. More recent SDOs, as well as older ones that have performed a thorough updating of their IPR policies, will require that disclosure must be made on a claim by claim basis. While claim-by-claim disclosures may be more useful to working groups, they also impose greater burdens on the organizations making disclosures and also introduce more room for inadvertent errors and omissions.

The timing and formality of Disclosure Obligations varies among SDOs. While some SDOs require mandatory (*members shall*) or voluntary (*members may*, or *members are encouraged to*) disclosure as soon as a member becomes aware of an essential claim that it owns or controls, other SDOs require that disclosure be made during a formal review period or "call for patents", usually around the time that a draft standard is posted for final working group comments or prior to voting. Many SDOs also require informal disclosure in response to a statement that is read prior to the start of every in-person and telephonic working group meeting. This is intended to provide early warning of potential essential claims owned by others as well as those owned by members participating in the working group. Those making disclosure are typically not held responsible if their assumptions prove to be false.

What must be Disclosed? When disclosure is encouraged or required, the disclosure obligation almost always applies to any essential claim that a participant is aware of. However, many SDOs only extend disclosure obligations to patent claims that the owner is unwilling to make available on a FRAND basis. In other SDOs, disclosure is required only if the participant is unwilling to license on a royalty-free basis. In an SDO with only a few working groups, this obligation may be imposed both on members and non-members of the relevant working group. As SDOs expand the range of their activities, however, they will usually limit this requirement to working group members only if it begins to negatively impact the SDO's recruiting and retention efforts. This is because a typical member may be interested in participating in only a few working groups, and would not wish to be responsible for disclosing (or licensing) patents relevant to other, unrelated working groups. Once again, this result reflects a balancing of burdens (performing frequent patent reviews in connection with standards in which a member has no interest) relative to the benefits (desiring to participate in the development of only some of the standards that the SDO creates).

International Filings. Disclosure requirements usually cover patents registered anywhere in the world (even when the SDO is a national rather than an international organization). Some SDO policies also cover patent applications in addition to issued patents. Many policies require that the specific claims deemed to be essential must be identified, as well as the portion(s) of the draft standard that would result in infringement if implemented. Because patent applications are not public in all jurisdictions, a policy may provide that disclosure can be more general to the extent necessary to avoid the disclosure of trade secrets when an application is not yet public.

Blanket Disclosures. When specific disclosure is not required, an SDO may require what is sometimes (rather misleadingly) referred to as "blanket disclosures." A "blanket licensing" requirement is a more appropriate label for two reasons: first, no disclosure is required at all, and second, what is involved is a conditional licensing commitment rather than disclosure. For this reason, such commitments will be discussed in greater detail below in the context of licensing.

While the brief overview above lays out the most common practices and suggests a means whereby an organization can evaluate whether or not the disclosure rules of a specific SDO will be acceptable, it does not cover every term relating to disclosure that can be found in SDO IPR Policies. Managers, and their legal counsel, should therefore be alert to additional, and different requirements.

Licensing

While many SDOs recognize that it may be inevitable that a standard will include technology covered by patent claims, they may also wish to avoid, to the extent possible, approving a standard covered by patents unless anyone that wishes to implement the standard can do so for a reasonable royalty or other fee (or, for some SDOs, for no fee at all), that all other licensing terms will also be reasonable, and finally, that all would-be implementers will be treated equally. Such terms are referred to as being "fair, reasonable and non-discriminatory," or FRAND, or simply RAND (these two terms being synonymous for all practical purposes).

SDOs do not have the ability to require non-participants to license essential patent claims on a FRAND (or any other) basis, although many do address the topic of how to react when a non-member owner of an essential claim declines to commit voluntarily to license such claims. But because participants are in a position to make contributions to a standard under development that are covered by their own patent claims, most SDOs view it as most important to address patent licensing by participants .

The goal of patent licensing commitments is to provide certainty in the marketplace regarding the availability of licenses for essential claims covering a standard. This goal can sometimes best be realized by requiring all members that participate in a given working group (and sometimes all members) to state whether they will make their essential claims available without charge (on terms that are otherwise FRAND), or at rates and other terms that are FRAND. Some IPR policies offer the alternative of stating that the participant will not assert its essential claims at all, often referred to as a "non assertion covenant". Some SDOs do not give their members a choice regarding which of these approaches to take, and require one or another for all members. Some SDOs permit members to declare that they are unwilling to license certain patent claims (i.e., to "opt out" of the licensing obligation for those claims), which usually results in the relevant working group deciding to avoid or work around those claims.

Duration of Commitments. Patent licensing commitments, once made, should be perpetual and irrevocable once the standard has been adopted. Such commitments, however, do not typically extend to portions of the same standard that change when the standard is revised, unless the participant is still subject to the IPR policy. As earlier noted, SDOs are increasingly requiring that a party to whom an SDO participant assigns a covered patent be bound by the same licensing obligations as the original participant.

Rules Applicable to Contributions. An important distinction is recognized in many SDO policies between a member's patent claims that cover its own contributions to the SDO, and its claims that cover other members' contributions. One reason for this distinction is that a contributor is often motivated by an expectation that it will achieve a direct or indirect benefit from the inclusion of its technology or methodologies in a standard, especially in preference to the inclusion of an alternative from one of its competitors. Another reason is that if a participant in a working group were permitted to withhold a license later in the development process, it could result in a long delay while the group backtracks to pursue another course. In order to prevent such game playing (sadly, companies have been alleged to join a working group to impede rather than work towards its success), IPR policies typically state that a contribution once made cannot be withdrawn, and that a licensing commitment made at the time of that contribution cannot later be changed (for example, from royalty-free to royalty-bearing). Some SDOs go further, and only accept contributions accompanied by commitments to license any underlying essential claims on a royalty-free basis.

A participant's obligation to grant licenses may be less with respect to portions of a standard that it did not contribute. In this case, the concern is to protect the participant if its technology has been included in a standard without its encouragement or consent. With respect to that material, other licensing options (usually including withholding a license entirely) are available, and a decision most often need not be made until the end of the development process.

Licensing Options. Except where disclosure obligations or opportunities are ongoing, the disclosure of essential claims and making of licensing commitments usually occur during a formal review period at the end of the standards-development process. That period is often 30 to 45 days long, but can occasionally run longer or shorter. The period is intended to permit a member to perform whatever degree of patent portfolio review and business analysis it may wish before making a binding commitment to license its essential claims, and if so, whether with or without payment, if that decision is permitted by the SDO. Sometimes a participant may elect different alternatives for different patents, resulting in the possibility that it might agree to license some of its essential claims for free, some for a fee, and some not at all. In order to ensure uniformity of the commitments made, many SDOs only permit a member to make licensing commitments using a standardized "check the box" template to identify its essential patent claims, affected standard sections and desired licensing option(s).

Blanket Commitments. As noted above, many SDOs permit members to make "blanket" commitments to license their essential patent claims on FRAND or royalty-free terms, but do not require them to disclose specific claims. There are two reasons that have been advanced to justify this practice. The first is that if the member is willing to license an essential claim on a FRAND basis, there is insufficient benefit to the marketplace to learn whether in fact the member actually owns an essential claim, and if so, whether it would decide to charge a fee. The second, and less obvious, reason is that when a member is permitted to agree to license without disclosure, it is less likely (at least in theory) to review its patent portfolio in detail to determine whether or not it owns any essential claims. Were disclosure required, it is sometimes said, more essential claims would be discovered and more licenses would be required. On the other hand, some SDOs require specific disclosure of every essential claim, whether or not the participant has committed to license them, to inform working groups and potential implementers about the potential patent landscape of various technical directions, and to enable them to assess the advisability of working around patent-heavy approaches. A manager should therefore determine what the practice, or policy, of his or her own company would be when evaluating whether to join an SDO that requires disclosure of essential claims in order to seek compensation when licensing those claims.

Use of Information. SDOs vary in what they do with the information disclosed, and licensing obligations made, during the standards development process. Some restrict access to this data to their members, while others post the information on a public part of their Web sites. In recent years, pilot efforts have been made in Europe and the U.S. to index such information, not only for the benefit of standards implementers, but also to make it available to patent examiners for use as prior art, thereby helping raise the quality of new patents that are issued (NRC 2013). Managers evaluating a given SDO should therefore not only be aware that SDO practices in this regard vary, but also that they may change in the future.

FRAND Definition. IPR policies rarely define what FRAND is intended to mean, leaving disagreements on this topic to be resolved in court, which is occurring with increasing frequency. Policies will often, however, clarify what is intended in two cases, stating that "defensive suspension (or revocation)" and/or requiring a "reciprocal license," are considered to be consistent with a FRAND licensing obligation (ABA 2007). The former term means that an entity with a FRAND licensing obligation may revoke a license if the licensee later asserts one of its own patents against the licensor, thus leveling the playing field between the two companies. Most of these policies permit defensive suspension of a FRAND license if the asserted patent claims would be essential under the same standard as the licensed claims. A few SDOs have gone farther, permitting suspension if the licensor is sued for patent infringement at all, but this is uncommon and may raise antitrust/competition law issues. In some cases, all SDO members with essential claims covering the same standard may be permitted to suspend or revoke their licenses if the licensee asserts its patents against any SDO member.

Where reciprocal licensing is permitted, the licensor is permitted to require that the licensee grant a license back of its own patents to the licensor. Typically an IPR policy states that this requirement extends only to the licensee's essential claims covering the same standard, but some companies are known to request greater coverage from more patents. Some IPR policies provide that if a member is bound by a royalty-free licensing obligation, it may charge a royalty if a licensee is only willing to extend a royalty-bearing license back.

As earlier noted, regulators have urged SDOs to provide greater guidance on what a FRAND obligation is intended to mean, and to urge, or require, disputes on this topic to be resolved through alternative dispute resolution mechanisms, such as arbitration (Lemley and Shapiro 2013, Contreras and Newman 2014). One major SDO that has recently updated its IPR policy to address some of these concerns is the IEEE-SA. However, the amendment process is typically long and contentious, and few other SDOs have successfully met this challenge to date. Managers should nevertheless be alert to the possibility that expanded definitions, and perhaps terms relating to FRAND dispute resolution, may begin to appear with more regularity in the future.

Reach of Obligations. Managers employed by companies with large and complex corporate structures should pay particular attention to the extent to which the obligations under an IPR policy may extend to affiliates of their company. Typically, the obligations imposed by the policy will apply in one of two ways. The most common approach is to capture the patents owned only by the entity that becomes a member and its own direct and indirect subsidiaries. A smaller number of SDOs apply their policies to the entire corporate family of their members and their aggregate patent portfolios. When a small subsidiary company joins such an SDO, all of its direct and indirect parent companies and their respective subsidiaries and other affiliates are affected. If the SDO has thought the situation through, it may require the ultimate parent company to enter into a written commitment to be bound by the terms of the IPR policy, often by countersigning the membership application or agreement. For a variety of reasons, a meaningful number of companies find that they are unable to obtain such a signature. Some SDOs have mechanisms in place to grant exceptions in such circumstances. Managers should therefore pay particular attention to an IPR policy's definition of words such as "Affiliate" and "Related Party."

Obligations will also typically apply to patents that are "controlled" by a member. In this context, "control" usually means that a patent is either owned or co-owned by the member, or licensed by the member from a third party with the right to grant sub-licenses. Often, the policy will limit licensing obligations by excluding patents when the member is required to obtain the permission of a co-owner, or to pay a third party a royalty, if the patent is sublicensed. These exclusions are particularly important under IPR policies that include a blanket licensing option, since a member may otherwise (still) need to perform a careful patent review before deciding whether or not it can safely agree to this obligation.

Essentiality. Managers should also determine whether essential claims arise under all, or only some parts, of a standard. Some policies will restrict the definition of "essential" claims to "required" or "mandatory" elements of a standard, but not to extensions or other optional elements. When all elements are included, a licensing obligation is therefore incurred that extends to parts of a standard from which the member may gain no benefit. Relatedly, a policy may state that a member's licensing obligation applies only when an implementer implements all of the required elements of a standard, or it may instead state that the obligation will arise even in the case of a partial implementation. In the case of an interoperability standard, this may mean that a member may receive no benefit at all when a firm implements less than all of the standard, but may still have an obligation to provide a license at a below market rate, or even for free. Managers should carefully review the definition of "essential" or "necessary" in the SDO IPR policy, as it could have a significant impact on which patents are subjected to a licensing commitment.

Penalty terms. Some IPR policies include terms that specify the consequences of a failure to return a disclosure form within the required time period. Often, the penalty is that the member will be obligated to grant a FRAND license with respect all of its essential claims under the standard in question. As a result, when an SDO has a blanket disclosure option, few of its members may bother to return a disclosure and election form at all; as a practical matter, the licensing obligation is the same whether or not the form is returned. In a few SDOs, however, failing to return such a form may result in an obligation to license the member's essential claims under the relevant standard on a royalty-free basis, which is a far more serious penalty.

Most IPR polices are vague regarding the consequences for failure to disclose a particular essential claim, and some SDOs treat an undisclosed claim as though no form had been returned at all. Other SDOs do not differentiate between essential claims that are owned by members or third parties if they are asserted after adoption of a standard. Typically, in such cases the owner of the essential claim will be contacted and asked if it will commit to grant a FRAND license. If the request is refused, the situation may be referred to the Board or the Technical Committee for "appropriate action." In some cases, SDOs will impose penalties on participation or, in the most extreme cases, eject a member from the SDO if repeated violations have occurred.

Members that discover that they have failed to disclose an essential claim that they would wish to withhold or on which they wish to charge a royalty for may find themselves in an uncomfortable position, as they may be accused of bad faith if they later assert the claim. This can have serious consequences of two sorts: in the extreme case, potential licensees may complain to enforcement agencies that the omission represents a violation of law (as occurred in the case of Dell Computer, referred to above), or may raise a defense that the non-disclosed patent(s) are unenforceable when the patent holder seeks to enforce them. Second, the violation of an SDO's policies, if discovered, can give rise to penalties and sanctions within the SDO itself. These may include public reprimand, suspension of participation rights, rejection of the violator's affected technical contribution(s), automatic royalty-free licenses of the non-disclosed patents and, in extreme cases, expulsion from the SDO. But even when none of these formal sanctions occur, the reputation of a company may be damaged, resulting in a lack of trust when it continues to participate in the same and other SDOs. For this reason, some firms may elect to live quietly with the consequences of their failure to identify an essential claim in a timely fashion and simply refrain from asserting the newly discovered claim against implementers of the standard in question, despite the risks.

Waivers. Another important issue for managers to note in SDO policies is whether participants are required to waive other rights that they may have. For example, IEEE-SA recently imposed a requirement that participants waive the right to seek an injunction preventing an infringer from making products that comply with the SDO's standards except in certain limited circumstances. Other SDOs are considering similar measures. The theory behind such requirements is that FRAND commitments, which are generally required by such SDOs, mandate that the patent holder agree to grant a license to any implementer of the standard. In light of such a requirement, the patent holder should not have the ability to enforce its patent to prevent an implementer from using the patented technology, so long as the implementer is willing to take a license on that basis. Though such waivers are arguably consistent with many of the objectives and policies of SDOs, they can also have a serious impact on the value of an enterprise's patent portfolio. While terms of this nature are currently very rare, they may become more frequent in the future. Thus, the manager evaluating an SDO's IPR policy should take careful note of provisions that require participants to waive rights afforded to them under the law.

Patent Transfers. In today's technology marketplace, patents are changing hands with increasing rapidity. It is inevitable that some patents transferred individually or in large portfolio sales, corporate mergers and other transactions will have been subject to commitments regarding licensing or non-assertion in the standards context. The law is not entirely settled regarding the effect on such a commitment of the transfer of the underlying patent. Do such pledges automatically travel with patents, binding their new owners, or are they personal commitments binding only on the original promisor unless specifically included in patent transfer documents?

This question was central to the U.S. FTC's 2008 action against Negotiated Data Solutions (N-Data), the successor to a patent essential to IEEE-SA's fast Ethernet standard.[9] The original patent owner, National Semiconductor, committed that it would license the patent to all implementers of the standard for a flat fee of $1,000. National eventually sold the patent, and after a second such transfer, the patent was held by N-Data. N-Data, which was aware of National's original $1,000 licensing commitment, nevertheless announced that it would seek dramatically higher royalties. The FTC brought an action against N-Data, alleging that its disregard for National's original commitment was an unfair method of competition in violation of the FTC Act. The action was settled with N-Data agreeing to honor National's original pledge.

The issue arose again in 2011, when the trustee of bankrupt Nortel Networks, a significant contributor to telecommunications and computer networking standards, proposed the sale of its remaining assets, including approximately 4,000 patents, "free and clear" of many existing encumbrances.[10] Several product vendors, together with IEEE-SA, intervened in the case, arguing that Nortel's "free and clear" sale could invalidate several standards-related licensing commitments that Nortel had made. Ultimately, the purchaser of the patents, a consortium including several large product vendors, agreed to abide by Nortel's prior licensing commitments.

To address the issue of patent transfers, a growing number of SDOs require that purchasers of patents agree to abide by commitments made by the previous owners of those patents. The goal of these requirements is not to prohibit the transfer of patents, but to ensure that the transferees of patents, whoever they may be, remain bound by the original owner's commitments. For example, some SDO policies now require that a participant that wishes to transfer a patent that is subject to a licensing commitment is responsible for ensuring that the transferee contractually agrees to abide by the commitment and to ensure that any subsequent transferee makes a similar commitment.

While such requirements may benefit the SDO and overall market of firms implementing a standard, they also make it more difficult to sell patents unless careful records of all incurred obligations are created. When they are, the existence of such obligations may decrease the market value of the patents involved. The enterprise that is considering joining an SDO with such requirements on onward transfers should reflect carefully on what impact, if any, such requirements could have on its ability to complete an important business sale that includes patents essential to the SDO's standards. The enterprise should also take such requirements into account when evaluating the reliability of potential purchasers of patents, as a violation of a patent licensing commitment by a subsequent owner could, under some circumstances, result in some liability for the original patent owner.

Patent Pools

Business managers should also be aware that in some cases a "patent pool" may be created in conjunction with (but usually independent of) an SDO. In such a case, the implementer of a standard may be able to acquire the right to license some or all relevant patent claims from a single source and for a single, package price. While such a facility can be useful in a situation where claims essential to the implementation of a standard are owned by multiple parties, only a relatively small number of such pools have been created due to the difficulty of rallying a sufficient percentage of patent owners to the cause and the up-front expense of identifying patents that are "essential" to the standard for inclusion in the pool.

CONCLUSION

The above issues represent the principal intellectual property considerations that arise under SDO IPR policies. A manager evaluating the pros and cons of participating in an SDO must be familiar with these issues and able to guide his or her enterprise accordingly. The above discussion will hopefully give the manager a suitable overview of these issues and the more important provisions of SDO IPR policies to assist in this important evaluation.

REFERENCES

ABA Section of Antitrust Law. (2011). *Handbook on antitrust aspects of standard setting* (2nd ed.). Chicago, IL: ABA Publishing.

Bekkers, R., & Updegrove, A. (2012, September). *A Study of IPR Policies and Practices of aRepresentative Group of Standard Setting Organizations Worldwide*. Symposium conducted at a meeting of the National Academies of Science. Retrieved from http://sites.nationalacademies.org/xpedio/groups/pgasite/documents/webpage/pga_072197.pdf

Biddle, B., Curci, F. X., Haslach, T. F., Marchant, G. E., Askland, A., & Gaudet, L. (2012). The expanding role and importance of standards in the information and communications technology industry. *Jurimetrics*, *52*, 177–209.

Blind, K., Bekkers, R., Dietrich, Y., Iverson, E., Kohler, F., Muller, B., . . . Verweijen. (2011). *Study on the interplay between standards and intellectual property rights (IPRs) (final report).* Berlin: Author. Retrieved from http://ec.europa.eu/enterprise/policies/european-standards/files/standards_policy/ipr-workshop/ipr_study_final_report_en.pdf

Brooks, R. G., & Geradin, D. (2011). Interpreting and enforcing the voluntary FRAND commitment. *International Journal of IT Standards and Standardization Research*, *9*(1), 1–23.

Cannady, C. (2013). *Technology licensing and development agreements*. Oxford University Press.

Carrier, M. (2009). *Innovation for the 21st Century – Harnessing the power of intellectual property and antitrust law*. Oxford University Press. doi:10.1093/acprof:oso/9780195342581.001.0001

Cary, G. S., Nelson, M. W., Kaiser, S. J., & Sistla, A. R. (2011). The case for antitrust law to police the patent holdup problem in standard setting. *Antitrust Law Journal*, *77*(3), 913–945.

Contreras, J. L. (Ed.). (2007). *Standards development patent policy manual*. Chicago, IL: ABA Publishing.

Contreras, J. L. (2013). Fixing FRAND: A pseudo-pool approach to standards-based patent licensing. *Antitrust Law Journal*, *79*(1), 47–97.

Contreras, J. L. (2015a). Patent pledges. *Arizona State Law Journal*, 47(3).

Contreras, J. L. (2015b). A market reliance theory for FRAND commitments and other patent pledges. *Utah Law Review*, *2015*(2), 479-558.

Contreras, J. L. (2015c). A brief history of FRAND: Analyzing current debates in standard-setting and antitrust through a historical lens. *Antitrust Law Journal*, 30(1), 39-120.

Contreras, J. L., & Hernacki, A. T. (2014). Copyright termination and technical standards. *University of Baltimore Law Review*, *43*, 221–254.

Contreras, J. L., & Newman, D. L. (2014). Developing a framework for arbitrating standards-essential patent disputes. *Journal of Dispute Resolution*, *2014*, 23–50.

Contreras, J. L., Roth, H., & Lewis, M. (2011). Toward a rational framework for sustainable building materials. *Standards Engineering*, *63*(5), 1–17.

Cotter, T. F. (2014). Comparative law and economics of standard-essential patents and FRAND royalties. *Texas Intellectual Property Law Journal*, *22*, 311–363.

Ernst, D. (2012). American's voluntary standards system – A "best practice" model for innovation policy? *East-West Center Working Papers, Economics Series*, *128*, 1-57.

European Commission Directorate-General for Enterprise and Industry (EC). (2014). *Patents and Standards: A Modern Framework for IPR-Based Standardization*. Author.

Farrell, J., Hayes, J., Shapiro, C., & Sullivan, T. (2007). Standard setting, patents, and hold-up. *Antitrust Law Journal*, *74*(3), 603–670.

Gilbert, R. J. (2011). Deal or no deal? Licensing negotiations in standard-setting organizations. *Antitrust Law Journal*, *77*(3), 855–888.

Herman, M. K. (2010). Negotiating standards-related patent licenses: How the deal is done (Parts 1 & 2). *Landslide*, *3*(2-3), 31-41.

Kesan, J. P., & Hayes, C. M. (2013). FRAND's forever: Standards, patent transfers, and Licensing Commitments. *Indiana Law Journal (Indianapolis, Ind.)*, *89*, 231–314.

Layne-Farrar, A. (2010). Nondiscriminatory pricing: Is standard setting different? *Journal of Competition Law & Economics*, *6*(4), 811–838. doi:10.1093/joclec/nhq009

Layne-Farrar, A., Llobet, G., & Padilla, J. (2014). Payments and participation: The incentives to join cooperative standards-setting efforts. *Journal of Economics & Management Strategy*, *23*(1), 24–49. doi:10.1111/jems.12046

Layne-Farrar, A., Padilla, A. J., & Schmalensee, R. (2007). Pricing patents for licensing in standard-setting organizations: Making sense of FRAND commitments. *Antitrust Law Journal*, *74*(3), 671–706.

Lemley, M. A. (2002). Intellectual property rights and standard-setting organizations. *California Law Review*, *90*(6), 1889–1980. doi:10.2307/3481437

Lemley, M. A., & Shapiro, C. (2007). Patent holdup and royalty stacking. *Texas Law Review*, *85*(7), 1991–2049.

Lemley, M. A., & Shapiro, C. (2013). A simple approach to setting reasonable royalties for standard-essential patents. *Berkeley Technology Law Journal*, *28*, 1135–1166.

Lichtman, D. (2010). Understanding the RAND commitment. *Houston Law Review*, *47*, 1023–1050.

Lundqvist, B. (2014). *Standardization under EU competition rules and US antitrust laws: The rise and limits of self-regulation*. Edward Elgar Publishing Limited. doi:10.4337/9781781954867

Maskus, K., & Merrill, S. A. (Eds.). (2013). Intellectual property challenges for standard-setting in the global economy. Washington, DC: National Research Council (NRC).

Michel, S. (2011). Bargaining for RAND royalties in the shadow of patent remedies law. *Antitrust Law Journal*, *77*(3), 889–911.

Miller, J. S. (2007). Standard setting, patents, and access lock-in: RAND licensing and the theory of the firm. *Indiana Law Review*, *40*, 351–396.

Samuelson, P. (2007). Questioning copyrights in standards. *Boston College Law Review*, *48*, 193–224.

Shapiro, C., & Varian, H. R. (1998). *Information rules: A strategic guide to the network economy*. Harvard Business Review Press.

Sidak, J. G. (2013). The meaning of FRAND, part I: Royalties. *Journal of Competition Law & Economics*, *9*(4), 931–1055. doi:10.1093/joclec/nht040

Sidak, J. G. (2015). The meaning of FRAND, part II: Injunctions. *Journal of Competition Law & Economics*, *11*(1), 201–269. doi:10.1093/joclec/nhv005

Skitol, R. A. (2005). Concerted buying power: Its potential for addressing the patent holdup problem in standard setting. *Antitrust Law Journal*, *72*(2), 727–744.

Swanson, D. G., & Baumol, W. J. (2005). Reasonable and nondiscriminatory (RAND) royalties, standards selection, and control of market power. *Antitrust Law Journal*, *73*(1), 1–58.

Updegrove, A. (2009). A concise introduction to free and open software. *Standards Today*, *8*(5). Retrieved from http://www.consortiuminfo.org/bulletins/aug09.php#feature

U.S. Dep't of Justice & Fed. Trade Comm'n. (2007). *Antitrust enforcement and intellectual property rights: Promoting innovation and competition.* Retrieved from www.usdoj.gov/atr/public/hearings/ip/222655.pdf

Vetter, G. R. (2007). Open source licensing and scattering opportunism in software standards. *Boston College Law Review*, *48*, 225–250.

KEY TERMS AND DEFINITIONS

Disclosure Obligation: A provision of some IPR policies requiring SDO participants to disclose, with some degree of specificity, SEPs that they hold covering the SDO's standards.

ETSI: The European Telecommunications Standards Institute, a European SDO responsible for many prominent wireless telecommunications standards.

FRAND: An acronym for "fair, reasonable and nondiscriminatory" license terms, which SDO participants are often required to offer to potential licensees of their SEPs. FRAND terms may be royalty-bearing or royalty-free.

IEEE-SA: The Institute of Electrical and Electronics Engineers Standards Association, a major international SDO which is responsible, among other things, for the 802.11 series of "Wi-Fi" wireless networking protocols.

Implementer: An entity that manufactures or sells a product, or offers a service, that conforms to a particular standard.

IPR Policy: A written document setting forth an SDO's rules and requirements concerning intellectual property (usually patents and copyrights) held by its participants.

Licensing Commitment: A provision of some IPR policies requiring SDO participants to commit to grant licenses of their SEPs to implementers of a standard.

Standards Development Organization (SDO): An organization that serves as a forum for interested parties to develop one or more technical interoperability standards. SDOs are often non-profit trade associations. The term "Standards Setting Organization (SSO)" is often used interchangeably with SDO.

Standards Essential Patent (SEP): A patent that will necessarily be infringed by a product or process complying with a particular technical standard. Some SDOs consider patents to be SEPs only if they cover mandatory portions of a standard, but others will consider any patent covering a mandatory or optional portion of a standard, or which is commercially necessary to utilize the standardized product, to be a SEP.

Voluntary Consensus Standard: A standard that is developed by private sector actors through open, consensus processes, the use and adoption of which is typically not mandated by a governmental authority.

ENDNOTES

1 *In re. Dell Computer Corp.*, 121 F.T.C. 616 (1996).

2 *Rambus, Inc. v. Infineon Technologies AG*, 318 F.3d 1081 (Fed. Cir. 2003), cert. denied, 124 S.Ct. 227 (2003); *In re. Rambus, Inc.*, 2006 WL 2330117, 2006-2 Trade Cas. 75364 at 53 (FTC, Aug. 2, 2006), rev'd, 522 F.3d 456 (D.C. Cir. 2008).

3 Berne Convention for the Protection of Literary and Artistic Works art. 9(2), Sept. 9, 1886, revised at Paris July 24, 1971, 25 U.S.T. 1341, 828 U.N.T.S. 221, available at http://www.law.cornell.edu/treaties/berne/overview.html.

4 Participants should be aware that if an IPR policy includes discriminatory terms, or if otherwise appropriate terms are applied in a discriminatory fashion, potential liability can arise under the antitrust and competition laws of relevant jurisdictions. Accordingly, at a minimum, any stakeholder should be able to eligible to participate in the SDO and/or obtain copies of its standards for a reasonable fee.

5 IETF Trust, Legal Provisions Relating to IETF Documents, Sec. 4 (Dec. 28, 2009), http://trustee.ietf.org/license-info/IETF-TLP-4.pdf.

6 The development of open source software is as much a movement as a method, inspiring strong opinions among many in the developer community. The community values that have arisen in the process complicate legal analysis because some licenses and licensing situations have assumed secondary meanings beyond what is explicit in the legal documentation.

7 *In re Dell*, 121 F.T.C. 616.

8 *Rambus. v. Infineon,* 318 F.3d 1081.

9 *In re Negotiated Data Solutions LLC*, FTC File No. 051-0094 (2008).

10 *In re. Nortel Networks*, Order Authorizing And Approving (A) The Sale Of Certain Patent And Related Assets Free And Clear Of All Claims And Interests, (B) The Assumption And Assignment Of Certain Executory Contracts, (C) The Rejection Of Certain Patent Licenses And (D) The License Non-Assignment And Non-Renewal Protections, No. 09-10138 (KG) (Del. Bcy., Jul. 11, 2011). Under Section 363(f) of the U.S. Bankruptcy Code, a bankruptcy trustee or debtor-in-possession may sell the bankruptcy estate's assets "free and clear of any interest in such property."

Chapter 11
IP and Electric Vehicles Standards:
Local Policies vs. Global Standards? Standardization Management in a Multi-Stakeholder Environment in China

Martina Gerst
Tsinghua University, China

Xudong Gao
Tsinghua University, China

ABSTRACT

This book chapter explores the interplay between IP and standardization in the case of New Electric Vehicles (NEV) in China. This case has been chosen because NEVs are an important part of urban E-mobility concepts and therefore currently on top of Government agendas in many countries in the world – also in China. NEV standardization takes place in an international multi-stakeholder environment, embedded in a rapidly changing, competitive and complex global environment, highly influenced by competing regional innovation policies. Today, we observe a technology-based competition over the inclusion of IPR in standards and the resulting capability to exert control over markets. Thus, national and regional innovation policies influence not only the standards development of electric vehicles in one particular market – in our case China - but also affect the standardization management of multi-national companies (MNC), for example in charging.

INTRODUCTION

Currently, China is the world's largest automotive market and highly interesting for Multi-National Automotive companies (MNC). China is poised to have reached about 25 million private cars and trucks sold in 2014 with an annual prediction of 30 million cars by 2020 (Wang, A., Liao, W.& Hein, A.P ., 2014). Like in Europe, the dependency on fossil fuels for internal combustion engines and growing

DOI: 10.4018/978-1-4666-9737-9.ch011

concerns of urban pollution made the China government acting, emphasizing the importance of New Energy Vehicles (NEV). Whereas a first time buyer of a combustion engine car has to participate in a lottery for a license plate, it is much easier to get one for a NEV (Wang, et al., 2014).

NEVs are part of electric mobility (e-mobility) concepts and are becoming now an integral part of many communities' sustainable energy concepts in the 21st century, gaining global importance particularly with regard to the building of sustainable mobile communities or even smart cities. Thus, due to an increasing demand of an expanding middle class requiring unlimited mobility, there is an urgent need to reduce greenhouse gas emissions. Therefore, NEVs can be one pillar to ensure sustainability in the transportation sector including this reduction and improving air quality (Gerst & Gao, 2013).

However, China is not just one of the largest and most attractive markets; it is also the most challenging one for MNCs manufacturing NEVs. The reasons for that are manifold. First of all, e-mobility is a cross section topic involving many stakeholders of different industries. In addition, the electric vehicle environment and market is still a very dynamic and volatile area with a number of ongoing challenges such as i.a. battery power reach, a lack of charging infrastructure (Fojcik &Proff, 2014). This is because NEVs have gained global importance as a new way of mobile urban life in what is called smart cities (EU, 2013). However, the dissemination of e-mobility is to a large extent based upon standardization of different interfaces, including vehicle engineering, energy supply, and the associated information and communication technologies (ICT) in NEVs. Thus, interoperability of NEVs' individual sub-systems is one of the central factors to gain market acceptance.

Furthermore, in the context of a still immature concept of electric mobility, interoperability standards serve as valuable enablers of innovation and are at the heart of a rising complexity in technology, laws and regulations, business organization and market structure. They do not only serve as enablers of innovation (Blind, 2013) but have also become a significant factor in the creation and shaping of markets (Gibson, 2007). They have also become a significant factor in international trade with considerable economic importance. Today, we observe a technology-based competition over the inclusion of IPR in standards and the resulting capability to exert control over markets. Thus, NEV standards should also be understood as a market regulation mechanism (Fomin, et al., 2011). This holds particularly in a Sino-European context where standardization and IP strategies as well as approaches to NEV standardization differ vastly; here, standards become a significant factor in the creation and shaping of markets. Thus, standards and standards development in this particular area also seem to be driven by other factors rather than by interoperability.

Since urban e-mobility concepts are currently on top of government agendas in many countries and regions in the world, this book chapter explores the interplay between IP and NEV standardization in China in a multi-stakeholder environment. Regarding this huge market for NEVs, the tensions between IP and standards are of relevance for MNCs as many of these tensions are rooted in the two different IPR and standardization systems (TCEITSRP, 2012). In practice, this situation may contribute to a dis-coordination of innovations related to NEVs, e.g. charging plugs, and may negatively influence innovations in the entire field. Moreover as previously mentioned, NEV standardization in general, not only in China, takes place in an international environment with many different players, embedded in a rapidly changing, competitive and complex global context, highly influenced by competing regional and local innovation policies like in China.

China has recently become very active in developing its own standards (e.g. TD-SCDMA) (Kwaka, Heejin &Fomin, 2011). It has also implemented its own innovation policy and derived IP strategies. China aims to position its standardization system as part of its national innovation strategy; standards are regarded as critical factor of the country's innovative capacity (Ernst, 2011). On the other hand, China as one of the major trade players and member of the World Trade Organisation (WTO), it has to comply with the agreement on 'Trade-Related aspects of Intellectual Property' (TRIPS agreement) (MacQueen, H. et al., 2008), and is supposed to recognize the Agreement on Technical Barriers to Trade (ATBT) (Ernst, 2011). Thus, the fundamental challenge for China will be to combine its leading role in international trade with its objective to strengthen its domestic innovation and Intellectual Property (IP) without violating WTO regulations. In addition, China together with Korea and Japan, has established Northeast Asian regional standard cooperation program called CJK (China Daily, 2012). This mechanism aims to focus especially on ICT because those technologies and the relevant standards are playing a major role not only in an e-mobility setting but also in the society that in the 21st Century is to a large extent dominated by information technologies.

Against this background, this chapter seeks to analyze 1) whether China's IP policies linked to national innovation policies in the technology field of e-mobility are stifling or promoting the development of NEV standards taking into consideration the global industry context, and 2) how this is affecting the standardization management of stakeholders such as the MNCs. This is of particular relevance for European industry stakeholders in China because here, the standardization environment is characterized by tensions triggered by different technology levels as well as by both new and well-established players with different sets of interests leading to different standardization dynamics. Consequently, the outcomes of standards setting often remain uncertain because they are subject to competing arrays of interests including driving and opposing forces (Gerst, 2013).

EV standardization highlights that technological and institutional changes create a dynamic environment in which China seeks establish its own standards regime to support its technological development and national innovation policy (Kennedy, Suttmeier & Su 2008). China has internalized the economic importance of IPR in national standards. Thus, international companies should be alerted because the Chinese – quite masterfully – keep the balance between promoting national interests and not violating WTO regulations. China has entered the established international standards landscape as a new player, aiming to shape the global economy. This could lead to a shift of power in international standardization, a situation which does not seem to be that terribly far-fetched (Jakobs & Gerst, 2013).

The framework of analysis draws upon the Social Shaping of Technology (SST) perspective, by explaining ICT innovations as history- and context-specific, actor-focused technological change processes (Williams & Edge, 1992). The SST framework provides the perfect tool to explore the complexity associated with innovation development by incorporating a multifaceted socio-technical perspective. Thus, the analysis will identify and look at the factors influencing NEV standardization.

Data was collected over a period of five years. Ethnographic observation was used as the major method for data collection, as the focus was on what people do rather than on what they say (Arksey & Knight, 1999). Observation was facilitated by the authors being in China during the period of data collection, which enabled them to obtain a deep understanding of China, the automotive market, stakeholders, and the broader context of electric mobility and role of NEVs (Myers, 1999). In addition, the paper is based on a German-China collaboration project in electric mobility, supplemented by industry and research literature and official policy documents (Ziegler & Gerst, 2012). Furthermore, applying the framework developed by Fomin and de Vries (2010), core stakeholders have been identified and interviews have been conducted with selected industry stakeholders in China.

With a focus on IP policies and stakeholders involved, the chapter will illustrate the current status quo in China and will outline if and how IP policies influence NEV standards management of MNCs.

After some background on the electric mobility multi-stakeholder environment, and the interplay between IP and standards, China's eco-system for NEVs is outlined, including an introduction into their standardisation system and strategy, as well as an introduction into their IP policies. Finally, the effects of China's IP policies on the standardisation management of MNCs and resulting strategies are discussed.

BACKGROUND

China is the world's largest automotive market since 2010 and highly interesting for European automotive companies. However, the growth rate has slowed down and customer expectations increased. In addition, the government has been driving automotive development for decades but more recently, with an increasing urbanisation going on, environmental concerns like air pollution, the government has indicated that it favours electric vehicles (Manyika, Remes & Woetzel, 2014). For sustainable communities, addressing the topic of transportation and its impact on the environment is vital, considering that mobility is essential for economic growth. On a global scale, the unprecedented growth in the global population has led to an expanding middle class, which is demanding increased mobility (SLoCAT, 2012). Due to this increasing demand, there is an urgent need to reduce greenhouse emissions. Therefore, NEVs can be one pillar to ensure sustainability in the transportation sector including this reduction and improving air quality (Cox 2012). However, the electric vehicle topic and respective technology aspects such as standardisation in general, not only in China, take place in an international multi-stakeholder context, embedded in a rapidly changing, competitive and complex global environment, highly influenced by competing regional innovation policies.

E-Mobility: A Complex Setting of Technology and Stakeholders

The availability of fossil fuels for internal combustion engines decreases while prices are on the rise (Hölk, 2012). NEVs are – entirely or partially – powered by electricity, typically stored in batteries. 1 They were invented at the end of the 19th Century, around the same time as combustion engines. Yet, NEVs were marginalised at the beginning of the 20th century (Van den Bossche, 2003). Today, however, they have become part of communities' electric mobility (e-mobility) concepts (EU, 2013) As illustrated in figure 1, the e-mobility environment consists of several layers, different components, and multiple stakeholders – implying multiple interests and multiple interfaces.

A NEV is a radical innovation that requires a new, cross-sector systems thinking. Accordingly, several features distinguish the standardisation processes in this field from those of other sectors where the main challenge is to coordinate activities in order to effectively meet demands (Gerst & Gao, 2013). NEVs, in contrast, are part of the still immature concept of sustainable mobility, embedded in a rapidly changing, competitive and complex global environment. This environment is influenced by competing regional innovation and IPR policies that reshape and intensify technology-based competition. Thus, economic success may well depend on the capacity to exert control over open architectures and interface standards, that nonetheless incorporate proprietary technologies.

E-mobility draws upon input from different industry sectors and from research. All stakeholders (including those from the automotive industry, the energy sector and the Information and Communication Technology (ICT) sector (BMBF, 2011) collaborate towards the development of the large number of ICT components and the necessary interfaces. The charging infrastructure is a highly relevant sub-

Figure 1. The electric mobility multi-stakeholder environment (According to figure 5 at Hölk, 2012).

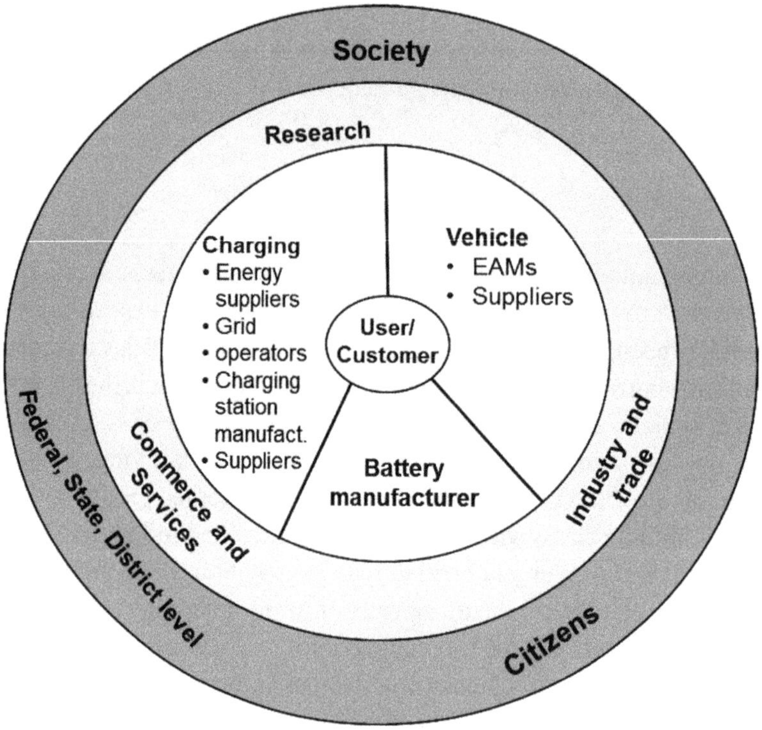

system with respect to NEV standardisation. Here, interoperability has to be ensured both between the individual components of NEVs and with the infrastructure. One current example are plugs; they define the interface between the vehicle and the grid, ensure interoperability and reduce the variety of different charging systems. This standardisation activity is a central pre-requisite for a broad take-up of e-mobility. Commissioned studies in the field clearly reveal some gaps for example in charging infrastructure standardisation and conformity assessment for NEVs as well as tensions between the stakeholders regarding harmonised standard (Ziegler & Gerst, 2012).

E-Mobility and Standards

When it comes to NEV standards, the focus is on interoperability and variety reduction standards. ICT interoperability standards are compatibility standards, based upon reference and similarity (TCEITSRP, 2012). Standardization in the field of e-mobility is characterized by several features distinguishing it from previous standardization processes. Standards in the electrical engineering/energy technology and in the automotive technology domain have thus far been considered as separate entities (FIA, 2011); there have only been very few attempts to look at them from a more integrated point of view.

Against this background, it is worthwhile to note that today standard setting in general has fundamentally changed from being a narrow 'technical' issue to a means of alignment of individual interests between the different players (Williams (2009). One of the main reasons is the fact that interoperability standards have become a significant factor in international trade (Gibson, 2009) with considerable economic importance and serve as valuable enablers of innovation (Jakobs & Blind, 2011). At a time when globally implemented technologies such as advanced Information and Communication Technologies (ICT) increasingly require compatible and harmonized standards to be fully effective, the role of standards is of increasing policy importance (Gibson, 2009).

Actually, the standard setting practice is subject to very different challenges for all stakeholders. Firstly, standard setting is characterized by an on-going and increasing convergence of IT systems and their global implementations, coupled with, and further accelerated by, the Internet. Hence, standard setting processes are embedded in a rapidly changing and complex standardization environment, driven by the growing importance of ICT and the globalization of markets on the one hand (Cox, 2011) and the respective national innovation policies on the other. Secondly, the standardization environment is characterized by tensions triggered by different technology levels as well as by both new and well-established players with different sets of interests leading to different standardization dynamics (Jakobs, 2011). Consequently, the outcomes of standards setting often remain uncertain because they are subject to competing arrays of interests including driving and opposing forces (Jakobs, 2011).

However, companies have understood that for a broad take-up of e-mobility standardization of different interfaces of an NEV is crucial. The challenge here is to continuously coordinate and integrate the diverse activities of stakeholders from different sectors in order to effectively meet customer demands. This holds particularly for European MNCs since decades being one of the industrial stakeholders benefitting from the robust China auto market.

The Interplay between Standards and Intellectual Property (IP)

Standards play an important role in technology-driven markets and may serve as catalysts of innovation, as leveraging and diffusion tools for IPR, particularly patents, or to influence markets (Blind, 2009). The traditional view was that standards and innovation contradict each other, thus hampering the integration of standardisation into innovation policy and innovation management. These days, the relation between standards and innovation is viewed differently. In principle, the former are seen as a common platform for the latter, especially in Research and development (R&D). Thus, IPRs can provide the creator with the temporary right of exclusive exploitation (Blind, 2009). Therefore, the prospect of incorporation of IP into standards and its associated wider dissemination may well be tempting for IPR holders.

However, the commercial success and economic impact of R&D results will only be realised when transferred into innovative products. Here, standards serve as a mechanism to support technology transfer (Blind, 2009). Due to the recursive interdependence of R&D and standards, high economic efficiency can be achieved if publicly funded R&D results become public goods via standards (Blind & Gauch, 2009). Unlike patents, standards are available to everyone at reasonable costs, and are more likely to be implemented as they represent a consensus reached amongst the stakeholders (Blind, 2009).

From an economic point of view, IPR are understood as 'appropriation mechanisms', playing at least three different roles in promoting technological diversity: 1) providing an *incentive* for R&D activity, 2) *diffusing* economically useful information and, 3) promoting a desirable level of *co-ordination* for R&D activity. Thus, the integration of IPR into standards has 1) an incentive, 2) a disclosure, and 3) a coordination function (Blind, 2009).

1. The possibility to include patents allows the IP holder to leverage a temporary monopoly based on the awarded IPR, thus generating additional incentives for investment in R&D (McQueen et.al, 2008). However, expanding the temporary into a permanent monopoly could restrict not only com-

petition but also innovation (Blind, 2013). In addition, this might cause a lock-in into an inferior, outdated standard. Even open and IPR-free standards may create incentives for R&D if they serve as platform standards upon which innovation may be based.

2. As outlined by Swann, standards influence the dissemination of innovations and technologies (Blind, 2009) and promote the diffusion of incorporated IPR through licensing. Patent pools may be used to reduce transaction costs for both patent owners and standard implementers; additional licensing revenues are generated for the former and licensing costs are reduced for the latter (Bekkers, et.al, 2012). Therefore, incentives are increased and the diffusion of the standard and the included IPR is promoted (Blind, 2009).
3. The coordination function avoids not only parallel developments of technologies but also supports the transition from old to new technologies It also helps reduce inefficiencies in transition periods of new technologies (Blind, 2009).

Therefore, patents, the central IP in the area of electric mobility, are essentially an incentive-oriented appropriability mechanism linked to a diffusion-oriented disclosure mechanism (Blind et.al, 2002). Standards and patents are mechanisms for knowledge creation and diffusion. However, the major difference is that patents are used to secure private property whereas standards diffuse technology in the public interest (Lea & Hall, 2004). Nevertheless, it has been argued that the inclusion of IPRs can only have a negative effect on technical standards development and usage due to this inherent conflict between the two. Globalisation of markets, the fast pace of technology change, and the convergence of technologies have contributed to the way on how patents and standards are used and have increasingly led to conflicts (Blind et.al, 2002). In consequence, the balancing of private and public knowledge is a constant challenge for SDOs, government agencies, and industry because any such conflicts may result in higher costs. Particularly in ICT, the intensity of patenting and the development of standards have increased dramatically (TCEITSRP, 2012). This situation especially holds for China. Taking into consideration the focused innovation policy strategy for home grown IP/patents, the question is arising whether those particular IP policies targeting e-mobility embedded in a global industry context, are stifling or promoting the development of NEV standards.

CHINA'S ECO-SYSTEM FOR ELECTRIC VEHICLES

The Standardisation System of China and Standards Development

The European approach and perspective towards the Chinese standardisation system can be characterised as more cooperative than competitive – a fact that may be rooted in the EU's co-regulation tradition (Meyer, 2012). Figure 2 illustrates the relations between the Chinese organisations involved in standardisation.

The standards system in China is administered by the General Administration of Quality Supervision, Inspection and Quarantine (AQSIQ). AQSIQ is a ministerial level administration directly under the supervision of the State Council. AQSIQ administers both the Certification and Accreditation Administration of China (CNCA), supervising all certification and accreditation activities throughout China, and the Standardization Administration of China (SAC), established by the State Council in 2001, in charge of National Standards (Jakobs & Gerst, 2013).

Figure 2. Overview of China's Standardisation System based on the Standardisation Law of 1988[2]

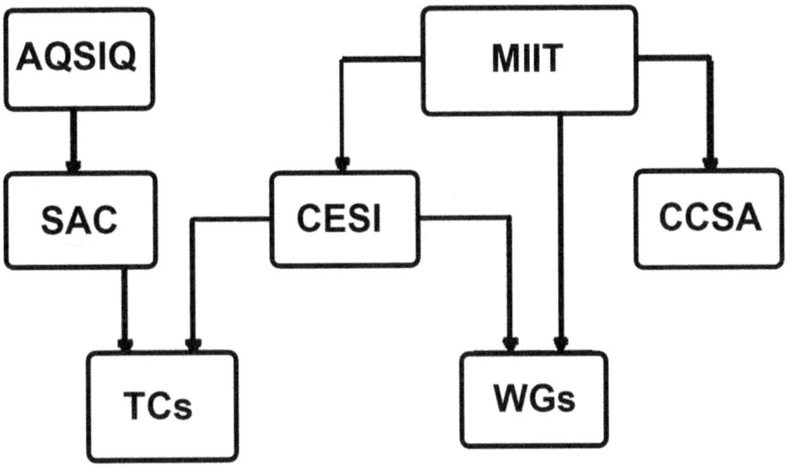

AQSIQ: General Administration of Quality Supervision, Inspection, and Quarantine
MIIT: Ministry of Industry and Information Technology
SAC: Standardization Administrative of China
CESI: China Electronics Standardization Institute
CCSA: China Communications Standards Association

On the international level SAC represents China in ISO and IEC. Chinese National Standards are owned by SAC, whilst all other standards are supervised by SAC and ideally also registered with them. In addition to SAC, a number of other entities are also involved in ICT standardisation (Jakobs & Gerst, 2013). China Electronics Standardization Institute (CESI) is a professional institute for standardisation in electronics and ICT. It is administered by the MIIT. Among other activities, CESI manages 51 national mirrors groups of IEC Technical Committees (TC) and Sub-Committees (SC) and ISO/IEC Joint TC1 (JTC)/SCs. China Communication Standards Association (CCSA) is the Chinese member of the Global Standards Collaboration, the association of national/regional SDOs active in the area of telecommunication. Like CESI, it is administered by MIIT (Jakobs & Gerst, 2013).

The Chinese standardisation system includes both, voluntary and compulsory standards and distinguishes four levels of standards, all of which should be registered with SAC: national, industry, local, and enterprise standards (Jakobs & Gerst, 2013).

In general, the standardisation systems in both China and Europe are modelled upon the ISO/IEC process which is based on the consensus amongst the interested stakeholders such as industry, users, interest groups, public authorities (Sherif, 2003). However, mostly researchers from universities and public research institutes are dominating the standards development work, whereas trade associations and enterprises are comparably unimportant (Rongping & Zhuoliang, 2005). The members of the WGs participate as individuals and do not have to represent any position other than that of their own expertise (Jakobs & Gerst, 2013). Regarding the integration of standards consortia, the preferred approach of China is to work rather through institutionalized standardization organisations than involving consortia (Suttmeier et al., 2006) which has been critisised for excluding managerial innovation (Wang et al., 2010).

As far as China is concerned, Stuttmeier's & Xiangkui's (2004) observation that "In 2001 and 2002 China thus began to reform its standards regime with an eye toward building standard setting into its national research and development programs as a priority objective. This has now resulted in a system in which policy purposes for the standards regime – expressed through laws, administrative directives, and policy statements – are increasingly integrated with a research and development (R&D) network". Today, the draft standardisation strategy (MST, 2012) puts a very strong emphasis on the link between R&D and standardisation. What's more, it makes concrete recommendations how this link shall be established and strengthened, respectively.

In China, standard setting has been included into the national research and development programs as a priority objective since reforms of the standards system started in 2001 and is build-upon the strong IP policies introduced in paragraph 3.3.

The Standardisation Strategy of China

In 2012 the Chinese Ministry of Science & Technology released a Draft 12th Five-Year Special Plan on Technical Standards for comments (MST, 2012). The document observes that "*Technical standards are the technical basis of social and economic activities, a strategic resource for national development and a core element for international competitiveness*". The strategy identifies how technical standards are to be used and considered in planning, R&D, Advanced Development, Testing and Certification, Intellectual Property, and a host of other areas.

Research and Development in support of standardisation is a major focus of the strategy. This holds specifically for the creation of stronger links between R&D and standardisation, in both directions. 'Next Generation Information Technology Industry' is the second of the identified 13 major areas where science and technology shall support the development of new technical standards. Dedicated R&D efforts shall also help to promote (superior) national standards in the international arena. This also holds for the development of industry and 'alliance' standards and their subsequent elevation to national standards. 'Capacity building' is another major aspect of the strategy. This refers to an improved education on standards-related matters, including both technical and non-technical (e.g. procedural) aspects and also to an improved information flow between the different parties involved in standards setting and implementation (Jakobs & Gerst, 2013).

Overall, the strategy addresses all aspects relevant to the development and use of standards. A particularly strong emphasis is put on the integration of education, R&D and standards development. Standards play an important role in technology-driven markets and may serve as catalysts of innovation, as leveraging and diffusion tools for IPR, particularly patents, or to influence markets (Blind, 2009). The traditional view was that standards and innovation contradict each other, thus hampering the integration of standardisation into innovation policy and innovation management (Blind, 2009). These days, the relation between standards and innovation is viewed differently. In principle, the former are seen as a common platform for the latter, especially in Research and development (R&D). Thus, IPRs can provide the creator with the temporary right of exclusive exploitation. Therefore, the prospect of incorporation of IP into standards and its associated wider dissemination may well be tempting for IPR holders (Blind, 2009).

In Europe and China, standards are used as market access tools. Together with IPRs they provide an incentive for the creation and diffusion of knowledge. In order to increase the number of IP, China has issued an array of innovation policies.

China's IP Policies

China's detailed innovation Policy framework is complex; figure 3 offers a simplified view.

1. The framework for China's innovation policies was initiated in 2006 with its 15-year National Medium- and Long-Term Program for Scientific and Technological Development (2006–2020) (MLP) (Stevenson-Yang & DeWoskin, 2005). In order to become an "innovative society" by 2020, and adopting a systemic approach, the Government aims to integrate and coordinate the policies and institutions in law, finance, management, education, and foreign trade (McGregor, 2012) in order to strengthen the country's capabilities in science and technology (Kennedy et al., 2008). The focus of the MLP is on what is often referred to as "indigenous innovation" (Cao et. al., 2006), i.e. incorporating national IP into national standards to demonstrate technological sovereignty, national power and international influence (Kennedy et al., 2008). Hence, the filing of patents and the development of standards are considered important "outputs" in research, and a company's IP and standards record influences its eligibility for technology policy privileges Moreover, the plan also calls for China's industrial enterprises to become the core of the national innovation system (NIS). Therein, standards development is a key task, supported by special R&D programs for standards (Kennedy et al., 2008). Besides a number of quantitative targets for the scientific and technological development, the R&D program with sixteen megaprojects aims to trigger the catch-up process by using, assimilating and improving foreign high technology in order to produce indigenous IP and standards (Suttmeier & Yao, 2011). To enforce indigenous innovation, China has implemented a number of industrial policies (OECD, 2008), as well as special national strategies in education, human resource and IP development (Suttmeier & Yao, 2011).

Figure 3. Overview of China's innovation Policy Framework[3]

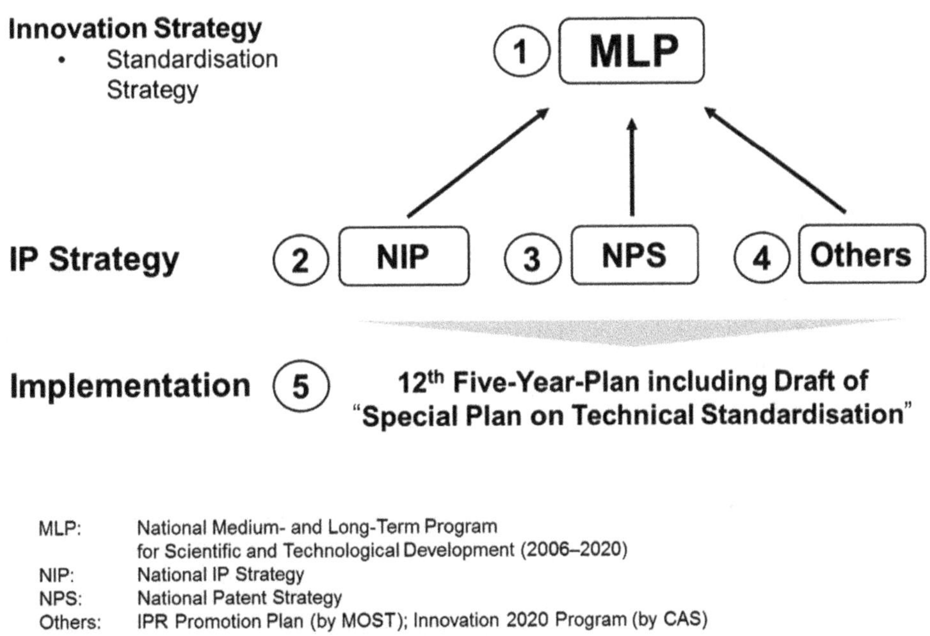

2. Key policy measures and strategic goals of the National IP (NIP) strategy of 2008 [4] address patents, trademarks, copyrights, and trade secrets. NIP emphasises the protection, creation and use of IP, especially for standards (Suttmeier & Yao, 2011). The measures also include the improvement of the IP regime, like preventing IP misuse or abuse. Furthermore, the strategy emphasises an IPR culture, education of IP professionals, and training for civil servants (SIPO, 2013). Also, the importance of international cooperation and exchange of information on IP matters are listed in the 'Promotion Plan for the Implementation of the National Intellectual Property Strategy in 2013'. Quantitative benchmarks are introduced to evaluate the progress and to reinforce IP protection, e.g. through the implementation of an evaluation tool.
3. The National Patent Strategy (NPS) of 2010 also stresses the importance of IP for the country.[5] It focusses on government activities, but also on market forces, emphasising the importance of protecting IPR owners and the public interest. This is supported by the establishment of a "patent service industry" (Suttmeier & Yao, 2011).
4. Other related initiatives include the "IPR Promotion Plan (2010)" by the Ministry of Industry and Information Technology (MIIT) to support copyright-intensive industries and the "innovation 2020 Program" by the Chinese Academy of Sciences (CAS), a major R&D programme emphasising national IP and technical standards, as well as innovations programmes launched by local governments (Suttmeier & Yao, 2011).
5. The 12th Five-Year-Plan (FYP) from 2011(APCO, 2010) focuses on the "emerging strategic industries programme", covering IT and NEVs, among others.[6] The plan aims to stimulate knowledge-intensive developments, exp and the community of IP stakeholders, form patent alliances among companies, and simplify the IP transfer from university or R&D institutions to industry (Suttmeier & Yao, 2011). In 2012, the Ministry of Science & Technology (MOST) released a draft "Special Plan on Technical Standards" for comments (MST, 2012). The draft understands technical standards as strategic resource for national development and as a core element for international competitiveness". It further notes that national IP is an important component of standards in order to play a vital role in international cooperation as well as in competitions between standards (MST, 2012).

STANDARDISATION MANAGEMENT OF MNC IN CHINA

The Effects of Local IP and innovation Policies on the Standardisation Management of MNCs

innovation policies are promoting the development of IP and standards in new technology fields. However, local ones may have a negative effect on standards development and management in a global context that is characterised by harmonised standards to ensure interoperability. Thus, IP and standardisation management of MNCs marketing NEVs in China are shaped by four factors. This part of the chapter explores to which extent they influence a respective strategy in standardisation management.

Innovation and IP in the Automotive Industry in China

Not only in China but also in Europe IP affects standardisation through various policies and their associated strategies that are embedded in the respective standardisation systems. However, both regions vary considerably regarding the emphasis put on the importance of national/regional IP. China explicitly stresses this importance for its economic development. In contrast, European IP policies and strategies rather implicitly focus on strengthening the internal market and its global market competitiveness, with no strong ties to standardisation, except for ICT. Worthwhile to take into consideration that Europe and China differ considerably with respect to their economic situation; Europe is a highly industrialised region and China is still an emerging economy with an annual growth rate of more than 7% per year (Reuters, 2015). In Europe and China, standards are used as market access tools. Together with patents they provide an incentive for the creation and diffusion of knowledge. As described before, in China, the strong link between R&D and standardisation is a major focus of the standardisation strategy (Jakobs & Gerst, 2013). However, the commercial success and economic impact of R&D results will only be realised when transferred into innovative products. Here, standards serve as a mechanism to support technology transfer (Blind, 2009). Due to the recursive interdependence of R&D and standards, high economic efficiency can be achieved if publicly funded R&D results become public goods via standards (Blind & Gauch, 2009). Unlike patents, standards are available to everyone at reasonable costs, and are more likely to be implemented as they represent a consensus reached amongst the stakeholders (Blind, 2009). Thus, the integration of IP into standards matches all three functions outlined in the paragraph before, the incentive, disclosure, and coordination function.

Outlined in the last paragraph, NEV policies and standardisation strategy are in line with the 12th FYP, aiming to establish a scientific, systematic, open, and adjustable NEV standards system which fully meets research, industrialisation and commercialisation requirements and which represents an important technical support tool for the e-mobility industry (Ziegler & Gerst, 2012). The idea behind this is to transform a large number of achievements and advanced experiences into standards and to get involved in relevant international standard activities. Hence, China's standards development attempts aim to transform the country from a follower to a standard setter in the automotive sector. The emphasis will be on the coordination of a key framework of national standards with both industry-specific and enterprise standards.

Since the first standards for NEVs were released in 2001, they have become the basis for major projects like the State '863' program[7] launched by the MOST (Earley et.al. 2011). The "Renewable energy automotive manufacture and product access management", released by the MIIT in June 2009, stipulates that NEVs have to meet the existing standards for market access. In addition, NEV standards are supposed to play also an important role in manufacturing (Ziegler & Gerst, 2012). Since the 9th 5-Year Plan, 57 NEV standards have been published, seven standards are waiting for approval, 19 new standards are under development, and 45 standards are under preliminary research (Ziegler & Gerst, 2012). From a Chinese point of view, this is obviously a success of the implementation of the numerous strategies; from a European point of view this large number of national standards has been criticised as contradicting the WTO agreement, and absolutely not being in line with harmonised international standards, and therefore impeding market access.

Moreover, as illustrated in figure 1, NEVs are created by a network of stakeholders (Powell & Grodal, 2005); they comprise different components that need to interoperate. Together, they are part of a larger network that also includes the charging infrastructure. Interoperability standards serve as the basis of this network. Accordingly, IPR in those standards receive increasing attention as they could cause various problems including competition issues (Lemley, 2002). These concerns stem from the character of those standards: They should include state-of-the-art technology (Mueller, 2002), which may well be based on patents, but should at the same time be available for implementation (Meir, 2012).

This has led to an increasing number of conflicts to be solved at the policy level (competition, IPR and standardisation policies) (TCEITSRP, 2012), and in other IP contexts (e.g. patent pooling, licensing schemes) (Bekkers et al. 2006). With regard to China as a huge market for NEVs, the tensions between IP and standards are of relevance for MNCs as many of these tensions are rooted in the two different IPR and standardisation systems (TCEITSRP, 2012). In practice, this situation may contribute to a discoordination of innovations related to NEVs, e.g. charging plugs, and may negatively influence innovations in this field that will be further discussed in the following chapter.

However, the national approach adopted by China is in line with China's efforts to conquer the global NEV markets with their own national standards. For China, NEVs are "THE" opportunity to counter the Western prominence in combustion engine technology. As described, the implementation of this IP strategy has led to contradictions and tensions on the way towards "indigenous innovation" efforts and IP development, due to the complex government-industry relationship (Suttmeier & Yao, 2011). To encourage patent applications by companies a variety of incentives and subsidies have been introduced. The result is overwhelming; the implementation of the IP strategy has indeed led to a filing of more than 1.6 Million patents with the State Intellectual Property Office (SIPO) in 2011 (Prudhomme, 2012). Thus, it seems the incentives have brought China closer to its goal of being a creative nation.

However, observers question the explosion of patent applications arguing that quality is more important than quantity. This claim is supported by the fact that utility models and design patents[8] are frequently applied for. These types of patents are considered as "junk" by foreign stakeholders due to their frequently low quality (Prudhomme, 2012). These observations have been confirmed by a study in 2012, finding that the Chinese patent policies hamper innovations rather than encouraging them. Moreover, the Chinese legal community is concerned about a degradation of the overall IP system because utility model and design patents are not examined for substance and thus allow abusive assertions for strategic purposes, for example in bringing legal action against Western companies (Suttmeier & Yao, 2011). Whereas the influence of IP in European standardisation is institutionalised via the ESOs' IPR rules, in accordance with international harmonised standards, China is still keen to incorporate national IP first into national, and then into international standards as there is evidence in the not to be neglected number of first home-grown NEV standards.

Although China's innovation policies are producing more patents and standards and thus confirming that those mechanisms are serving as R&D incentive, ensure dissemination of innovation, it becomes obvious that in terms of coordination, tensions occur with regard to international industry stakeholders pursuing a harmonised standardisation strategy and related management. The view from Europe is generally that standards and standardisation processes in China are barriers to market entry (TCEITSRP, 2012). One reason is the complex system of mandatory and voluntary standards with different regional validity. Moreover, some standardisations processes are shaped or even controlled by public policy and finally, the strong links between government and Chinese firms are found to be difficult. Interestingly, these views are largely mirrored by the Chinese views on the European system. This might be due to a mutual lack of information and knowledge that the EC and the ESOs try to overcome, e.g. through the EU-China Standardisation Platform (CESIP, 2013) whereas the European automotive industry has started to develop their own standardisation strategies to keep and defend their market share in the China market.

In general, vehicles are characterised by an increasing complexity of interoperable, modular systems. The industries' R&D and manufacturing are embedded in a global innovation and supply chain network. NEVs and the standardisation of their charging infrastructure require full interoperability between systems, thus increasing the economic impact of international standards. This confirms research results on China's standards and innovation policies and their evolving role in the global knowledge economy, undertaken by the National Bureau of Asian Research projects (Kennedy et al, 2008).

Although China has been the world's largest automotive manufacturing and sales market since 2009 and despite the fact that the automotive industry is one of China's "pillar industries", the slowly developing domestic R&D and innovation capabilities[9] are key barriers to its international competitiveness (APCO, 2010). Even extensive foreign investments through joint ventures (JV) [10] between European and Chinese companies have not improved the situation much. These investments aimed to exploit the situation by establishing a technology transfer platforms to locally develop technology and IP. One of the major issues in these JVs, often complained about by the European partners, is the insufficient protection of the foreign partner's IPRs; typical violations include reverse engineering and copying (Tang, 2009). Hence, MNCs in China have taken steps to protect proprietary technology, product design, or development. According to an MNC manager, one of the measures is *to "compartmentalise knowledge, so that only few employees have a complete overview of complex core technology."* Another one is to incorporate IP protection in internal 'codes of conduct' as a core part of a company's innovation culture". [11]

When talking to foreign companies, the perceptions of managers regarding innovation and IP policies vary considerably (Suttmeier & Yao, 2011). In numerous publications the US stress that China remains the world's worst IP infringer (Okun, 2010). In contrast, European voices are more restrained. However, even the European business community is concerned about the way in which the Chinese IP strategy is implemented as part of China's industrial and technology policies; but this is usually paraphrased in an 'IP-free' language. For example, the comment on the recent Five-Year Plan, although recognising the need for reforms, states that it still lacks a "meaningful implementation" which leaves room for interpretation (EUCCC, 2012).

Overall, for the majority of foreign companies doing business in China the industrial policies (outlined in the chapter before), still seem to be more an annoyance than a serious threat (Suttmeier & Yao, 2011). This has been confirmed by a study, initiated by the European Commission that investigated IP and trade secret issues under 'China's Compulsory Certification' procedure. Interestingly despite complaints, the study found that no tangible problems of know-how loss or concrete IP-related irregularities occurred (Gerst, 2010). EU stakeholders were rather hesitant answering IP-related questions.

However, one of the major concerns of the NEV community, represented by the European Automobile Manufacturers' Association (ACEA), is the implementation of the 'Administrative Regulation on Access for NEV manufacturers and Products' (ACEA, 2009). It stipulates that the IPR of at least one of the three core technology components of an NEV[12], i.e. the "modification right and usufruct of design" have to be transferred to the Chinese JV partner (ACEA, 2009). This Regulation seems to confirm the techno-nationalist approach to support indigenous innovation (Suttmeier & Xiangkui, 2004). European managers have expressed deep concerns concerning this paragraph (which is – interestingly – listed in the annex of the document). They observed that they *"have no choice to withhold the transfer of IP and therefore will select the least developed technology where the company as well would like to see further research progress and could realise purchasing benefits"*.[13] However, it has been argued by others that this technology transfer approach of China has to be understood in the context of its economic situation, i.e. its attempt to further its economic development (Ernst, 2011).

On the other hand, as already mentioned, the US tends to see China's approach to develop indigenous innovations as potentially damaging to free trade and competition while exercising 'mercantilist innovation policies' focusing on 'home grown' innovations (Atkinson, 2012). Again, although representatives of

MNCs agree with this point of view, they would not mention it publicly. Moreover, the Chinese Government requires MNCs to set up local NEV R&D centres if they intend to expand their China business, in another attempt to strengthen national IP and standards. Accordingly, all MNCs have established R&D facilities at Chinese universities, e.g. sponsoring PhDs or funding projects. All those R&D activities complement public R&D platforms and national research test centres (Tagscherer, 2012).

The MNCs also raise concerns regarding the barriers to move test items or prototypes into and out of China (ACWG, 2012). As automotive R&D takes place in global engineering networks, prototypes developed in China may have to be moved to other R&D centres outside the country and perhaps back again. Unfortunately, Chinese customs regulations prohibit or restrict the import and export of those products as well as of test equipment, creating long delays and thus hampering innovations. This seems to motivate the MNCs to carry out NEV research in countries without such customs restrictions. A manager states *"the long administrative and often not very transparent processes are one of the reasons why China leading edge technology is not developed in China."* [14]

China's Standardisation and Market Access Strategy

A study by Bresnitz and Murphree (2013) found that Chinese strategies to support innovation and technology do not necessarily follow one clear objective, due to biased interests of stakeholders and the different institutions involved (Bresnitz & Murphree, 2013). Moreover, embedding indigenous IP into standards has only recently started and is not considered as the real challenge for Western companies. This challenge would rather be to influence the mechanisms how "embedded IP in standards is monetised and licensed" (Bresnitz & Murphree, 2013). Their research observes a trend in China for inexpensive licensing of embedded national IP in technology standards.

China has entered the international standards arena as a new 'player', aiming to shape the global economy, using technological and institutional changes to create a dynamic environment in which China seeks to establish its own standards regime to support its technological development and national innovation policy (Kennedy et al, 2008). China follows a multi-track approach. As part of the international standardisation system it has the long-term perspective to incorporate patentable technology into global standards in order to strengthen its bargaining power and reduce royalty fees. In parallel, it uses its economic power to try and influence the international system itself by creating IP essential to global standards (Ernst, 2011).

Against the background of increased globalisation and the equally increasing complexity of technology, China links its government-centred standardisation strategy with a more open market-led approach (Ernst, 2011). Therefore, Chinese delegates are increasingly populating international SDOs' working groups, most notably in ITU-T (Levin, 2010). Chinese policy makers have realised that standards are crucial for market success. Therefore, Chinese policy circles share the view that only standard making companies may be considered as international leaders in technology (Breznitz & Murphree, 2013). However, MNCs in China struggle in the competition between the national and the international standards regime, the latter requiring the harmonisation of standards. The MNCs follow a rather conservative approach in standardisation; they participate in all international formal SDOs without any motivation to push own IPR into a standard. Rather, they aim to extend or create markets and develop through standards.

In the automotive sector, the promotion of "green technology" is high on the agenda; the government encourages local companies to further develop their technological capability, including IP and the respective standards, particularly with regard to own brands and to increased exports (Collis & Donnelli,

2012). Thus, the number of newly developed or approved national standards for NEVs is huge. In addition, international standards have been modified to an extent rendering them incompatible to the original (Gerst & Gao, 2013b). One example is the Type 2 plug used in China that has not been designed according to IEC 62196, one of the issues for MNCs addressed in the BMWi study (Ziegler & Gerst, 2012).

A major concern for NEV MNCs is homologation, the local vehicle registration necessary for market access. It has been found that different procedures exist, and that it remains unclear to the MNCs whether standards or regulations are effective (Ziegler & Gerst, 2012). From the European industry perspective, managers argue that "standards are understood as voluntarily established, representing newest technology whereas regulations are seen to define a minimum level of requirements to be fulfilled". The issue has been addressed in the recent position paper of the EU Chamber of Commerce in China (EUCCC, 2012). The paper recommends a simplification of the process that currently involves a number of uncoordinated regulators, and the adherence to the UN/ECE 1958 Geneva Agreement. In addition, regarding an international standardised physical NEV infrastructure with interoperable interfaces, an international export-oriented regulatory framework including homologation requirements and cost is favoured by the MNCs (Filipovic, 2013).

The Distribution of Power between the Different Stakeholders in Global Standard-Setting

The potential market size for NEVs in China is huge in comparison with other regional markets; the Government has realised its economic power and seems to be ready to not only question but also to compete against what Fang has called the "Western game" (Suttmeier & Yao, 2011). Fang emphasises the economic success of China and thus sees no reason to adapt to the Western IP and standardisation model. This is supported by Drezner who argues that economic powers are primary stakeholders in the landscape of global regulatory standards (Drezner, 2002). On the other hand, the so-called "techno-nationalist" policies promoting indigenous innovation are an adaption to the Western model. Eventually, this may lead to duplicate standards, negatively influencing global trade (Updegrove, 2007). However, it remains to be seen whether international stakeholders consider this–as a zero-sum or a positive-sum game. In fact, the international community needs to understand that technology-based competition does not have to be zero-sum game (Suttmeier & Yao, 2011). From US standardisation China has learned how companies can use standards to drive the development and direction of an industry sector, from the EU they learned how standards can be used to support a single market, and Germany demonstrated how standards can support industrial export policy (Jakobs & Gerst, 2013).

The Role of Inter-Governmental Relations

Finally, good inter-governmental relations between Europe and China facilitate MNC business in China and have contributed to an improvement of IPR protection and to the understanding of IP and standards related matters on all levels – politically and in practice. In 1985, the EU and China signed a strategic partnership agreement for trade and cooperation (EEAS, 2013). Today, China is the EU's 2nd largest trading partner (Trade 2013). The EU is represented by a permanent delegation in Beijing, and each year numerous political events take place, e.g. the EU-China summit. Further cooperation projects include the IPR SME helpdesk or the EU SME centre to provide legal and market access services, respectively, to European SMEs. Furthermore, research projects such as the China EU IT Standards Research Partnership analysed strategies of public policy and standardisation approaches adopted (TCEITSRP, 2012). In addition, individual EU Member States like Germany have launched special technology cooperation projects such as the Sino-German cooperation on NEVs (Schaefer, 2011). The objective of such projects is to get relevant government institutions and industry stakeholders involved to better understand market and technology, and to identify further projects in policy, technology and science (Ziegler & Gerst, 2012).

Resulting Standardisation Management Strategies

Regarding standardisation and its management, one has to notice that standardisation has gradually become a strategic instrument, particularly in the NEV area. In the EV case, the European perception of China's industry policy is an obvious discrimination of foreign car companies while developing home-grown and internationally incompatible standards (Meissner, 2014) that has been analysed before. For international car companies involved, the task to develop and implement NEV standards and to ensure their conformity even on an international level is challenging. Also, it introduces additional risks because automotive cycles, although already shortened, still require some planning. They have to deal with different standardisation approaches comprising international and national standards, policies and regulations. Thus, market access for NEVs seems to be driven by economic interests of the stakeholders rather than by interoperability.

China's electric vehicle market is far from being a walk in the park for Western automotive companies. Among the many challenges of selling NEVs in China, there is still a lack of public-charging infrastructure and Chinese charging standards differ from any other standard in the world (Tillemann, 2015). It has not been easy going in China thus far for electric-car maker Tesla Motors. Local governments where some of the current pilot projects are implemented, having a great deal of power over everyday business transactions in the country, have little incentive to assist Tesla in setting up a proprietary charging network (De Feijter, 2015).

The European perception of the current status quo of NEV standardisation reveals that in China MNCs are confronted with the national approach to NEV standardisation and homologation and that concerns are more focussing on standards as market access tools than on IP incorporated into standards. This has been confirmed by the BMWi study (Ziegler & Gerst, 2012). The study reveals that major differences in charging modes, in the charging infrastructure itself, and in communication systems exist, e.g. for plugs. One reason is that the MNCs are favouring existing international standards. The idea is to use already available standards from automotive or electrical engineering sectors and to develop them further in order to save R&D cost, to focus on large scale production, and most notably to save market access time (Jakobs & Gerst, 2013a).

In 2011, all MNCs agreed on standards for charging modes and plugs, e.g. the Combo plug[15] capable of AC and DC charging, or the IEC Type 2. Although this was also discussed with manufacturers from US and Japan[16], the agreement did not include the Chinese automotive industry. Moreover, China was not part of the international understanding on preferred future charging modes. Thus, China developed its own standards for AC and DC connector types, which are based on international standards but contain significant modifications.

However, the MNCs insisted to use their "club standard" in China, stressing the importance of common standards as market acceptance mechanism that contribute to a wider dissemination of NEVs due to the positive network effects, the reduction of variety of plugs, and increased interoperability of systems, confirming Blind (2009). Interestingly, and due to the good Sino-German relations, German car makers have convinced the Chinese NEV manufacturers to agree on one common, high technology plug. Eventually, the governmental institutions agreed to re-open standardisation of the plug design. An observer stated," *it seems that there will be compromise to achieve a common plug"*.[17]

According to an MNC manager, "the real issue is not the common plug which is not really the most relevant product feature of a NEV but the regulatory uncertainty concerning certification in China where it is currently not clear which plugs are accepted". He further stated "if we have no other choice, we will of course adapt to the Chinese plug required for market entry". [18] Meyer confirms this behaviour, noting that competitive pressure forces MNCs to participate in different standardisation systems (Meyer, 2012). However, due to the existing different regional standard setting approaches, companies often have to face not only the ambiguity of standards developed but also have to deal with some very strict local technical specifications.

With regard to charging another important area where MNCs still see major differences between Europe and China, influencing the standardisation management are testing and homologation processes[19]. The elements to be standardized include charging poles, wall boxes (or home-chargers), cables and plugs. Currently, there is no existing global set of standards for the charging infrastructure for electric vehicles. Thus, current market access regulations are based on a combination of existing standards for vehicles in general with some specific NEV requirements and reliance on automotive manufacturers' own testing specifications. Since it is unclear which regulations and standards will be applied, the ground for any meaningful and comprehensive certification scheme is not yet in place. For an MNC and its standardization management this means that if it intends to sell an NEV in China, there is no type approval available with a specific focus on charging. In China the charging infrastructure for NEVs is currently not part of the China Compulsory Certification System (CCC). China as Europe currently has no specific homologation requirements for Charging Infrastructure in place: The basic principle is that this infrastructure must be safe for consumers, comply with low voltage regulations and ensure EMC compatibility. Hence, in terms of testing procedures, which are required for homologation of cars in the respective markets, there is still a gap to fill. Therefore, both MNCs and Chinese car manufacturers are concerned that such testing rules will become a major obstacle for the commercialization of NEVs if not handled properly; current markets are largely segmented by different rules for infrastructure and related testing processes. Whilst integration of processes within Europe and within China is foreseeable, an alignment of testing processes in Europe and China would need a clear political will on both sides.

The example of the charging infrastructure shows that there is indeed a basic global understanding on some key elements – nevertheless the standardization work is yet to be completed. It is thus pre-mature to compare testing requirements or to develop certification schemes. Moreover, the fact that certification and approval requirements for infrastructure are still incomplete and consequently hamper the market entry of this new technology remains a major concern for many stakeholders. In particular, an agreed harmonized set of standards for compliance with infrastructure requirements is still missing. Whilst a global alignment of standards for the charging Infrastructure of NEVs seems to be slowly on the way, this cannot be said for testing and homologation processes in Europe or China. It is thus imperative that cooperation on conformity assessment is taken seriously, and that such work has to be undertaken in close cooperation between governments and industry (Gerst & Gao 2013b).

Thus related to standardization management of MNCs in China, the international car community in China forge alliances to draft and present common position papers or to organise joint workshops to inform Chinese suppliers and even sometimes government organisations about ISO/IEC standards.

In addition, confirming the view of Kennedy et al. (2008), a very dense cross-company network has been established where EV representatives and technicians exchange information on a regular basis and maintain good personal relationships (Gerst & Gao, 2013).

With respect to participation in standard setting, China does not yet allow foreign companies to participate in Chinese standardisation if Chinese IP is involved (Kennedy et. al. 2008). However, foreign car makers have established special departments in their businesses in China in order to better understand rules and regulations, participate in industry consortia, and do standard lobbying at industry associations or governmental institutions. As an MNC manager stated *"we are very often asked about our industry and standardisation experience by professors of R&D institutions which have to draft standards without having experience. This is a chance to early introduce existing experience from different R&D methodologies and address the need for international standards"*. [20]

MNCs often have pointed out that the major issues are rather the regulatory uncertainty affecting market access overall than having a common plug standard. Moreover, the automotive industry has to adapt to the national standards if this is the way to market access[21]. However, in general, MNCs consider the Chinese system to be inadequate for global markets. Meyer confirms this behaviour, noting that competitive pressure forces international companies to participate in different standardisation systems (Meyer, 2012). However, due to the existing different regional standard setting approaches, companies often have to face not only the ambiguity of standards developed but also have to deal with some very strict local technical specifications.

On the other hand, although MNCs are concerned about national IP embedded in technology standards, the community in general does not seem to be overly interested in actively participating in standard development of Chinese consortia (Bresnitz & Murphree, 2013). The reason might be unclear licensing terms and the fact that participation would force companies to license their IP. Participation is primarily to collect information, demonstrate goodwill and follow market trends – rather more defensive purposes (Kennedy et. al., 2008). Although MNCs are concerned about national IP embedded in technology standards, the community in general does not seem to be overly interested in actively participating in standard development of Chinese consortia (Breznitz & Murphree, 2013). The reason might be unclear licensing terms and the fact that participation would force companies to license their IP. Participation is primarily to collect information, demonstrate goodwill and follow market trends – rather more defensive purposes (Kennedy et. al., 2008). MNCs in China, although confronted with a national IP and standards regime, focus on an international standardisation strategy and apply lobbying and alliance mechanisms in order to support their strategy.

To sum up, market access of NEVs is high on the agenda of MNCs and governments. However, the speed and success of the intake of the new technology in both regions will depend on a variety of complex and intertwined factors. Amongst others, those factors include: the availability of a reliable and customer friendly charging infrastructure; the capacity of the automotive industry to tackle technical challenges such as cruising range, battery durability and reliability; trust in overall car safety; the ability of industry and governments to agree on a globally harmonized approach towards standards, common interfaces, vehicle topologies and architecture, homologation and certification processes. The uncertainty regarding the existing NEV standards world and the degree of global convergence of standards for infrastructure does not yet allow a substantial discussion on harmonization of testing and homologation processes which is a topic for all countries.

Regarding standardization and its management, one has to notice that standardization has gradually become a strategic instrument, particularly in new areas of technology such as the e-mobility field. For the stakeholders involved, particularly the MNCs, the task to develop and implement e-mobility standards and to ensure their conformity on an international are challenging. Also, it introduces additional risks because automotive cycles, although already shortened, still require some planning. They have to deal with different standardization approaches comprising international and national standards, policies and regulations. Market access for NEVs seems to be driven by economic interests of the stakeholders rather than by interoperability.

Some of the NEV technology is still not very mature and in the area of charging huge infrastructural investments have to be made in the future. We may expect to see the emergence of new business relationships and business models offering value added services. New service configurations, e.g. in the battery field, require standards to ensure the necessary interoperability for re-charging. Further research will need to be carried out to analyze, for example, how stakeholders such as grid operators or manufacturers of the charging stations will influence NEV standardization management.

CONCLUSION

This chapter has explored the interplay between national IP policies and international standards and their development in the relatively new field of electric vehicle technologies. This case has been chosen because electric vehicles are an important part of smart city concepts and therefore currently on top of Government agendas in many countries in the world, not only in Europe but also in China.

The NEV standards case highlights that technological and institutional changes create a dynamic environment in which China seeks establish its own standards regime to support its technological development and national innovation policy (Kennedy et.al, 2008). China has internalised the economic importance of IPR in national standards.

The analysis found that although China uses national standards as market access tools, trying to include national IP into them, the European automotive stakeholders do not seem to be worried, despite numerous IPR issues in the past. Although confronted with a national IP and standards regime, they keep pursuing their international IP and standardisation strategy, shaped by the four factors identified:

- Regional/national innovation and IP policies,
- Standardisation and market strategies,
- Distribution of power between the different stakeholders in global standard setting,
- Degree of inter-governmental cooperation,

Furthermore, this is supported by alliance building and lobbying on the industry and governmental level in China. However, this attitude may have negative long-term effects if Chinese manufacturers manage to cover NEV technology with their own IPR. To this end, they strongly support knowledge transfer from R&D to standardisation, something the EU still needs to render.

China has internalised the economic importance of IPR in national standards. Thus, the EU and European companies should be alerted because the Chinese – quite masterfully – keep the balance between promoting national interests and not violating WTO regulations. China has entered the established international standards landscape as a new player, aiming to shape the global economy. China is also part of further regional standardisation programmes in Asia such as the CJK. Thus, the fundamental challenge for China is to combine its leading role in international trade with its objective to strengthen its domestic innovation and IP without violating WTO regulations and commit to commonly agreed mutual standards with its cooperation partners.

Analysing the NEV case, it became obvious, there is a potential risk with country-specific IP and standards that the coordination function of standards is cancelled that this eventually, may lead to duplications and consequently negatively influences global trade (Updegrove, 2007).

However, NEVs are still very much driven by high technology, with MNCs as leaders. Moreover, China's auto market has already started to consolidate with a more moderate growth by halve to 7 percent (Reuters, 2014) which may cause internal economic issues that need to be solved first. On the other hand, this could lead to an even stronger focus on national IP and standards in order to strengthen domestic economic growth, which has so far been neglected in favour of its export strategy. A shift of power in international standardisation, a situation which does not seem to be that terribly far-fetched, may be possible in the near future (Jakobs & Gerst, 2013). China - as a latecomer – has started to exercise influence on long-standing major players like US or Europe. However, it has been argued that no individual Asian country will have the power to lead international ICT standardisation in the foreseeable future (the way the EU and the US are leading today). Yet, an alliance of these countries (particularly China, Japan and South Korea) might well assume a leading role. To this end, it had been further argued, it will be necessary for these countries to avoid – and learn from – the mistakes made by the EU and the US (Jakobs & Gerst, 2013). The CJK standards cooperation programme mentioned might be a first step in the right direction. However, it needs to be seen to which extent outcomes may have an influence on the existing global standards environment.

REFERENCES

ACEA (2009). *Translation of MIIT document 'Administrative Rules on Access for New-energy Vehicle Manufacturers and Products' for NEV policies MNCs have to comply with the regulations and comments, No. (44 of GCY [2009]), (for internal use only)*. ACEA.

ACWG. (2012). *Auto Components Working Group Position Paper, 144-145, by EUCCC*. Retrieved from http://www.europeanchamber.com.cn/en/publications-archive?categoryId=3

APCO. (2010). *China's 12th Five-Year Plan, "How it actually works and what's in store for the next five years", December 2010*. Retrieved from http://www.apcoworldwide.com/

Arksey, H., & Knight, P. (1999). *Interviewing for Social Scientist*. London: Sage Publications.

Atkinson, D. R. (2012). *Enough is enough: Confronting Chinese innovation Mercantilism*. The Information Technology & Innovation Foundation.

Bekkers, R., Iversen, E., & Blind, K. (2006). *Patent pools and non-assertion agreements: coordination mechanisms for multi-party IPR holders in standardization*. Paper for the EASST 2006 Conference, Lausanne, Switzerland.

Bekkers, R., Iversen, E., & Blind, K. (2012). Emerging ways to address the reemerging conflict between patenting and technological standardization. *Industrial and Corporate Change*, *21*(4), 901–931. doi:10.1093/icc/dtr067

Blind, K. (2009). *Standardisation: a catalyst for innovation*. Retrieved from http://hdl.handle.net/1765/17558

Blind, K. (2013). *The Impact of Standardization and Standards on innovation, Compendium of Evidence on the Effectiveness of innovation Policy Intervention*. Retrieved from http://www.innovation-policy.org.uk/compendium/

Blind, K., & Gauch, S. (2009). Research and Standardisation in Nanotechnology: Evidence from Germany. *The Journal of Technology Transfer*, *34*(3), 320–342. doi:10.1007/s10961-008-9089-8

Blind, K., Thumm, N., Bierhals, R., Hossein, K., Iversen, E., van Reekum, R., & Rixius, B. (2002). *Study on the Interaction between Standardisation and Intellectual Property Rights*. Final Report to the DG Research of the European Commission (EC Contract No G6MA-CT-2000-02001).

BMBF. (2011). *Fortschrittsbericht der Nationalen Plattform Elektromobilität (Dritter Bericht)*. Retrieved from http://www.bmbf.de/de/22327.php?hilite=elektromobilit%E4t

Breznitz, D., & Murphree, M. (2013). *The Rise of China in Technology Standards: New Norms in Old Institutions*. Retrieved from http://origin.www.uscc.gov/sites/default/files/Research/RiseofChinainTechnologyStandards.pdf

Cao, C., Suttmeier, R. P., & Simon, D. F. (2006, December). China's 15-Year Science Plan: Mapping Research and innovation Strategies for the 21st Century. *Physics Today*, *59*(2), 38–43. doi:10.1063/1.2435680

CESIP, China Europe Standardization Information Platform. (2013). Retrieved from http://www.eu-china-standards.eu/

China Daily. (2012). *Milestone investment agreement is signed from China Daily*. Retrieved from http://www.chinadaily.com.cn/business/2012- 05/14/content_15282436.htm

Collis, C., & Donnelly, T. (2012). Joint Ventures and the Development of the Chinese Automotive. *International Journal of Automotive Technology and Management*, *12*(4), 318–319. doi:10.1504/IJATM.2012.051358

COM. (2007). *581 final*. Communication from the Commission to the European Parliament, the Council, the European Economic and Social Committee and the Committee of the Regions, The European Interest: Succeeding in the age of globalisation. Retrieved from http://eur-lex.europa.eu/Result.do?T1=V5&T2=2007&T3=581&RechType=RECH_naturel&Submit=Search

COM. (2008a). *133 final*. Communication from the Commission to the Council, the European Parliament and the European Economic and Social Committee, Towards an increased contribution from standardisation to innovation in Europe. Retrieved from http://ec.europa.eu/enterprise/policies/european-standards/standardisation-policy/policy-activities/innovation/

COM. (2008b). *465 final*. Communication from the Commission to the European Parliament, the Council and the European Economic and Social Committee, An Industrial Property Rights Strategy for Europe. Retrieved from http://ec.europa.eu/enterprise/policies/innovation/policy/intellectual-property/index_en.htm

COM. (2010). *546 final.* Communication from the Commission to the European Parliament, The Council, The European Economic and Social Committee and The Committee of the Regions Europe 2020 Flagship Initiative Innovation Union. Retrieved from http://ec.europa.eu/research/innovation-union/index_en.cfm?pg=keydocs

COM I. (2011). *311.* Communication from the Commission to the European Parliament, the Council, and the European Economic and Social Committee on 'A strategic vision for European standards: Moving forward to enhance and accelerate the sustainable growth of the European economy by 2020', final. Retrieved from http://eur-lex.europa.eu/LexUriServ/LexUriServ.do?uri=COM:2011:0311:FIN:EN:PDF

Cox, W. (2012). *Reducing Greenhouse Gases from Personal Mobility: Opportunities and Possibilities.* Available at http://reason.org/files/reducing_greenhouse_gases_mobility_development.pdf

Drezner, D. W. (2002). *Who rules? The Regulation of Globalization, Previous versions of the paper were presented at the 2001 American Political Science Association and other events.* Retrieved from www.danieldrezner.com/research/whorules.pdf

Earley R., Kang, L. &Green-Weiskel, L. (2011). *Electric vehicles in the context of sustainable development in China.* The innovation Center for Energy and Transportation (iCET), Beijing Los Angeles New York, Background Paper No.9, CSD19/2011/BP9, 7-8.

EEAS. (2013). Retrieved from http://eeas.europa.eu/china/

EPO (European Patent Office). (2007). *Scenarios for the Future, March 2007.* Retrieved from http://www.epo.org/topics/patent system/scenarios-for-the-future/

Ernst, D. (2011). *Indigenous innovation and Globalization: The Challenge for China's Standardization Strategy.* La Jolla, CA: UC Institute on Global Conflict and Cooperation.

EU Mobility and Transport Website. (2013). Retrieved from http://ec.europa.eu/transport/themes/urban/vehicles/road/electric_en.htm

EUCCC Study. (2012). *Executive summary.* Retrieved from http://www.europeanchamber.com.cn/en/publications-position-paper

FIA. (2011). *Federation Internationale de l'Automobile.* Brussels: Towards E-Mobility the Challenges Ahead.

Filipovic, E. (2013). *Mapping Electric-Mobility: Standards Infrastructure for Market uptake.* Paper presented at SIIT 2013 Conference, Sophia-Antipolis, France. Retrieved from http://ieee-siit.org/program/

Fojcik, T. M., & Proff, H. (2014). *Accelerating market diffusion of battery electric vehicles through alternative concepts.* Paper presented at the 20th GERPISA International Colloquium: Structuring new automotive industries, restructuring old automotive industries and the new geopolitics of the global automotive sector, Krakow, Poland.

Fomin, V. V., & de Vries, H. J. (2009). How Balanced is Balanced Enough? Case Studies of Stakeholders' (Under-) Representation in Standardization Process. In *Proceedings of the 6th International Conference on Standardization and innovation in Information Technology SIIT 2009.* Aachen: Wissenschaftsverlag Mainz.

Fomin, V.V., Su, J., & Gao, P. (2011). Indigenous standard development in the presence of dominant international standards: the case of the AVS standard in China. *Technology Analysis & Strategic Management, 23*(7), 748.

Gerst, M. (2013). *IP Issues in Technology Standards in a Multi-Stakeholder Environment – Analysing New Energy Vehicle Standardization in Europe and China.* (Unpublished MS Thesis).

Gerst, M. (2010). *CNCA CCC IP Support Study commissioned by the IPR2 project of the European Commission: IPR and Trade Secrecy related issues in the Chinese CCC certification process.* (not published).

Gerst, M., & Gao, X. (2013a). Electric Vehicle Standardization Management. In M. Zelm et al. (Eds.), *Enterprise Interoperability: I-ESA'12 Proceedings.* John Wiley & Sons, Inc. doi:10.1002/9781118561942.ch63

Gerst, M., & Xudong, G. (2013b). Innovation Management of Electrical Vehicle Charging. Infrastructure Standards in the Sino-European Context. In *Proceedings of the 2013 ITU Kaleidoscope Academic Conference. Building Sustainable Communities.* Kyoto, Japan: ITU. Retrieved from http://www.itu.int/en/ITU-T/academia/kaleidoscope/2013/Documents/K-2013_Proceedings.pdf

Gibson, C. S. (2007). Globalization and the Technology Standards Game: Balancing Concerns of Protectionism and Intellectual Property in International Standards. Suffolk University Law School Legal Studies Research Paper No. 07-39, 2007. *Berkeley Technology Law Journal, 22*, 6.

Hölk, S. (2012). Die deutsche Normungs-Roadmap Elektromobilität – Version 2. Arbeitsgruppe 4 „Normung, Standardisierung und Zertifizierung“ der Nationalen Plattform Elektromobilität (NPE), Gemeinsame Geschäftsstelle Elektromobilität der Bundesregierung (Hrsg.). Berlin.

Interest. (2005). *D02 – Report on the results of the survey among researchers.* Retrieved from http://www.nets.rwth-aachen.de/Interest/D02.pdf

Jakobs, K. (2011). The European Union's ICT StandardizationPolicy - Changes Ahead!? In *New Applications in IT Standards: Developments and Progress.* IGI Global.

Jakobs, K., & Blind, K. (2011). The European ICT Standardization Policy – Recent Developments, Current Situation, and a Brief Outlook. In *Proceedings of the 16th EURAS Conference. EURAS contributions to standardization research* (vol. 5). EURAS.

Jakobs, K., & Gerst, M. (2013). *How to Perceive the (Future) Role of China in ICT Standardisation – A European Perspective.* Paper presented at the 8th IEEE Conference on Standardisation and innovation in Information Technology. Retrieved from http://ieee-siit.org/program/

Kennedy, S., Suttmeier, R.P. & Su, J. (2008). *Standards, Stakeholders, and innovation: China's Evolving Role in the Global Knowledge Economy.* NBR Reports, 4.

Kwaka, J., Leeb, H., & Fomin, V. V. (2011). Government coordination of conflicting interests in standardisation: Case studies of indigenous ICT standards in China and South Korea. *Technology Analysis & Strategic Management, 23*(7), 789-806.

Lea, G., & Hall, P. (2004). Standards and intellectual property rights: An economic and legal perspective Information. *Economic Policy, 16*, 71.

Lemley, M. (2002). Intellectual Property Rights and Standard Setting Organizations. In *California Law Review (1889 2002), 1900: While standardization has great economic value in many markets, group standard setting also poses some potential threats to competition*. Retrieved from http://papers.ssrn.com/sol3/papers.cfm?abstract_id=310122

Levin, A. (2010). *Bridging the Standardization Gap: China and the ITU-T*. Retrieved from http://www.china-eu-standards.org/dissemination/slides/Levin.pdf

MacQueen, H. (2008). *Contemporary Intellectual Property: Law and Policy* (2nd ed.). Oxford University Press.

Mair, C. (2012). *Openness, intellectual property and standardization in the European ICT sector*. Retrieved from https://papers.ssrn.com/sol3/papers.cfm?abstract_id=2065271

Manyika, J., Remes, J., & Woetzel, J. (2014). A productivity perspective on the future of growth. *The McKinsey Quarterly*.

McGregor, J. (2012). *China's Drive for 'Indigenous innovation' A web of industrial policies*. Global Regulatory Cooperation Project for US Chamber of Commerce. Retrieved from http://www.uschamber.com/reports/chinas-drive-indigenous-innovation-web-industrial-policies

Meyer, N. (2012). *Public intervention in private rule-making: the role of the European Commission in industry standardization*. (PhD thesis). The London School of Economics and Political Science. Retrieved from http://etheses.lse.ac.uk/236/

Meyer, N. (2012). *Public intervention in private rule-making: the role of the European Commission in industry standardization*. (PhD thesis). The London School of Economics and Political Science. Retrieved from http://etheses.lse.ac.uk/236/

MST. (2012). *12th Five-Year Special Plan on Technical Standards – Draft for Comments*. Obtained privately.

Mueller, J. M. (2002). *Patent Misuse through the capture of industry standards*. Retrieved from http://papers.ssrn.com/sol3/papers.cfm?abstract_id=1346546

Myers, M. D. (1999). Investigating Information Systems with Ethnographic Research. *Communications of the Associations for Information Systems, 2*(23), 1–20.

OECD. (2008). *Review of innovation Policy China*. Retrieved from http://www.oecd.org/sti/inno/oecdreviewsofinnovationpolicychina.htm

Okun, D. T. (2010). China: Intellectual Property Infringement, Indigenous innovation Policies, and Frameworks for Measuring the Effects on the U.S. Economy. Investigation No. 332-514. U.S. International Trade Commission, USITC Publication 4199 (amended), November 2010.

Powell, W., & Grodal, S. (2005). Networks for Innovators. In The Oxford Handbook of innovation. Oxford.

Prudhomme, D. (2012). *Dulling the Cutting-Edge: How Patent-Related Policies and Practices Hamper innovation in China, study of the European Chamber of Commerce in China*. Academic Press.

Reuters. (2014). Retrieved from http://www.reuters.com/article/2014/10/25/us-china-autos-idUSKCN0IE04M20141025

Reuters. (2015). Retrieved from http://www.reuters.com/article/2015/01/20/us-china-economy-idUSKBN0KT04920150120

Rongping, M., & Zhuoliang, W. (2005). *The Role of Standards in National Technology Policy in China*. Retrieved from http://strategicstandards.com/files/China.pdf

Schaefer, M. (2011). *Grusswort zur "Sino-German Electric Vehicles Science and Technology Cooperation" am 24*. Retrieved from http://www.china.diplo.de/Vertretung/china/de/__pr/2011/reden-bo-2011/111024-emobilitaet-ps.html

SESEC Project. (2013). Retrieved from http://quality-partnerships.cn/standards-in-china/sesec-project/

Sherif, M. H. (2003). When Is Standardization Slow?. *International Journal of IT Standards and Standardization Research*, *1*(1), 19–32. doi:10.4018/jitsr.2003010102

SIPO. (2013). Retrieved from http://english.sipo.gov.cn/news/official/201303/t20130326_789188.html

SLoCAT@ Rio + 20. (2012). *Expanding the Use of electric mobility: Options for sustainable urban transport*. UN Department of Economic and Social Affairs (DESA). Available at http://www.slocat.net/content/expanding-use-electric-mobility-options-sustainable-urban-transport

Stevenson-Yang, A., & DeWoskin, K. (2005, March). China Destroys the IP Paradigm. *Far Eastern Economic Review*, 9-18.

Suttmeier, P., & Yao, X. (2011). *China's IP Transition: Rethinking Intellectual Property Rights in a Rising China*. NBR Special Report, 7.

Suttmeier, R.P. & Xiangkui, Y. (2004). *China's Post-WTO Technology Policy – Standards, Software, and the Changing Nature of Techno-Nationalism*. NBR Special report no. 7.

Suttmeier, R. P., Yao, X., & Tan, A. Z. (2006). *Standards of Power? Technology, Institutions, and Politics in the Development of China's National Standards Strategy*. Retrieved from http://kms2.isn.ethz.ch/serviceengine/Files/RESSpecNet/106854/ipublicationdocument_singledocument/DEC48CA8-82BF-4097-AEBF-3C47AA85F045/en/10_Apr06.pdf

SWD. (2012). *458 final*. Commission Staff Working Document, Towards enhanced patent valorisation for growth and jobs. Retrieved from http://ec.europa.eu/enterprise/search/index_en.htm?q=SWD%282012%29+458+final

Tagscherer, U. (2012). Electric mobility in China – A policy review. Fraunhofer ISI Discussion Papers innovation Systems and Policy Analysis No. 30, Karlsruhe, February 2012, 8.

Tang, R. (2009). *The Rise of China's Auto Industry and Its Impact on the U.S. Motor Vehicle Industry*. Report for Congress Prepared for Members and Committees of Congress.

TCEITSRP. (2012). *Deliverable 3: Review of State of Art: China and EU*. Retrieved from http://www.china-eu-standards.org/deliverables/index.htm

Trade. (2013). Retrieved from http://ec.europa.eu/trade/policy/countries-and-regions/countries/china/

Updegrove, A. (2007). *Government Policy and "Standards – Based Neo-Colonialism"*. Retrieved from www.consortiuminfo.org

Van den Bossche, P. (2003). *The electric vehicle: raising the standards*. (PhD dissertation). Vrije Universiteit Brussel.

Wang, A., Liao, W., & Hein, A. P. (2014). *Bigger, better, broader: A perspective on China's auto market in 2020*. McKinsey & Company.

Wang, P., Wang, Y., & Hill, J. (2010). *Standardization Strategy of China: Achievements and Challenges*. East-West Center Working Paper, Economics Series No. 107.

Williams, R. (2009). *Common ICT Standards or Divided Markets: The EU and China*. China-Eu Standards Project Workshop held in conjunction with the I-ESA Conference.

Williams, R., & Edge, D. (1992). The Social Shaping of Technology: a Review of UK Research Concepts, Findings, Programmes and Centres. In New Technology at the Outset. Campus-Verlag & Westview.

Ziegler, K., & Gerst, M. (2012). *Sino-German Cooperation on Standardisation and Conformity assessment for New Energy Vehicles*. Feasibility Study for BMWi (German Ministry of Economy and Technology) and CNCA.

KEY TERMS AND DEFINITIONS

Automotive Industry: Companies and activities involved in the manufacture of motor vehicles, including most components, such as engines and bodies, but excluding tires, batteries, and fuel. The industry's principal products are passenger cars (Encyclopaedia Britannica).

Chine Eco-System for NEVs: The Standardisation System of China and Standards Development, the Standardisation Strategy of China, and China's IP Policies.

Electric Mobility: Represents the concept of using electric powertrain technologies, in-vehicle information, and communication technologies and connected infrastructures to enable the electric propulsion of vehicles.

Innovation: Result of a process where a new product or service is created.

MLP (Medium-And Long-Term Program for Scientific and Technological Development): Framework for China's innovation policies was initiated in 2006.

MNC: (Multi-National Company): A corporation that has its facilities and other assets in at least one country other than its home country.

SST (Social Shaping of Technology): Social Shaping of Technology (SST) perspective, by explaining ICT innovations as history- and context-specific, actor-focused technological change processes (Williams & Edge, 1992).

ENDNOTES

1 In this chapter, NEVs include the categories: hybrid electric vehicles (HEVs), battery electric vehicles (BEVs), fuel cell electric vehicles (FCEVs) for passenger cars.

2 Standardization Law of the People's Republic of China and its Implementing Regulations, 1988, http://www.sac.gov.cn/sac_en/AffairsOpening/LawsandRegulations/201011/t20101123_4214.htm, (last accessed April 4th 2013).

3 Own illustration for simplifying the complexity of the different policies.

4 For detailed information of the Chinese National IP strategy 2008, see Government website, Promotion Plan for the Implementation of the National Intellectual Property Strategy in 2013 http://english.gov.cn/2008-06/21/content_1023471.htm, (last accessed July 28th 2013).

5 For further details in English of the National Patent Development Strategy (2011-2020), SIPO, see http://www.ipwatchdog.com/2011/11/06/key-considerations-for-patent-strategies-in-china/id=20241/, (last accessed July 29th 2013).

6 Decision of the State Council on Accelerating the Fostering and Development of Strategic Emerging Industries, (No.32 [2010] of the State Council), http://www.lawinfochina.com/display.aspx?lib=law&id=8570, (last accessed July 29th 2013).

7 A key funding mechanism for research in the research and application of new technology in China is the State High-Tech Development Plan, also known as the "863 Programme" (named as such for having been founded and finally endorsed by Deng Xiaoping in the third month of 1986). In recent years, this fund has mostly supported the development of electric vehicles.

8 In China, the patent system grants three types of patents: utility model patent, design patents and invention patents. "Invention patents can be granted to both products and processes, and must meet a standard for novelty (not part of the "prior art," i.e. not openly known to the public abroad or in China before their filing date), "inventiveness," and practical use as determined by a review called a substantive examination. Utility models can be granted on the shape and/or structure of a product, and do not undergo a Substantive Examination but are required to be novel, meet a far lower level of "inventiveness" than invention patents, and must meet criteria for practical use/functionality. Invention patents and utility models enjoy basically the same level of legal protection during their lifetimes. Design patents are granted on the appearance of a product that makes it particularly recognisable, do not undergo a substantive examination nor have to meet any technical or functional thresholds but must be distinct from prior designs, and should not conflict with prior rights like copyrights or trademarks."

9 The lack of core design and R&D facilities has meant that domestic firms had low levels of IP. For more details see Collis, C., Donnelly, T., "Joint Ventures and the Development of the Chinese Automotive", International Journal of Automotive Technology and Management 12 (4), 318-319.

10 An activist government policy has liberalized the Chinese automotive sector in some key respects, but it requires foreign manufacturers to undertake joint ventures with local partners in order to obtain market access.

11 Those statements were made during a conversation with an MNC manager who preferred to remain anonymous.

12 According to the industry, NEV three core technology components are: on-board energy system/battery, power transmission system, and control system of the new-energy vehicle. One of them,

the "Manufacturer shall well know and master the core technology: technical principle, structure, function and performance requirements, control method, communication and data exchange, failure mode, security risk, evaluation test and diagnosis and solution of main default patterns".

13 This statement was made during a conversation with an MNC manager who preferred to remain anonymous.

14 This statement was made during a conversation with an MNC manager who preferred to remain anonymous.

15 The German-US Combined Charging System (CCS) is a universal charging system for fast loading requiring only one vehicle interface enabling the customer to charge with all sorts of different available charging modes. For further definition, see Ladestecker in der Elektromobilität, http://www.e-stations.de/stecker.php#, (last accessed August 5th 2013).

16 Europe, USA, Japan Will Harmonize Electric Vehicle Regulations, http://ens-newswire.com/2011/11/22/europe-usa-japan-will-harmonize-electric-vehicle-regulations/, (last accessed August 5th 2013).

17 The statement was made during an informal meeting, following up on the BMWi study.

18 The statement was made during a conversation related to the Sino-German certification study, the manager wanted to be not named.

19 Homologation is based on regulations and compulsory standards which may refer to standards and technical specifications as means to comply with the regulations. Homologation processes may include some or all of the following elements: approval of documentation, type testing, factory and production inspections, certification, registration and licensing, post-registration surveillance. Homologation rules may include a description of processes, applicable standards, and of the testing methodology to be used.

20 ibid.The statement was made during a conversation related to the Sino-German certification study, the manager wanted to be not named.

Chapter 12
Standardization Commercial Strategy: An EU Antitrust Perspective

Justin Pierce
University of Lund, Sweden

ABSTRACT

This contribution offers an overview of recent developments in European Union antitrust regulation of single-firm strategies in standardization. The law has moved cuttingly towards a rigid enforcement model securing the foundations and future benefits of standardization. There remains a paradoxical gap between effective single-firm strategy and antitrust enforcement. Whereas firms are maximizing their strategies to a position of strength and maximum security within standardization, the law is acting is suspicious of such activities. The law currently applied to standardization strategies and especially those that involve essential intellectual property represents a strong warning to firms strategizing in t field of standardization. In the main the law has focused on single-firm strategies and largely on the exclusion by a dominant firm of rivals from access to or use of essential technologies protected by intellectual property. This contribution reviews the antitrust repercussions for single-firms in standardization activities under EU antitrust rules.

INTRODUCTION

Company strategy is fundamentally about forms of decision making and their rationales and impacts at a firm level. Business strategy as a field of study often has a holistic view, assessing the drivers, strategies, rationales and repercussions of decision-making strategies through a strategy or economic lens, the impact of law is often less considered. This may as pointed out by Denrell, et al. (2003) be a result of the predominance of strategy analysis through the prism of economics focusing on those strategies which optimize economic surplus and raise profits or enhance the market position of the firm. In some sectors, the focus of company strategies has come under more scrutiny and investigation than others, one such sector been the so-called high technology or technology-enabled markets.

DOI: 10.4018/978-1-4666-9737-9.ch012

Although technology is prevalent in most sectors and markets, it plays a special and more centralized role in markets that have an intense degree of innovative activity. Since the 1980's a characterization of these markets are the high degree of intellectual property strategizing and particularly the use of patents (Economides, N. (1996), Katz & Shapiro (1994), Kortum & Lerner (1999)). In other words, they are populated by firms with significant research and development (R&D) activities and which rely heavily on the intellectual property (IP) system (Iversen, E. (1999)). In general, and at the risk of over simplification, these markets are comprised of sectors where technology plays a central role in the competitive process. Therefore, the strategic decision making and management in these markets naturally sees a focus on the way in which decision maker's use or benefit from technology or technologies. In recent years, especially at a European level, there has been debate and examination of these markets from an economic, strategy and general business perspective (Gifford & Kudrle, (2011). The focus of many of these studies has rested on the strategic positioning and decision making within companies and the link between strategies, innovation, and intellectual property (Blind, K. & Thumm, N. (2004)). More recently, and as standardization has become a strategy focus, understanding the benefits and dangers for companies from engaging, strategically, in standardization activities and commercial planning have formed focal points for researchers.[1] Standards and standardization in technology enabled markets potentially offer great synergies to companies especially where their technology plays a predominant role in the standard. As Swann (Swann (2010)) points out this is especially the case where these markets are overlaid with Network Effects, especially where such effects are 'direct Network Effects'. In these circumstances the Network Effect can be seen as creating value in the market, for example, the 3G mobile network enables communication between handsets, leading to a direct Network Effect for consumers.

Consumers are able to communicate with each other and interact by connecting to the network. Companies are able, by incorporating the 3G technology into their products, to access the pool of consumers. Moreover, it heightens competition on the market, and the use of one standard over another can add to the competitiveness of consumer products. 3G enabled devices were far more likely to be attractive to the consumer than the slower 2G option. It is here the strategic incentive for companies involving themselves in standardization can be drawn out (Bekkers, R., Duysters, G. & Verspagen, B. (2002)). Especially the inclusion of owned intellectual property into these standards potentially delivering not only a technological advantage but also a continuous revenue stream.

From a strategic point of view, simply put, it is better to be in the standard setting than simply following the standard, especially if firms strategically place themselves and incorporate their technologies.[2] To date, much of the research into standardization strategy, especially in sectors marked by Network Effect or reliant on Interoperability information, makes use of economics as a framework for understanding decision making, typically focusing on the wealth and market share objectives that private companies have. Surprisingly, however, is the lack of evaluation of the impact of the law. While many often raise, in a superficial way, the potential legal issues they are certainly not dealt with in any great depth. An approach that is surprising, especially in the field of Antitrust, which is, in many ways the flipside of the strategy coin.[3] Whereas strategy is often related to the growth of profitability, competitiveness and market share, Antitrust economics is concerned with the dangers of market power, specifically what it is, how it is sustained and what are its effects. The natural effect of this from a regulatory perspective is the action, through Antitrust, that is taken is to prevent the accumulation of market power and to prevent certain business strategies runs against the realization of firm strategies. Take as a general example, loyalty rebating or predatory pricing strategies of single-firm's been examples of where Antitrust steps restricting business strategizing.[4] Outside of these examples commercial strategy can and often does, especially when successful, raise the concern of the regulator and opens up for a strategy, however commercially sound, to be assessed and evaluated through a legal prism to evaluate whether increased market power and market success is potentially unlawful.[5]

In the European Union (EU) the field of standardization has become one of the focal points for the EU regulator, especially where firms are perceived to be using intellectual property or market share strategically to enhance their own market position or to influence the way the market is structured.[6] To highlight the potential legal impact on a firm's strategy in standardization this contribution takes four recent cases and rationalizes each from a business perspective before contrasting the legal outcome. By doing so it highlights the intersection between standardization strategizing and Antitrust enforcement in the EU to show the importance of Antitrust considerations in single-firm strategizing in standardization. In this respect, it first provides an overview of the business strategy literature illustrating that many of the goals of any strategy are profitability, competitive advantage, and market share. Setting the scene for the remaining parts of the chapter in examining how Antitrust authorities in the EU have stepped in to regulate standardization as well as illustrating the danger of accumulated market power through *De Facto Standards*.

The structure of the remainder of the paper involves a four-part discussion. Firstly, an overview of the predominant literature on the rationale and understanding of why businesses strategize and their protected freedom within the EU to do so. Followed secondly by an overview of how the EU regulator approaches commercial strategies, especially in the context of standardization. Thirdly is the evaluation of the four chosen case studies which is illustrative of the distinctive legal analysis that occurs in competition matters and how strategic objectives can infringe the law. In concluding it is shown that the case-by-case application of competition rules has blurred the lines of lawful and unlawful behavior when it comes the strategizing of single-firms to maximize their returns or positions in the field of standardization.

COMMERCIAL STRATEGY AND ANTITRUST

An overview of the business strategy literature illustrates that some of the main goals of any strategy are profitability and competitive advantage. For others such as Robert Grant (2008), these taken together are mechanisms for creating value for the company and the shareholders, financially and competitively. For others such as Smit (2004), the key to delivering such value rests on the ability of firms to generate a competitive advantage in the relevant market. For the likes of Porter (1996), (2004) the relevant market and the competitive structure of this market including the competitive pressures influences strategy. Williams (1998) asserts that a contributing factor is the effective use of the individual capabilities of the company and their ability to deploy these capabilities to achieve competitive advantage. At their heart these have a common goal in so far as it relates to the strengthening or establishing of a firm in a market and leading to a greater level of competitiveness and market share or profitability. Porter (2004), for example, in building on his classical 'five forces' sees competitiveness through strategy in relation to the five forces as been a focal point in designing an effective strategy.

The extent to which companies are free to compete depends greatly on the larger economic and legal frame which is premised on a market conceptualization. This chapter makes no claim to the market type but it is necessary to consider the nature of the European market in the EU context to have a better appreciation and understanding of EU Antitrust and regulatory actions. The governing law of the EU is the treaties and they are authoritative of the type of economic framework firms find themselves operating in. As a starting point for the European Community (European Union since 2009) the model was and remains based on a system of free competition, this was a central objective of the original Treaty of Rome (1957). This objective has remained a constant through the various treaty reforms and growth of the EU. The Treaty on the Functioning of the European Union (TFEU) in Article 119 TFEU states that the economic activities of the EU and its Member States should be "conducted in accordance with the principle of an open market economy with free competition." This principle and approach is given further effect throughout the TFEU, for example Article 120 TFEU requires Member States and the EU to "act in accordance with the principle of an open market economy with free competition, favoring an efficient allocation of resources." Moreover, Protocol 27 of the TFEU states as follows:

The High Contracting Parties, considering that the internal market as set out in Article 3 of the Treaty on European Union includes a system ensuring that competition is not distorted, have agreed that: to this end, the Union shall, if necessary, take action under the provisions of the Treaties, including under Article 352 of the Treaty on the Functioning of the European Union.

In support of the economic goal of a competitive and free market the TFEU adds the competition provisions, these are Article 101 and 102 TFEU. Both of these are of equal importance but for the purposes of this paper Article 102 TFEU and the decisions thereunder forms the focus. Where Article 101 TFEU deals with agreements between firms Article 102 TFEU focuses on placing certain restrictions on the unilateral conduct of firms. It provides that "any abuse by one or more undertakings of a dominant position within the common market or in a substantial part of it shall be prohibited as incompatible with the common market insofar as it may affect trade between Member States." The text goes on to offer the following examples of an "abuse": "(a) directly or indirectly imposing unfair purchase or selling prices or other unfair trading conditions; (b) limiting production, markets or technical development to the prejudice of consumers; (c) applying dissimilar conditions to equivalent transactions with other trading parties, thereby placing them at a competitive disadvantage; (d) making the conclusion of contracts subject to acceptance by the other parties of supplementary obligations which, by their nature or according to commercial usage, have no connection with the subject of such contracts." In principle Article 102 TFEU is the provision most likely to be invoked when considering individual-firm strategies and it is the provision that is most specifically aimed at strategic market behaviors. As such it is unsurprising that this is the Article that has been most predominantly used by the EU regulator in the field of standardization.

The autonomy of firms is in principle a fundamental component of the EU free market. In this sense free is equal to the freedom to act in the market to the extent that such action does not damage the competitive structure of the market. This is clearly illustrated in the EU Charter of Fundamental Rights (Charter). Article 16 of the Charter provides a positive freedom to conduct a business that companies can rely on. However, this is by far an unconditional right and the second paragraph of Article 16 requires any conduct of a business to be in accordance with the law, this includes conformity with the competition rules. It is generally accepted that complete freedom in the market provides the proverbial forbidden fruit, meaning a failure to regulate is potentially as damaging as over or bad regulation. As such it is universally recognized that without the appropriate regulatory institutions armed by the legal system with effective rules and norms of regulation and enforcement powers the successful creation of a market economy is doomed to failure. At an EU level, considerable responsibility for market regulation falls under the watchful eye of the European Commission, Directorate General of Competition. [Commission or DG Comp.]. Within their array of duties, the regulation and stabilization of the EU market is a central function. This underlies an inherent tension in the free market model, in so far as the reality of the free is far from what one may presume free to be. How the regulator balances this is critical to the ability of a company to strategize. This is especially pertinent in the modern day marketplace and the advancements in the studies of strategy that are applied at a business level. It is fair to say that on balance, commercial strategy has become an art itself and is marked by a complex process of determining the actions that need to be carried out in order to achieve the commercial purpose. There is more focus on the stabilization of the company, and therefore it is not so much about short-term market power or short-term gains it is more about medium/long term stability and value. As such strategy has become more dynamic, and strategies are more likely to be characterized by the specific value adding factors that are identified instituted by the company than are otherwise attributable to the specific market. This presents a regulatory conundrum in so far as it is necessary to decide the manner in which a market is to be regulated. In high technology and network enabled markets there is no consensus on the most effective manner of regulation and the lines of permissible strategies affecting market behaviors is often blurred (Gandia & Tourancheau (2015)). There also remains uncertainty over the type of behavior associated with companies strategizing in the field of standardization that will engage the competition rules. One such problem is associated with the use of property, especially intellectual property for a Rights-Holder to gain an advantage in the process.

STRATEGIES AND INTELLECTUAL PROPERTY

As alluded to above, the European regulatory authority has recently taken a strong interest in cases involving standardization and the strategies used by firms to obtain a strategic position within the standardizing processes. In a line of investigations, the extent to which standards are strategic commercially and have the potential to affect the European market and thus raise concerns of the regulator has become evident. Investigations stemming from market power resulting from firm *De Facto Standards* to the strategic licensing or purchasing of essential intellectual property in standardization for competitive advantage has led to strong enforcement against firms (Emanuelson (2012)).

The driver behind these investigations is to protect the competitive structure of the European market and to restrict unduly damaging behavior to the standardization processes. Additionally, protecting the integrity and advantages of standard setting activities in the EU internal market. In this context, the investigated behavior is strongly linked to strategy and goes hand in glove with the increased commercial awareness been shown by firms of the advantages of standardization participation. Additionally, the strong powers of the EU Commission and the open-ended texture of the EU legal provisions means that the level of proof required to establish anti-competitive behavior is easily surmountable. There is a growing trend to use commercial strategies as evidence in a regulatory investigation of the intention to restrict or impede competition. Under this set of rules, the effect on the market is not a necessary hurdle to be proven by the Commission. It suffices to demonstrate the potential effect in the future, whether exercised or not, a business strategy may have on the market and the competitive structure of the market is sufficient to raise objections.

The imposition of the Antitrust norms is, or at least should be, a major consideration for commercial strategy, especially in areas of intellectual property where the repercussions may be severe, both in fines and in imposed licensing terms. Especially given the recent Antitrust investigations that have specifically targeted technological market leaders or incumbents owning essential technologies, where their market share or activity is surround by or could cause a *De Facto Standard* the scrutiny has intensified. As will become clear from the later parts of this paper the Commission has not concentrated on a specific business strategies that dominant firms put in place in order to maintain and increase their market power. Rather the regulatory scrutiny has widened and has included actions against rebating to tying, to Interoperability knowledge, as well as technology licensing. This line of investigations is interesting legal trend given that it is a move away from more traditional Antitrust interventions that have typically focused on incumbent's practices and not the effect on the research and development, incremental innovation, or Interoperability considerations.

This shift in focus is potentially concerning from a business angle as it means that the rules on competition are been extended into a new sphere of a mapped single-firm strategy on increasing return or extraction of profits from innovation. Potentially meaning that a company's strategy of innovation, culminating in intellectual property filings and protections or process know-how through research efforts are opened up to regulation where the innovation or knowledge is overly successful or essential to the market. Here the interaction between the successful innovator and the business practices linked to the exploitation or enforcement of the fruits of their research and development can be subject to competition provisions with potentially negative effects.

It is becoming evident from the emerging line of Commission Antitrust decisions that the intellectual property and the Interoperability strategies of market leading firms in high-tech industries are particularly susceptible to Antitrust intervention. The impact of the application of the law on these companies has serious implications for further investment and disclosure of their innovations in the future. The threat of 'appropriation' of technology through Antitrust decisions marks a clear and present danger to innovation and prohibitive of further investment into areas of focus for regulators. It is suggested that the drive of Antitrust regulation at market leaders is an underlying need for knowledge dissemination and access to Interoperability information. Especially where there was a need for substantial investments in industries that are network reliant, such as telecom. The case studies that follow will show that intellectual property and licensing strategies in these markets are open to Antitrust rules and the remedies may require the sharing of protected technology with the market.

In this respect, the motivation of industry leaders to participate in standardization presents a logical and potentially fruitful avenue through which firms can gain exploitation rights whilst avoiding the pitfall of becoming a *De Facto Standard* holder. The standardization platform permits the licensing and exploitation of patents. While it may require the Rights-Holder to grant a license to standards users, it still presents them with a market advantage of been a predominant technology. As the inventor and the licensor remain in command of information that may not be disclosed in the patent or the standard. Meaning that the 'first mover' advantage still exists although to a much lesser extent. However, these companies are only able to benefit from standardization if they can participate in and have their technologies included in the standard.

Standardization and the Antitrust Safe Harbor

There has been a rise of standardization focused investigations and guidance from the Commission in recent years. Specifically from an EU perspective and the pro-competitive benefits that accompany effective standardization outputs has led to a substantial revision of the Commission's competition law 2010 Horizontal Cooperation Guidelines (Horizontal Guidelines, Official Journal C33, 2.2.2011, p. 20).

Unsurprisingly, mainstream economic analyses and Commission-sponsored studies, find that the Commission has rightfully reached the conclusion that standardization is deserving of special Antitrust focus, as such finding its way into the Antitrust provisions of the EU. However, it remains questionable whether it is desirable or necessary for such direct Antitrust linkage, although this discussion is not the focus of this paper. On the back of the studies mentioned above, an assumption is drawn that standards provide efficiency gains and can be directly allied to innovation and standards are further attributable to competitiveness. The efficiency gains argument stems from the harmonized effect of the standards, especially when the principle of conformity underpins them. Put simply standards are catalysts for further market integration, and this connects to the economies of scale and wider consumer choice and ultimate cost savings to the consumer. In much the same way, the link to innovation derives from the belief that standards unlock information that would otherwise be protected under inaccessible intellectual property or know-how. Moreover, they present a forum exempting or legitimizing a behavior, i.e. knowledge exchange, market strategy and revealing of future developments, these would otherwise likely attract attention from even the most docile of enforcement agencies (Geradin, D. (2007)).

The dissemination, discussion and knowledge exchange that occurs through standards promotes and invites, if not radical, then certainly incremental innovation (Carrier M. (2009)). The upshot of these taken as pieces of a whole is that standards drive competition both in terms of participation in the standards formation but also in technological competition deriving from innovation advancement. In fields of Interoperability, companies should view the establishment of standards as a platform to prevent market lock-in, from a consumer point of view this activity leads to wider choice and better Interoperability. From an EU standpoint, this drives the economic integration vision and contributes to the single market and European competitiveness. As Granieri and Renda, A. point out if more justification for promoting standards was necessary the economic advantages include reduction in production costs, focused research and development activities and overall product or service efficiency.

Despite all the good things associated with standards the Guidelines identify that there is a risk that standard setting leads to negative effects on competition and, in those markets where standards are vital to entry, such negative effects can be an impediment to effective competition. A number of examples are laid out in the standardization chapter of the Guidelines and include the potential for abuse of the standardization process with companies hiding behind standardization to disguise information exchanges, allowing them to 'semi-cartelize' by supporting each other's technological solutions for inclusion in a standard to the exclusion of a competitor or another market player is one example. Additionally, the reverse is possible. Allowing predominant players to use their technological weight and advancement to exclude disruptive innovators and to allow a protection of royalties for the future and even worse this may be premised on less effective technological solution been adopted into the standard.

The intrusion of Antitrust investigation into the field of standardization is not necessarily a new phenomenon; contrary to the sudden popularity it is currently experiencing in the legal sphere with lawyers treating it as a new field of discovery. As early as 1992 the Commission intervened in regulating the potential danger of standardization abuse at an institutional level.[7] Likewise outside of competition law, private standardization through trade associations has been controlled when a fundamental freedom has been engaged. So too has EU law intervened when standard setting organizations have taken on quasi-private/public duties.[8] It is true that in recent years, standardization cases under Antitrust rules have increased, and the focus has rested with network and Interoperability standards and has moved away from the institutional angle to focus more on the activity of participants, the firms. To over-generalize so as, not to flog a dead horse, most of the recent cases and complaints have concerned the involvement of intellectual property rights, and more specifically patent licensing and disclosures in the standard-setting process. It is not doubted that this results from (a) an increased commercial awareness and savvy in the obtaining and strategically using IP and (b) with the increase in innovation, technology orientated standards that incorporate IP has risen to the surface. In reality these are two sides of the same coin, when we speak of the advantages both competitive and economic it is not only beneficial to use standards to release information otherwise restrained by IP, but necessary if we are to see EU competitiveness grow.

On the anti-competitive side of the coin, intellectual property law provides the avenue for undesirable effects to be regulated and if not prevent then certainly punished. Whether or not the current approach produces an optimal enforcement system or not is a matter of debate, but not one that this paper explores in detail. It is nonetheless prudent to point out the EU seems to be of the traditional U.S. view of corporate strategic behavior as summed up by Donald Turner as the tradition is that judges view each business practice with suspicion, always wondering how firms are using it to harm consumers (Areeda, P. & Turner, D. (1978)). This view is evident in a number of the case studies that follow and, as will be highlighted, the strategic use of patents and licensing of essential technologies is fast becoming fertile ground for Antitrust intervention.

In fact at the EU level a string of controversial cases has arisen out of the question of how far the legal rights of a property holder are limited as a direct result of their involvement in standardization and a commitment to license on fair, reasonable and non-discriminatory terms (FRAND). In particular the questions that are raised are to what extent a Rights-Holder can refuse a license and secondly, whether the right to enforcement through an application for an Injunction against a user of the standard is permissible. It is here that the potential incompatibility between IP strategies and standardization is at its greatest. While IP locks others out, standards aim to make the same information readily available on pre-determined terms and rates and where a dispute arises the Rights-Holder will act by falling back on their strict legal rights. Under Directive 2004/38 this is a responsibility of the national courts tasked with the case, and they are obliged to provide remedies, of which one is injunctive relief. On the other side of the coin, we have seen that the Commission is concerned with the blocking of access to information that is necessary in order to innovate. Hence the introduction of instruments such as the technology transfer regulation and the distrust of technology pooling and certain settlement clauses, such as payment to not challenge IP. All of which can be argued for from a strategy point of view, pooling technologies may be an effective market share stabilizer, or can be strategically used to prevent a new comer into the market. Likewise settlements and agreements to no challenge obligations for a payment of some allows a strategic decision that ultimately protects the property.

Nonetheless, the case examples highlight the dangers of an overly aggressive and successful intellectual property strategy from an Antitrust perspective. The first example is a case that involved an investigation into *IBM* following complaints that *IBM* had refused to disclose the information necessary for third-part suppliers to attach 'interface' peripherals to its system/370 mainframe computers.[9] *IBM* relied on the fact that the information requested was protected by IPRs. However, the Commission found that this refusal constituted a breach of the EU competition rules and an abuse of a dominant position. The result of which was to require *IBM* to release the necessary interface information to facilitate Interoperability and to open the market to greater competition. In *IBM*, the opening of the Interoperability information led to the 'plug and play' of products. As such there is an argument to validate the Antitrust intrusion, especially where it leads to a breakthrough in the market that would otherwise be prevented as a result of a firm's licensing strategy.

In EU Antitrust law there are other infamous examples of the regulator moving in the direction of unlocking IP information to grant access to firms wishing to make use of the information in new markets or been able to compete effectively on an existing market. In *Magill*, a case involving access to IP protected TV programming information, the IP owner's respective refusals to provide programme-scheduling information to third parties prevented the appearance of a new product, namely a comprehensive weekly television guide.[10] As such the IP owner was compelled to allow access to the otherwise IP protected information on FRAND terms. Two more infamous examples in the cases of *IMS Health* and *Microsoft* exist, in *IMS* the refusal to provide access to information needed for a downstream market has held to be unjustifiable. In *Microsoft*, a failure to provide access to necessary Interoperability information for Windows servers was also indefensible. In *IMS*, a secondary downstream market was created and in *Microsoft* the Interoperability information shared with the market led to higher levels of competition, greater Interoperability and more innovation. To the extent that these decisions are reconcilable with sound licensing strategy or with IP law is debatable. Nonetheless, the Union courts have 'backed' the Commission and it seems that there is institutional harmony on the point. While many of these types of cases have been criticized, especially the last two, there is a common thread running through the cases, even more so if we put it in the light of standardization. As already pointed out the Commission's position is that one of the more valuable facets of standardization is the information sharing and unlocking of the intellectual property information necessary for innovation and implementation of a standard. Given the importance of the telecommunications and high technology sectors it is hardly surprising that Standard Essential Patents (SEP) investigations by the Commission, a sector that is characterized by Interoperability and Network Effects, has been the greatest cause for Antitrust concern.

The EU has, since the mid-80's, been focused on and pursuing a common market in telecommunications equipment and services to make for a more competitive Europe.[11] The role of standards in this goal has been constant and given that standard setting in the sector is unique in the sense that it is the standard that creates the Interoperability and networking beforehand rather than the standard following the market. Put another way, the standard needs to be agreed and implemented prior to technological advancement, failure to do so leads to a gap in the market as was the case with the 4G standard.

Meaning that there is an increasing importance of standards acting as platforms for technological advancement in the mobile telecommunications market. Therefore, any tensions or delays assume a particular importance, especially where these are caused by the deliberate licensing strategy of a firm. The sensitivities in this area combined with the goals of information sharing, innovation and market integration may well explain the line of cases relating to SEPs in the field of standardization. There has been special attention paid to the IP use, from IP hold-up to IP ambush and then more recently following a global 'mobile war' on the failure to license and apply for injunctive relief.[12]

A further motivation for the Commission to be involved is that there is great disparity at a national level on the ease or difficulty in obtaining an exclusionary Injunction for IP holders. Combined with the length and delay of a so-called 'willing licensee' been able to challenge an Injunction in what is a fast moving market could cause detrimental damage to the excluded party.[13] Whether the application of the law to cases that intersect the fields of standardization, intellectual property and commercial strategy is desirable or effective remains a point of contention, the reality is Antitrust has moved into this area and there is little evidence of it abating, it is therefore important from a firm point of view that care is taken when formulating strategies related to these areas.

CASE-LAW ANALYSIS

The four cases selected as examples of this intersection between standardization, intellectual property and strategy will demonstrate that the strategy selected by companies is directly linked to the anti-competitive concerns raised by the regulator.

Microsoft, for example, has been said to operate an "embrace, extend, and extinguish" strategy, which has a long play goal of establishing an industry standard with widespread adoption before cutting out interoperable competitors to cement the Microsoft market position. This type of behavior is seen as anti-competitive despite the commercial reality of its success.

In *Google*, the acquisition of a patent portfolio from Motorola Mobility was a strategic purchase to balance the threat from competitors to the Android platform development. As a strategy of protection, this was still susceptible to the Antitrust rules with a particular regulatory focus on the Commission receiving commitments that Google would not generate a foreclosure strategy, such as refusing to license, causing an impediment to effective competition.

The *Intel* case involved the pricing and rebate system introduced to increase and protect existing sales. In Intel, this was found to amount to the implementation of a strategy aimed at foreclosing a competitor, AMD, from the market for x86 CPU microprocessors on which Intel was considered to have an industry standard (*De Facto Standard*). A finding that was made despite evidence being adduced that AMD had enjoyed great growth in sales during the time of the Intel strategy.

Finally in *Samsung* the strategic decision of Samsung not to grant a license to use Standard Essential Patents and threaten to enforce their strict IP rights against usage of the patents without a license was questioned. The interim remedy, a restricting Injunction amounted to a threat capable of distorting licensing negotiations and lead to anticompetitive licensing terms that the licensee would not have accepted absent the seeking of the threat of enforcement. Each of these cases is examined in more detail below, the objective and purpose of which is to illustrate the manner in which the Antitrust rules have been applied in the EU to restrict business practices which can often be attributed to direct strategies.

Case 1: Microsoft

The case originated from an application for annulment against a Commission decision.[14] The Commission having found Microsoft infringed Article 102 TFEU by refusing to supply Interoperability information to its competitors as well as other breaches of the competition law that are not considered in this contribution. One of the grounds of the Commission decision was based on Microsoft limiting technical development to the prejudice of consumers by refusing to supply Interoperability information to competitors. The Commission ordered Microsoft to license on reasonable and non-discriminatory terms the relevant Interoperability information to its competitors. Justifying the order, the Commission considered that the possible negative impact of an order to supply on Microsoft's incentives to innovate was outweighed by its positive impact on the level of innovation of the whole industry. The arguments put forward by Microsoft, among others, was that a refusal to supply Interoperability information to a competitor who could otherwise not compete cannot constitute an abuse of a dominant position within the meaning of Article 102 TFEU. Microsoft further pointed to the fact that the information was protected by intellectual property rights or at the very least constituted trade secrets.

From a strategy angle one could assume that Microsoft's actions to not share the Interoperability information was designed to achieve a market position and obtain *a De Facto Standard*. This ultimately that they were not only able to earn a higher profit margin than competitors, but also to sustain a strong market position over a significant Antitrust is evident and the extent that a strong strategy in terms of restrictive licensing and minimal sharing is successful. From a legal perspective the starting point remains that Antitrust should not protect a market for the sake of protection, in other words, the law is there to protect the competitive market structure and not to protect individual competitors. Meaning that there should be competition on the merits. From a strategy perspective, it means that any business plan will need to take into account the need to compete on the merits on the market. For example, the introduction of a protected innovation onto the market that benefits and does not exploit consumers but leads to competitors from being excluded is legitimate. Here the introduction of the innovation must be permissible if Antitrust law is not significantly going to impair the incentive to innovate. Leading scholars in the field of Antitrust such as Hovenkamp (2012) provides the example in the development of the handheld digital electronic calculators. Following the introduction of the innovation in a decade's time the development of handheld digital electronic calculators destroyed the market for slide rules, a product that had been around since the seventeenth century. Not only did electronic calculators perform much better, but slide rule manufacturers were not in a position simply to switch their plants over to calculators. For the most part, they had no experience or manufacturing capability for electronic devices. Although the balance arises in the situation where, as in Microsoft, the market share has led to an industry standard, whether intentioned or not.

It seems that the 'standard' is a catalyst for the evaluation in so far as in normal circumstances the position in law is that a refusal by a dominant company 'to license a third party to use a product covered by an intellectual property right cannot in itself constitute an abuse of a dominant position within the meaning of Article 102 TFEU. It also follows from previously decided cases that a *De Facto Standard* is sufficient to be considered exceptional for purpose of the law and if on the back of this it is shown that: (a) the refusal relates to a product or service indispensable to the exercise of a particular activity in a neighboring market, (b) the refusal is of such a kind as to exclude any effective competition on that neighboring market, and (c) the refusal prevents the appearance of a new product for which there is potential consumer demand. The incumbent IP holder will be required to make available their technology, albeit on FRAND. In *Microsoft* the court affirmed that in order to be able to compete viably with Windows workgroup server operating systems, competitors' operating systems must be able to interoperate with the Windows domain architecture on an equal footing with those Windows systems, this was based on the assumption that the Interoperability with the Windows domain architecture was indispensable.

If we consider that Windows was, in fact, a standard then the judgment begins to make more sense, especially when the position within the EU in respect of standardization is that access to the standard must be made available. In other words, a standard cannot lock out, intentionally, other market players from using it. If we put this in Antitrust terms, what is meant is that a standard should never be operated in way that allows any party to the standard to exercise such control so as to lock out competition.

In *Microsoft*, the Commission considered whether the refusal at issue gave rise to a "risk" of the elimination of competition on the work group server operating systems market. In response to this potential "risk", Microsoft relied on the existing case law position requiring that the Commission establish that the refusal to license is "likely to eliminate all competition'. However, the Court refused to accept the Microsoft interpretation on the basis that it would become overly restrictive and would lead to a situation whereby the regulator would need to wait for the harm to occur before acting. Much criticism has been levelled at the General Court's judgment in Microsoft on the grounds that the previous case law tests of indispensable and restricting the appearance of a new product, were not the only parameters which determined whether a refusal to license an intellectual property right is capable of causing prejudice to consumers within the meaning of (Article 102(b) TFEU). More than this the Court brought in, the potentiality of an adverse effect on innovation generally. This as a potential abuse has now found its way into the Guidance Paper of the Commission and states that: 'The Commission considers that consumer harm may, for instance, arise where the competitors that the dominant undertaking forecloses are, as a result of the refusal, prevented from bringing innovative goods or services to market and/or where follow-on innovation is likely to be stifled.'[15] Whilst it may well be difficult to reconcile *Microsoft* with the general perception of freedom to operate and the right to strategically protect and acquire market position, if one is to evaluate the judgment and Commission decision in the case from a standardization angle and the pro-market benefits of standards then the breaking out of the Interoperability information seems logical and consistent with the practice of the Commission. Moreover, in later cases such as *Google* we see the Commission clearly laying out its stall for the difficulty it associates between intellectual property and standardization in cases where essential information is inaccessible. Nonetheless, the *Microsoft* approach has been heavily criticized. It is, however, illustrative of the difficulty that Antitrust enforcement has with been able to deal with business practices that are otherwise firmly grounded in commercial strategy. The overriding problem in *Microsoft* was the fact that there was not necessarily behavior that was traditionally unlawful rather there was an underlying danger to innovation resulting from a *De Facto Standard.*

Case 2: Google/Motorola

Although not an Antitrust case as such but rather a merger decision it involved the acquisition of a patent portfolio containing a number of encumbered patents to various standards. The same approach in *Microsoft* is seen in the decision of the merger in respect of the non-foreclosure or exclusion of competitors from the market when it comes to essential information that permits other companies to innovate, access and use a standard or otherwise allow the market to mature to the point of providing sufficiently competitive restraint. The decision stemmed from the European Commission receiving notification of a proposed concentration by which Google Inc. acquired sole control over Motorola Mobility Holdings, Inc. Under the agreement and Plan of Merger signed on 15 August 2011, Google acquired the whole of Motorola Mobility.[16] As part of the transaction, Google acquired Motorola Mobility's patent portfolio, approximately 7-8.000 patents. Google had strategically aimed at Motorola Mobility as the patent portfolio, focused mainly on wireless communication hardware and to some extent high-definition television (''HDTV''), related directly to Google's strategy to enter the mobile technology market. A difficulty arose in the transaction in so far as Motorola Mobility owned a number of patents that it had declared to be standard-essential patents

In particular, Motorola Mobility had a significant portfolio of cellular SEPs including patents relating to LTE, 3G and 2G as well as WCDMA-UMTS, GSM- GPRS, CDMA, WiFi, WiMAX, MPEG-4 Visual, HDTV, Battery etc. These patents read on wireless standards, smartphones, video, HDTV and batteries. A concern for the Commission was at the time of notification Google had no declared SEPs. The acquisition of Motorola Mobility's SEPs for which there was no substitute was considered to present Google with additional market power with regard to the standardized technology.

A question arose as to whether Google was restricted by the previous Motorola Mobility commitments of SEP's to various standards. In particular, whether Google would be constrained by the FRAND commitment which has been given by Motorola Mobility. Interestingly, the Commission considered that even if Google was restrained by the FRAND commitment this was not a guarantee that Google holder would not abuse its market power. Although a FRAND commitment may influence Google's incentives to significantly impede effective competition, it remains true that the company would still have some ability to do so. The regulator is essentially reserving the scope of the Antitrust provisions by considering that Google could still strategically use the FRAND commitments strategically to exclude other market players. This is clear from the Commission going on to consider that Google's ability to do so would ultimately be constrained to an extent by the fact that FRAND is an obligation upon which inter alia courts, arbitral tribunals or competition authorities may rule. This was especially important to the Commission in respect of Google strategically using Injunctions to prevent use of the SEP's in the future, here the regulator considered that the courts are the ultimate decision-makers on whether Injunctions should be granted and any SEP holder needs to convince a court before it can obtain an Injunction. This is a direct statement of constraining IP strategies and in some ways it restricts the property rights of SEP owners from been able to use Injunctions as weapons against users of a standard. An Injunction strategy was at the heart of DG Comp.'s concerns in Google. One of the dangers identified was that despite the limitations imposed by courts and other bodies, it is likely that Google would retain some ability to use its SEPs to significantly impede effective competition. For example, Google could certainly threaten to seek or seek Injunctions at any time. Moreover, depending on the national court in question, an Injunction may be granted without a detailed examination of whether FRAND and Article 102 TFEU have been respected, leaving Google free to enforce the Injunction. Thus the Commission considered that Google would have some ability to use its SEPs to significantly impede effective competition post-merger. Much of the concern in this merger was the fact that Motorola Mobility had already sought Injunctions against Apple and Microsoft on the basis of its SEPs prior to the merger. There was a danger that Google would continue the strategy and do so more strategically in the future. Ultimately the merger was allowed, with certain commitments, even though the Commission considered that Google was likely to have some ability post-transaction, to seek and enforce Injunctions, and in doing so to significantly impede effective competition, including by potentially extracting onerous licensing terms from good faith potential licensees to which they would otherwise not have agreed, be it in respect of royalty levels or other license terms. On the back of this the Commission sought assurances from Google that the strict legal right of the Rights-Holder to pursue an Injunction against a competitor for accessing the information from the standard would not occur. Interestingly for the Commission the original encumbrance of the patents was such as to stretch over the merger. Evidentially a strategic decision by Google to acquire through merging the patent portfolio of Motorola also allows and places Google, in the future, in a strategic position in standard discussions. Especially in those sectors that involve telecom or other high Network Effect markets. This coincides with their Android challenge against Apple, whilst one might argue that Motorola had already given up their information whereas in Microsoft the decision was not to release the information changes little from a legal point of view.

Again we see that the Commission focus is to ensure access to information and safeguard markets from the operation of corporate strategy, such as licensing exclusions or Injunction claims against competitors.

Case 3: Samsung

This was a case that originated via complaint from Apple to the Commission surrounding the licensing practices, or lack thereof, of Samsung. The Commission took the view that *Samsung*, by seeking injunctive relief against Apple based on Samsung's 3G UMTS SEPs in several EU Member States may breach Article 102 TFEU, which prohibits the abuse of a dominant position.[17] In the Commission's view standard-essential patents are patents protecting a technology which is essential for the implementation of an industry standard developed by a standard-setting organization. Accordingly, it is technically not possible to make a standard-compliant product without using the technology protected by the SEP. Interestingly there was little discussion over the question of substitutability or a more detailed definition of the SEP's. Seemingly by the mere ownership of SEPs related to mobile telecommunications standards and having committed to license these SEPs on FRAND terms there was an immediate assumption of anti-competitive behavior. Put another way, the ownership of a SEP brings with it an assumption of market power.

Whether this is compatible with the generally theory of intellectual property or not is part of a larger debate that is not addressed in this paper. The basis of a Commission investigation was the assertion made by Apple that they were willing to enter into a licensing agreement on FRAND terms for Samsung's SEPs. It was found that *Samsung* had started judicial proceedings seeking Injunctions against Apple to prevent Apple from accessing and using the patent information contained in the standard. Under these specific circumstances, the Commission took the preliminary view that Samsung's behavior amounted to an abuse of a dominant market position prohibited by EU Antitrust rules. According to the Commission recourse to Injunctions may unduly distort FRAND licensing negotiations and allow SEP holders to charge royalties or impose licensing terms which a licensee would not agree to absent the threat of having its products excluded from the market. To address these concerns, Samsung committed, for a period of five years, not to seek any Injunctions on the basis of any of its SEPs, present and future, that relate to technologies implemented in smartphones and tablets against any company that agrees to a particular licensing framework. Such a licensing framework consists of: (i) a negotiation period of up to 12 months and (ii) if no agreement is reached, a third party determination of FRAND terms by either a court or an arbitrator, as agreed by the parties. If the parties cannot agree on either submitting to court or arbitration, the parties will have to submit to arbitration. This is in line with the 2010 Horizontal Guidelines requiring FRAND Commitments be made by participants who wish to have their IPR included in a standard in order to ensure effective access to the standard. These FRAND commitments prevent Rights-Holders from refusing to license, or requesting excessive fees after the industry has been locked-in to the standard. Therefore *Samsung* is quite easily reconcilable with the Commission's program of restraining the locking in of information and excluding competitors to the detriment of the market and more strikingly in markets where innovation is a key. It also brings with it some concern, the main been the introduction of a new legal test for SEP holders wishing to apply for injunctive relief and that is the so called willing licensee test. Under this test a SEP holder can seek an Injunction for an infringements against a party where a party (a) breaches the license agreement, (b) is unlikely to be able to pay, (c) refuses per-standard reciprocity in cross-license negotiations, or (d) is engaged in SEP-based litigation against the SEP holder. A prospective licensee cannot be deemed "unwilling", however, on the ground that it challenges validity, infringement, or essentiality of the SEP in question. This imposes a strong limitation on the traditional patent rights under IP law. SEP holders should be aware that strategizing under an IP strategy with SEP's is subject to a new and different set of rules.

Case 4: Intel

The background to this case relates to two main investigations into *Intel* by the Commission following complaints of abuse.[18] The first brought by Advanced Micro Devices (AMD) in 2003 and the second in

2005. Following a number of on-the-spot investigations the Commission issued a Statement of Objections concerning Intel's conduct vis-à-vis five major Original Equipment Manufacturers (OEMs) namely: Dell, HP, Acer, NEC and IBM. On top of this a further complaint was made by AMD in 2006 claiming that Intel had engaged in exclusionary marketing arrangements and other practices with Media-Saturn-Holding GmbH (MSH), a European retailer of microelectronic devices. Following another investigation, in 2008, the Commission issued a supplementary Statement of Objections (SSO) concerning Intel's conduct in relation to MSH and Lenovo. The General Court found that the market share of Intel resulted in a degree of dominance. Interestingly was the fact that the Commission and the Court both narrowed the product market to the x86 CPU's meaning that this resulted in a 'standard' within the CPU market. In addition to this and as a result of the fact that Intel had embarked on an aggressive pricing strategy and plan to prevent disruptive innovation activities from eroding their market share by the main competitor, AMD, they had committed an abuse of their market position. Although there is no direct evidence of the strategy been directed towards this objective, it is not a far leap in faith given the strong relationship between Intel and Professor Christensen. Christensen had previously advised Intel on the dangers of disruptive technology.

This type of technology considers that companies who operate in high-end novel markets are able to maximize their innovation efforts on the top tier of their market, i.e. the most profitable and undertake sustaining innovation to maintain this segment of the market. Examples of which are mainframe technologies, telephony and vehicle industries. Considering that upon their introduction the original innovators have captured a high end of the market, in that the average company or consumer would not be able to afford these innovations (Christensen & Eyring (2011)). However, this focus on sustaining innovation at this level neglects the bottom end of the market, i.e. the everyday day potential user. This opens the door for disruptive innovation. According to Christensen (1997), an innovation that is disruptive allows a whole new population of consumers at the bottom of a market access to a product or service that was historically only accessible to consumers with a lot of money or a lot of skill. The personal computer market being a good example of this, when introduced they were outrageously expensive but as this market has been disrupted by new innovations we now find ourselves in a position where computers are affordable and freely available. The focus of Christensen's theory can, at the risk of oversimplification, be said to focus on those areas of the market that are at the high end of innovation. Meaning that these are areas where technology is new and as yet the competitive structure of the market has not altered in such a way so as create a natural competitive environment. Intel had structured their pricing and supply strategies in such a way so as to prevent the launch of products that could threaten their market share from the bottom end up.

The fact that there were significant barriers to entry and expansion in the x86 CPU market arising from the necessity of high sunk investments in research and development, intellectual property and production facilities necessary to produce x86 CPUs which one could argue was the while point of the Intel strategy to protect the market share they had built. Based on Intel's strong (must-stock) brand status and the resulting product differentiation the Court found that Intel had created a barrier to entry. It is easily arguable that Intel had created at least close to a *De Facto Standard* for the industry. The Commission reached the conclusion that Intel's conditional rebates and payments induced the loyalty of key OEMs and of a major retailer, the effects of which were complementary in that they significantly diminished competitors' ability to compete on the merits of their x86 CPUs. Intel's anticompetitive conduct thereby resulted in a reduction of consumer choice and in lower incentives to innovate. One can argue that again the application of Antitrust rules in Intel were such as to free up the market for further innovation. It is also interesting that it is one of the few cases where the Court has found a potential stifling of innovation been anti-competitive.

In summary of the cases considered a clear trend is developing which sees the EU Antitrust provisions moving into a system that is aimed at unlocking information and breaking out the dominance of single-firms where this dominance is based on intellectual property or a *De Facto Standard*. This is especially noticeable in market sectors which are characterized by Interoperability and Network Effects.[19] However, there remain question marks over the application of the existing framework tests when one looks for some form of market certainty for companies to strategize and behave in a given market.

CONCLUSION

In conclusion, there is a clear need, given the obvious trend of Antitrust involvement in company strategies for the regulator to clarify the scope of application of the Antitrust rules to single-firms. This is especially the case if the Commission desires to encourage and stimulate industry participation in standardization activities in the future where a level of legal certainty is necessary so not to allow the law to become counterproductive. Currently in the field of standardization the use of the Antitrust provisions are haphazard and lack any certain principles against which to assess the current legal status with any certainty. This is especially the case where Commission and the EU courts have dealt with cases of market leaders in fast moving markets.

Currently there is a possibility of a paradoxical situation, when commercial actors who have been encouraged to exercise their freedom to operate their businesses and a devised strategy becomes a success there may an infringement of law for such success. In the context of standards and standardization, the use of a strategy to increase an existing market share could prove fatal [Intel]. The failure to share or increase Interoperability with a De Facto Standard generated through market strategy could be unlawful [Microsoft]. Whereas the acquisition of IP as a strategy to compete against larger players in a particular market may be encumbered where the IP is subject of a standard, thus reducing the value of the IP [Google]. Finally, the strategic participation in standardization and the inclusion of Standard Essential Patents into an approved standard results in a loss of enforcement rights of the committed IP by the owner [Samsung]. In all of these cases, the companies concerned owed a separate and higher level of responsibility in their decisions to ensure their behavior was responsible for the market.

The takeaway from the above is the restriction the law, at an EU level, places on the strategic market activities of companies is unquantifiable. Evidently, once a firm gains market power, or is integral to the standardization process or a completed standard it becomes incumbent upon them to behave in a way that other companies are not required to. To act in a manner that is not necessarily in the best interests of their value creation but to rather act in a way that takes the market into account as well. Failure to do so is likely to result in the firm been subject to Antitrust scrutiny when undertaking commercial practices. A key area that this style of legal enforcement ignores is the impact of such a broad approach on the future research and development. An overly robust enforcement policy is likely to drive market leaders out of the standardization process altogether and a constant fear of been a market leader in respect of innovation.

This seems to sit uneasily with the policies, objectives and aims of the EU in climbing the competitiveness scale through research, development and innovation. The same level of uncertainty exists on companies that are disruptively innovating. How long will their market share be allowed to exist, before the Commission deems it necessary that the information be subject to a forced FRAND licensing arrangement? Ultimately, a balanced legal enforcement would see the law been able to adapt to new technologies, short lived market power resulting from inventive behavior and taking into account the market structure in relation to disrupting innovations and particular market costs associated with innovating.

REFERENCES

Blind, K., Bierhals, R., Iversen, E., Hossain, K., Rixius, B., Thumm, N., & van Reekum, R. (2004). Interaction between Standardisation and Intellectual Property Rights. European Commission Technical Report EUR 21074 EN, Seville.

Blind, K., & Thumm, N. (2004). Intellectual property protection and standardization. *International Journal of IT Standards and Standardization Research*, *2*(2), 61–75. doi:10.4018/jitsr.2004070105

Bohannan, C., & Hovenkamp, H. (2012). *Creation without restraint: promoting liberty and rivalry in innovation*. Oxford, UK: Oxford University Press. doi:10.1093/acprof:oso/9780199738830.001.0001

Denrell, J., Fang, C. & Winter, S. G. (2003). The Economics of Strategic Opportunity. *Strategic MGMT. J.*, *24*, 977-78.

Economides, N. (1996). *The Economics of Networks*. *International Journal of Industrial Organization*, *14*(6), 673–699. doi:10.1016/0167-7187(96)01015-6

Faull, J., Nikpay, A., & Taylor, D. (2014). *The EU law of competition*. Oxford, UK: Oxford University Press.

Ghemawat, P. (2002). Competition and Business Strategy in Historical Perspective. *Business History Review*, *76*(37), 37–38. doi:10.2307/4127751

Grant, R. M. (2004). *Contemporary strategy analysis: concepts, techniques, applications* (5th ed.). Oxford, UK: Blackwell Publishing.

Hamel, G., & Prahalad, G. K. (1996). *Competing for the future*. Boston: Harvard Business School Press.

Iversen, E. (1999). Standardization and intellectual property rights: conflicts between innovation and diffusion in new telecommunications systems. In K. Jakobs (Ed.), *Standards and Standardization: A Global Perspective* (pp. 80–101). Idea Group Publishing.

Johnson, G. & Scholes, K. (2005). *Exploring corporate strategy: text and cases*. Academic Press.

Kortum, S., & Lerner, J. (1999). What is behind the recent surge in patenting? *Research Policy*, *28*(1), 1–22. doi:10.1016/S0048-7333(98)00082-1

Kovacevicj, A. (2009). *Google's Approach to Competition*. Google Public Policy Blog.

Porter, M. E. (2004). *Competitive strategy: techniques for analyzing industries and competitors*. New York: Free Press.

Posner, R. (2013). *Antitrust Law* (9th ed.). University of Chicago Press.

Pound, J., Lehn, K. & G. Jarel. (1986). Are Takeovers Hostile to Economic Performance? *Regulation, 10*, 25-30.

Prusa, T., & Schnitz, J. Jr. (1991). Are New Firms an Important Source of Innovation? Evidence from the PC Software Industry. *Economics Letters, 35*(3), 39–342. doi:10.1016/0165-1765(91)90155-E

Rumelt, R. P., Schendel, D., & Teece, D. J. (1991). Strategic Management and Economics. *Strategic Management Journal, 12*(S2), 5–29. doi:10.1002/smj.4250121003

Salzberger, E. M. (2012). *Law and economics of innovation*. Cheltenham, UK: Edward Elgar. doi:10.4337/9781784714345

Schumpeter, J. (1943). *Capitalism, Socialism and Democracy*. Unwin University Books.

Shapiro, C., & Varian, H. (1999). Information Rules: A Strategic Guide to the Network Economy. Harvard Business Press.

Smit, J. (2004). *Het drama Ahold*. Amsterdam: Balans.

Spence, A. M. (1984). Cost Reduction, Competition, and Industry Performance. *Econometrica, 52*(1), 101–112. doi:10.2307/1911463

Swann, P. (2000). *The Economics of Standardization. Final Report for Standards and Technical Regulations*. Manchester, UK: Directorate Department of Trade and Industry, University of Manchester.

KEY TERMS AND DEFINITIONS

Antitrust: These are a set of laws that promote competition and protect consumers from anticompetitive business practices.

De Facto Standard: A product or protocol which has not been ratified by any official standards body but which nevertheless has a large market share and is followed by the market.

Frand: A level of royalty that can be expected from a Standard Essential Patent included in a standard, meaning licensing on fair, reasonable and non-discriminatory terms.

Injunction: A court order sought by a patent holder requiring a 3rd party be restrained from performing a particular act, such as using protected technology without been granted a license to do so.

Interoperability: A product or system, whose interfaces are completely understood, to work with other products or systems, present or future, without any restricted access or implementation.

Network Effect: Describes a network that attracts a number of users to use a good or service through a common network. Once a network reaches critical mass of users firms will have access to large pools of consumers in the network.

Rights-Holder: Refers to the intangible rights granted through legal ownership to a firm or an individual protecting the products of human intelligence and creation.

Standard Essential Patent: Patents protecting a technology which is essential for the implementation of an industry standard developed by a standard-setting organization.

ENDNOTES

1 From the EU regulators perspective this is certainly the case, see Commission Communication COM (2008) 133 final "Towards an increased contribution from standardization to innovation in Europe" and the recent DG Enterprise Public Consultation on Patents and Standards [2015]. Retrieved July 13, 2015, from http://ec.europa.eu/enterprise/policies/industrial-competitiveness/industrial-policy/intellectual-property-rights/patents-standards/index_en.htm

2 It is widely accepted that Sony's Betamax video standard that competed against and ultimately lost to JVC's VHS standard was not a result of the superiority of one technology over the other but rather the slick marketing of VHS and a more organized and accessible licensing of the VHS standard over Betamax. Licensing strategies potentially provide single-firms with revenue but also market position. Take for example the licensing strategies of standard essential patents in the telecommunication standards, and Ericsson as an example of technology licensing in standards as a business model, Retrieved July 13, 2015, from http://www.ericsson.com/res/thecompany/docs/press/backgrounders/ipr_licensing_press_backgrounder.pdf

3 In this paper there is no distinction drawn between antitrust and competition law and the terms are used interchangeably.

4 See DG Competition enforcement priorities Retrieved July 13, 2015, from http://eur-lex.europa.eu/legal-content/EN/ALL/?uri=CELEX:52009XC0224(01) at paragraphs 23-37.

5 See for example the Case C-457/10 P, AstraZeneca v Commission, Judgment of 6 December 2012, the EU Commission fined AstraZeneca for abusing its dominance using a strategy of misleading statements to obtain supplementary protection certificates that extended its patent exclusivity. This is a case example illustrating the far reach of the EU competition rules where firms enter into a particular strategy to extend their market share.

6 See DG Comp Competition Brief of June 2014 where standard essential patents are highlighted as an area of concern and focus for the EU regulator, Retrieved July 13, 2015, from http://ec.europa.eu/competition/publications/cpb/2014/008_en.pdf

7 See the investigation of ETSI in the Sun investigation by the Commission, Press Release 12 December 2005, IP/05/1565

8 See the case of Case C-171/11 Fra.bo SpA v Deutsche Vereinigung des Gas- und Wasserfaches eV (DVGW) – Technisch-Wissenschaftlicher Verein, judgment of 12 July 2012 where the German standard setting body was held to be a public body and susceptible to the EU free movement rules. See also the application of the competition rules to an international body set up through international cooperation Case T-155/04 SELEX Sistemi Integrati v Commission [2006] ECR II-4797

9 IBM v Commission (60 & 190/81R) [1981] ECR 1857, [1981] 3 CMLR 93

10 Magill (RTE and ITP *v* Commission) (C-241 & 242/91P) [1995] ECR I-743.

11 See for example Commission Communication "on the role of European standardization in the framework of European policies and legislation," COM (2004) 674 final, see also the classic Directive for the control of standardization (technical standardization) under Council Directive 98/34/EC laying down a procedure for the provision of information in the field of technical standards and regulations (OJ L 204, 21.7.1998).

12 For the locking out of industry competitors see the Commission Decision of 14 October 2009 in Case 39.416, Ship Classification. Retrieved July 13, 2015, from http://ec.europa.eu/competition/antitrust/cases/index/by_nr_78.html#i39_416 (Last visited 16 July, 2015).

13 For a recent opinion on this point see the Opinion of Advocate General Wathelet in Case C-170/13 Huawei Technologies Co. Ltd v. ZTE Corp., ZTE Deutschland GmbH, who gives credence to the willing licensee test as a standard to be applied in SEP cases when considering a failure to license.

14 T-201/04 Microsoft v Commission [2004] ECR II-4463

15 See paragraphs 88-90 of Communication from the Commission - Guidance on the Commission's enforcement priorities in applying Article 82 of the EC Treaty to abusive exclusionary conduct by dominant undertakings (Text with EEA relevance) OJ C 45, 24.2.2009, p. 7–20

16 Google/Motorola Mobility (M.6381) (13 February 2012)

17 Samsung-enforcement of ETSI standards essential patents (COMP/39939)

18 Intel *v* Commission (T-286/09)

19 The EU Commission has recently identified a class of investigation and cases they have carried out as been information communication technologies cases. Retrieved July 13, 2015, from http://ec.europa.eu/competition/sectors/ICT/cases.html

Section 6

Looking at Innovation

Chapter 13
Standardization, Innovation, and Organization:
A Contingency Perspective

Nizar Abdelkafi
Fraunhofer MOEZ, Germany & University of Leipzig, Germany

Sergiy Makhotin
Fraunhofer MOEZ, Germany & University of Leipzig, Germany

ABSTRACT

Companies that intend to leverage standardization to drive innovation have to set up their organizations appropriately. Data from four German companies reveal different ways for organizing the standardization activities. Companies may allocate a high level of resources by dedicating a specialized department; they may assign the standardization task to top managers; they may treat standardization within occasional projects; or create sophisticated solutions to facilitate the use of standards. Contingency theory is used to narrow down the number of factors that influence the organization of the standardization function. There are three relevant contingency factors: competition, size, and strategy. Company's age, technological complexity and volatility of customer demand are found to be less important. Instead, maturity of technology and the period of time, during which the company has been actively engaged in standardization work seem to be better contingencies. Another finding is that innovation appears to be a by-product and not the main driver of standardization work.

INTRODUCTION

Standardization and innovation are interrelated issues (DIN, 2000). Their mutual relationships have been discussed intensively in the innovation management and standardization literature. The innovation management literature argues that companies winning the dominant design battle can achieve economic success (Suárez & Utterback, 1995; Utterback & Abernathy, 1975), since the dominant design develops to a standard on the marketplace, providing the company with an advantage against its competitors (Swann, 2000). Hence, companies' strategies and business models that aim to establish a dominant design are at

DOI: 10.4018/978-1-4666-9737-9.ch013

the focus of the innovation management literature that deals with standards. Standardization literature, however, looks at innovation from a different perspective. It particularly focuses on those standards that are initiated by standards developing organizations (SDOs). These so-called committee standards are virtually neglected in innovation management. The literature on standardization shows, for instance, how these standards can be established during the product development phase, even before a prototype has been created (Byrne & Golder, 2002), or investigates the economic benefits of standardization and standards from the macro-economic viewpoint (Blind, Jungmittag, & Mangelsdorf, 2011).

Research on standardization and innovation at the company level is still scarce. We believe that standards and standardization are conducive to innovation, in spite of the controversy of the literature (Wright, 2012). The underlying assumption of this work is that if companies intend to support their innovation process with standards and standardization, they have to integrate their standardization activities in their organizations appropriately. Hence, they have to embed the standardization function in their activities in a way to benefit innovation. Depending on certain contextual factors (contingencies), however, companies are expected to adopt different organizational structures. In other words, there is no single formula that fits well to all eventualities. Consequently, contingency theory can be used to generate new insights into the overlapping field of standardization and innovation. This work is, therefore, an attempt to apply a theory from organization science to understand the conditions under which standards and standardization can support innovation. Hence, we derive two major research questions:

- How do companies organize their standardization activities?
- What are the relevant contingency factors?

The chapter is structured as follows. The next section deals with the theoretical underpinnings of this research and draws upon the current understanding of the interdependencies between standardization and innovation. It also introduces contingency theory as a suitable frame for the analysis of how companies organize their standardization activities. The section after the literature review provides information on the research design, and describes four case studies conducted in German small and medium-sized companies. In the research findings, we identify different ways of organizing the standardization function and apply contingency theory to find out the relevant factors that make companies implement different types of organizations. In addition, we conclude that innovation appears to be a by-product of standardization and standards, but not the primary objective of companies. The final section summarizes the main findings, discusses them, and proposes directions for future research.

LITERATURE REVIEW

This work integrates three fields of research: standardization, innovation, and organization theory. The relationship between standardization and innovation has not been directly linked to organization theory, in particular to the contingency approach. In the following, we deal with the literature on organization in the context of standardization, discuss the relationships between standardization and innovation, and then establish their connections to contingency theory.

Organizational Structure of Standardization within Companies

The body of research on how companies organize their standardization activities is limited in scope. Most of the identified work in this regard deals somehow with organization, but does not specifically focus on it. ISO, for example, developed a methodology to demonstrate the benefits of standards in various companies (ISO, 2012, 2011). This methodology, which is based on the generic value chain model by Porter (1985), uncovers how business functions are affected by standardization. Although primarily not focused on organization, the case study of Nanotron, a young ICT company based in Berlin, Germany, provides key insights into strategy, attitude of management towards standardization, in addition to the qualitative and quantitative impacts of standardization on the company (ISO, 2011).

The INTEREST project (Integrating Research and Standardization) delivers some insights into the organization of the standardization function (ISI, 2007, 2006). Standardization is considered a key issue as "[t]he way enterprises organize standardization and research activities involve[s] central aspects of the way firms and, ultimately, the way industries develop" (ISI, 2006, p. 71). On the basis of five case studies conducted with commercial companies, two organizational modes of standardization are distinguished: coordinated and self-organized. The coordinated mode implies that standardization activities are coordinated by an internal department or organizational layer, whereas self-organized mode is characterized by ad hoc or project-related standardization activities (ISI, 2006, p. 73).

The question of whether an organizational structure fits strategy, technology, or another contingency is, however, not the focus of the case studies that have been conducted so far. Nevertheless, without explicitly relating standardization to contingency theory, these case studies demonstrate that typical factors found in contingency research such as strategy, managers' perception of standardization as well as technology and environment are of major importance for the organization of standardization activities within companies.

Innovation and Standardization: Current Understanding

This work essentially focuses on those standards that are developed by national and international organizations at such as ISO and DIN (formal standards or committee standards). According to Egyedi & Blind (2008), who adapt the ISO/IEC definition, a committee standard is "a document established by cnsensus that provides, for common and repeated use, rules, guidelines or characteristics for activities or their results, aimed at the achievement of the optimum degree of order in a given context" (p. 3). The use of committee standards is voluntary unless required by legal regulation. Standardization denotes the process, by which committee standards are developed. Standardization also designates the work of SDOs, which moderate the consensus process that can involve individuals, companies, public organizations and other institutions.

Standardization is linked to innovation (e.g. Abdelkafi & Makhotin, 2014), and it is important to explain what we understand under the term "innovation". Hauschildt (1997) defines innovation as invention and exploitation. This is a broad definition that includes all types of innovations, be it radical or incremental. It seems, however, unlikely that standards and standardization give rise to groundbreaking innovations in the Schumpeterian sense (Abdelkafi & Makhotin, 2014). The creation process of radical innovations is messy and unstructured, and the odds of having standardization and standards leading to breakthrough innovations are low. In fact, we are aware of many innovation definitions, but in our research we are interested in all what is "new", no matter its characteristics, provided that it is enabled by standardization and standards. For further innovation definitions, it can be referred to Tidd & Bessant (2013), Reichwald & Piller (2009), OECD & Eurostat (2005), and Cohen & Levinthal (1990).

Research on innovation and standardization has gained a remarkable momentum over the last two decades (Choi, Lee, & Sung, 2011). A significant share of work can be traced back to researchers such as Blind, Egyedi, Sherif, de Vries, and Swann who introduced new ideas to this field and criticized the common belief that standardization is associated with rigidity and inflexibility. It has been shown that the mechanisms by which innovation and standardization are interrelated are complex and require further scrutiny.

The effects of standards and standardization have been scrutinized from the macro- and micro- perspective. For example, the DIN-survey studying the economic benefits of standardization consists of two parts: benefits for business and benefits for the economy as a whole (DIN, 2000; Blind et al., 2011). The study uncovers the significant contribution of standards and standardization to national economies and their impacts on the competitiveness and strategies of companies (DIN, 2000). Nevertheless, the interactions between the micro- and macro-levels have been mostly neglected. Based on an extensive literature research, Swann (2000) concludes that standards give rise to an infrastructure that shapes the environment outside the firm and enables innovation within the firm. Thus, the studies by DIN (2000) and Swann (2010, 2000) can be seen as the initial impetus for researchers to study the environmental and technological change that are triggered by standards and the economic benefits for companies when they actively pursue standardization.

Egyedi & Blind (2008) argue that standards in the ICT sector do change over time, leading to so-called standards dynamics. "Standards are revised, extended, replaced, succeeded, reinstated, etc." (Egyedi & Blind, 2008, p. 4). The change of standards can constitute an opportunity or a threat. Change "accompanies innovation in science and technology", but also requires adaptation from companies (Egyedi & Blind, 2008, p. 6). Because technology is identified as one of the change drivers, Egyedi & Sherif (2010) relate the timing of standardization to the technology S-curve and technology performance. The transition from an existing to a new technology with higher performance is supported by three types of standards: anticipatory, enabling and responsive standards. Though these findings originate from ICT, some aspects can be transferred to other sectors (Abdelkafi & Makhotin, 2014).

Abdelkafi & Makhotin (2014) note that the relationships between standardization and innovation are twofold: getting innovations to standards and using standardization and standards to derive innovations. It is straightforward that innovations can redefine the state-of-the-art by giving rise to new standards. The other direction, which closes the loop, is, however, less evident. Abdelkafi and Makhotin (2014) demonstrate that the use of committee-based standards and involvement in standardization processes can support innovation. The authors make a good case in this regard, but do not deal in-depth with the context factors. In other words, it is unknown under which conditions this innovation potential in standards and standardization unfolds. Abdelkafi & Makhotin (2014) rather focus on the opportunities that originate from standards and standardization to drive innovation. By conducting interview-based research, they identify six innovation potentials. In addition, they differentiate between standardization and standards in their analysis. Standards can be conducive to efficient and targeted innovation and business model innovation. They can also stimulate innovation by triggering impulses to innovation (e.g. when standards are updated) and by enabling innovations through exceeding the requirements of standards. The standardization process can foster the innovation process, as it supports innovation communication, enables the absorption of innovations, or stimulates innovation through the participation in the standardization process (Abdelkafi & Makhotin, 2014). These opportunities for innovation can thrive, however, only under specific conditions. Some conditions can help companies better leverage the innovation potentials in standardization than others. Because of this, contingency theory can deliver good insights to identify the factors that have an impact on the company's capability of reaping the innovation benefits due to standardization.

Contingency Theory as a Suitable Frame

The contingency approach can be fitted into a simple formula: "At the most abstract level, the contingency approach says that the effect of one variable on another depends upon some third variable" (Donaldson, 2001, p. 5). It basically postulates that there is no best way to organize (Weill & Olson, 1987) due to the variety of internal and external influencing factors. "The contingency approach attempts to understand the interrelationships within and among organizational subsystems as well as between the organizational system as an entity and its environments. It emphasizes the multivariate nature of organizations and attempts to interpret and understand how they operate under varying conditions" (Szilagyi & Wallace, 1980, p. 178; based on Kast & Rosenzweig, 1973).

The understanding of contingency approach has evolved since its rise from social research in the early 1950ies, resulting in various definitions, application domains and controversial discussions (Sousa & Voss, 2008; Gerdin & Greve, 2004; Bourgeois, 1985; Mintzberg, 1979; Child, 1972; Duncan, 1972; Pugh, Hickson, Hinnings, & Turner, 1969). Contingency theory research has dealt with many organizational issues such as management structures, leadership, human resource management, and decision-making. Since this research aims at interpreting and understanding the interdependencies between standardization and organizational structure, we only focus on structural contingency theory (Donaldson, 2001; Pfeffer, 1982).

Structural contingency theory is rooted in early contingency research (e.g. Tushman, 1979; Ouchi, 1977; Miles, Snow, Meyer, & Coleman, 1978; Child, 1972), but still provides a high potential for application in new fields (Donaldson, 2001) such as standardization research. According to Donaldson (2001, p. 7), a contingency is "any variable that moderates the effect of an organizational characteristic on organizational performance." The contingency factors that are frequently mentioned are: environment, age, size, technology, uncertainty and complexity. Donaldson (2001) identifies three "core commonalities" of prior research. First, there is a relationship between contingency and organizational structure. Second, contingency change leads to adaptation of the organizational structure. Third, fit between contingency and the organizational structure affects performance.

Thus, contingency theory postulates a trivariate relationship among contingency, structure, and performance (Donaldson, 2001; Woodward, 1965). The concept of fit matches the company's performance (high or low) with the state of the interrelation (congruence) between contingency and structure (fit or misfit). Considerable research has focused on the concepts of fit and performance (Burkert, Davila, Mehta, & Oyon, 2014; Volberda, van der Weerdt, Verwaal, Stienstra, & Verdu, 2012; Roth & Morrison, 1992; Venkatraman & Prescott, 1990; Drazin & Van de Ven, 1985), but also provided the ground for a controversial scientific debate about the limitations of the contingency approach such as the lack of clear definitions or measurement techniques (Ginsberg & Venkatraman, 1985; Venkatraman & Camillus, 1984; Tushman, 1979; Longenecker & Pringle, 1978; Wooton, 1977).

Despite the criticism, structural contingency theory constitutes a suitable frame for this research. Standardization literature can be viewed through the lenses of this theory. Research has dealt so far with related issues, yet without direct connection to the contingency approach. For instance, in the literature on standards dynamics (Egyedi & Blind, 2008; Egyedi & Sherif, 2010) environment and technology seem to be relevant contingencies, as a change in the environment or technology leads to change in the standards and standardization activities. Based on expert interviews from different sectors, Abdelkafi & Makhotin (2014) conclude that the environmental characteristics have a strong impact on the standardization activities of a firm. In standardization literature, the company size has been frequently addressed, since authors disagree on who is actually profiting from standardization and who is rather suffering from it: SMEs or large companies (Blind & Thumm, 2004; Swann, 2000).

SMEs frequently complain about the extensive costs of standardization. We believe, however, that high expenditures are actually due to the misfit between contingency and structure. The company's benefits from standardization are also difficult to measure. Additionally, there is usually a time lag between the investment made in standardization and the harvested gains. De Vries (2006), for instance, studied four companies that were able to extend their market shares and achieve cost advantages through participation in the standardization process. Nevertheless, the companies do not achieve these advantages in the same way. Hence, the concepts of fit and performance from structural contingency theory can shed more light on standardization.

In this research, the performance indicator that we focus on is whether companies are capable of using standards and standardization to support their innovation process. Consequently, we are not comparing different levels of innovation among companies. The case studies described in the following are all in the state of fit. In a previous study by Abdelkafi & Makhotin (2014), these companies are considered high performing, as they are able to leverage the innovation benefits of standardization. Though their performance with regard to innovation is high, these companies do not organize the standardization function in the same way. Hence, two questions are important: What are the types of organization of the standardization function that are encountered in practice and what drives companies to organize differently? Contingency theory provides a suitable theoretical frame. First, it enables us to narrow down the number contingency factors that are possibly relevant and then to posit that the structure of the standardization function depends on specific contingency factors that we propose to determine.

STANDARDIZATION AND ORGANIZATION: FOUR CASE STRUDIES

Research Design

Our research design consists of two consecutive phases. The first phase is a research study conducted by the authors on behalf of the German Institute for Standardization DIN (Deutsches Institut für Normung e. V.) between April 2012 and December 2013. This study focused on the identification of the innovation potentials in standardization and the development of recommendations for practitioners and policy makers, so as to achieve a better leverage of these—mostly latent and unexploited—potentials. The research was exploratory in nature and based on interview methodology. 40 semi-structured interviews were carried out with practitioners from German companies, in particular SMEs, and 20 interviews with diverse institutions such as universities, associations, and chambers of commerce. The qualitative interviews were analyzed by means of content analysis, resulting in different codes, sub-categories and main categories (Budd, Throp, & Donohew, 1967; Lindkvist, 1981). Although the interview guide was designed to capture the relationships between innovations and standardization, only 7% of the transcript sections pointed actually to innovation potentials, hence leaving us with rich data. The context-related categories were particularly useful to analyze the contingency factors. For instance, these categories fit well into the components defined by Bourgeois (1985) to characterize the contingency factor "environment", which are: actors such as customers and suppliers, and attributes such as complexity and dynamism.

The interview-based study was essential for the discovery of how standards and standardization can drive innovation. Hence, the objective was not to validate the hypothesis whether companies become more innovative, but to demonstrate that the strategic use of standards and standardization can be supportive to innovation, while uncovering the forms that the innovation potentials in standardization can take. Nevertheless, the companies that have exploited the benefits of standards and standardization seemed to not follow the same pattern of behavior.

To further investigate this pattern, we launched a second research phase, which consists in conducting case studies. Case studies constitute an appropriate methodology to explore how best-practice companies internally organize their standardization activities. With regard to contingency theory, the case studies provide a fertile ground for the identification of contingencies and for the analysis of how companies achieve organizational fit to leverage the innovation potentials in standardization.

Research Sites

This work applies the multiple case study approach (Yin, 2003). A comparative case study analysis consists of a small number of cases that are scrutinized in a qualitative manner (Dul & Hak, 2008). From our database of companies that initially participated in the first research phase, four case studies are selected based on the following requirements:

- Successful exploitation of standards and standardization as an effective means to drive innovation;
- Different organizational structures of standardization;
- Willingness to participate in this study.

The first selection criterion rules out companies that are unable to leverage innovation potentials of standardization. The second criterion means that companies should exhibit different organizational structures of their standardization activities. Only a few companies from our sample have a department that is exclusively dedicated to standardization work. Frequently, standardization activities are assigned to one person or a group of people as a secondary task. The third criterion reflects the openness of the companies to share their experiences and knowledge with the researchers.

Data Collection

The generation of relevant data for each case study has relied on primary sources (interviews with 3 to 5 key informants) and secondary data (press releases and articles in practitioners' journals). In this way, the data sources are triangulated to increase the reliability level of the research (Yin, 2003). All interviews were carried out face-to-face in the time period between December 2012 and August 2014 with the participation of both authors. The interview durations are between 35 and 100 minutes. All of them were recorded and transcribed. Because the companies have different organizational structures of standardization, our key informants occupy different hierarchical levels: middle management such as managers in the areas of product development, sales, costs, quality and standardization as well as CEOs (table 1).

The semi-structured interviews were based on an interview guide with two main parts. The first part is based on our prior work (Abdelkafi & Makhotin, 2014) and captures data on the company itself and its products, the positions and tasks of the key informants as well as the company's involvement in standardization activities. The second part contains questions that address standardization issues from the contingency perspective. This part reflects frequently encountered issues in contingency theory such as external environment, structure, size, age, strategy, technology, in addition to fit and company's performance.

Data Analysis

The interview transcripts and the secondary data are coded by means of software for qualitative content analysis (MAXQDA) (Schreier, 2012; Mayring, 2010). Coding refers to the process of assigning text sections to categories, which designate frequently mentioned issues by the informants (Gibbs, 2007). Thus, the categories reflect key contents of the interviews.

The first step of data analysis deals with the identification of relevant categories. Toward this aim, we apply a deductive-inductive approach; potential categories are derived from the literature and others from the interviews. An online search is conducted via the databases "web of science" and "Scopus" to find publications that relate standardization to contingency theory and organization. For instance, the

Table 1. Participants in the semi-structured interviews

	Company Name	Position of Interviewee
Phase 1	OptSys	Strategic Innovation Manager
	Lafur	CEO (Product Development and Cost Engineering)
	ElecS	Executive Director (CEO)
	MechC&I	Quality Manager
Phase 2	Lafur	Head of Product Development and Design
		Cost Engineering Manager
		Product Manager
		Quality Manager
	OptSys	Applications Manager
		Sales Manager
		Strategic Innovation Manager*
	ElecS	Standardization Manager 1
		Standardization Manager 2
	MechC&I	Head of Product Development
		Quality Manager*
		Designing Engineer/Risk Analysis Manager
		CE Coordinator/Technical writer

*Same interviewee as from the phase 1.

search for ["organizational structure" standardization case][1] in Scopus returned 103 publications. After going through the titles and abstracts, only 13 publications were classified as potentially relevant. None of them, however, could serve the purpose of this research. The same procedure has been repeated by varying the combinations of search words (more than 20 queries in total), but research that relates the issues of standardization, organization and contingency could not be identified. Therefore, we decided to consider these streams of research separately.

We consulted the literature that provides a good overview on contingency theory and identified frequently mentioned issues: environment, structure, size, age, manager's perception, performance, fit, technology and strategy (e.g. Matyusz, 2012; Sousa & Voss, 2008; Graubner 2006; Donaldson 2001). Subsequently, we looked in the standardization literature for explicit statements related to these issues. When many statements are found, the contingency factor is assumed relevant for standardization and therefore retained as a category for the coding process. In addition, sub-categories are defined to better address the particularities of standardization. For example, in contingency theory, structure relates to the overall organization. Since we are interested in the organization of the standardization function, this category is divided to two sub-categories: company structure and standardization-related structure. Beside the categories found in contingency theory, we add two categories identified by Abdelkafi & Makhotin (2014): application of standards and active involvement in the development of standards (table 2). The interviews were coded by one author and then separately controlled by the other. In the case of disagreement, discussions were initiated to reach a consensus.

Case Study Description

In the following, the case studies conducted with the German companies are described. Because of anonymity reasons, we do not provide the real, but rather pseudo names of the companies. These companies are called: ElecS, Lafur, OptSys, and MechC&I. In addition, some companies are operating in industries, in which the company's name can be guessed easily, if one knows the product. In such cases, the product is also not disclosed.

Table 2. List of categories

Main Category	Sub-Category
Environment	External environment
Structure	Company structure Standardization-related structure
Size	Company size Standardization unit size
Age / Time	Company age Duration of standardization activities Timing of standardization activities
Performance	Impact on innovation Other impact factors (e.g. financial impact)
Technology	-
Strategy	-
Standardization activities	Application of standards Active involvement in the development of standards

ElecS

ElecS is a producer of electrical systems that has grown to a size of 600 employees over the last six decades. The company is operating in a highly competitive business field, in which multinationals are the main competitors. Production is driven by the orders of the customers, who get customized solutions according to their specific requirements.

Standardization constitutes a crucial part in the company's philosophy. Over more than 40 years, ElecS could accumulate a considerable amount of know-how. Standards are relevant for achieving safety, ensuring customers' trust, and covering the legal aspects in case of accident. The company integrates standardization activities in its organizational structures and has strong connections to standardization bodies. One of the company's top managers has been the chairman of a standardization body for several years.

Currently, ElecS has an organizational unit that exclusively deals with standardization issues on national and international levels. The standardization department is composed—at the time during which the study has been carried out—of four persons: two full-time employees dedicated to standardization work and two part-time employees. The head of the department participates regularly in diverse committee standard meetings and reports directly to the company's management board. One of the major tasks of the department is to monitor new technologies and to proactively initiate new committees by submitting new applications to SDOs. The department has developed and implemented processes to capture systematically the internal knowledge required during the committee meetings. This knowledge reflects different perspectives and serves to protect the company's interests. For instance, the department can ask the company's engineers to comment an early version of the standard. Reversely, the critical information gained during the standardization-related work is brought to the company and diffused during internal meetings with the management board. The agenda of these meetings usually contains a heading called standardization, so that the head of the standardization department can report on his work and inform the participants on major issues from the committees. The standardization department acts, therefore, as a conduit between the committees organized by SDOs and the company.

Lafur

Lafur is a company that manufactures laboratory equipment. It is over 60 years old and employs around 700 people. Lafur offers a high variety of products and services. It serves two customer groups with different requirements: public institutions (equipment for educational services) and industry firms, mainly pharmaceutical companies. Lafur produces customized products for the final customer (lot size of one). Within the value chain, so-called planners occupy an intermediate position between the company and the final customer. These planners provide services that consist in designing the laboratories for the final customer and consequently influence the set of suppliers with whom Lafur has to work.

Lafur regularly creates customized products to satisfy the needs of its technology-oriented customers by using the principle of modularity in design. Hence, the final customers are the main innovation triggers. Because of the constant pressure to innovate due to changing customers' requirements, the company considers itself an innovation leader. These innovations can drive the development of new standards or the update of current standards, resulting in an improvement of state-of-the-art. The company indirectly benefits its cost-oriented customers from the public institutions. As public institutions usually request the fulfillment of specific standards, they can get better products. Thus, innovations made for the private sector are translated to standards that subsequently benefit the public sector.

OptSys

OptSys is a small 20-years old company employing around 70 people. The company produces high-tech optical systems for the measurement and visualization of surfaces on a nanometer-scale. The devices developed by the company consist of hardware and software. The technical parameters as well as the mathematical procedures and algorithms are the critical parts in the standardization endeavors, since they affect the comparability of the results of the measurements. For instance, in the absence of standardized procedures, the competitor's software would generate distorted visualizations and hence wrong measurements, if applied to analyze the output data of OptSys devices. For OptSys, this problem was the main impulse that triggered the initiation of the development of a new ISO standard.

Standardization-related tasks at OptSys are part of the activities of the strategic innovation manager. The company is aware of the importance of standardization, especially at the international level, to achieve market diffusion of its products. The company's production is customer-driven, and the devices are customized according to the customers' needs. Customers' specifications can lead to major innovations. For instance, the desire of a metallurgical multinational company to optically measure surfaces of big and heavy items that cannot be transported to the laboratory has given rise to a new product: the portable scanner. Later on, this portable scanner has been used in other industries such as in automotive companies for the control of car painting. It is noteworthy that the portable scanner may not have been developed, hadn't the customer requested this innovation. The competitive environment of OptSys consists of many big players. OptSys could achieve a competitive advantage because it can implement the customer's requirements very quickly.

MechC&I

MechC&I is a mechanical engineering company with about 150 employees that makes quality products for the high-end niche market with about two competitors worldwide. Over the last ten years, the company went through a considerable growth phase. Recently, the company has diversified its business model, as it does not only sell its own products, but also rents them to clients. Furthermore, it provides reparation services for its own and competitors' products. Because of its continuous growth, the company is currently going through a restructuring process, resulting in changes of the organizational structure. As the company produces customized products, it rarely receives two identical orders. Innovation happens as a result of the customer's request for specific features, leading the company to carry out product adaptations. Therefore, innovation is mostly incremental.

MechC&I has not been so far involved in formal standardization processes initiated by SDOs. It is only a user of standards. About 50 standards are relevant for the company's products and processes. A centralized system operated by the R&D department has been recently installed for the storage and management of all standards used by the company.

Customers may ask for the fulfillment of certain standards that the company has not worked with so far. Thus, to comply with customers' needs, the company has to adapt its products according to these standards. In addition, some innovations can emerge because customers may ask for products that do not satisfy certain standards. Consequently, innovation can happen during the process of reconciling the customers' requirements and these standards (e.g. Swann, 2000).

FINDINGS

Organizing Standardization Activities within Companies: A Contingency Perspective

The case studies show different approaches how companies internally organize their standardization-related activities.

Organization of the Standardization Function

Companies allocate different levels of resources to standardization-related work. ElecS is the only organization in our sample that has a unit dedicated to standardization. The company has recognized the relevance of standardization a long time ago. Lafur does not dedicate a department to standardization, but the head of research and development participates intensively in standard-setting processes. He communicates to his colleagues the relevant information from the standardization committees by organizing regular internal meetings. This anchors the transfer of knowledge to the whole organization. Lafur considers its involvement in standardization a critical success factor. In contrast to ElecS and Lafur, OptSys is operating in a less standardized industry. The company works itself on occasional projects. To save resources, however, it may delegate the participation in standardization committees to third parties who represent its interests. OptSys itself initiated the development of a new international standard to harmonize software in its industry. MechC&I exhibits the lowest level of involvement in standardization processes. The company faces competition from only two foreign companies worldwide. Hence, the initiation of standard development is not attractive at the national level. Even internationally, standardization may not be interesting because only a few companies are serving the niche market. The company signalizes the importance of working with standards through its investment in a centralized IT system for the storage, management, and retrieval of standards.

Contingency Factors

In the following, we analyze how the internal organization of the standardization activities is affected by these contingency factors: environment (customer demand and competition), size, age, technology, and strategy. Recall that all companies in the sample occupy leading positions in their industries, are innovative, and produce customized one-of-a-kind products.

Environment

All the market environments are characterized by unstable customer requirements and relentless quest for innovation. The structure of competition, however, is different among the four industries. ElecS competes with big players on the market, and its growth from a company with 30 employees in 1975 to a company with 600 people by 2014 is actually due to the company's strong involvement in standardization processes, as illustrated by its former CEO:

We found more applications for our products during the discussions initiated in the committee meetings. [...] The benefit of standardization is that the awareness level of our applications has increased considerably. With this increased awareness, the market demand for our devices has grown, leading to a higher turnover. (ElecS)

For ElecS, standardization contributed to growth and discovery of new markets, helping the company to prevail in a highly competitive market. Like ElecS, Lafur's external environment is characterized by many competitors. What make the competitive situation even harder are the so-called planners, which are positioned in the value chain between laboratory equipment producers and final customers. Due to its participation in standardization committees, Lafur got relevant insights into new developments, enabling the company to achieve an information advantage. Similar to ElecS and Lafur, OptSys evolves in a market with many competitors. MechC&I, however, has only two competitors worldwide. It does not participate in any standardization-related work and is, therefore, the only company within the sample with no standardization activities. Hence, the level of competition in an industry seems to have an influence on the level of involvement in standardization-related work (Acemoglu, Gancia, & Zilibotti, 2012); the higher the competition, the higher the level of involvement in standardization. Thus, to achieve a competitive advantage, companies can actively look for participation in standardization committees (Abdelkafi & Makhotin, 2014).

Size

ElecS (600 people) and Lafur (more than 700 people) are the largest companies in the sample; OptSys (70 people) and MechC&I (120 people) are the smallest ones. ElecS allocates the highest level of resources to standardization, followed by Lafur, OptSys (about one man-month a year is dedicated to standardization activities), and finally MechC&I with the lowest level of resource allocation. ElecS and Lafur do more work on standards and standardization than OptSys and MechC&I. To compensate for the lack of resources, OptSys involves third parties that represent its interests in standardization committees. Consequently, the level of resource allocation seems to correlate with size: larger companies allocate more resources than smaller companies. Thus, size has an impact on how a company organizes its standardization activities and how much resources are allocated to these activities. Note, however, that this finding is intuitive, as larger companies do have more resources and can virtually allocate more efforts to standardization than smaller companies. The quotient "Resources dedicated to standardization relative to the size of the company" can provide insights into the intensity of the standardization efforts per each employed person. This, however, does not change the fact that the absolute company size is the critical factor that influences how companies organize their standardization functions.

The direction of the causation between the level of resources allocated to standardization and size is unclear: are the companies with the highest involvement in standardization activities capable of growing more over time, thus achieving a relatively big size, or are the companies with relatively big sizes those that can justify their active involvement in standardization-related activities? Nevertheless, regardless of the company size, all key informants from the companies that participate in standard-setting processes acknowledge that they would actually require more resources to better deal with the standardization tasks:

We are both full-time standardizers. But our capacities are not enough to participate in too many standardization committees… (ElecS).

Age

Company age does not seem to be a relevant contingency factor. MechC&I is the oldest company in the sample (95 years old), but exhibits the lowest level of involvement in standardization activities. OptSys (20 years old) is the youngest one and has the second lowest level of involvement—in terms of resource expenditure—in standardization processes. ElecS and Lafur are about of the same age (60 years old). They are neither the oldest nor the youngest in the sample. Nevertheless, both companies allocate a higher level of resources to standardization-related work.

The length of the period of time, during which the company has been doing standardization work, seems to explain better the level of resource allocation to standardization activities. For instance, ElecS, which has been working actively on standardization for more than 40 years, has a dedicated department. OptSys, which is involved in specific standardization projects, is rather a new player in the field. Hence, the level of allocated resources appears to be dependent on how long the company has been involved in standardization activities during its history.

Technology

The complexity of the technology does not seem to be a relevant contingency factor (Gauch & Blind, 2014; Acemoglu et al., 2012). The technology of Lafur is of moderate complexity (equipment producer), as compared to the other companies in the sample, but the company still shows a high level of involvement in standardization. The technologies by ElecS and MechC&I have a higher level of complexity, but both companies exhibit different levels of involvement in standardization. OptSys has the technology with the highest complexity (optical measurements on nano-scale, involving software and hardware components). Its current technology is mature, making the transition to a new technology very likely in the near future:

We have reached a high level of technological maturity. Only a few technical improvements are possible […]. A revolution might come from other technologies […] in ten years or so. (OptSys)

OptSys dedicates in absolute values fewer resources to standardization than do Lafur and ElecS. In spite of this and the high complexity of the technology, OptSys seems to do well in standardization. Recall that all companies in the sample are in the "fit" state. Consequently, the complexity level of technology does not appear to have an influence on how standardization is organized internally. This result is predictable since standardization activities and the availability of standards, per se, are not dependent of the complexity of the technology. Take, for instance, mechanical engineering. Standardization efforts are intensive, and this industry has produced a rich repertoire of standards. However, high-tech sectors such as nano-technology with a higher level of complexity than mechanical engineering still require an intensification of standardization efforts.

The maturity level of technology may actually better explain how companies organize their standardization function. "The consequences of standardization … are not invariant with respect to questions of timing in the development of the industry and the maturity of its technology" (David & Steinmueller, 1994, p. 220). We expect that the resources allocated to standardization efforts are dependent on the technological lifecycle, beginning with low level of resource allocation at the inception phase, then increasing during growth, stagnating in the maturity phase, and finally going down during the decline phase of the technology. The data collected from the case studies do not allow us to check for the impact of technological maturity and to recognize a specific pattern, as all technologies in our sample exhibit a high maturity level.

Strategy

The strategic importance of standardization in all four companies is not only explained by the resources allocated to standardization-related work, but also by the position of this function in the company's organization. ElecS has a dedicated unit that directly reports to the top management board. At Lafur, a top manager (member of the executive board) does the standardization work. Standardization at OptSys is assigned to the strategic innovation manager. At MechC&I, there is no standardization-relevant activities, but the maintenance and update of the centralized IT system for the storage and management of standards is carried out by a product development engineer under the responsibility of the product development manager. Concerning whether there is a written strategy for standardization, even ElecS did not formulate one (during the time at which the study has been carried out). One key informant at ElecS notes:

We have a standardization strategy, but this strategy is not formulated in a written form. (ElecS)

The standardization strategy of the company determines the level of allocated resources. For instance, in its early days, ElecS was working on new standards, as it was broadening its product line and improving further its technology. ElecS was able to set the standards requirements high for durability testing, in order to take a big advantage from its superior technology, putting the main competitor in a difficult situation. However, this aggressive role changed over time. The current endeavors of ElecS are rather focused on maintaining its position and defending itself against standardization attempts at national and international levels that might be disadvantageous for its own products. All in all, the intensity of participation in standardization for ElecS remained high over the years.

Without assigning a lot of resources to standardization, OptSys aims at changing its market environment by setting a new standard that might shift the focus of competition to hardware. The work is done on a project-basis and carried out by a single engineer. Although it is unclear if OptSys engages in the future in a similar standardization endeavor, the impact on the niche market, in which the company is active, is high. Thus, these observations of past developments at ElecS and OptSys indicate that their intentions and objectives behind their standardization endeavors change over time.

Innovation as a By-Product of Fit

The companies in the sample are highly heterogeneous, operate in different technological fields, and are of different sizes. Their industries induce diverse environments for standardization activities. ElecS interprets standardization as one of the key elements of its strategy. ElecS is constrained by the illiberality of its environment (Child, 1972), since standards are indispensable for safety reasons. Lafur strives for being the first in the marketplace that comes up with a new standard or adapts to a new version of an existing standard. Hence, it actively shapes standards and contributes to the variability of the standardization landscape in its industry. OptSys, however, has other motives; it strives for harmonization of algorithms and mathematical procedures in the whole market, so that innovation capabilities in the industry no longer depend on software, but only on hardware. Obviously, OptSys faces the challenge of reducing complexity in its environment through its open innovation approach (Chesbrough, 2003). Though OptSys has to freely reveal proprietary knowledge, the company still maintains an information advantage by being the main driver of the standard-setting process. MechC&I is more reactive than the other companies in the sample, since it focuses on fulfilling specific standards, which can be prescribed by the customers themselves. Innovation usually emerges after solving eventual conflicts between the customers' requested features and the requirements of the standards. Both Lafur and ElecS stress their roles as pioneers in standard-setting, whereas OptSys attempts are rather oriented on the unification of different players to agree on one software standard, and MechC&I is only focused on the use of standards.

These case studies as well as our previous research study (Abdelkafi & Makhotin, 2014) confirm the existence of innovation potentials in standardization. For instance, ElecS found new applications for their products thanks to standardization, an obvious example for market innovation (e.g. Reichwald & Piller, 2010). Lafur achieved an information advantage that enabled the company to launch a product that is conform to the new standard much faster than competitors. OptSys supports its advantage in hardware innovations by making nanoscale measurements less dependent on the software and the algorithms with which it works. MechC&I gets inspired from the standards prescribed by their customers to provide them with innovative solutions. In this regard, requirements-oriented committee standards can be much more helpful than solution-oriented standards. However, although standards and standardization lead to innovation, all companies did not have innovation as an objective in mind, when they participated in standardization processes and worked with standards. The real drivers are rather different. Companies wanted to achieve an information advantage; to discover new applications; to improve their competitiveness level; or to conquer new markets. Innovation was not the main indicator of performance that the companies focused on.

Consequently, innovation appears to not be the main driver that pushes companies to intensify their standardization endeavors. Consequently, we posit that innovation is a by-product of fit between the environment and the way a company organizes internally its standardization activities. Take the example of Lafur. Competition is high, size is relatively big, and the complexity of its technology is moderate compared to the other companies in the sample. The company chose an organizational form of standardization, in which a top manager from the executive board pushes standardization activities. In this way, the company could release its innovation faster than other competitors in the market. Obviously, it would have missed the opportunity, if it has ignored standardization. Lafur would be still able to capture this innovation opportunity, if it has had a department dedicated to standardization. But the total cost position would be much worse than in the current situation because of the additional resources required to operate a standardization department.

IMPLICATIONS, CONCLUSIONS AND DIRECTIONS FOR FUTURE RESEARCH

This paper marks a first attempt to anchor contingency theory in standardization research. It contributes to extant literature by applying structural contingency theory to analyze and explain why companies organize their standardization work the way they do. The research design is based on a qualitative case study approach. The multiple case study analysis uncovers many ways how standardization work can be organized within companies. ElecS dedicates an entire department to standardization activities and has devised several routines and processes to ensure that other departments in the company get relevant information from the standardization department and vice versa. Though the other companies do not have dedicated departments, they still deal with standardization appropriately. Lafur allocated the standardization work to one member of the top management board, who reports on his activities in regular management meetings. OptSys deals with standardization within the scope of occasional projects and gets the support of third parties who represent its interests in standardization committees, whereas MechC&I is not active in standardization, but sets value on the management of and easy access to standards for engineers by installing an adequate IT system.

Hence, companies should not necessarily dedicate a lot of resources to the standardization activities. Instead, they have to adjust the level of resources, depending on relevant contingency factors. Competition, size and strategy have a large influence on how companies organize their standardization activities. Higher competition is found to drive companies to allocate more resources to standardization. Companies can use standardization as a strategic weapon to outpace competitors and even to exclude them from the market, for instance when the standard includes technical features that cannot be fulfilled by competitors. Thus, the allocation of high level of resources through the creation of a dedicated department or the allocation of standardization tasks to top managers can increase the likelihood that the company is participating in major standardization committees. A low competition level may lead companies to assign fewer resources to standardization, as found in the case of MechC&I. Size also seems to be a relevant contingency factor. Nevertheless, the cause-consequence relationship can be ambiguous: does standardization work facilitate growth (as in the case of ElecS) and hence increase the company size, or does larger size, which is equivalent to higher abundance of resources, leads to the allocation of more resources to standardization? The third relevant contingency factor is strategy. OptSys is a case in point and its strategy is crystal clear. The company pushes the standardization of parameters, mathematical procedures and algorithms to move the focus of competition to hardware, at which OptSys can excel and achieve a competitive advantage. Furthermore, whereas the company's age and the complexity of technology were not retained as relevant contingencies, we expect that the length of the time period, during which the company has been active in doing standardization and maturity level of technology are more likely to be relevant contingency factors. All companies in the sample are operating a one-of-a-kind production, characterized by an unstable customer demand. Thus, the volatility of customer specifications constitutes a common factor to all case studies. This contingency does not seem to have an influence on the organization of the standardization function, although it can be considered the main factor that drives all the companies to be innovative.

With respect to innovation, we have argued that innovation is a by-product of fit between the internal organizational structure for standardization and the environment. Companies that participate in standardization primarily envision goals that are different from innovation. Nevertheless, they are able to leverage some innovation benefits on their way due to their involvement in standardization and use of standards. In a previous research, we have demonstrated that companies can augment their innovative abilities through standards and standardization. Because of the latent innovation potentials in standards and standardization, SDOs should promote these potentials more aggressively, in order to raise the awareness level of companies. Innovation can constitute, in fact, a primary objective, not only a by-product, of the company's engagement in standardization.

The data used in this research stems from four in-depth case studies. Our research objective is rather exploratory in nature and does not aim at generalization. Nevertheless, our findings provide valuable insights in order to conduct more quantitative research in the area of standardization and organization. Research on the link between both streams of research can be more intensified in the future. It can be also very promising to study the relationships between standardization and dynamic capabilities of the company. Dynamic capabilities highlight "organizational and (strategic) managerial competences that can enable an enterprise to achieve competitive advantage, and then semi-continuously morph so as to maintain it" (Teece, 2007, p. 1346). Because companies involved in standardization activities can maintain a better overview of recent advancements in their field of technology, it can be assumed that they can achieve high level of competitiveness over a long period of time. Hence, managing standardization becomes an important competence that can lead to a higher performance, also in terms of the support of the company's innovation activities. This issue can constitute an adequate topic for research to investigate more in-depth the connections of standardization to the company's internal organization.

REFERENCES

Abdelkafi, N., & Makhotin, S. (2014). Seizing opportunities for the support of innovation through committee standards and standardization: Insights from German SMEs. *International Journal of IT Standards and Standardization Research*, *12*(2), 38–56. doi:10.4018/ijitsr.2014070103

Acemoglu, D., Gancia, D., & Zilibotti, F. (2012). Competing engines of growth: Innovation and standardization. *Journal of Economic Theory*, *147*(2), 570–601. doi:10.1016/j.jet.2010.09.001

Blind, K., Jungmittag, A., & Mangelsdorf, A. (2011). *The economic benefits of standardization: An update of the study carried out by DIN in 2000.* Retrieved September 26, 2014, from DIN German Institute for Standardization e. V.: http://www.din.de/sixcms_upload/media/2896/DIN_GNN_2011_engl_akt_neu.pdf

Blind, K., & Thumm, N. (2004). Interrelation between patenting and standardisation strategies: Empirical evidence and policy implications. *Research Policy*, *33*(10), 1583–1598. doi:10.1016/j.respol.2004.08.007

Bourgeois, L. J. III. (1985). Strategic goals, perceived uncertainty, and economic performance in Volatile Environments. *Academy of Management Journal*, *28*(3), 548–573. doi:10.2307/256113

Budd, R. W., Throp, R. K., & Donohew, L. (1967). *Content analysis of communications*. New York: Macmillan.

Burkert, M., Davila, A., Mehta, K., & Oyon, D. (2014). Relating alternative forms of contingency fit to the appropriate methods to test them. *Management Accounting Research*, *25*(1), 6–29. doi:10.1016/j.mar.2013.07.008

Byrne, B. M., & Golder, P. A. (2002). The diffusion of anticipatory standards with particular reference to the ISO/IEC information resource dictionary system framework standard. *Computer Standards & Interfaces*, *24*(5), 369–379. doi:10.1016/S0920-5489(02)00057-0

Chesbrough, H. W. (2003). *Open innovation: The new imperative for creating and profiting from technology*. Boston: Harvard Business School Press.

Child, J. (1972). Organizational structure, environment and performance: The role of strategic choice. *Sociology*, *6*(1), 1–22. doi:10.1177/003803857200600101

Choi, D. G., Lee, H., & Sung, T.-. (2011). Research profiling for 'standardization and innovation'. *Scientometrics*, *88*(1), 259–278. doi:10.1007/s11192-011-0344-7

Cohen, W. M., & Levinthal, D. A. (1990). Absorptive capacity: A new perspective on learning and innovation. *Administrative Science Quarterly*, *35*(1), 128–152. doi:10.2307/2393553

David, P. A., & Steinmueller, W. E. (1994). Economics of compatibility standards and competition in telecommunication networks. *Information Economics and Policy*, *6*(3-4), 217–241. doi:10.1016/0167-6245(94)90003-5

De Vries, H. J. (2006). *Standards for business - How companies benefit from participation in international standards setting*. Retrieved August 12, 2014, from http://www.iec.ch/about/globalreach/academia/pdf/vries-1.pdf

DIN. (2000). *Economic benefits of standardization: Summary of results*. Final report and practical examples. Retrieved September 26, 2014, from DIN German Institute for Standardization e. V.: http://www.din.de/sixcms_upload/media/2896/economic_benefits_standardization.pdf

Donaldson, L. (2001). *The contingency theory of organizations. Foundations for organizational science*. Thousand Oaks, CA: SAGE Publications. doi:10.4135/9781452229249

Drazin, R., & Van de Ven, A. H. (1985). Alternative forms of fit in Contingency Theory. *Administrative Science Quarterly*, *30*(4), 514–539. doi:10.2307/2392695

Dul, J., & Hak, T. (2008). *Case study methodology in business research*. Oxford, UK: Butterworth-Heinemann/Elsevier.

Duncan, R. B. (1972). Characteristics of organizational environments and perceived environmental uncertainty. *Administrative Science Quarterly*, *17*(3), 313–327. doi:10.2307/2392145

Egyedi, T., & Sherif, M. (2010). Standards dynamics through an innovation lens: Next-generation ethernet networks. *IEEE Communications Magazine*, *48*(10), 166–171. doi:10.1109/MCOM.2010.5594692

Egyedi, T. M., & Blind, K. (2008). *The dynamics of standards*. Cheltenham, UK: Edward Elgar.

Gauch, S., & Blind, K. (2014). Technological convergence and the absorptive capacity of standardisation. *Technological Forecasting and Social Change*, 1–14.

Gerdin, J., & Greve, J. (2004). Forms of contingency fit in management accounting research—a critical review. *Accounting, Organizations and Society*, *29*(3-4), 303–326. doi:10.1016/S0361-3682(02)00096-X

Gibbs, G. (2007). *Analyzing qualitative data*. Los Angeles, CA: SAGE Publications.

Ginsberg, A., & Venkatraman, N. (1985). Contingency perspectives of organizational strategy: A critical review of the empirical research. *Academy of Management Review*, *10*(3), 421–434.

Graubner, M. (2006). *Task, firm size, and organizational structure in management consulting: An empirical analysis from a contingency perspective*. Wiesbaden, Germany: Deutscher Universitats-Verlag.

Hauschildt, J. (1997). Innovationsmanagement: Vahlens Handbücher der Wirtschafts- und Sozialwissenschaften (2nd ed.). Munich: Vahlen.

ISI. (2006). *INTEREST Integrating research and standardisation: D04 - Report on Case Studies.* Retrieved October 11, 2014, from http://www-i4.informatik.rwth-aachen.de/Interest/documents.html

ISI. (2007). *INTEREST Integrating research and standardisation: D 10 – Final Report.* Retrieved October 11, 2014, from http://www-i4.informatik.rwth-aachen.de/Interest/documents.html

ISO. (2011). *Economic benefits of standards - International case studies - Volume 1.* Retrieved March 13, 2015, from International Organization for Standardization: http://www.iso.org/iso/benefits_of_standards.html

ISO. (2012). *Economic benefits of standards - International case studies - Volume 2.* Retrieved March 13, 2015, from International Organization for Standardization: http://www.iso.org/iso/benefits_of_standards.html

Kast, F. E., & Rosenzweig, J. E. (Eds.). (1973). *Contingency views of organization and management.* Chicago: Science Research Assoc.

Lindkvist, K. (1981). Approaches to textual analysis. In K. E. Rosengren (Ed.), *Advances in content analysis*. Beverly Hills, CA: Sage.

Longenecker, J. G., & Pringle, C. D. (1978). The illusion of Contingency Theory as a general theory. *Academy of Management Review*, *3*(3), 679–683.

Matyusz, Z. (2012). *The effect of contingency factors on the use of manufacturing practices and operations performance.* (PhD thesis). Cornvinus University of Budapest.

Mayring, P. (2010). *Qualitative Inhaltsanalyse: Grundlagen und Techniken* (11th ed.). Weinheim: Beltz. doi:10.1007/978-3-531-92052-8_42

Miles, R. E., Snow, C. C., & Meyer, A. D., & Coleman, H. J., Jr. (1978). Organizational strategy, structure, and process STOR. *Academy of Management Review*, *3*(3), 546–562. PMID:10238389

Mintzberg, H. (1979). *The structuring of organizations.* Prentice-Hall Inc.

OECD & Eurostat. (2005). Oslo manual: Guidelines for collecting and interpreting innovation data (3rd ed.). Paris: Organisation for Economic Co-operation and Development; Statistical Office of the European Communities.

Ouchi, W. G. (1977). The relationship between organizational structure and organizational control. *Administrative Science Quarterly*, *22*(1), 95–113. doi:10.2307/2391748

Pfeffer, J. (1982). *Organizations and Organization Theory*. Marshfield, MA: Pitman.

Porter, M. E. (1985). Competitive advantage: Creating and sustaining superior performance. New York: Free Press.

Pugh, D. S., Hickson, D. J., Hinnings, C. R., & Turner, C. (1969). The context of organization structures. *Administrative Science Quarterly*, *14*(1), 91–114. doi:10.2307/2391366

Reichwald, R., & Piller, F. T. (2009). Interaktive Wertschöpfung: Open Innovation, Individualisierung und neue Formen der Arbeitsteilung (2nd ed.). Wiesbaden: Gabler.

Roth, K., & Morrison, A. J. (1992). Business-level competitive strategy: A contingency link to internationalization. *Journal of Management, 18*(3), 473–487. doi:10.1177/014920639201800303

Schreier, M. (2012). *Qualitative content analysis in practice*. Thousand Oaks, CA: SAGE Publications.

Sousa, R., & Voss, C. A. (2008). Contingency research in operations management practices. *Journal of Operations Management, 26*(6), 697–713. doi:10.1016/j.jom.2008.06.001

Suárez, F. F., & Utterback, J. M. (1995). Dominant designs and the survival of firms. *Strategic Management Journal, 16*(6), 415–430. doi:10.1002/smj.4250160602

Swann, P. (2000). *The economics of standardization: Final report for Standards and Technical Regulations Directorate Department of Trade and Industry*. Retrieved July 17, 2014, from https://www.gov.uk/government/uploads/system/uploads/attachment_data/file/16506/The_Economics_of_Standardization_-_in_English.pdf

Swann, P. (2010). *The economics of standardization: An update*. Retrieved May 17, 2015, from https://www.gov.uk/government/uploads/system/uploads/attachment_data/file/16509/The_Economics_of_Standardization_-_an_update_.pdf

Szilagyi, J. A., & Wallace, M. (1980). *Organizational behavior and performance*. Santa Monica, CA: Goodyear.

Teece, D. J. (2007). Explicating dynamic capabilities: The nature and microfoundations of (sustainable) enterprise performance. *Strategic Management Journal, 28*(13), 1319–1350. doi:10.1002/smj.640

Tidd, J., & Bessant, J. R. (2009). *Managing innovation: Integrating technological, market and organizational change* (4th ed.). Chichester, UK: Wiley.

Tushman, M. (1979). Work characteristics and subunit communication structure: A contingency analysis. *Administrative Science Quarterly, 24*(1), 82–98. doi:10.2307/2989877

Utterback, J. M., & Abernathy, W. J. (1975). A dynamic model of process and product innovation. *Omega, 3*(6), 639–656. doi:10.1016/0305-0483(75)90068-7

Venkatraman, N., & Camillus, J. C. (1984). Exploring the concept of "Fit" in Strategic Management. *Academy of Management Review, 9*(3), 513–525.

Venkatraman, N., & Prescott, J. (1990). Environment-strategy coalignment: An empirical test of its performance implications. *Strategic Management Journal, 11*(1), 1–23. doi:10.1002/smj.4250110102

Volberda, H. W., & Weerdt, N. (2012). Contingency fit, institutional fit, and firm performance: A metafit approach to organization–environment relationships. *Organization Science, 23*(4), 1040–1054. doi:10.1287/orsc.1110.0687

Weill, P., & Olson, M. H. (1989). An assessment of the contingency theory of management information systems. *Journal of Management Information Systems, 6*(1), 59–85.

Woodward, J. (1965). *Industrial organization: Theory and practice*. London: Oxford University Press.

Wooton, L. M. (1977). The mixed blessings of contingency management. *Academy of Management Review*, *2*(3), 431–441.

Wright, C., Sturdy, A., & Wylie, N. (2012). Management innovation through standardization: Consultants as standardizers of organizational practice. *Research Policy*, *41*(3), 652–662. doi:10.1016/j.respol.2011.12.004

Yin, R. K. (2003). *Case study research: Design and methods* (3rd ed.). Thousand Oaks, CA: SAGE Publications.

KEY TERMS AND DEFINITIONS

Case Study: A research methodology that investigates in-depth one (a single case study) or many (a multiple case study) subjects of study in a qualitative or quantitative manner. Case studies represent a suitable methodology when the researcher cannot separate the phenomenon to be studied from the context. Different data collection methods such as interviews, archives, and observations can be used to compile a case study.

Committee Standard: A document established by consensus within a committee, which consists of representatives of relevant stakeholder groups and is organized by a formal/recognized Standard Developing Organization (SDO). Committee standards aim at the achievement of an optimal degree of order, as they provide guidelines, rules and reference points for common and repeated use in a given context. Their use is voluntary unless required by legal regulation.

Contingency Factor: Any moderator variable (e.g. environment, technology) that affects certain aspects of an organization (e.g. organizational performance).

Contingency Theory: A theory postulating that under different circumstances different solutions may prove effective; hence there is no one-size-fits-all solution to decisions related to organizational structure, leadership, and company strategy. The theory emphasizes that the success of an organization depends on its relationship with the environment.

Organizational Fit: A concept that matches the organizational performance (high or low) with the state of the interrelation (congruence) between contingency and structure (fit or misfit).

Innovation: Is defined as a combination of invention and exploitation. There are many types of innovation in the literature: product, process, paradigm and position innovations; radical and incremental innovation; or new-to-the-world, new-to-the-market, new-to-the-firm innovation.

Organizational Structure: The way an organization arranges its resources (e.g. human, time and monetary resources) so that it can perform its activities and meet its objectives. The structure of an organization is characterized by hierarchies, rights, duties, processes, roles, information flows, etc.

Standardization: The process, by which committee standards are developed. Standardization also designates the work of Standards Developing Organizations (SDOs), which moderate the consensus process that can involve individuals, companies, public organizations, and other institutions.

Structural Contingency Theory: A branch of contingency theory, which focuses on how certain contingency factors such as environment, size and strategy affect certain aspects of organizational structure (see contingency theory and organizational structure).

Technology S-Curve: A graphical representation of the technology lifecycle that describes how a technology progresses and evolves over different phases in terms of performance or revenues. There are four phases in the lifecycle of a technology: emergence, growth, maturity, and decline.

ENDNOTES

1 The search word "case" was used to identify publications that might contain case studies.

Chapter 14
Standardization of Information and Financial Innovation:
Lessons from Mortgage Securitization

Antonios Kaniadakis
Queen Mary University of London, UK

ABSTRACT

Mortgage securitization markets emerged as an extension of the primary mortgage lending markets. This created the need for standardization of information across these two contexts that would enable a collective and universal understanding of credit risk and its management. The securitization industry, however, instead of developing standardization management strategies that would support this vision, it rather chose to implement an organizing vision that was centered around operational efficiency and profit-making supported by a focus on functional specialization. The outcome was the fragmentation of the securitization supply chain via vertical disintegration, which undermined the unity of the risk analysis process. This chapter argues that the effects of technological standardization on innovation in the mortgage industry should be explored beyond a narrow focus on efficiency and profit in relation to an individual organization's business strategy; but rather within an extended scope that includes broader social and policy contexts that guide innovation.

INTRODUCTION

Not that long ago the securitization industry and its practices were practically unknown to the broader public. In fact, in 1998, Business Week was reporting on "A $2.5 trillion market you hardly know" (Business week, 1998), referring to securitization. This very profitable industry had to a large extend remained the domain of a few players who had managed to assemble the technical knowledge and capabilities to practice it. As the 2007-8 financial crisis started spreading allover the world, however, this rather unknown industry was suddenly at the center of everyone's attention. How was it created? What were the innovations necessary for its emergence and evolution? In this chapter I shall show that stan-

DOI: 10.4018/978-1-4666-9737-9.ch014

dardization of information across contexts was a driving force for innovations that led to the emergence of securitization. I shall talk more specifically about the mortgage industry as a special case which was heavily implicated in the 2007-8 global credit crunch.

A typical definition of securitization is "The process through which an issuer creates a financial instrument by combining other financial assets and then marketing different tiers of the repackaged instruments to investors. The process can encompass any type of financial asset and promotes liquidity in the marketplace." (Investopedia, accessed 15.05.2015)

Although this seems like a rather technical definition, simply it means that issuers (mortgage lenders, investment banks, building societies etc) pool together financial assets (mortgage loans, credit card loans, auto loans) which then divide in slices (tranches) depending on how much risk they carry, and then they sell these slices to third party investors in the capital markets. Capital markets constitute secondary markets as opposed to the primary ones, in which the underlying assets were bought and sold. Although securitization is a centuries-old financial innovation, as a practice of selling loans in secondary markets (Riddiough, 2011), in its current form and scale, especially in the context of the mortgage markets that were implicated in the US housing bubble that initiated the credit crunch, its origins are placed in the 1970s.

One of the main characteristics of securitization is that it is based on creating linkages between multiple layers of context by which credit risk can travel from primary to secondary markets. During this process, risk is transformed from something to be dealt with as part of a risk analysis process aimed at improving decision-making (i.e. lending decisions) in the mortgage lending market, into *raw material* used to create a different type of financial product aimed at the capital markets. By making this risk transfer possible, securitization as a financial innovation creates linkages between an individual homebuyer with the vast resources of the global capital markets (Business Week, 1998). The creation of these linkages gives securitization a ubiquity, as if a person has a mortgage, or a credit card loan, or an auto loan, the chances are they are part of it whether they know it or not.

The complexity of securitization and the way it makes possible the "enmeshing" of multiple layers of context of social and economic life into a single financial product, makes it quite a controversial topic. In fact, the way risk can be interpreted outside the specific context of the credit arrangement that created it (e.g. a borrower paying their mortgage in monthly installments) and acquire a broader, generic meaning, has raised debates as to how beneficial securitization is as a financial innovation. On the one hand, it has been seen as "one of the most important and abiding innovations to emerge in the financial markets since the 1930's." (Leon T. Kendall quoted in Business Week, 1998), while on the other, it has been termed as "financial weapons of mass destruction" (Buffett, 2002). This ambiguity surrounding securitization raises questions in regards to the role of financial innovation in economic and social life, and also suggests that this type of financial innovation may be pointed towards alternative possible directions. These alternative paths for financial innovation are created through a collective effort by industrial actors, in their attempts to enter the industry and "plug into" its value chain. As I will show here, part of these efforts was the collective attempt of the industry to standardize credit information across primary and secondary mortgage markets. Although the need for *standardization of information* was initially aiming at finding a collective and standardized way to understand and interpret credit risk, as the industry spread to private sectors, standardization of information formed the basis for operational efficiency based on functional specialization. This had the opposite effect on the understanding of risk as it resulted to its compartmentalization.

The empirical material discussed in this chapter, is drawn from existing published research from a variety of disciplines focusing on the mortgage industry. I combine published findings with some of my own unpublished material and insights, which I have been developing via a long-term ethnographic engagement with mortgage industry players for the past five years. Although reference is made to empirical material, by no means should this chapter be treated as a pure research piece. The chapter should also not be treated as purely theoretical or conceptual work or as an attempt to provide a complete and accurate history of the mortgage industry. Instead, it should be read as an effort to set the grounds for the construction of a narrative around the relationship between technological standardization and innovation in the context of the mortgage industry. Although this is not a purely conceptual piece, concepts from ICT innovation studies, technological standardization, and economic sociology will be used to establish analytical links between processes of technological standardization, their motivation and possible effects on innovation and on society at large.

By examining the role and effects of technological standardization on the innovations that triggered the emergence and evolution of the mortgage securitization industry, the main message and main contribution of this chapter is the following: Standardization management in mortgage finance, if understood as the efforts to balance the positive and negative effects of standardization on innovation, is gradually moving[1] away from a pure business logic around competitiveness, efficiency, profitability. Instead its scope as an academic discipline and as a professional practice is expanding to include social and public policy discourses that generate a new type of accountability; not only towards the firm and/or the organization one represents but also towards stakeholders at the level of the industry and towards the broader public. In other words, the particular historical circumstances and idiosyncrasies of this industry surely put standardization management in a unique context on which lies the future of the industry itself and questions on the role of the financial system in society more generally.

The chapter is structured as follows: In the next session I present a theoretical discussion of the relationship between technological standardization and innovation. This discussion is setting the scene on which to explore the developments in the mortgage industry. In the following section, then, I discuss the mortgage industry and the need for standardization of information across primary and secondary mortgage markets. This is where standardization becomes the aim of innovation in a collective industry effort to establish mortgage securitization markets. In the section after that I discuss the effects of these efforts on techno-organizational and business innovations. The chapter concludes with a discussion on the changing nature of standardization management in the mortgage industry.

TECHNOLOGICAL STANDARDIZATION AND INNOVATION

A major theme in the standardization literature is its relationship with innovation. The question that is most commonly asked by both practitioners and academics is whether standardization enables or constrains innovation (Swann, 2010). It only takes one look at the relevant literature on technology standards to realize that such a question may be proved quite difficult and tricky to answer. It is no surprise, then, that the relationship between standardization and innovation has been seen as a "paradox" (de Vries, 2006).

The difficulty in understanding the effects of standardization on innovation (and visa-versa) seems to lie on the fragmented state of knowledge on technology standards, resulting from different ways that scholars approach standardization and also from diverse particular empirical contexts. Indeed, technological standardization has been studied at many different levels: the level of the standard (Tassey,

2000), the level of a technology or technology platform (Loch & Hubermann 1999; Economides & Katsamakas, 2006), the level of a firm (Suarez, 2004; Teece, 2007; Benner & Tushman, 2003), the level of public policy (Greenstein & Stango, 2006) and the level of a broader innovative terrain, usually conceptualized either as an industry (Arthur, 1996) or as a technological community (Hargrave & Van De Ven, 2006) or a technological paradigm (Dosi, 1982). All these studies, depending on their focus and mandates, have produced rich diverse accounts on technological standardization but at the cost of creating a fragmented knowledge landscape on the management of standardization and its implications for innovation. This led to an uneven conceptual and theoretical development. As Narayanan & Chen (2012) put it "…the richness of the literature came at the cost of the fragmentation caused by the inconsistent use of terms and the lack of communication between scholars operating at different levels of analysis or treating standards sometimes as exogenous and sometimes as endogenous variables" (2012: 1375). Such a fragmented knowledge landscape with theoretical, conceptual and empirical inconsistencies would indeed make it difficult to produce any comprehensive accounts on the relationship between technological standardization and innovation.

Despite the inconsistency and fragmentation, however, scholars have still managed to produce integrative accounts, which led to some level of consensus based on the realization (whether it is founded on empirical evidence or on conceptual articulation) that standardization can, in fact, *both enable and constrain innovation*. As a result, accounts founded on this basis have produced theorizations of technological standardization that tend to put at the center of analysis the balancing of the tensions between its enabling and constraining effects on innovation. Authors have theoretically addressed this balance in different ways: David (1995), for instance, talks about the "flux between freedom and order", while Hanseth et al (1996) talk about the "tension between standardization and flexibility" in relation to innovation in information infrastructures. Economic theorists examine this question in relation to economic growth and have found that "standardization is both an engine of growth and a potential barrier to it" (Acemoglu et al, 2012). Finally, social shaping of technology approaches have focused on discussing the "mutual influence between standardization and innovation" (Jakobs, 2006).

In the context of this two-way dynamic relationship between technological standardization and innovation, the *focus* of standardization management, both as an academic field and as professional practice, would be to explore/address the dynamics and mechanisms by which *industrial actors* (this includes producers of standards, users, regulators, policy makers, professional associations etc) engage in *efforts to balance the tensions between the positive and negative effects of technological standardization to innovation* (however they are temporally defined and understood). This will include the consideration of the implications for the organization each of these actors happen to represent - and as I shall argue here, for society at large. I shall show, for instance, that the question is not only whether standardization enables innovation but also how this innovation could lead to socially destructive consequences as it happened with the 2007 global financial crisis.

I use the term industrial actors as opposed to just "firms" or "companies" because besides producers and users of standards, I also include regulators, policy makers, the State, professional associations and other intermediary organizations and institutions involved in technological standardization. This is necessary to understand the mortgage industry, as it constitutes a "technological field" (Pollock & Williams, 2011) that is shaped by a variety of different and heterogeneous actors operating at the political, economic, and technological spheres. It is the relationships amongst these actors that shape mortgage finance and give it its form. Technological standardization, therefore, has to be seen within the context of interactions amongst these industrial actors.

For this reason, the most suitable theoretical approach to examine technological standardization in mortgage finance is a "collective action approach" (Markus et al, 2006; 2008; Hargrave & Van de Ven, 2006). Collective action approaches understand technological standardization as a "dialectical process in which a coalition of partisan actors espousing conflicting views confront each other and engage in political behaviors to create and change standards" (Hargrave & Van De Ven, 2006; Narayanan & Chen, 2012: 1389). Industrial actors that have a stake in the technological field (i.e. mortgage finance) collectively engage in the creation of technological standards and also institutional structures that affect innovation (Narayanan & Chen, 2012:1389). The meaning of the term "collective" may include both actions and interactions founded on competition but also on a more collaborative basis (Oshri & Weeber, 2006). Authors that have studied standardization and innovation in mortgage finance have already been doing so from a collective action approach (Markus et al, 2006; 2008; Poon, 2009). Collective action approaches to technological standardization, as discussed next, also are more compatible with networked and communitarian approaches to innovation.

Indeed, a collective action view of technological standardization fits better with a more "networked" (Swan & Scarbrough, 2005), "distributed" (Rammert, 2003) and "communitarian" (Swanson & Ramiller, 1997; Wang & Ramiller, 2009) view of innovation. Such views of innovation provide the analytical space in which technological standardization as collective action can become more visible. In such an understanding of innovation, I will employ concepts from Economic Sociology such as "cognitive frames" (Beunza & Stark, 2004) to describe the emergence of different decision-making contexts in mortgage lending that created the need for technological standardization in the first place. I will use the concept of "calculative agency" (Callon, 1998; Barry & Slater, 2002; Callon & Muniesa, 2005; Preda, 2006) to explain the role of standardization in the process of transferring the data from one decision-making context to another. Finally, I use the concept of "organizing visions" (Swanson & Ramiller, 1997) to explain the different paths that the impact of technological standardization on innovation may take, depending on how industrial actors involved in technological standardization "collectively" engage with it.

In the context of technological standardization in mortgage finance, and for the purposes of this chapter, the concept of "innovation" has a dual meaning: 1) *Innovation as the process of technological standardization in itself*, that is, the collective and individual innovative efforts by various industrial actors to achieve technological standardization (either driven by competition, or collaboration, or compliance, and so on); and 2) *Innovation as the effects of technological standardization on financial innovation* (development of new innovative financial products, new techno-organizational and business models for finance organizations – business innovation, the emergence of new secondary markets). These two views will give us different interesting flavors of the relationship between standardization and innovation and will paint a more complete picture of the challenges involved in the role of technological standardization in the emergence of mortgage securitization.

STANDARDIZATION OF INFORMATION AND ICT INNOVATION IN MORTGAGE SECURITIZATION MARKETS

In mortgage finance, technological standardization processes must be understood in the context of "standardization of information" across contexts. It was this type of standardization that was necessary for the emergence and evolution of mortgage securitization as a new type of market and also as a business possibility for many mortgage lenders. Standardization of information across contexts would indeed enable lenders to transfer the assets (i.e. mortgage loans) that they were warehousing and "plug into" the emerging mortgage securitization chain (Kaniadakis & Constantinides, 2014). A big part of this transition was to integrate and standardize risk calculation practices across two decision-making contexts: a) lenders' decisions to grant loans to individual customers; and b) decisions on which of these loans to securitize and which not. Standardization of information would make possible a common understanding of risk and the criteria by which decisions were made in these two different contexts. For example, the risk involved in a single loan and that involved in a pool of loans would be represented by the same metric (Poon, 2009).

As I shall show here, however, standardization of information was not simply depended on a single organization's willingness to adopt a certain standard (i.e. a platform like FICO); it was rather a collective effort, an industry-wide movement that involved technological, organizational and business innovations by financial actors. To better understand the broader context of standardization of information in mortgage markets, I shall discuss below how these markets work.

Mortgage Markets Explained: Securitization as an Idea and as a Techno-Organizational Framework

Mortgage markets can be distinguished between primary and secondary. In the primary mortgage markets, borrowers (individuals or legal entities) obtain loans from mortgage originators, while in the secondary mortgage markets, existing mortgage loans are bought and sold in the context of a different kind of relationship between the originator and an investor (or multiple investors) (Cummings & DiPasquale, 1997; Kaniadakis & Constantinides, 2014). In the primary mortgage market there is a *mortgage origination* process by which borrowers apply for a mortgage loan while the lender initiates an *underwriting* process by which they check whether the borrower has both the ability and willingness to repay the loan. During the underwriting process an amount of information is generated (i.e. borrowers income) which helps guide the decision to give a loan or not to specific applicants. Finally, there is the production of the legal documentation necessary to create the mortgage and transfer the ownership of the residential property (Cummings & DiPasquale, 1997; Kaniadakis & Constantinides, 2014). In the secondary mortgage markets multiple such mortgage loans are *pooled* and *bundled* together and are sold to investors (Kaniadakis & Constantinides, 2014). In other words, individual mortgage loans were used as income-generating assets that were packaged and put into a pool and subsequently securities that are claims on that income were sold to investors (MacKenzie, 2011).

Mortgage securitization started out as a public policy innovation in the 1970s United States, in the context of a political campaign of the US Federal Government to seek additional means to finance the housing needs of the American people. For these purposes, the U.S. Government manufactured a network of government-sponsored industrial actors (the so-called GSEs) that were competing with each other in buying individual mortgages from lenders, re-package them into mortgage-backed securities (MBS) and sell them to third party investors (Fligstein & Goldstein, 2010). More specifically, as Fligstein & Goldstein explain: "the US government reorganized the Federal National Mortgage Association (Fannie Mae) as a quasi-private organization, called a government sponsored enterprise (GSE), to lend money and hold mortgages. They also created a new entity, the Federal Home loan Mortgage Corporation (Freddie Mac) to compete with Fannie Mae and a government owned corporation to insure those mortgages against risk of default, the Government National Mortgage Association (Ginnie Mae)" (Fligstein & Goldstein, 2010).

Although mortgage securitization emerged as a "social policy innovation" as part of the U.S. Government's budgetary politics (Quinn, 2009) and although it was reinforced by the deregulation of savings and loans banks (Fligstein & Goldstein, 2010), it gradually evolved beyond the arena of the U.S. federal public policy and into other national, business, and techno-organizational contexts. This means that mortgage securitization exited the narrow socio-political boundaries of the US housing policy and as it was diffused in different contexts, it gradually evolved into something that was simultaneously an *idea*, a *technology* and a *business model* for fundraising purposes (Wainwright, 2009). Indeed, soon after its invention, private actors adopted the practice and, in 1977, the first private label (non-government sponsored) U.S. MBS was issued by the Bank of America. Then, from 1985 onward, banks began securitizing other types of consumer debts, such as auto loans, credit card receivables, and so on (MacKenzie, 2011). By the 1990s, mortgage securitization practices spread from the US to the UK and Europe.

The *idea* of securitization travelled to Europe relatively freely, as North American investment banks started promoting it. However, the innovative landscape around securitization was fundamentally different than in the U.S. Indeed, as the idea of securitization spread in Europe and the UK, it mobilized context-specific technological, economic, social, cultural and legal forces (Wainwright, 2009; Kaniadakis & Constantinides, 2014) that determined the composition of the collective of industrial actors responsible for the development of the securitization markets. For example, in contrast to the U.S. markets, the U.K. Government was not directly involved in the development of a secondary market for U.K. mortgages; the market was rather the outcome of financial institutions (Wainwright, 2009) such as mortgage lenders, investors, credit rating agencies, and so on. What is interesting is that securitization spreads to the UK and Europe not as a policy to assist home ownership, but rather as a *business innovation* that would enable mortgage lenders to engage the capital markets to raise capital, instead of relying on savings and deposits from individual customers. Although securitization was used in the UK exclusively as a funding tool until the late 1990s (Wainwright, 2009), it seemed to offer more incentives and motives for the financial actors considering to adopt it, apart from access to funds. One of the main benefits of securitization for lenders had nothing to do with lending or borrowing, but with the fact that it allowed them to easily accelerate the circulation of their capital and circumnavigate capital adequacy requirements (Leyshon & Thrift, 2008) since it allowed mortgage debt to be removed from a lender's balance sheet (Wainwright, 2009). Within this new innovative landscape, turning the *idea* into a *techno-organizational framework* and *a product* was a question of how European industrial actors would address this as an innovative effort.

In summary, in different socio-economic and political contexts, the transition, or rather the extension, of a primary mortgage loan business into a secondary securitization market involved different types of motivation and incentives. In the US it initially emerged as a public policy innovation but as it spread in the UK and Europe, mortgage securitization emerged in the context of techno-organizational and business innovations on behalf of private lenders as well as other intermediary actors and regulatory players involved in the new industry. It is in this context where standardization of information across contexts in the mortgage markets becomes relevant as it represents the innovative efforts of mortgage lenders to adopt securitization as a business innovation.

The Need for Standardization of Information Across Primary and Secondary Mortgage Markets

For U.K. mortgage lenders with large amounts of pre-existing legacy assets (e.g. individual mortgage loans) and established lending practices, adopting securitization as a business and operational model was particularly challenging for various reasons. Firstly, the strategic decision to adopt securitization was not straightforward but involved a lot of internal organizational politics between different sources of decision-making (Kaniadakis & Constantinides, 2014). The reason for this is that, in the 1990s, securitization was still a new innovation in the UK and most of the adoption risk fell on the shoulders of early adopters (Fincham et al., 1994); additionally, the housing crash of the 1990s led to heavy losses by the centralized lenders, something that made U.K. bankers doubt securitization's viability as a useful tool. Securitization's reputation as a celebrated tool for fundraising was thus thrown into disrepute (Wainwright, 2009).

Secondly, for U.K. lenders, to turn securitization from an idea into a techno-organizational framework and a new business, they had to engage in a slow and mundane process of techno-organizational transformations. Indeed, as soon as the strategic (and political) decision was made to enter the emerging securitization chain, banks had to develop new information infrastructures (Kaniadakis & Constantinides, 2014), adopt standardized tools and platforms (Markus et al, 2006; Poon, 2009) that would allow them access to the new markets and to diverse sources of funds. It is these ICT changes and innovations that are most relevant to understanding standardization as a technological innovation in itself.

Historically, as discussed earlier, the primary mortgage markets pre-exist while the secondary mortgage securitization markets emerged as their extension. The coming of securitization in Europe, therefore, initially appealed to those industrial actors that were already active in the primary mortgage loan markets and thus had reserves of "legacy" mortgage assets on which to try securitization on. More specifically, mortgage lenders that already had an established mortgage origination infrastructure started undertaking technological and organizational innovations aimed at extending their existing practices into the secondary markets. The main objective was to develop ICTs that would enable these mortgage lenders to "transfer" their pre-existing "legacy assets" onto the emerging secondary markets (Kaniadakis & Constantinides, 2014); in other words, to engage in ICT innovations with the purpose of "plugging into" the securitization supply chain. Transferring the existing assets in the secondary markets, however, proved to be challenging. The issue that was raised was related to the "informational quality and integrity" of those assets (Kaniadakis & Constantinides, 2014). More specifically, since existing mortgage loans that were held in mortgage lenders' data warehouses were underwritten over a period of many years and under different temporal circumstances, their eligibility to be considered as assets fit for securitization at a present time, was put under question. Their quality was challenged by the very nature of securitization practice, the main task of which was to perform a risk analysis on behalf of potential investors. Indeed, the changing historical circumstances under which each individual loan was underwritten may render the information available on a specific loan as inaccurate or misleading. Inaccurate and misleading data would affect the calculations involved in the risk analysis and credit rating process, which would determine how risky an investment in one of the securitization products would be for a potential investor.

What effectively happened with securitization, was that the transition from primary to secondary mortgage markets created a whole different managerial decision-making context, a whole new underwriting process that constitutes, I argue, a separate "cognitive frame" (Beuza & Stark, 2004) in which information on an individual's credit worthiness (ability to repay the loan), become relevant. More specifically, the decision-making context in the secondary mortgage markets referred to the choices on which of the existing loans can or cannot be selected for securitization purposes. Therefore, although in the primary markets, underwriting referred to the calculations done to determine a customer's credit worthiness and the decision to be made was whether to grant a loan to an individual customer or not, in the secondary markets, underwriting referred to the practice of selecting which individual loans to include in a securitization transaction. This process is often referred to as "pooling". It has been shown that pooling large numbers of home mortgages reduces the amount of information needed to understand their value, so instead of compiling idiosyncratic information about each individual home and borrower, a lender need only use aggregate information about means and variables of pool mortgages (Carruthers & Stinchcombe, 1999; Kaniadakis & Constantinnides, 2014). Pooling, then, renders the value of a loan more generally knowable and encourages those with little particular knowledge of the individual mortgage to invest their money (Carruthers & Stinchcombe, 1999; Kaniadakis & Constantinides, 2014). However, in order for an individual loan to make it into the pool, it has to meet certain criteria and standards.

It is in this context where data and information that were generated during the initial mortgage underwriting process are put into question and where the issue of standardization of information across different contexts becomes relevant. The question of ICT standardization, therefore, takes a specific form, which is to standardize loan information across the two distinct cognitive frames reflecting the calculations and decisions that need to be made in the primary and secondary mortgage markets. ICT innovations in securitization by mortgage lenders, therefore, were carried out with the specific purpose to achieve standardization of information across primary and secondary market contexts.

Standardization of Information and the Creation of Calculable Objects

Calculative Agency and the Role of ICTs as Standardizer of Information

As shown in the section above, the need for standardization of information was in fact addressing the need for calculating financial risk across contexts. To put it differently, what posed the need for standardization of information across primary and secondary market cognitive frames was the intention to make the assets that were being transferred calculable (Preda, 2006) within the new securitization context. Indeed, in order to effectively calculate financial risk, you need a standardized way to define and measure it. Standardization of information in the context of mortgage securitization, then could be understood as what Economic Sociologists call "calculative agency" (Callon, 1998; Barry & Slater, 2002; Callon & Muniesa, 2005; Preda, 2006). Calculative agency involves the practices of *framing*, *disentanglement* and *performativity* (Preda, 2006). Framing designates the practice through which technology creates calculable objects and separates them from non-calculable ones. Disentanglement is the marking of boundaries between what is relevant and non-relevant with respect to calculability. Finally, performativity refers to the designation of the status of technology as a set of intervention tools in market transactions (Preda, 2006). Framing and disentanglement, then, as a central process in ICT innovation, may be understood as a preparation of the data before they are used for the performance of calculation of financial risk. In their recent paper, Kaniadakis & Constantinides (2014), they present empirical evidence of how this "preparation" of the data happened. Indeed, they present a case study of a UK mortgage lender with existing mortgage assets that wanted to enter the securitization markets. They show that the specific

Table 1. Underwriting in primary and secondary mortgage markets

	Primary mortgage markets	Secondary mortgage markets (Securitization)
Underwriting	Decision to lend individuals who apply for mortgage loans	Decision on which loans to select and transfer to the securitization markets.

bank developed a separate organizational unit to conduct this preparation by validating the data on the existing assets before they are considered for securitization. This re-examination of the legacy assets had no aim to question the lending decision in the cognitive frame of the primary mortgage markets; this would be pointless as such a decision might have been made years ago. The aim was rather to determine whether the data that were available and accompany each asset were reasonably accurate (Kaniadakis & Constantinides, 2014).

Standardization in this context does not refer to the compression of information into data but rather to a set of rules, conventions and tools by which data are transferred across contexts (Preda, 2006). As Preda (2006) explains, this type of standardization plays a major part in the rationalization of decision making and the constitution of financial transactions as separated from broader social ties and obligations. In this way, trust and authority are dissociated from individuals and transferred to the technology: trustworthy data are data produced or recorded by an authoritative technology, which can be transferred across heterogeneous contexts without losing their properties. The role of ICT then was to become the standardizers of data across contexts and as they are transferred to open predictable and reliable paths of economic action across these contexts (Preda, 2006). The innovative efforts of industrial actors, therefore, were centered around the development of ICTs capable of achieving standardization of information across contexts. This is where we may understand standardization as a goal in itself. In other words, standardization of information was part of a broad innovative effort of mortgage lenders to create new calculable objects out of pre-existing ones.

Standardization of Information as a Collective Action

The levels of accuracy of the data on the existing assets were not determined based on criteria set internally and independently by the mortgage lender; rather the framing and disentanglement involved in the standardization of information across cognitive frames was a collective process involving external actors, such as, credit rating agencies (CRA) and communities of potential investors. More specifically, for mortgage lenders, one way to standardize data across primary and secondary markets was to determine the extent of which their internal individual mortgage underwriting criteria based on which each of their loans was underwritten, match the criteria that are commonly accepted in the broader securitization market. One way of doing so was to use the same templates that CRAs used to evaluate securitization transactions. If the underwriting information of a single loan was providing information that would match what CRAs required in order to rate a transaction, then this loan would be a better candidate to enter securitization. Another way to make sure that the information requirements of your assets were up to the standards of the broader market, is through politicking and networking with investors before you start preparing for a deal. Indeed, interacting with investor communities is useful to prepare the ground before one attempts to proceed with a securitization transaction. Mortgage lenders, then, would meet with potential investors, explain to them what kind of assets they have, how they are thinking of structuring a securitization deal and so on, and this way, they would prepare the ground for an upcoming transaction.

This brings us to the discussion of an understanding of technological standardization as a *collective action* and a *movement* involving industrial actors that constitute the market. Existing research on mortgage securitization in the US confirms an understanding of ICT standardization as a collective action or a movement (Markus et al, 2008; Poon, 2009). With reference to specific ICT-based tools for securitization practices, such as, mortgage underwriting (i.e. Automated Underwriting tools) and credit scoring (i.e. FICO scores), researchers explain standardization as a process whereby technical apparatuses are adopted and reframed according to different contexts and actors' interests (Markus et al., 2008). Poon (2009), for instance, shows the successive "movements" and "translations" that took place as the FICO infrastructure spread through the U.S. mortgage industry and was shaped as a standardized platform to represent credit risk by interactions amongst industry actors. Similarly, Markus et al, (2008) describe the diffusion of standardized automated underwriting tools in the US mortgage industry as a succession of multiple "technological action frames". These are defined as "multi-dimensional composite understandings – constituted and circulated in language – that legitimate high levels of investment for potential users, and form the core ideas about how a technology works and how a future based on its use should be envisioned" (Iacono & Kling, 2001 in Markus et al, 2008). Markus et al argue that several framings of automated underwriting emerged over time, which were the result of interactions between the technology and various social (industrial) actors. Such actors (e.g. banks, lenders, CRAs, investors, IT companies, regulators, etc) actively produce and maintain a technological action frame by giving it a certain meaning. They talk therefore for the "productivity master frame" where the adoption of standardized automated underwriting tools would enhance productivity; about the "democratization master frame" where AU tools would turn credit scoring into a fairer and more objective process; and the "anytime-anywhere master frame" whereby AU tools were bringing underwriting to the hands of consumers (Markus et al, 2008).

This research suggests that the effects of technological standardization on innovation are "negotiated" by the various stakeholders that participate in the market. Industrial actors as market participants are collectively engaged in a process by which they develop visions and discourses around how standardization may benefit business practices and enable innovation. It seems that standardization of information across primary and secondary mortgage markets was itself a part of such a technological action frame. More specifically, the adoption of standardized tools that were performing parts of the securitization chain, such as mortgage underwriting (Markus et al, 2008; Poon, 2009) along with other ICT-related innovations, such as, developing new information infrastructures (Kaniadakis & Constantinides, 2014), especially in the early days of securitization, were done with the purpose of standardizing information across different cognitive frames. The purpose of standardization of information across mortgage markets was to establish universal, industry-based representations that would help *understand and express risk across these contexts.*

So far I have discussed how the emergence of securitization as an extension of the primary mortgage lending markets has created the need for standardization of information across primary and secondary market contexts. The reason for this need is that primary and secondary markets represent two distinct administrative and decision-making contexts, which I explained as two different 'cognitive frames' for the calculation of credit risk. Then, I discussed the ICT innovations that were involved in the standardization of information as mortgage lenders started transferring their pools of existing mortgage loans to plug them into the emerging securitization chain. I propose an understanding of this process as a calculative agency, which involves not only the performance of a calculation by an authoritative technology but also the preparation for this calculation, as explained by the processes of framing and disentanglement. Finally, I explained that calculative agency, as it has been understood in this context, is not an independent internal process but rather a collective effort by various market participants involved.

The standardization of information as discussed here has contributed to the flattening of the calculative spaces between primary and secondary markets. Indeed, before 1995 the description of each individual loan through underwriting was done with an entirely separate set of tools, metrics, and vocabulary than those used to describe a securitized pool of loans as a composite whole (Poon, 2009:664). Standardization of information, however, made possible the techno-organizational and administrative integration of primary and secondary markets which were placed on the same calculative platform (ibid:665). Single loans and pools of loans, therefore, could be expressed by the same metric (ibid: 671), a fact that reinforced the importance and relevance of single loan information for securitization. As a result, underwriting tools that produced credit scores for individual consumers expanded to become a central technology of securitization as they were diffused and adopted by securitizers and rating agencies (Poon, 2009).

THE EFFECTS OF STANDARDIZATION OF INFORMATION ON BUSINESS AND ORGANIZATIONAL INNOVATION

Two Different Organizing Visions

In the previous sections I discussed how in order for the mortgage securitization markets to emerge, standardization of information across the primary and secondary markets was necessary. ICT innovations were central to this process as they led to the establishment of market infrastructures on which to understand risk and transfer loan information. Standardization of information, then, gave ICT innovation a more specific context and purpose. The purpose was to be able for actors in the primary markets with an existing infrastructure of loan origination, to plug into the emerging securitization chains and thus gain access to capital markets. ICT innovation, therefore, contributed to the standardization of information across primary and secondary markets by providing both mechanisms for single loans to be "transferred" to the secondary markets (Kaniadakis & Constantinides, 2014) and also provided integrated "calculative platforms" on which the risk of both single loans and a pool of loans could be expressed and understood (Poon, 2009).

I argue at this point, however, that the very process of information standardization that was necessary for securitization to exist, was one of the main reasons that led to the collapse of securitization markets in 2007-8. The implications of such an argument would mean that the scope of standardization management would go well beyond balancing positive and negative effects of standardization on innovation with respect to a single firm (i.e. to improve productivity, or competitiveness, etc); it would rather extend to the foundations of the mortgage market itself and also to society at large.

As hinted in the previous sections, standardization management processes (i.e. innovative efforts towards standardization of information) were not motivated and guided by a culture around securitization as an accurate and reliable risk analysis; they were rather motivated by a culture of securitization as a product development process, as a new supply chain and as a work organization issue. Indeed, as securitization spread all over the world, it was adopted by organizations not simply as an innovative way to manage risk but rather as a business model to support fund-raising. The establishment of common ways to understand and represent risk, via standardization of information, therefore, was not done with the purpose to make the markets and the financial system more resilient towards risk, but rather in the context of developing a business model and products that would allow lenders to enter the emerging securitization markets.

Theoretically, this can be explained in the context of two co-existing "organizing visions"[2] (Swanson & Ramiller, 1997) that became visible around securitization. One was referring to *securitization as a risk management innovation*, while the other was referring to *securitization as a money-making business innovation*. It seems that the industry chose to pursue the latter. I am not arguing here that there is something inherently wrong with exploring a new type of business innovation; the way it was done organizationally, diminished the possibility of using securitization as an innovative way to handle risk and undermined the standardization of information movement that gave rise to the whole industry.

The first organizing vision of securitization being a risk management innovation was mainly expressed in the political rhetoric of powerful global financial institutions. The IMF Global Financial Stability report of April 2006, for example, just one year before the break out of the global financial crisis, states:

There is growing recognition that the dispersion of credit risk by banks to a broader and more diverse group of investors, rather than warehousing such risk on their balance sheets, has helped to make the banking and overall financial system more resilient. Over the last decade, new investors have entered the credit markets, including the credit risk transfer markets. These new participants, with differing risk management and investment objectives (including other banks seeking portfolio diversification), help to mitigate and absorb shocks to the financial system, which in the past affected primarily a few systemically important financial intermediaries. (IMF Report, 2006)

In this view, securitization is understood as a positive and socially beneficial innovation that would lead to the democratization of finance, enhance market efficiency and financial stability (Shiller, 2003). This vision was based on the nature of securitization as a risk management technique, more specifically on its ability to transfer risk and disperse it amongst a large number of investors and thus smoothing out its negative effects (Franklin & Gale 1994; Duffie 2008). ICTs would have a central role in the organizing vision of securitization as a risk management method, as explained by Claudio Ciborra:

This new way of looking at, and practicing, finance as a science for managing risk 'democratically' gives digital technologies an overarching importance and a new role. They become 'grid technologies', that is an information infrastructure that allows the calculation of indexes and units of accounts, so that risks are quantified and can be traded, pooled and shared on global markets by large numbers of individuals. Digital technologies diffuse and enmesh with the structure of markets under this encompassing grid infrastructure, creating virtual enterprise networks and affecting more than ever the personal lives of workers, managers and consumers. (Ciborra, 2007)

As we can see, therefore, as ICTs "enmesh" with the structure of financial markets, they create information infrastructures whereby the generation and transmission of standardized information would enable a more collective and democratic management of financial and credit risk. Nevertheless, this organizing vision was far from being realized. The securitization industry chose instead to focus their innovative standardization management efforts around securitization not as this "wonderful" innovation that would benefit societies and make the financial system more resilient to systemic risk threats, but as a vehicle for speculation and profit, which ultimately led to the credit crunch. Paul Krugman (2007) in a New York Times article graphically explains this contrast:

The innovations of recent years — the alphabet soup of C.D.O.'s and S.I.V.'s, R.M.B.S. and A.B.C.P. — were sold on false pretenses. They were promoted as ways to spread risk, making investment safer. What they did instead — aside from making their creators a lot of money, which they didn't have to repay when it all went bust — was to spread confusion, luring investors into taking on more risk than they realized. (Krugman, 2007)

In the context of securitization as a business model and a platform for speculative profit seeking, the standardization of information made securitization actors to shift their focus from risk analysis to efficiency, product development and profit. The standardization of underwriting information of individual loans across primary and secondary markets, then, was not happening with the purpose to produce more accurate risk calculations and improve the risk management process; it was rather happening with the purpose to create the necessary resources that would be used for the production of complex financial products; and also with the purpose of making this product development process more cost effective and more efficient operationally. In other words, standardization management in mortgage securitization was indeed seen in the context of the newly developed business strategies of mortgage lenders. Nevertheless, as shown in the following section, the individual organization's business strategy context was proved to be rather narrow in terms of guiding innovations in the industry.

Intra-Organizational Partitioning: Undermining the Unity of Risk Analysis

The effects of standardization of information on the realization of the vision of securitization as a speculative profit-making tool are explained by Jacobides (2005). He argues that the standardization of information led to an "intra-organizational partitioning" in the securitization industry and a fragmentation of its supply chain (Jacobides, 2005). More specifically, he explains that the very nature of securitization itself is based on the separation of ownership of a loan (or an asset more generally) from its origination and servicing (479). Indeed, standardization of information which made possible a "universal" understanding of loan characteristics and a customer's credit worthiness, led to vertical disintegration of the securitization value chain, as emerging intermediary actors started developing their business models around a small, specific part of the whole process of securitization. This way, tasks, such as, mortgage origination and warehousing were gradually de-coupled and interactions between them were reduced initially within the same organizations and later followed by the vertical disintegration of the whole industry (480). For instance, as Jacobides explains:

If a mortgage banker produced more loans than anticipated, the company need not worry about not being able to sell them to an investor; the flexible arrangements alleviated this concern. Thus, a part of the company could just focus on origination without having to worry about the next step in the value chain. (Jacobides, 2005: 480)

Indeed, as mortgage lenders grew, they developed internal organizational structures that were disassociated from each other. In other words, mortgage lenders were undergoing organizational innovations that led to a functional specialization around different steps of the securitization value chain. Gradually, these functional units grew apart and developed distinct cultures that made a single holistic understanding of the risk analysis process quite impossible. MacKenzie (2011), for instance, explains how practices of evaluating securitization products (i.e. credit rating) were transformed into organizational routines that became the province of different parts within the same organization (MacKenzie, 2011: 1779). He explores how these different organizational units were developed around different product lines. In a bank, for instance, one unit was devoted to the evaluation of asset-backed securities (ABSs) like residential mortgage securitizations, while another unit would be devoted to collateralized debt obligations (CDOs)[3]. In the case of the rating agencies, to mention another example, both ABSs and CDOs fell within the remit of their structured finance departments, but the latter had separate groups dealing with each (MacKenzie, 2011: 1779). When the more complex products of ABS CDOs (which are CDOs with ABSs nested within them, so to speak) were fashioned, "the decision as to how to evaluate them was thus also a decision about how their evaluation should be mapped onto the organizational structure of rating agencies" (ibid).

Beunza & Stark (2004) also explain the spatial distribution of different securitization practices (risk analysis, evaluation) across different "desks" of a trading room. As a result, safeguarded by distinct cultures, the organizational units devoted their efforts on the particular function/practice they were in charge of, while ignoring the rest of the steps that were necessary for a complete risk analysis process. Decisions and investment behaviours that were largely governed by one such specific task were made highly profitable, while other cognitive peripheral functions were left as "epistemic orphans" (MacKenzie, 2011: 1779).

These developments in the internal organizational structures and the partitioning of different tasks paved the way for the vertical disintegration in the whole industry that was more prominent in the separation between the origination and servicing of a mortgage loan and the simultaneous emergence of new intermediaries that would undertake a specific task (e.g. mortgage servicing) (Jacobides, 2005). Each of these new intermediary actors pursues their own interests while the new intermediary markets developed their own rules and cultural aspects.

The standardization of information across primary and secondary mortgage markets, which was initially necessary for several pre-existing mortgage lenders to enter the emerging securitization chain, gradually led to an organizational partitioning of securitization practices and subsequently to the vertical disintegration of the whole value chain. Although the standardization of information enabled the transfer of assets from one context to another and seemed like the industry had a way to collectively understand, transfer and manage credit risk, a focus on operational efficiency of product development that was organized around specialized functions within organizations and around emerging new intermediary markets, shifted the initial attention from the organizing vision of risk management to the (business strategy-informed) one about speculation and profit making. Securitization as a techno-organizational framework within industrial actors and a product, therefore, took a very specific form that undermined the unity of the risk analysis process. The paradox here is that although standardization of information made possible a collective understanding of risk across primary and secondary mortgage markets, on the other hand this universal understanding was subsequently broken down into multiple, organizational, cultural, functional, operational, spatial realities, which made it impossible for the financial markets to contain. One the one hand we have a really unique innovation with a lot of potential that would revolutionize the way markets function and the way they deal with risk; on the other hand, these new potential were not realized because the new innovation took a more "traditional" techno-organizational form, based around functional specialization, operational efficiency, competition and so on.

In the following section, I offer a discussion on the implications of standardization management for broader social relations that extend the scope of narrow management planning around goals on productivity, effectiveness, competitiveness etc. to include aspects that might seem initially inconceivable.

DISCUSSION AND CONCLUSIONS

Bringing Securitization Back to its Social Context

It is now time to revisit my initial discussion on the paradoxical relationship between standardization and innovation. The story of the financial innovations in mortgage markets is indeed a story that unfolds in a context of technological standardization. Technological standardization in this industry should be understood both as a goal of innovation that needs to be pursued (i.e. the need for standardization of information) but also as an enabling factor for industrial actors in the mortgage industry to further develop innovative business models and techno-organizational frameworks around securitization.

Although standardization was an innovative force in mortgage markets, I showed in this chapter that to understand standardization as an enabler or constraint on innovation is not enough. One has to look into the different types of innovation that standardization can enable, to understand alternative potential routes for innovation and explore what would be the implications for the individual firm, for the industry and for society at large. Before pursuing innovative goals, then, standardization management practitioners would need to envision alternative innovative paths, in the form of organizing visions, and evaluate their potential impact.

Different such organizing visions developed around securitization. Industrial actors, however, decided to utilize the innovation possibilities that were made available by standardization within a specific organizing vision around securitization as a profit-making tool. For this reason, they centered their innovative efforts around making securitization more efficient and more profitable. Thus, a more "traditional" model of functional specialization was pursued both at the level of organization and of the supply chain, which in turn undermined the unity of securitization as a risk management process. Whether the industry decided to go this way as a result of an impulse or reflex or whether is was the result of the "greed of the bankers", or some other reason, should be the focus of future research.

Although standardization of information aimed at and succeeded to a certain extent to create a common understanding of credit risk and credit worthiness, industrial actors utilized this understanding not as a way to collectively manage risk, but to use it as raw material for the development of new types of products. The collective spirit surrounding the standardization of information process, therefore, was not directed towards a communitarian and democratic way to handle risk. Its objective was rather to establish a common language around which to set up new fund-raising securitization infrastructures. Standardization of information across contexts as an enabler of innovative ways to collectively manage and mitigate financial risk remained a political rhetoric. What happened in reality is that it triggered the speculative instincts of finance organizations who devised new ways to profit from transferring and selling that risk. To make this process more efficient, industrial actors were drawn to mainstream organizational models of functional specialization, which gradually led to the partitioning of the risk management process. Although industrial actors collectively constructed a common understanding of risk, on the other hand, from a business innovation point of view, they "selfishly" remained attached to old principles around efficiency and competitiveness. This happened at the cost of removing credit and financial risk from its social/societal context and treat it in the context of new product development.

Post financial crisis reforms have addressed this issue are targeted towards bringing securitization back to its social policy context. These reforms are mainly centered around the notion of transparency, which puts standardization management in a more extended socio-political context. Indeed, transparency is now a big part of the glossary of mortgage securitization and has become the new organizing vision for industrial actors and a new frame on which to understand and approach standardization of information. Standardization of information in this context is not only related to the transition from primary to secondary mortgage markets; it rather refers to the re-integration of the securitization supply chain and the re-unification of credit risk assessment. In other words, standardization of information targets the different organizational and inter-organizational contexts that were created with the disintegration of the value chain. It refers to relationships between originators and servicers, credit rating agencies and investors, financial organizations and regulatory bodies, and so on. As mortgage securitization enters a third more mature phase after 2010, the relationships between securitization industrial actors and the role of each one of them in the securitization value chain are being re-visited. This new securitization landscape is not just technological, or just political, or economic. It is all these things together.

New needs for standardization of information are created and new challenges for standardization management strategy development. The case of mortgage securitization teaches us that it is not enough for a standardization management strategy to simply be aligned with the organization's business strategy. The nature of the securitization practice is such that if organizations develop their standardization management strategies only to support a single business strategy, that might have a negative impact over-time. Indeed, it might create industry-wide systemic issues with unfathomable consequences for the industry itself but also for society at large.

I argue here that standardization management in mortgage finance gradually left the scope of managerial decision making and became highly politicized. Indeed, it became the subject of public debates and a public policy issue. From the visions that gave rise to the occupy social movements globally to the motivations for introducing new initiatives by Central banks to standardize transaction reporting and monitoring, the topic of standardization of information seems to slowly being removed from the frames of competitiveness, efficiency and productivity and instead placed within a political discourse around informational transparency. I argue here, therefore, that standardization management in the context of innovations in the mortgage industry, should be seen in a broader socio-political context. Standardization management professionals, therefore, should not only consider choices around the adoption of certain standards and what would be the impact on the company; they should rather practice their profession within a broader socio-political terrain. Indeed, standardization management should aim beyond improving operational efficiency, organizational performance and competitiveness of a single company, traditional objectives of ICT innovation and standardization. Standardization management strategies should also be aligned with the broader social policy visions that aim at guiding innovations in a specific industry.

I showed how securitization historically emerged in the US as a "social policy innovation" as part of the US government's budgetary politics (Quinn, 2009) and it gradually evolved beyond the arena of the US federal public policy and into other national, business and organizational contexts. As soon as securitization became a business then it was exposed in a wide variety of strategies, incentives, rationales and action frames. In this chapter I discussed mainly two: One was an understanding of *securitization as a risk management process* that would democratize finance and make the financial system more resilient to risk threats. This was an organizing vision that was projected by policy circles and advocates of securitization and was giving ICTs a certain role as technologies that would enable such a collective risk management process. The other vision was about using *securitization as a speculative tool* to make profits and exploit "information asymmetries" (Akerlof, 1970). These two visions offered two alternative ways in which securitization could be turned from an idea into a techno-organizational framework and a product. As shown in this chapter, however, the risk management vision was rather hard to become realized.

Figure 1. Standardization management "in-context"

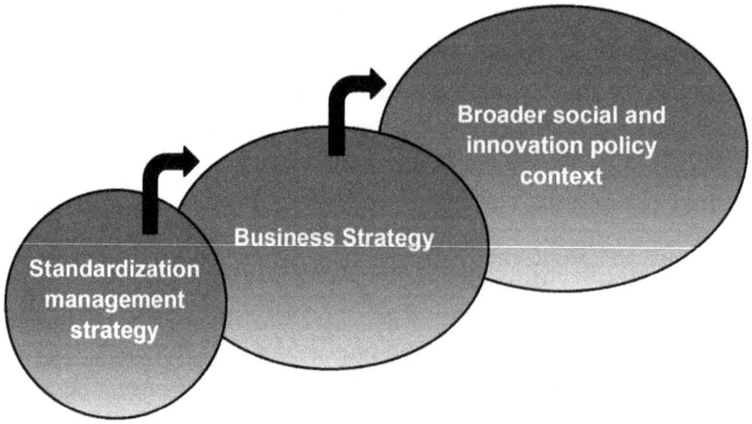

Indeed, although initially, the aim of standardization management was to integrate primary and secondary mortgage markets and create entry incentives for lenders, it later shifted its focus: From trying to maintain the unity of risk analysis across context towards making product development more efficient and more profitable. The outcome of this efficiency is reflected on the exotic, complex structures of securities-related products (ABS, CDO, etc). Standardization management strategies, therefore, took on board the "efficiency" and "speculation" organizing vision with the consequences that were discussed earlier both in regards with organizational structures and cultures and with the way the whole industry was organized. This means that standardization management was in fact aligned with a business strategy logic that was guiding the actions of market participants. Mortgage securitizers were indeed undergoing technological and organizational innovations in order to improve their operational practices and their competitive advantage in the market but while doing so, they forgot to care about reliable risk analysis. The risk management-focused organizing vision, then, was pushed out of the way and securitization participants focused on efficiency and speculation.

I argue, therefore, that standardization management should not just simply be aligned with the business strategy of a single organization but also with broader, industry-wide, social and innovation policy issues that guide the development of an industry. If at the very context of invention of securitization there was the vision that standardization of information would help risk management, then any standardization management strategies should use this as a point of departure.

Standardization management strategies should therefore not simply support an organization's narrow, independent and "selfish" interests for profit and competition, but also take into consideration systemic and longer-term effects at the level of the industry. Standardization managers will have to engage in coordinated actions with other industrial actors, with policy makers, with regulators and explore what would be the implications of specific standardization efforts to the industry, its practices and to society as a whole. Standardization efforts in the mortgage industry should therefore move away from a purely "business" logic and be brought back to a socio-political context, such as the one that securitization initially emerged from. The question that remains, however, is whether the socio-politization of standardization management will manage to enhance the resilience of the financial system and become a catalyst for fairer social policies by enabling multiple relationships of accountability amongst industrial actors, company stakeholders, the state and the broader public, or will it set the basis for industrial actors to create the next big financial meltdown? History will tell.

REFERENCES

Acemoglu, D., Gancia, G., & Zilibotti, F. (2012). Competing engines of growth: Innovation and standardization. *Journal of Economic Theory*, *147*(2), 570–601. doi:10.1016/j.jet.2010.09.001

Akerlof, G. A. (1970). The Market for 'Lemons': Quality Uncertainty and the Market Mechanism [The MIT Press]. *The Quarterly Journal of Economics*, *84*(3), 488–500. doi:10.2307/1879431

Arthur, W. B. (1996). Increasing returns and the new world of business. *Harvard Business Review*, *74*, 100–109. PMID:10158472

Barry, A., & Slater, D. (2002). Introduction: The technological economy. *Economy and Society*, *31*(2), 175–193. doi:10.1080/03085140220123117

Benner, M. J., & Tushman, M. L. (2003). Exploitation, exploration, and process management: The productivity dilemma revisited. *Academy of Management Review*, *28*, 238–256.

Beunza, D., & Stark, D. (2004). Tools of the trade: The socio-technology of arbitrage in a Wall Street trading room. *Industrial and Corporate Change*, *13*(2), 369–400. doi:10.1093/icc/dth015

Buffett, W. (2002). *Letter to Berkshire Stakeholders*.

Callon, M. (1998). Introduction: The embeddedness of economic markets in economics. In M. Callon (Ed.), *The Laws of the Markets*. Blackwell. doi:10.1111/j.1467-954X.1998.tb03468.x

Callon, M., & Muniesa, F. (2005). Economic markets as calculative collective devices. *Organization Studies*, *26*(8), 1129–1250. doi:10.1177/0170840605056393

Carruthers, B., & Stinchcombe, A. (1999). The social structure of liquidity: Flexibility, markets, and states. *Theory and Society*, *28*(3), 353–382. doi:10.1023/A:1006903103304

Ciborra, C. (2007). Digital Technologies and Risk: A Critical Review. In Hanseth & Ciborra (Eds.), Risk, Complexity and ICT. Edward Elgar Publishing. doi:10.4337/9781847207005.00007

Cummings, J., & DiPasquale, J. (1997). *A primer on the secondary mortgage market*. Boston: City Research.

David, P. A. (1995). Standardization policies for network technologies: The flux between freedom and order revisited. In R. Hawkins, R. Mansell, & J. Skea (Eds.), *Standards, Innovation and Competitiveness: The Politics and Economics of Standards in natural and Technical Environments*. Cheltenham, UK: Edward Elgar.

de Vries, H. J. (2006). The Paradox of Standardization and Innovation. *ISO Focus*, *3*(February), 40–43.

Dosi, G. (1982). Technological paradigms and technological trajectories: A suggested interpretation of the determinants and directions of technical change. *Research Policy*, *11*(3), 147–162. doi:10.1016/0048-7333(82)90016-6

Duffie, D. (2008). *Innovations in Credit Risk Transfer: Implications for Financial Stability*. BIS Working Papers, No. 255.

Economides, N., & Katsamakas, E. (2006). Two-sided competition of proprietary vs. open source technology platforms and the implications for the software industry. *Management Science, 52*(7), 1057–1071. doi:10.1287/mnsc.1060.0549

Fincham, R., Fleck, J., Proctor, R., Scarbrough, H., Tierney, M., & Williams, R. (1994). *Expertise and Innovation: Information Technology Strategies in the Financial Services Sector*. Oxford, UK: Clarendon Press.

Fligstein, N., & Goldstein, A. (2010). The anatomy of the mortgage securitization crisis. In M. Lounsbury & P. M. Hirsch (Eds.), Markets on trial: The economic sociology of the U.S. financial crisis: Part A (Research in the Sociology of Organizations, Volume 30) (pp. 29-70). Emerald Group Publishing Limited. doi:10.1108/S0733-558X(2010)000030A006

Franklin, A., & Gale, D. (1994). *Financial innovations and risk sharing*. Cambridge, MA: MIT Press.

Greenstein, S., & Stango, V. (2006). *Standards and Public Policy*. Cambridge University Press. doi:10.1017/CBO9780511493249

Hanseth, O., Monteiro, E., & Hatling, M. (1996). Developing information infrastructure: The tension between standardization and flexibility. *Science, Technology & Human Values, 21*(4), 407–426. doi:10.1177/016224399602100402

Hargrave, T. J., & Van De Ven, A. H. (2006). A collective action model of institutional innovation. *Academy of Management Review, 31*(4), 864–888. doi:10.5465/AMR.2006.22527458

Iacono, S., & Kling, R. (2001). Computerization movements: The rise of the internet and distant forms of work. In J. Yates & J. Van Maanen (Eds.), *Information technology and organizational transformation: History, rhetoric, and practice* (pp. 93–136). Thousand Oaks, CA: Sage Publications. doi:10.4135/9781452231266.n4

IMF. (2006). *Global financial stability report: Market development and issues, April 2006*. IMF.

Jacobides, M. G. (2005). Industry change through vertical disintegration: How and why markets emerged in mortgage banking. *Academy of Management Journal, 48*(3), 465–498. doi:10.5465/AMJ.2005.17407912

Jakobs, K. (2006). Shaping User-side Innovation Through Standardisation – The Example of ICT. *Technological Forecasting and Social Change, 73*(1), 27–40. doi:10.1016/j.techfore.2005.06.007

Kaniadakis, A., Constantinides P. (2014). Innovating Financial Information Infrastructures: the transition of legacy assets to the securitization market. *Journal of the Association of Information Systems, 15*(5).

Krugman, P. (2007, December 3). Innovating Our Way to Financial Crisis. *New York Times*.

Leyshon, A., & Thrift, N. (2008). The capitalism of almost everything: The future of finance and capitalism. *Theory, Society and Culture, 24*(7-8), 97–115. doi:10.1177/0263276407084699

Loch, C. H., & Huberman, B. A. (1999). A punctuated-equilibrium model of technology diffusion. *Management Science, 45*(2), 160–177. doi:10.1287/mnsc.45.2.160

MacKenzie, D. (2011). The credit crisis as a problem in the sociology of knowledge. *American Journal of Sociology*, (May): 2011.

Markus, M. L., Dutta, A., Steinfield, C. W., & Wigand, R. T. (2008). The Computerization Movement in the US Home Mortgage Industry: Automated Underwriting from 1980 to 2004. In K. Kraemer & M. Elliott (Eds.), *Computerization Movements and Technology Diffusion: From Mainframes to Ubiquitous Computing* (pp. 115–144). Medford, NY: Information Today.

Markus, M. L., Steinfield, C. W., Wigand, R. T., & Minton, G. (2006). Industry-wide IS standardization as collective action: The case of the US residential mortgage industry. *Management Information Systems Quarterly*, *30*(Special Issue), 439–465.

Narayanan, V. K., & Chen, T. (2012). Research on technology standards: Accomplishments and challenges. *Research Policy*, *41*(8), 1375–1406. doi:10.1016/j.respol.2012.02.006

Oshri, I., & Weeber, C. (2006). Cooperation and competition standards-setting activities in the digitization era: The case of wireless information services. *Technology Analysis and Strategic Management*, *18*(2), 265–283. doi:10.1080/09537320600624196

Pollock, N., & Williams, R. (2011). Who decides the shape of product markets? The knowledge institutions that name and categorize new technologies. *Information and Organization*, *21*(4), 194–217. doi:10.1016/j.infoandorg.2011.08.001

Poon, M. (2009). From New Deal Institutions to Capital Markets: Commercial consumer risk scores and the making of subprime mortgage finance. *Accounting, Organizations and Society*, *34*(5), 654–674. doi:10.1016/j.aos.2009.02.003

Preda, A. (2006). Sociotechnical agency in financial markets: The case of the stock ticker. *Social Studies of Science*, *36*(5), 753–782. doi:10.1177/0306312706059543

Quinn, S. (2009). *Towards a market for mortgage-backed securities: Credit lending, the Federal budget, and the politics of debt management*. University of California Group of Economic History, Working Paper.

Rammert, W. (2003). Two Paradoxes of Fragmented Knowledge Production: Combining Heterogeneous and Cultivating Non-Explicit Knowledge. *SASE Annual Conference*, Aix-en-Provence, France.

Riddiough, T. J. (2011). Can securitization work? Economic, Structural and Policy considerations. *Journal of Portfolio Management*, *37*(5), 24–38. doi:10.3905/jpm.2011.37.5.024

Shiller, R. J. (2003). *The New Financial Order: Risk in the 21st Century*. Princeton, NJ: Princeton University Press.

Suarez, F. F. (2004). Battles for technological dominance: An integrative framework. *Research Policy*, *33*(2), 271–286. doi:10.1016/j.respol.2003.07.001

Swan, J., & Scarbrough, H. (2005). The Politics of Networked Innovation. *Human Relations*, *58*(7), 913–943. doi:10.1177/0018726705057811

Swann, G.M.P. (2010). *The economics of standardization: An update*. Report for the UK Department of BIS.

Swanson, E. B., & Ramiller, N. C. (1997). The Organising Vision in Information Systems Innovation. *Organization Science*, *8*(5), 5. doi:10.1287/orsc.8.5.458

Tassey, G. (2000). Standardization in Technology-Based Markets. *Research Policy*, *29*(4-5), 587–602. doi:10.1016/S0048-7333(99)00091-8

Teece, D. J. (2007). Explicating dynamic capabilities: The nature and microfoundations of (sustainable) enterprise performance. *Strategic Management Journal*, *28*(13), 1319–1350. doi:10.1002/smj.640

Wainwright, T. (2009). Laying the foundations for a crisis: Mapping the historico-geographical construction of residential mortgage-backed securitization in the UK. *International Journal of Urban and Regional Research*, *33*(2), 372–388. doi:10.1111/j.1468-2427.2009.00876.x

Wang, P., & Ramiller, N. C. (2009). Community learning in information technology innovation. *Management Information Systems Quarterly*, *33*(4), 709–734.

Week, B. (1998). *A $2.5 trillion market you hardly know*. Available at: http://www.bloomberg.com/bw/stories/1998-10-25/a-2-dot-5-trillion-market-you-hardly-know

KEY TERMS AND DEFINITIONS

Credit Risk: Is the risk of loss of principal or loss of a financial reward stemming from a borrower's failure to repay a loan or otherwise meet a contractual obligation.

Mortgage Underwriting: Is the process a lender uses to determine if the risk of offering a mortgage loan to a particular borrower is acceptable.

Primary Mortgage Markets: Are those where borrowers and mortgage originators come together to negotiate terms and effectuate mortgage transaction. Mortgage brokers, mortgage bankers, credit unions and banks are all part of the primary mortgage market.

Secondary Mortgage Markets: A large percentage of newly originated mortgages are sold by their originators into the secondary market, where they are packaged into mortgage-backed securities and sold to investors such as pension funds, insurance companies and hedge funds.

Securitization: Is the practice by which issuers pool together various types of assets such as residential mortgages, commercial mortgages, auto loans or credit card debt obligations, they divide them in slices depending on how much risk they carry and then they sell these slices to third party investors in the capital markets.

ENDNOTES

1 I was a bit uncertain as to whether I should write "moving" or "should be moving". The difference between the two at this stage is still unclear.

2 The organizing vision refers to the formulation of a vision around how to organize structures and processes in regards to an information systems innovation.

3 CDO is a structured financial product that pools together cash flow-generating assets and repackages this asset pool into discrete tranches that can be sold to investors. A collateralized debt obligation (CDO) is so-called because the pooled assets – such as mortgages, bonds and loans – are essentially

debt obligations that serve as collateral for the CDO. The tranches in a CDO vary substantially in their risk profile. The senior tranches are relatively safer because they have first priority on the collateral in the event of default. As a result, the senior tranches of a CDO generally have a higher credit rating and offer lower coupon rates than the junior tranches, which offer higher coupon rates to compensate for their higher default risk (Investopedia: accessed 19 Feb, 2015).

Section 7

Two Sides of the Same Coin

Chapter 15
Predatory Strategies in Standards Wars:
On Creating Fear, Uncertainty, and Doubt

Tineke Mirjam Egyedi
Delft Institute for Research on Standardization, The Netherlands

Anique Hommels
Maastricht University, The Netherlands

ABSTRACT

In standards wars, FUD (Fear, Uncertainty and Doubt) is sometimes created to weaken the opponent's market position. Little is known about these strategies, their use in committee standardization settings and how to respond to them. This chapter explores this phenomenon. It (1) identifies various FUD strategies, (2) their context of emergence, and (3) their effect on the dynamics of a standards war in a historical case study: the European standards war on digital mobile radio communication in the 1990s. The study highlights the need to distinguish 'FUD as perceived' from 'FUD as intended'. FUD strategies and case-specific characteristics of their emergence are illustrated. The chapter shows that perceived FUD polarizes and entrenches positions of warring parties thereby affecting the course of the standards war. The authors conclude that, given its impact, reflection by corporate standardization managers on (perceived) FUD, preclusion, counter-strategies and the downscaling of standards wars is warranted.

INTRODUCTION

In the early 1990s, several European member-states decided to replace their analogue radio systems for the police, ambulance and fire brigades by advanced digital systems. Their decision coincided with the aim of the Schengen Agreement to create a Pan-European network for public safety that would allow the police to use the same handset across national networks during 'cross-border surveillance and hot pursuit' (Schengen, 1990). It also concurred with activities of the European Telecommunications Standards Institute (ETSI). Supported by the European Commission, ETSI had started work on a standard for digital mobile radio communication in 1988. One of the objectives was to define a common air interface that

DOI: 10.4018/978-1-4666-9737-9.ch015

would ensure that any standard-compliant terminal (e.g., handset) could run on any European network regardless of the network supplier. A number of countries considered the Tetra standard, as it was called, to be a promising means for achieving the Schengen aims.

However, in April 1994, the period in which ETSI was approving the air interface technology (TDMA) of the Tetra standard, an incompatible competing technology (FDMA) was gaining support among European countries. The latter technology had already been developed by a French company called Matra Communications and implemented in 1992 for the French gendarmerie. To improve access to the European market, FDMA advocates also attempted to get their technology formally acknowledged as an ETSI standard. Initially they seemed to succeed. Early 1995, however, ETSI dropped FDMA from the work program. This was a conspicuous decision, according to Bekkers (2001, p. 380), given the number of suppliers supporting FDMA and FDMA being the preferred choice of a number of public safety network operators. It was the beginning of a standards war.

Standards wars are fought in markets as well as in and between standards committees (Besen, 1991). The TDMA-FDMA war foremost took place in a standardization setting. It involved a committee standard of a formal standards body (i.e., an ETSI specification based on TDMA technology) as well as a de facto standard (i.e., FDMA technology implemented in products by Matra) that later became a consortium standard (i.e., the FDMA-based technical specification of Tetrapol Forum). It was a 'rival revolutions' type of standards war (Shapiro, 1999). In such wars, two factors determine the stakes and the dynamics. First, the rival technologies are *incompatible* (Shapiro, 1999). This was the case, for example, in technology wars between Blu-Ray and HD-DVD (Gallagher, 2012), e-purse systems (Vries, 2006) and proprietary platforms (West, 2003), to name a few de facto standards wars. Incompatible technologies were also at stake in wars between committee standards, for example, in the field of wireless telecommunication (Lee, 2006) and for DVD formats (Dranove, 2003).

A second and related defining factor in standards wars is the effect of rival technologies on *network externalities*. The term refers to an increase in value of the network with every new connected network user (Farrell, 1985). Rival networks based on incompatible technologies or standards cannot exploit each other's externalities. For example, owners of a certain smart phone cannot usually make use of the infrastructure of a different smart phone for spare parts and customer services. In past studies, the workings of network externalities was a main reason to presume that in IT markets only one competing technology would be able to survive ('winner-takes-all'). While more recent work indicates that there may be room for two or more standards (Singh, 2009), the presence of network externalities has been shown to heighten the stakes in standards wars. Given these stakes, "[w]hen it comes to standards wars, traditional principles of strategy, while helpful, are not enough" (Shapiro, 1999), and some companies will turn to predatory market strategies.

The predatory strategies addressed in this chapter are those that create Fear, Uncertainty and Doubt (FUD), a category referred to by Pfaffenberger (2000). To our knowledge, no other studies have as yet analyzed standards wars on the use of predatory market strategies and FUD strategies, in particular. Perhaps understandably so. The term 'FUD' is foremost referred to in business magazine and trade journal articles to capture a class of industry strategies that are perceived to be unfair and used to publicly denounce certain market practices. That is, it is used in a decidedly normative and non-scientific way. The challenge for scientific research on FUD strategies is how to take in and address the inherent subjectivity involved in attributing unfairness. This point is revisited when defining FUD strategies in the next section.

Trade literature on standards wars indicates that companies have difficulty knowing how to deal with FUD strategies. When feeling cornered, companies risk retaliating with defensive counter-strategies that are ineffective and/or may unwittingly escalate problems. Our assessment is that corporate standardization managers lack certain insights that are necessary to consider pre-emptive action and recognize alternative strategies. In the following, we hope to feed them with insights that help them in their decision making. We aim to shed more light on:

1. How may FUD strategies manifest themselves?
2. In which context do they emerge?
3. How do they affect the dynamics of standards wars?

Given the scope these questions, we limit ourselves to studying standards wars involving committee standardization, an area underrepresented in scholarly literature. We expect the occurrence of committee standards wars to increase in tandem with the rising number of standards consortia (Egyedi and Hawkins, 2010). Furthermore, given the explorative nature of our questions, we present the findings of a historical case study on the use of FUD strategies in the committee standards war introduced above (Yin, 2009).

Our data were primarily gathered by means of extensive archival research on exchanged letters, policy documents, etc.[1] from three archives[2] and secondary literature. Twenty in-depth interviews with key actors were held to enrich our general understanding of the case. Five of them are referred to in this chapter as they have immediate relevance for answering our research questions and help illustrate and contextualize our archival data[3].

Below, we first review scholarly literature on predatory strategies and standards wars for insights that help define FUD behavior and its effects. Next, the case of the European standards war on digital mobile radio communication is further introduced. Used FUD strategies are identified and discussed. We conclude by revisiting our research questions and reflecting on lessons learned for corporate standardization managers.

FUD STRATEGIES IN STANDARDIZATION

In the 1970s IBM was accused of using FUD to ward off new market entrants (Pfaffenberger, 2000). Gene Amdahl remarked after he left IBM to found his own company, the Amdahl Corporation: “FUD is the fear, uncertainty, and doubt that IBM sales people instill in the minds of potential customers who might be considering Amdahl products” (Raymond, 2009). Drawing a parallel with Microsoft’s reactions to the rising popularity of the Linux platform, Pfaffenberger (2000) defines FUD as “a marketing technique that a market-dominating firm employs to blunt a competitor’s first-to-market advantage”. (We will re-address his definition later on.) He lists techniques that may be part of a FUD campaign such as “press releases designed to confuse costumers about the merits of the new product, and benchmark tests – generally rigged in the market-dominating firm’s favor – that raise questions about the new product’s performance (…)” and the predatory preannouncement of vaporware (i.e., a product which does not and may never exist) “that can be timed to steal the momentum from a competitor’s technologically superior product” (Pfaffenberger, 2000).

Systematic studies of FUD strategies used in standardization settings are lacking. But, as certain parallels can be drawn with strategies used in highly competitive markets and de facto standards wars, some insight can be gained from technology management, economic, standardization and marketing literature in this field. In such situations, Shapiro and Varian (1999) recommend two basic marketplace tactics, i.e., pre-emption (being first to market and learn from positive feedback) and expectation management. Regarding pre-emption, where a committee standard is based on an existing technology, prior market experience with the technology offers clear benefits (first-mover advantage). In line, the benefit of early experience with implementing a standard tempts some companies to prematurely implement draft standards (Jakobs, 2008).

As for managing the expectations of consumers and competitors (Shapiro and Varian, 1999, p.273), this may include the use of predatory preannouncements (Haan, 2003; Wu, 2004; Bayus, 2001) and vaporware, tactics that also fall within the remit of FUD strategies. Product preannouncement refers to a "formal, deliberate communication[s] that release information about a product well in advance of the product's actual introduction" (Wu, 2004). Predatory preannouncements and vaporware can be very successful in creating uncertainty in markets among competitors and consumers (Haan, 2003; Bayus, 2001). They also play a role in standardization settings. The preannouncement of new standards or standards versions is a common occurrence. During the DVD recordables standards war, for example, both rival consortia regularly announced new standard versions with increased recording capacity and speed. This led to a race in new standards versions (Gauch, 2008). Preannouncement of a new standards version can undermine the momentum for a competitor's standard's launch. If the preannouncement signals that the newcomer will not survive competition and that customers are to 'hang on' in anticipation of a new version, it can also deter newcomers from market entry (Haan, 2003; Bayus, 2001).

In respect to standards wars in and between standards committees, the procedures of standard bodies and consortia, however fair and impartial, cannot safeguard against the use of predatory strategies (e.g. Oshri, 2006; Wegberg, 2004). Examples of (perceived) misuse of the formal rules are staging a standards vote during the holiday season, when most participants will be absent, and defecting during the final vote after lengthy but seemingly fruitful negotiations in a situation where 100% consensus is required to proceed (http://www.noooxml.org/irregularities). Furthermore, increasingly lawsuits are used to settle standardization disputes. Fear of lawsuits sometimes paralyzes standards processes and may cause companies to withdraw from the standards process (Egyedi, 2001).

Pfaffenberger's (2000) description of FUD as "a marketing technique that a market-dominating firm employs to blunt a competitor's first-to-market advantage" assumes that FUD is intentional and therefore involves predatory strategies. however, it is often difficult to distinguish between strategic and innocent action (Haan, 2003; Bayus, 2001). Possibly FUD is raised unintentionally. And, in the current study, how should we methodologically address behavior that is intended to create FUD but is not perceived as such by the targeted party? A clearer definition is needed.

We use the word *affect*, that is, a psychological term for feeling, experience or emotion. To make possible a more systematic study of FUD, the *affect* of Fear, Uncertainty and Doubt that results from certain behavior must be conceptually de-coupled from *intentionality*. This implies two basic lines of enquiry: one focusing on the perspective of the sender ('FUD as intended' in short) and the other on the receiver ('FUD as affected' in short). In general, companies are understandably reticent about providing outsiders information on their market strategy. The same applies to their standardization strategy. In addition, the intention of creating FUD is not a societally acceptable standardization strategy, as trade journals indicate. As a result, data on intentional FUD is hard to come by. In this case study, we therefore focus on 'FUD as affected' (see instances A and B in Table 1). We only refer to 'FUD as intended' if self-reported and if perceived as such. [4]

Figure 1 illustrates that certain *behavior* may or may not give rise to the *affect* FUD, whether intended or not (see Figure 1).

Consequently, we define 'the strategy of creating Fear, Uncertainty and Doubt in standardization' as: a market strategy *perceived* to be designed to create unease among competitors and raise concern among users about the consequences of selecting a competitor's standard, by exerting pressure - psychological, legal or otherwise – and disseminating negative and/or incorrect information about the standard, the standards process or the standard's likely market uptake, as a result of which the credibility of the competitor and/or the competing standard is undermined.

FUD strategies feed uncertainty among standardizers and market players. Typically, uncertainty undermines competition in network-based markets (Farrell, 1986), to which our case also belongs. It leads to hold–ups of investments (Williamson, 1979). Producers will try to postpone investments for fear of investing in the 'losing' system and having to write off sunk costs (i.e., costs that are specific and irreversible and cannot be retrieved). The same hesitations exist on the side of consumers. They fear being stuck with an incompatible system and becoming an 'angry orphan' (David, 1987). Uncertainty arising from creating FUD in the standards setting or in the market causes market stagnation.

The case study's primary source for perceived FUD behavior and its effect are three public and private archives and data from interviews with former stakeholders. Key data are instances in the archived docu-

Table 1. Focus of the case study is on perceived FUD strategies (A and B)

FUD Strategies	Perceived	Not perceived
Intended	A	C
Unintended	B	D

Figure 1. Key elements of a conceptual model for studying FUD strategies. It distinguishes behavior from how behavior is experienced (affect) and what it leads to (effect).

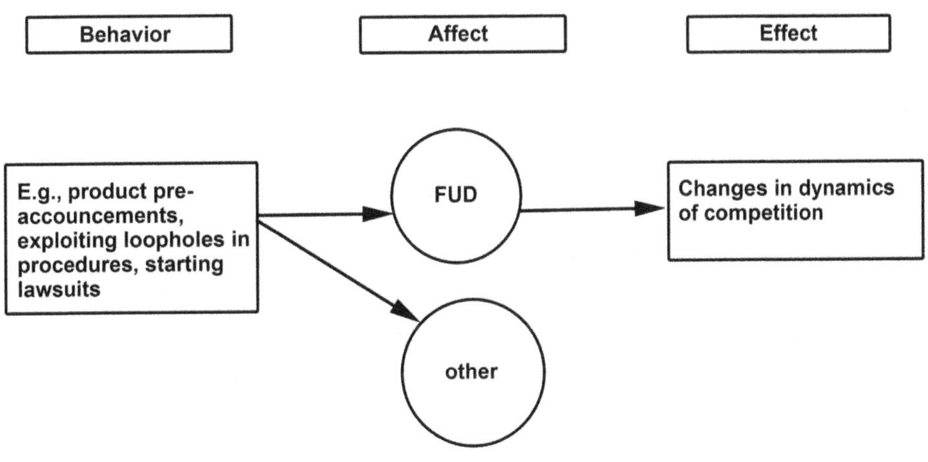

ments and interviews in which stakeholders themselves frame behavior as creating FUD. In our case, the archives almost exclusively made reference to the FUD strategies of one party - in the perception of the other party. Different causes may underlie this asymmetry. e.g.: methodological causes (inaccessibility of certain archives and interviewees) or case-related (e.g., asymmetry in *documented* FUD or different actor perceptions about what is a fair strategy). Whatever the cause(s), our aim is not to retrospectively choose sides in the standards war, but to contribute, empirically, to FUD studies by identifying (new) FUD strategies and, analytically, by exploring the conditions under which such behavior may emerge.

INTRODUCING THE CASE: TETRA VERSUS TETRAPOL

Although initially intended for a wider market (Bekkers, 2001), Tetra eventually became adopted as a standard for public safety networks in Europe under the influence of the Schengen Agreement. The police officials, who played a leading role among the potential users of Tetra, and the national governments of the - at that time five - Schengen countries (Belgium, the Netherlands, Luxemburg, Germany and France) saw it as an opportunity to meet a number of core objectives of the Schengen Agreement (signed June 14, 1985), a feeling that was shared more widely when the Schengen Agreement became officially adopted as EU policy in the Treaty of Amsterdam of 2 October 1997.

In the late 1980s, incompatible systems made direct connection between cross-border police units impossible. To overcome this, the Schengen Agreement outlined a short-term strategy (ad hoc improvements and bilateral agreements for police cooperation) and a long-term strategy aimed at developing a "uniform police radio communication system using common frequencies for exclusive use by the police" (#1). The latter required a common European communication standard like Tetra. The expectation was that "standardization would lead to a better quality of the cooperation between the emergency services, increased efficiency and increased effectiveness in case of large-scale emergencies." (#2)

The mobile digital radio system which ETSI started developing in 1988 was called MDTRS (Mobile Digital Trunking Radio System). In 1992, it was renamed Tetra (Trans European Trunked Radio, and later, when its possible use outside of Europe became clear, TErrestrial Trunked RAdio). In March 1991, the MDTRS specifications were prognozed to be ready at the end of 1992. The first tests were expected to be done in 1993/1994, and early 1995 the first systems would be operational.

The expectation of a smooth and quick standards process proved too optimistic. The first problems became apparent in 1993 and involved a competing technology that was later to be called Tetrapol. Tetrapol had been developed on request of the French government by the French company Matra Communications. Based on an FDMA access mode, this technology was the building block of the Rubis radio network for the French Gendarmerie (1992) and Acropol (1995) for the national police forces and fire brigades. The French put much effort into convincing other European countries of the advantages of the FDMA air interface technology underlying Tetrapol. But FDMA was incompatible with the TDMA air interface technology that was already being standardized in ETSI and was to be approved in 1994. Most TDMA proponents opposed standardizing two incompatible technologies. However, support for FDMA was growing, and in 1993 a number of operators and manufacturers (Bosch, Matra, Alcatel etc.) proposed the inclusion of the FDMA standardization in ETSI's work program. FDMA was argued to be more suitable for high coverage and low traffic density areas such as the countryside. TDMA was thought to better address high density communication and low coverage typical of urban areas. Given these complementary advantages, the FDMA proponents argued, it was appropriate to standardize both air interface technologies. Intense debate among suppliers and member-state representatives followed.

The beginning of the standards war can be pinpointed to early 1995, the period in which the ETSI committee decided to drop work on FDMA. (Hence forth the TDMA and FDMA technologies are referred to as a *pars pro toto* for Tetra and Tetrapol, respectively.) Companies such as Matra and Motorola played a pivotal role in the ensuing standards war. Matra left ETSI and decided to market its system under the name 'Tetrapol'. By then, it had more than two years of experience with the network of the French gendarmerie (Bekkers, 2001, p. 385) and had started rolling out Acropol for the French national police force and fire brigades. That is, it had a first-mover advantage in the market for public safety networks, which had the Schengen countries and the Police Cooperation as potential customers. In addition, the value of Matra's Intellectual Property Rights (IPRs) on the Tetrapol technology depended on the outcome of the standards war. The French government – representing a large and therefore influential member-state, geographically central to Europe and bordering with several other countries - promised to play the role of launching customer. Because it had high sunk costs and was locked into Matra's market. It had a direct stake in the standards war, even apart from having national industry interests like virtually all countries (e.g., Germany and the US).

Motorola's reasons to promote Tetra as a standard for public safety radio communication was that it wanted to re-use its technologies – and IPRs - in the development of Tetra. It had, for example, worked on a TDMA system that could serve in an adapted form as the technological base for the Tetra specification. The company already had an FDMA system in their portfolio (APCO 25), but it wanted to use this exclusively for the American market. Early on in the development of Tetra (1994), it publicly announced its intention to bring Tetra to the market. This raised the expectations of other manufacturers and end-users. It was the first company to invest heavily in Tetra and install Tetra test systems (on the Island of Jersey). In sum, both Matra and Motorola had a good starting position and ample reason to wage a war.

IDENTIFYING FUD IN THE CASE

In the following, occurrences in the Tetra-Tetrapol standards war are discussed that were perceived as creating FUD. The narratives are based on primary archival documents and, where possible, were confirmed by interviewees.

Confusion about Standards Names

In 1993, French and German technology suppliers (Bosch, Matra, Alcatel etc.), supported by the Schengen Telecom group, urged for a second, FDMA-based Tetra trajectory. The distinction Tetra 6 (FDMA) and Tetra 25 (TDMA) was introduced (i.e., based on the channel width of 6.25 kHz and 25 kHz, respectively). To outsiders, the names Tetra 6 and Tetra 25 suggested two standards options with little difference. Since both were called Tetra, the question of their (in)compatibility was not likely to be raised.

What happened before was that soon after the ETSI standards committee had renamed MDTRS to Tetra in 1992, Matra registered the name 'Tetrapol'. (Interview [5], former chair of ETSI Tetra committee). The name appears in the archives from 1993 onwards. As support for Tetra 6 within ETSI died down and was dropped in 1994 from its work program (#3), Matra increasingly used 'Tetrapol' to refer to the Rubis and Acropol networks.

Matra had chosen this name deliberately to put the FDMA-solution on the same plane as Tetra (Interview [1] Chairman of Tetrapol Forum). The similarity created much confusion. It positioned Tetrapol as a Tetra version developed for the police instead of a competing solution (Interview [2], Dutch representative in Schengen Telecom). Tetrapol made a 'claim by name' for a connection with Tetra (#4). By implication, ETSI, being the developer of Tetra, also sanctioned Tetrapol, which gave credibility to Matra's proprietary system. "This is deliberately misleading to users" (#5).

Over the years, the Schengen Telecom Group became an increasingly important user-player. Late 1993, it decided that an open ETSI standard would be needed to create a multi-vendor market. The Schengen Telecom group delegated, as it were, the technical decision making to ETSI. The term 'ETSI standard' became conditional for acquiring the group's support. Therefore, when ETSI decided not to pursue the Tetra 6 option, the Schengen Telecom group closed the door on Tetrapol (#6).

Confusion about the Status of Tetrapol

The Tetrapol proponents, not having achieved direct access to ETSI standardization, then explored indirect means of becoming ETSI 'accredited'. In December 1995, Tetrapol Forum expressed its wish to submit the Tetrapol technology to the so-called PAS (Publicly Available Specification) procedure in order to become a recognized ETSI standard. ETSI had introduced this procedure earlier that year "(...) to accelerate standards-making in Europe through [sic] the recognition of standards other than those produced formally within ETSI itself. (...)" (#7). Approval of Tetrapol via the PAS procedure would accredit it as an ETSI standard and provide it a status equal to that of the Tetra standard.

Tetrapol Forum submitted a formal request to ETSI on May 29 1996, and presented its specification on June 27 1996 (#8). Shortly afterwards, the ETSI Board decided that Tetrapol was eligible for the PAS procedure. Next in the PAS procedure, the ETSI committee was to evaluate if Tetrapol met the PAS criteria, one of which was that the PAS should not cover the same scope as an existing or proposed ETSI specification. (#30) An extra committee meeting was scheduled for 21-22 October 1996. But just prior to the meeting, Tetrapol Forum withdrew its request because the meeting had been announced too late and it had had too little opportunity to effectively lobby for votes (#21). Possibly it had guessed that behind the scene Tetra supporters, from government representatives to Tetra-manufacturers, had been lobbying successfully for a rejection of the Tetrapol PAS application. In a letter, Tetrapol Forum wrote:

The option of transferring the TETRAPOL P.A.S. to a Technical Specification might be more appropriate than the present proposal to transpose to an I-ETS. (...) In conclusion, we formally withdraw our request of 29 May 1996 for transposition to a P.A.S. (...) We further confirm our intention of restarting the procedure with (...) the new working rules for ETSI. (#9)

The letter created confusion and uncertainty. It unsettled ETSI members, first, because the loose use of the word 'P.A.S.' sharply contrasted with ETSI's specified use of the term 'PAS'[5] ("(...) any manufacturer or organization can publicly offer a specification and call it a PAS but this is nothing to do with ETSI." (#10)

Second, it raised the possibility of Tetrapol becoming a Technical Specification, which could create confusion and hurt the interests of TDMA proponents almost as much as a second Tetra-variant would.

With the current PAS-procedure the chances are almost zero that Tetrapol will become an EN (European Standard – the name for an open European Standard). It might sooner become a TR (Technical Report) (...). The status of a TR is low. But we are very likely to run the risk that those not well-acquainted with ETSI procedures (i.e. most people!) will not understand the difference between an EN and a TR. Matra may phrase it as that Tetrapol has been accepted by ETSI (#11)

Third, announcing that this PAS effort would not be the last one prolonged uncertainty about Tetrapol's status and Tetra's position in the market (#12).

Meanwhile, the European Commission was gradually changing its position. It started with Mr. Adoux from DG IV (Competition), noting that the Commission "(…) favor[s] competing standards but not conflicting standards. (…)." (#13) A few months later, the European Commission wanted "to speed up the standardization of telecommunications solutions which are already available on the market" (#14). This was an important impetus for Tetrapol Forum to re-start the Tetrapol PAS procedure on 21 January 1997. This time, before the voting on the Tetrapol PAS started on April 26 1999, a delegate of the European Commission informed the ETSI General Assembly that the Commission wanted to promote competition and that they should not block this (Interview [1]). Nevertheless, ETSI rejected the Tetrapol PAS, with only 37.5% of the votes in favor. The Presidency of the EU Council interpreted the outcome in a way that matched the Commission's views: "Until now, attempts to achieve a Common European Standard for radio communication have generally resulted in the adoption of two basic approaches, namely the TETRA *standard* and the TETRAPOL *technology solution*, both of which will now be used in future by the Member-States." (#15) The Commission hereby recognized Tetrapol as a 'de facto' standard.

Over the years, Tetrapol's status remained a source of confusion which, according to Tetra proponents, was damaging Tetra's market position:

Members of EP TETRA remain extremely concerned over the continuing confusion in the market place surrounding the status of Tetrapol in the context of an ETSI deliverable. You will be aware of the large investment in the European Standard, Tetra, already made by manufacturers and the TETRA procurement plans of major user organizations and network operators. The confusion is undermining confidence in TETRA and creating damaging uncertainty. (#16)

Brian Oliver, chairman of the ETSI Tetra Committee, noted that although it was clear that ETSI did not want a second standard, the Tetrapol proponents repeatedly tried to label their technology as such in the years 1994 up to 2003 (Interview [5]). A letter that was sent by Tetrapol Forum to Colonel Van Peer, chair of the Schengen Telecom group, dated 4 December 1995, further fuelled the confusion. Therein Tetrapol Forum called Tetrapol a Publicly Available Specification (PAS) and an open technology. "It respects the seven criteria fixed by the European Commission and approved by ETSI. The technical specifications are already available on request (…) The PAS documents and the report about compliance with the seven criteria are sent to the Director of ETSI to officially confirm that Tetrapol is already a PAS and can be transformed into an ETSI European stanard." The letter implied Commission support and suggested that Tetrapol was already halfway to being accepted by ETSI. Confused, the Schengen Telecom group contacted the European Commission. In response, Mr. Richter (European Commission) denied that ETSI had received the Tetrapol PAS and that an evaluation had already taken place. He noted that the Tetrapol PAS "is clearly in conflict with the criterion four, as its scope is the same as the draft ETSI TETRA ETS, being developed under mandate from the Commission" (…) and that " it is not at all guaranteed that the proposed PAS will be presented at the next [ETSI] Technical Assembly." (#17) He ended his letter sincerely regretting that Tetrapol Forum's letter did not correctly represent known public facts.

Discrediting ETSI

Uncertainty-enhancing were also Tetrapol Forum's references to ETSI's lack of integrity. ETSI's "anti-competitive practices (…) sought systematically to prevent TETRAPOL from gaining [European] recognition (…)" (#21). Certain members were colluding "to jeopardize the chances of the PAS application getting a fair treatment." (#22) More subtly, the chairman of Tetrapol Forum wrote the following in a letter to the ETSI secretary-general with a cc to the Commission:

(…) TETRAPOL Forum suggested that ETSI proceed with a thorough and objective examination of the complementary issues between the two technologies. In this respect, it is our view that the evaluation process cannot be based on anti-competitive criteria (such as a total ban on any overlap between a PAS application and an existing or in development ETSI standard) but must be market-oriented. In particular, the standardization process cannot be allowed to result in the unnecessary elimination of a widely used and complementary technology in this particular field of telecommunications whereas, like in other telecommunications fields, valid solutions can be found that permit the continued coexistence of partly overlapping standards. (…) I am sure you will understand that TETRAPOL Forum cannot – and will not accept being involved in a standardization process unless it is fair, objective and transparent and that no unnecessary restrictions are imposed. (#23)

By discrediting ETSI, doubts were cast on its decision not to support Tetrapol, on the quality of the Tetra standard, and on the demand for and market uptake of ETSI standards.

Law Suits

Law suits took place and legal threats were made. Countries like the Netherlands, Belgium and the UK explicitly favored Tetra over Tetrapol in their procurement policy. Matra viewed this as hindering free market competition. It objected to the way the UK Home Office had handled the public procurement process for its national Public Safety Radio Communications Project (later PSRC-S or "Airwave") for emergency services in 1998 (#24). The UK had issued an open tender but had not considered bids from non-Tetra companies, according to Gary Spicks, journalist for Communications Week International. The following quote from a letter of the UK Home Office to ETSI confirms this:

The UK Home Office is currently in a procurement process for the Public Safety Radio Communications Project (PSRCP) which will provide the national, next generation emergency service radio system for England, Wales and Scotland. PSRCP is a billion UK pound procurement carried out through rigorous application of EU Procurement Procedures. Accordingly, the procurement specifies the open European Standard that meets the requirements and that standard is TETRA. (…) The introduction of a competing ETSI standard would lead to market fragmentation, reduced manufacturing volumes, increased unit costs and reduced manufacturer viability, as well as loss of interoperability and reduced choice of terminal suppliers for users. (#25)

The Home Office chose the BT-led consortium Quadrant, which included Nokia Oyj and Motorola, "to carry out a pilot study, which if successful is likely to lead to the full contract worth 1.5- 2 billion (…) over 15 years" (#26). Matra felt that "Tetrapol manufacturers were excluded from this competition on the grounds that Tetrapol was not an ETSI standard" (#27) and started a legal case against the UK Home Office. However, Matra's case was rejected on the ground that Matra had been too late to complain (Interview [4], UK Home Office). Whether the lawsuits and legal references (#29) were called for or not, they made people wary and cautious.

Perceived FUD Behavior and the Market

The market for European public safety networks was slow to develop. A notable cause for delay was the scope creep and increasing complexity of standards committee work in ETSI. Over the years the targeted market shifted and new requirements (e.g. of Schengen Telecom) were added (Bekkers, 2001). In addition, the emergence of a rival incompatible technology also held back market development (Shapiro, 1999). Matra, the owner of the 'proven technology', was severely side-tracked by ETSI's Tetra initiative, while Matra's Tetra-6 and two Tetrapol-PAS initiatives heavily side-tracked ETSI's Tetra committee and delayed the standard's publication. The competing technologies created uncertainty among suppliers. The implication for roaming between networks was unclear, and the fear was that the market would become too small to compete (Bekkers, 2001, pp. 401-402). Many suppliers postponed investing in commercial Tetra equipment for fear of being stuck with the 'losing' technology and/or needing to support both Tetra and Tetrapol. For the same reason, also users postponed adopting one of the digital technologies.

Market uncertainty was further fuelled and prolonged by the use of FUD strategies in the settings of standardization and government politics. The following behavior was perceived as such (see Table 2):

- Procedural pre-announcement (e.g., announcement that a technology will enter a standards process which will heighten its status)
- Similar naming (e.g., renaming a technology in a way that creates a direct - assumed positive - association with another technology)
- Terminology reinterpretation (e.g., reinterpreting a term in a different – often unintended – way);
- Discrediting (e.g., raising doubts about a competitor's integrity)

Table 2. Perceived FUD strategies in the Tetra – Tetrapol standards war

Behavior	Example	Affect
Procedural Pre-announcement	Announcement of Tetrapol undergoing ETSI's PAS process	Creating Uncertainty
Similar Naming	• Tetra 6 and Tetra 25 • Tetra and Tetrapol	Creating Confusion
Terminology Re-interpretation	• PAS or P.A.S. • standard or technical specification	Creating Confusion
Discrediting	Discrediting ETSI's workings, standards and standards' uptake	Raising Doubt
Misrepresentation	Inaccurate account of events	Raising Doubt
Legal Action or Pre-announcement thereof	About breech of EU procedures	• Creating Fear • Creating Uncertainty

- Misrepresentation (e.g., inaccurate accounts)
- Legal action or pre-announcement thereof (e.g., about possible incorrect reference to standards in public procurement)

The delays in standardization caused by this behavior made it difficult to create momentum for the Tetra market. On ETSI's side, extra time and effort was needed to respond to disinformation and incorrect claims, and to restore trust in the Tetra standard, its standard process and its market expectations.

CONCLUSION

In this chapter a step has been taken towards a more systematic study of FUD strategies. A key element therein has been to redefine Pfaffenberger's notion of 'FUD as intended' as 'FUD as Affected' to emphasize the subjective, perceived nature of FUD. This conceptual shift has immediate bearing on our answers to the three research questions and our recommendations to corporate standardization managers and Standards Development Organizations (SDOs).

Revisiting the Research Questions

Regarding the way (perceived) *FUD strategies may manifest themselves* in committee standards wars, FUD behavior observed in de facto standards wars also occurs in committee standards wars. Compare, for example, product pre-announcements in de facto standards wars with procedural pre-announcements in the case study, a FUD strategy specific for committee standards wars. Other FUD strategies that emerge from the study are similar naming, terminology reinterpretation, discrediting, misrepresentation and legal action (pre-announced or effectuated). The FUD strategies we have observed are case-bound and may not be evident in other committee standards wars. But the case is rich in examples, and, while new case studies are likely to extend our list of (perceived) FUD strategies, this initial list serves to help corporate standardization managers to recognize them.

In the case study, behavior associated with these FUD strategies led to confusion, uncertainty, fear and doubt. Whether the FUD that was created ('FUD as affected') was also intended, was sometimes confirmed by the stakeholders themselves. In other instances, behavior perceived as such may not have been intended. That is, unfairness is attributed to 'unintended FUD'. One may ask oneself if it is at all relevant to correctly attribute intentionality to behavior resulting in FUD since, one way or the other, the outcome is the same. Our expectation is that in most situations it is relevant. The case illustrates how unsettling FUD strategies can be if perceived as intentional by the receiving end. The documents and interviews testify that ETSI standardizers often felt cornered and forced into a reaction. Perceived intentionality appears to entrench positions, reduce willingness to take the opponent's perspective, narrow down perceived options for counter-strategies, hinder the identification of common interests, and cloud views on partial, intermediate and/or long-term compromises. In short, company standardizers run the risk of retaliating with defensive counter-strategies that are ineffective and escalate a standards war. That is, incorrect attribution of intentionality can be self-defeating.

In order to react in a more effective and measured manner, corporate standardization managers benefit from realizing that what may seem as a predatory strategy may be seen by others as a necessary survival strategy in a highly competitive market. To recall an earlier quote, "When it comes to standards wars, traditional principles of strategy, while helpful, are not enough" (Shapiro, 1999) A better understanding of context in which (perceived) FUD strategies emerge reduces the chance of attributing unfairness to 'unintended FUD'.

Regarding the second research question, the *context in which FUD strategies may emerge,* FUD strategies find a fertile ground in standards wars. To recapitulate, standards wars involve incompatible rival technologies (i.e., which cannot exploit each other's externalities) in highly competitive markets (i.e., high stakes). In our case study, it started out as a war between a company with an existing market and an implemented technology, on the one hand, and a group of stakeholders that wanted to enter a new market and standardize a new rival technology, on the other. The former company feared being ousted out of the market and was perceived to use unfair FUD strategies.

Initially, both warring parties had IPRs on competing technologies and therefore incentives to exploit them. The targeted market was relatively specialized but profitable, which intensified competition. One party had a clear first-mover advantage (i.e., expertise and implementation experience) and high vested interests. Several large companies were involved that also produced for other markets (i.e. they ran businesses that were partly independent of the area standardized) *and* could survive a protracted standards war. Influential customer-users were involved, national government agencies, with high stakes in standardization (e.g. high sunk costs and crucial interests in public safety) and some with clear links to their national industry (i.e., standardizing companies).

The literature on standards wars and the case study suggest that the risk of resorting to FUD strategies will always be high in highly competitive markets and standards wars. It escalates where a company's survival is at stake. Comparative research is needed to determine whether Standards wars that show similar characteristics also show similar predatory behavior.

With respect to the third research question, *how do FUD strategies affect the dynamics of technology wars in committee standardization?* the case illustrates that perceived FUD increases the number and accelerate the succession of conflicts. It polarizes and entrenches positions. It leads to competing standards and technology solutions, and to market fragmentation.

To deepen the discussion, the distinction between using FUD strategies in de facto and committee standards wars is likely to be relevant here. In the light of game theory (Axelrod, 1984), committee standardization can be viewed as a game with multiple rounds. The (formal) aim of the game is ongoing market coordination among stakeholders. Once a standard has been finalized, stakeholders are likely to meet again at a later stage to revise the standard or develop standards in adjoining areas. Using FUD strategies intentionally will Affect a company's reputation and influence future standardization negotiations. This issue may be less acute in de facto standards wars, where coordination is not at stake and the game may end after one round. In these situations, intentional FUD and other short-term predatory strategies may go unpunished, but such a 'scorched earth' strategy most likely precludes future collaboration.

Recommendations

Why is it important for corporate standardization managers to better understand, recognize and address (perceived) FUD strategies? Because they typically escalate and prolong standards wars. This, in turn, increases market uncertainty and possibly leads to competing standards. Corporate standardization managers will usually want to avoid these consequences. In economic terms, uncertainty about the outcome of a standards war means that producers will try to postpone investments for fear of investing in a losing system and having to write off sunk costs (i.e., costs that are specific and irreversible and therefore cannot be retrieved). For the same reasons consumers will postpone their purchases. The market will stagnate. Moreover, there is a high risk of ending up with functionally equivalent, competing standards that fragment the market. Smaller, fragmented markets are more risky for producers to enter and less worthwhile (decreased economies of scale). Competing standards also create difficulties for consumers, here: the emergency services (e.g. reduced market transparency, lack of interoperability and high transaction costs such as switching costs). (Egyedi, 2014)

Corporate standardization managers and user agencies, who are or may be on the receiving end of FUD strategies, should therefore be aware of their interest in gaining better insight in what they entail and how they work. Insight in the context in which (intentional) FUD strategies are more likely to emerge, may aid in anticipating and precluding their emergence. early analysis and warning allows stakeholders to explore solutions that transgress the immanent committee standards war. Is there a possibility for adequately addressing the vested interests of a company whose survival may be at stake? Can the company be compensated by means of intellectual property or other commercial assets? Or by government contracts in other areas? Is it feasible to outline long-term technology convergence and a standards trajectory? Could it be that the life-cycle of the company's and the competitor's products are so short that they will soon be by-passed by a new technology anyway?

Once FUD strategies are perceived to be used, counter-strategies will come into play. As noted, those on the receiving end of FUD need to be cautious about too easily attributing intentionality and remain aware of the existence of different perspectives and interests. 'FUD as perceived' may –but need not - equal 'FUD as intended'. FUD entrenches and polarizes positions. It deepens conflicts. Once this happens, both the contemplated range of solutions and the room for negotiation tend to narrow down to short-term defensive company interests, if our case is anything to go by. Whether there is leeway for more, needs more elaborate research.

The case study illustrates that effective counter-strategies against FUD created by, for example, similar naming, misrepresentation and legal threats are difficult to devise. This picture also emerges from descriptions of de facto standards wars. In committee standards wars, one may presume that a FUD strategy like discrediting an SDO is less likely if SDOs have a strong reputation. Transparent standards procedures and clear terminology may partly buttress an SDO against targeted misuse ('terminology reinterpretation'). To our knowledge, no studies exist that address FUD counter-strategies in a systematic way, let alone strategies that down-scale standards wars and thus reduce their impact on technology development and markets. In this area more research is direly needed.

ACKNOWLEDGMENT

This research was funded by the Dutch research council (NWO, project number 458-06-027). We sincerely thank the interviewees for sharing their ideas and sources with us and the holders of archives for providing us access.

REFERENCES

Axelrod, R. (1984). *The evolution of Cooperation*. Basic Books.

Bayus, B. L., Jain, S., & Rao, A. G. (2001). Truth or Consequences: An Analysis of Vaporware and New Product Announcements. *JMR, Journal of Marketing Research*, *38*(1), 3–13. doi:10.1509/jmkr.38.1.3.18834

Bekkers, R. (2001). *Mobile Telecommunication Standards. GSM, UMTS, Tetra and Ermes*. Boston: Artech House.

Besen, S. M., & Farrell, J. (1991). The role of the ITU in standardization: Pre-eminence, impotence or rubber stamp? *Telecommunications Policy*, *15*(4), 311–321. doi:10.1016/0308-5961(91)90053-E

David, P. A. (1987). Some New Standards for the Economics of Standardization in the Information Age. In P. Dasgupta & P. Stoneman (Eds.), *Economic Policy and Technological Performance* (pp. 206–239). Cambridge University Press. doi:10.1017/CBO9780511559938.010

Dranove, D., & Gandal, N. (2003). The DVD vs. DIVX Standard War: Empirical Evidence of Vaporware. *Journal of Economics & Management Strategy*, *12*, 363–386. doi:10.1162/105864003322309518

Egyedi, T. (2014). The Impact of Competing Standards: On Innovation and Interoperability for E-government. *PIK - Praxis der Informationsverarbeitung und Kommunikation, 37*(3), 211-215.

Egyedi, T. M. (2001). Why JavaTM was -not- standardized twice. *Computer Standards & Interfaces*, *23*(4), 253–265. doi:10.1016/S0920-5489(01)00078-2

Egyedi, T. M., & Hawkins, R. (2010). Standards Consortia. In W. Hesser, A. J. Feilzer, & H. J. de Vries (Eds.), *Standardisation in Companies and Markets* (3rd ed.; pp. 443–490). Hamburg, Germany: Helmut Schmidt University.

Farrell, J., & Saloner, G. (1985). Standardization, Compatibility and Innovation. *The Rand Journal of Economics*, *16*(1), 70–83. doi:10.2307/2555589

Farrell, J., & Saloner, G. (1986). Installed base and compatibility: Innovation, Product Preannouncements, and predation. *The American Economic Review*, *76*, 943–954.

Gallagher, S. R. (2012). The battle of the blue laser DVDs: The significance of corporate strategy in standards battles. *Technovation*, *32*(2), 90–98. doi:10.1016/j.technovation.2011.10.004

Gauch, S. (2008). + vs. –, Dynamics and effects of competing standards of recordable DVD media. In T. M. Egyedi & K. Blind (Eds.), The dynamics of standards (pp. 47-67). Cheltenham, UK: Edward Elgar.

Haan, M. A. (2003). Vaporware as a Means of Entry Deterrence. *The Journal of Industrial Economics*, *3*(3), 345–358. doi:10.1111/1467-6451.00204

Jakobs, K. (2008). The IEEE 802.11 WLAN installation at RWTH Aachen University: A case of voluntary vendor lock-. In T. M. Egyedi & K. Blind (Eds.), *The dynamics of standards* (pp. 99–116). Cheltenham, UK: Edward Elgar.

Lee, H., & Oh, S. (2006). A standards war waged by a developing country: Understanding international standard setting from the actor-network perspective. *The Journal of Strategic Information Systems*, *15*(3), 177–195. doi:10.1016/j.jsis.2005.10.002

Oshri, I., & Weeber, C. (2006). Cooperation and Competition Standards-Setting Activities in the Digitization Era: The Case of Wireless Information Devices. *Technology Analysis and Strategic Management*, *18*(2), 265–283. doi:10.1080/09537320600624196

Pfaffenberger, B. (2000). The Rhetoric of Dread: Fear, Uncertainty, and Doubt (FUD) in Information Technology Marketing. *Knowledge, Technology & Policy*, *13*(3), 78–92. doi:10.1007/s12130-000-1022-x

RaymondE. (2009). http://www.catb.org/~esr/jargon/html/F/FUD.html

Schengen. (1990). The Schengen acquis - Convention implementing the Schengen Agreement of 14 June 1985 between the Governments of the States of the Benelux Economic Union, the Federal Republic of Germany and the French Republic on the gradual abolition of checks at their common borders. *Official Journal L 239*, 22/09/2000 P. 0019 – 0062.

Shapiro, C., & Varian, H. R. (1999). *Information rules - A Strategic Guide to the Network Economy*. Boston: Harvard Business School Press.

Singh, R. M., & Dahlin, K. B. (2009). *The winner does not take it all: Survival of niche standards*. Paper presented at the EGOS 2009 "The Social Dynamics of Standardization", Barcelona.

Vries, H. J. (2006). Competing E-Purse Systems: A Standards Battle. *Journal of Cases on Information Technology*, *8*(1), 1–15. doi:10.4018/jcit.2006010101

Wegberg, M. (2004). Standardization Process of Systems Technologies: Creating a Balance between Competition and Cooperation. *Technology Analysis and Strategic Management*, *16*(4), 457–478. doi:10.1080/0953732042000295784

West, J. (2003). How open is open enough? Melding proprietary and open source platform strategies. *Research Policy*, *32*(7), 1259–1285. doi:10.1016/S0048-7333(03)00052-0

Williamson, O. (1979). Transactions-Cost Economics: The Governance of Contractual Relations. *The Journal of Law & Economics*, *22*(2), 233–262. doi:10.1086/466942

Wu, Y., Balasubramanian, S., & Mahajan, V. (2004). When is a Preannounced New Product Likely to be Delayed? *Journal of Marketing*, *68*(2), 101–113. doi:10.1509/jmkg.68.2.101.27792

Yin, R. K. (2009). *Case Study Research: Design and Methods* (4th ed.). London: Sage.

KEY TERMS AND DEFINITIONS

Affect: A psychological term for feeling, experience or emotion.

Committee Standardization: The process of developing standards in technical committees of formal standards bodies, consortia, professional organizations, (inter)governmental agencies, etc.

Committee Standards: Standards that have been created and documented for common and repeated use.

De Facto Standards: Popular products, technologies, etc. They dominate the market. Similar to popular committee standards (e.g. A4 paper format ISO 216 or Wifi IEEE 802.11g), de facto standards function as points of reference for consumers and producers; but, unlike committee standards, they were not created for that purpose.

FUD: Fear, Uncertainty and Doubt. Most often used in the context of someone 'creating FUD'.

FUD Strategy in Standardization: A market strategy *perceived* to be designed to create unease among competitors and raise concern among users about the consequences of selecting a competitor's standard, by exerting pressure - psychological, legal or otherwise – and disseminating negative and/or incorrect information about the standard, the standards process or the standard's likely market uptake, as a result of which the credibility of the competitor and/or the competing standard is undermined.

PAS: Publicly Available Specification, a term used by standards bodies for a specific, accelerated standards 'development' trajectory. The procedure allows market-relevant and mature specifications, which were developed outside of the standards body, to become a formally recognized standard.[6]

Predatory Strategies: Aggressive market strategies intended to undermine a competitor's position.

Product Preannouncements: A formal and deliberate communication that releases information about a product well in advance of the product's actual introduction (Wu, 2004).

Standards War: A situation in which two competing parties with incompatible standards (i.e. committee or de facto standards) and very high stakes fight for a dominant market share.

ENDNOTES

1 Archival sources are indicated by hashtag numbers: #1 etc. They are listed at the end of the chapter.

2 I.e., archive of the Schengen Telecom group (personal archive of H. Borgonjen), archives of the Dutch Ministry of the Interior on C2000, and the ETSI archive. They include policy reports, correspondence and minutes of meetings.

3 Interviewees are indicated by straight brackets and numbered: [1] etc. They are listed at the end of the chapter.

4 NB: Research question (iii) about the conditions under which FUD strategies arise, only addresses situations in which FUD-raising behavior was intended and perceived.

5 Two days later a letter was sent by the chairman of Tetrapol to the ETSI Technical Assembly noting that "TETRAPOL remains a Publicly Available Specification; the last sections will be made public this coming December (…) (#31)

6 For ETSI, see http://www.etsi.org/about/what-we-do/publicly-available-specifications-pas

APPENDIX

Interviewees

[1] Chairman Tetrapol Forum. (1991-present), Paris:17-4-2008.

[2] Former Head of the R&D Department of the Dutch Police, Former chairman of Schengen Telecom, former member of ETSI RES-6 representing the Netherlands, Odijk: 31-5-2006, 29-1-2009.

[3] Motorola Marketing department. Chairman of ETSI TC Tetra since 2002, Exeter: 26-11-2007.

[4] UK Home Office, Police ICT department, Chelmsford: 27-11-2007.

[5] Chairman ETSI Tetra committees between 1991 and 2002, Chelmsford: 27-11-2007.

Archival Sources

1. Minutes of a meeting to improve cross-border communication of the police, 11 - 12 July 1989, Bonn-Bad Godesberg, p. 3.

2. Final Report initial phase C2000 (the Dutch Tetra implementation).

3. Letter from ETSI RES to Schengen Telcom Group, 15 November 1994.

4. Letter from A.N. Kent, UK Home Office to K.H. Rosenbrock, ETSI DG, 26-08-1997.

5. Letter O. Lauridsen, September 1995, TETRA MoU Chairman, to M. Bangemann, European Commission.

6. Minutes meeting Schengen Telecom, 9 May 1994, Brussels.

7. Letter from D. Kynaston, Philips, to M. Bangemann, European Commission, 20 September 1995.

8. Note from the French Delegation to WG Telecommunications, 16 October 1996

9. Letter from the chair and vice-chair of Tetrapol Forum to the ETSI DG, K.H. Rosenbrock, 9-10-1996.

10. Committee document, Motorola, J. Jensen, 14 November 1995.

11. International consultation, Police, NL, 24-11-1997, p.2 [translated from Dutch].

12. Letter from H. Borgonjen, 14-10-1996.

13. Minutes of 23/24 October meeting from J. Jensen (Motorola, MoU) to H. Borgonjen and others, 25-10-1996, p. 4.

14. Letter from H. Azémard (Tetrapol Forum) to K-H. Rosenbrock (ETSI), January 24, 1997.

15. Note from Presidency [Council of the European Union] to Police Cooperation working group, 16 November 1999 (italics added)).

16. "TETRA- the one and only European standard", press release Nokia/Tetra, 16-9-1997.

17. Letter from Mr. Richter, European Commission, to Schengen Telecom group, 15-12-1995.

21. Letter from H. Azémard, Chairman Tetrapol Forum to F. Chumbinho, chairman of the Schengen Telecom Group, 2-4-1997.

22. Letter from H. Azémard, Chairman Tetrapol Forum to K.H. Rosenbrock (ETSI), 08-09-1997.

23. Letter from H. Azémard, Chairman Tetrapol Forum to K.H. Rosenbrock, Secretary-General of ETSI), 08-09-1997.

24. TETRA MoU document, Why is the Tetrapol PAS application contentious?, December 1998.

25. Letter from the UK Home Office to K.H. Rosenbrock, ETSI DG, 26-08-1997.

26. Letter from the UK Home Office to K.H. Rosenbrock, ETSI DG, 26-08-1997.

#27. Cogent Defence System's Response to The Radio Spectrum Review Consultation Paper, 2001, p. 6.
#29. Letter from the chair and vice-chair of Tetrapol Forum to the ETSI DG, K.H. Rosenbrock, 9-10-1996.
#30. Letter from M. Bangemann, member of the EC, to H. Dijkstal, Dutch Minister of the Interior, 29 April 1998.
#31. Letter from H. Azémard, chairman of Tetrapol to the ETSI Technical Assembly, 11-10-1996, ETSI/TA25(96)TD24.

Chapter 16
Human Action in the Loop: Ethical Considerations and Standardization

Hartwig Steusloff
FhG IOSB, DKE, Germany

Michael Decker
KIT-ITAS, Germany

ABSTRACT

Extremely complex systems like the smart grid or autonomous cars need to meet society's high expectations regarding their safe operation. The human designer and operator becomes a "system component" as soon as responsible decision making is needed. Tacit knowledge and other human properties are of crucial relevance for situation-dependent decisions. The uniform modeling of technical systems and humans will benefit from ethical reflection. In this chapter, we describe human action with technical means and ask, on the one hand, for a comprehensive multidisciplinary technology assessment in order to produce supporting knowledge and methods for technical and societal decision making. On the other hand—and here is the focus—we propose a system life cycle approach which integrates the human in the loop and argue that it can be worthwhile to describe humans in a technical way in order to implement human decision making by means of the use case method. Ethical reflection and even ethically based technical decision making can support the effective control of convergent technology systems.

HUMANS, TECHNOLOGY AND STANDARDIZATION

Humans act to reach their targets. They are able to justify these targets and they are able to use means, often technological means, to reach them. In this scenario, the human as the actor uses and controls the technical means and bears responsibility for the consequences of the action. This idea of technological progress was a kind of unwritten rule because there was societal agreement that the development of technological means for human action is a substantial pillar of human well-being. However, this changed in the second half of the last century. Major accidents (such as Chernobyl or Fukushima), environmental

DOI: 10.4018/978-1-4666-9737-9.ch016

consequences (e.g. of Bhopal), consequences for human health (e.g. of asbestos), cultural consequences (e.g. of pre-implantation technology) were described as side-effects of technical developments that seriously questioned the benefit of the technology itself (Grunwald, 2002).

The complexity of large technical systems and their implementation as discrete manufacturing factories or continuous/batch processing plants, including the interface systems supporting the interaction with technical infrastructures, transportation, energy supply, etc.), pose particular challenges for research in technology assessment. Obviously, this research is increasingly focused on non-technical and socially relevant aspects such as economics, legislation, social issues, and ethics.

Considering standardization, we find a similar situation: For decades and from the very beginning, standardization was, and up to a large extent still is, dealing with well-delimited fields of *technology* and the related product-oriented technical issues. As a result of the extreme multiplicity and variety of such fields and issues, a world-wide task force of millions of technical experts is co-operating in national and international technical committees (TCs), creating and maintaining the currently available set of international and national standards and technical rules. Representing the current state of the art in specific fields of technology, such standards, however, must take the form of multidisciplinary documents as soon as we start dealing with technical *systems* and their convergence of technologies which nowadays are becoming crucial for the implementation of complex and "smart" systems such as smart grids or electro-mobility.

With the standard ISO 26000 on Corporate Social Responsibility (2010), the earlier concentration on the standardization of technologies and materials started to become obsolete. A proof of this trend can also be observed in the foundation of CEN/TC 389 Innovation Management in 2008. Thus, human behavior issues and the support of (non-technical) management issues through international standards will pave the way for consideration of ethics in and for standardization.

The issue of convergent technologies is internationally understood to be increasingly important for the standardization of systems, of system components, and of interoperability. As an example we would like to shortly mention the German standardization strategy from the year 2003. As another representative example of such highly complex systems we refer to "e-mobility." The convergence of well-established and presently separated technologies such as the automotive technologies (DKE, 2012) and the "smart grid" technologies (DKE, 2011) must be understood as a socio-technical innovation. Technology Convergence results in a number of societal consequences such as:

- necessary adoption of road traffic codes
- legislation and regulation
- extension of safety-related specialist training of workshop staff

Thus, (electro-)mobility is a perfect field for considering technical aspects combined with societal perspectives, such as ethical, social, legal and/or economic issues relevant to standardization.

Multidisciplinarity is also required in the personal profiles of operators and decision makers during the design and the operational phases of complex systems. We need technology experts with an explicit readiness to deal with complexity and to take responsibility in unexpected operational situations of complex systems in real-time. In chemical production facilities, highly critical situations may arise from unexpected catalytic reactions. In such dangerous system states, the process operators may find themselves confronted with an unprecedented responsibility for making decisions at the right time. The operator's explicit and, perhaps much more importantly, his/her tacit knowledge becomes the "ultimate resort" for tackling such situations (cf. Figure 1). The effective application of human experts' broad span of multidisciplinary explicit and tacit knowledge "in the loop" in combination with his/her "unique properties" (cf. sect. 3) has often proved to be decisive in tackling uncontrollable and catastrophic accidents. In the majority of *real life scenarios* (later on referred to as "use cases" (cf. sect. 4), we have so far relied on automation subsystems dealing with foreseeable exceptions. With complex systems, however, we can never be sure that we have anticipated, during the system design phase (cf. Figure 2), all possible system states and their related scenarios of system behavior. Consequently, we must include the human expert with all his/her technical knowledge and experience as a final resort ("ultima ratio") for tackling such unexpected situations.

However, up to now human workers or operators are well separated from machines, robots or any other potentially dangerous objects in their environments. The new Machinery Directive (006/42/EG), issued by the European Commission in 2006, is mandatory within all EU nations. Industrial production enterprises involved in the project "INDUSTRY 4.0", however, have committed to the concept of physical cooperation between mobile production equipment and human workforce. The new EU machinery directive requires effective methods and provisions for establishing and maintaining the required safety levels, including the physical human-machine/-robot collaboration.

In this context, a careful and comprehensive risk analysis is a key provision for establishing a safe and productive environment within any enterprise. If we accept that there is no "zero risk" technical task,

Figure 1. From data to action—Closed Loop (Real Time) System Modeling concept

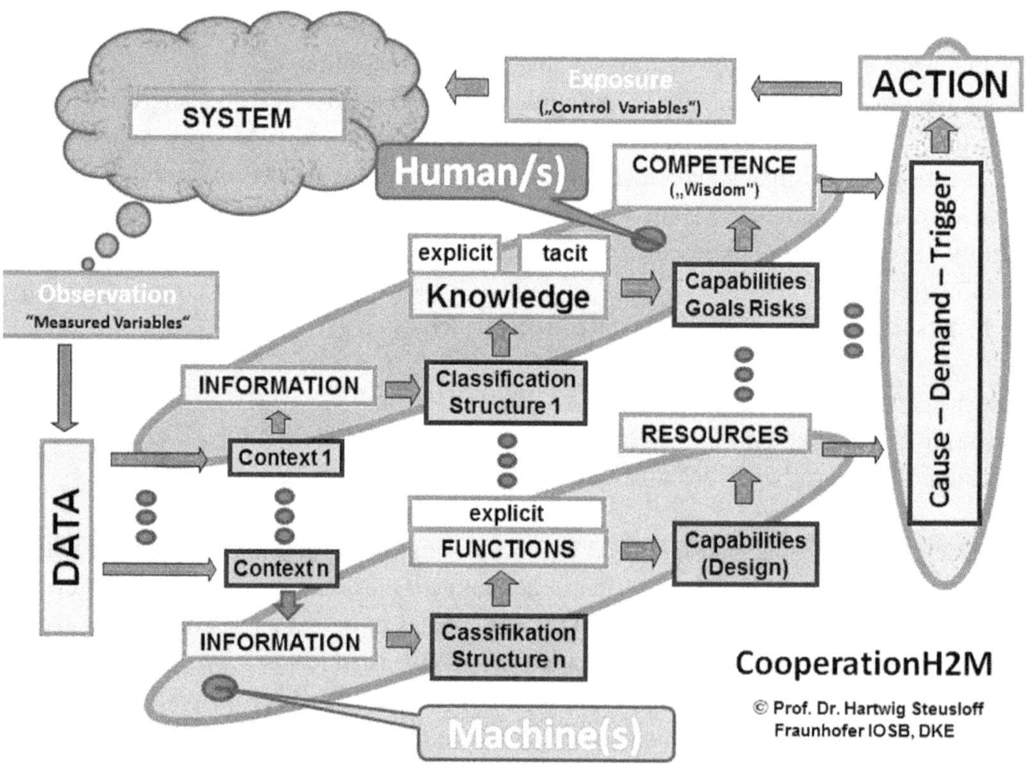

Figure 2. Human experts supporting the system life cycle

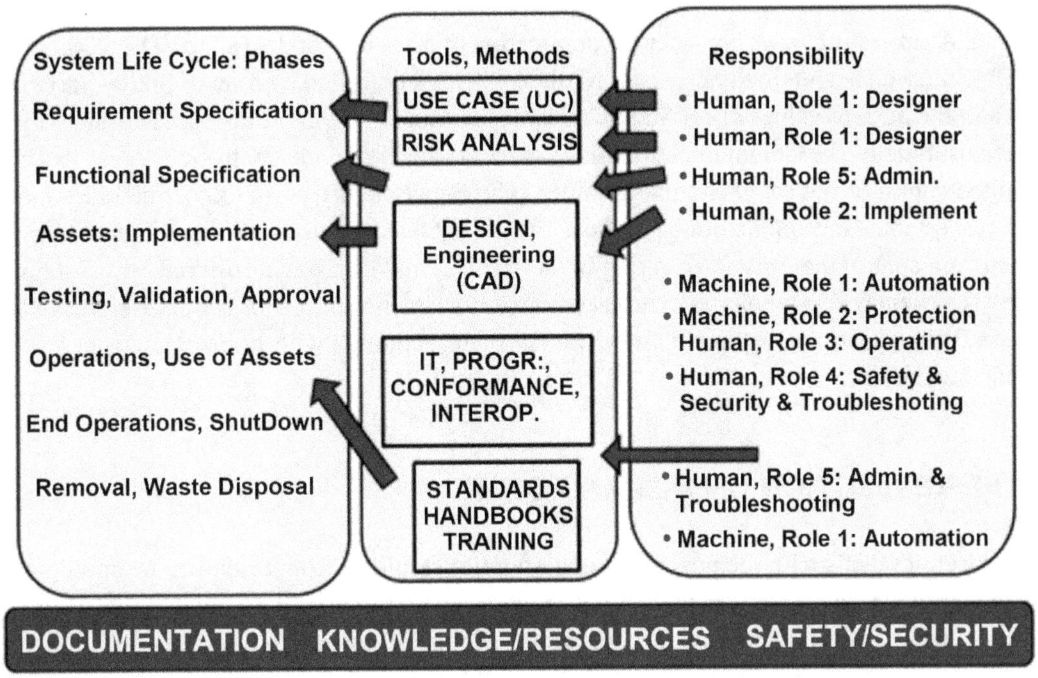

then the burden of technical risk to a worker is a first task to be tackled by ethical reflection already in the planning of a production activity, facility or service.

Currently most prominently discussed is human-machine interaction along the example of autonomous cars (Maurer et al., 2015). Here, it is discussed that the human "driver" takes over the steering wheel—and with that the responsibility—in situations the autonomous system is not able to deal with. If we assume that the technical systems are well designed, this does not happen very often. Nevertheless, the driver still needs to be concentrated all the time because the request of taking back control, i.e., responsibility, may happen at any time. This is not a very attractive division of labor: staying focused while doing nothing. Moreover, we can assume—if the systems are well designed—that the situation in which the technical system hands over to the human driver may also be difficult to deal with for humans. So it will take some time for the driver to orient and prepare in order to take the appropriate decision.

Let us mention an example from batch processing of the chemical industry. Critical situations may occur through unexpected catalytic reactions in reactors after switching to another product or recipe while using the same apparatus as for the last batch with different chemical reaction partners. Even if this apparatus is carefully cleaned, there remains a risk of unknown or unexpected micro-reactions which could end up in a situation not anticipated in the development of the new product and, for this reason, the automation functions may not be able to handle the situation. Again, the human expert will be the ultimate resource to avoid catastrophic consequences for life and environment.

At the end of the day, we want to employ the human expert as a system component, equipped with "unique" properties enabling him/her to turn unexpected, maybe safety-relevant system behavior into safe or at least controllable system states. Employing standardized system components during system design means establishing a unique and comprehensive modeling concept for each and any particular system component. As an example, we consider the application of an object model (VDI/VDE, 2003), based on sets of Objects, which contain Functions or Methods, being controlled by a set of Status Variables. The number and the current values of all Status Variables within a system represent its complexity and its overall status. Finally, the Communication Links of each Object provide controlled access to a confined or unlimited subset of the set of Status Variables. The Functionality of an Object includes messaging functions for establishing application-specific networks of Objects. Such sets of communicating Objects implement and control the Actions resulting in the services and/or products offered by a system. Figure 1 sketches a system with a human as system component and main source for action. Before we describe (in section 3) humans in a "technical" way, we start (in section 2) with human acting in general and with ethical reflection.

ETHICAL REFLECTION ON HUMAN ACTING

Technology ethics deals with the normative uncertainties resulting from scientific-technological progress. Ethics provides so-called orienting knowledge in order to support responsible decision making (Grunwald, 2013). In pluralistic societies, we are (or *should* be) prepared to tolerate different opinions, beliefs, etc. With respect to human actions this means that we define a broader spectrum of actions as being socially acceptable as we did a hundred years ago. This means, we accept a broader spectrum of justifications for a specific action. If we take individual human actions as the atomic particles of societal actions, it is a crucial element of the functioning of a society as a whole that most humans in the society mostly behave within this spectrum of societally acceptable actions. In our everyday lives, this usually works without any major conflicts. However, such conflicts nevertheless occur and then there are (again but on a different level) accepted societal rules of how to deal with these conflicts. Usually, this arguing process starts with a detailed justification of why a specific action is assumed to be adequate. Then an opponent needs to justify why he or she sees this action as inappropriate. If we take the normative parts of these justifications as moral justifications to be put forward by individuals or groups of society, then also these normative statements can get into conflict with each other.[1] Professional ethics can be described as the reflection on moral statements (Gethmann and Sander, 1999) by comparing them with respect to different approaches of ethical concepts such as, for example, deontological or teleological ethics, utilitarian or consequential ethics (Grunwald, 2013). Based on this reflection process, the different moral positions can be discussed and conclusions about the different and finally preferred options for action can be drawn.

As ethical reflection was introduced as reflection on moral justifications of human actions, it can also be applied to standardization processes as these are human actions as well. It is important to distinguish and understand the process of generating standards, the conditions of applying such standards, and the effect of including such standards in the specification of products and services, in legislation and regulation, and in daily actions of human individuals. Even if the use of standards is generally voluntary, the refusal to apply them could result in a loss of market success. Vice versa, the voluntariness of using standards is reduced as soon as, e.g., safety standards are referenced in legislation. And the expert "democratic" standards development process, although providing the basis for regarding the content of standards to represent the concurrent state of the art, may be influenced by special interests of stakeholders. Therefore, we can assume that with a new standard we always produce "winners and losers" in society with respect to, e.g., economic, environmental or risk-related advantages and disadvantages, and thus produce conflicts which need ethical reflection. The development of appropriate processes to include ethical reflection could turn out to be a valuable support for understanding and mediating such problems of standardization.

However, in this chapter we do not want to focus on ethical reflection in standardization processes on a general level, but on ethical reflection to be included in standardization itself. Operating and controlling complex technological systems is becoming a vital part of our societal well-being. Smart systems are taking over some of the "actions" formerly done by humans, and cooperation scenarios between humans and technological systems are arising. What are the differences between human acting and machine action?

It is a constitutive element of human beings to act autonomously and to be able to justify the respective action in a so-called "space of reasons," a notion introduced by Wilfrid Sellars (1997).[2] This means that humans make experiences and perceive their natural and cultural environment in a specific way, but this "empirical" awareness also needs an adequate syntactical form in the space of reasons: "The essential point is that in characterizing an episode or a state as that of knowing, we are not giving an empirical description of that episode or state; we are placing it in the logical space of reasons, of justifying and being able to justify what one says" (Sellars, 1956, p. 76). In the space of reasons, the classical epistemic, moral and aesthetic meaning of human capabilities and properties develops, which allows a "dense description" of a human being or a person referring to notions such as self-consciousness, experience, expression, planning of life, education, cultural practice, etc., to name just a few.

Obviously, this human way of autonomous acting in the space of reasons includes the idea that humans are able to spontaneously change their attitude, and with that, their action within a certain context of actions. This includes also the idea and consideration of second level desires and interests resulting in changes in the aims of action. This kind of higher level abstraction is of outmost relevance in the interaction between humans and machines. In order to avoid confusion: The "autonomous acting" of a machine, as it is sometimes called, takes place on a different, rule-based level, even if some kind of "decision making" is implemented in the system. Therefore, from an ethical point of view, the human part of the human-machine interface is in the focus. The following recommendation developed in an interdisciplinary technology assessment (Christaller et al., 2001, p. 220, translation by the authors) refers to the human-machine interaction starting from an ethical perspective, namely Kant's formula of humanity:

In contexts of robot application we should stick to the competence of setting goals of humans to avoid instrumentalization.

Thus, the technical realization of the competence to decide is a crucial aspect in shaping the human-machine interface. In order to enable humans to take over responsibility for the well-functioning of robots, the robots need to be controllable in the sense of transparency, predictability, and influenceability.

We recommend for all cases in which decision making is delegated to the robot that the people concerned are well informed about that and give explicit or implicit notice of acceptance in advance. Especially in the context of medical application and care taking, the refusal of acceptance should be taken as a veto.

This recommendation gives normative guidance for technology developers and, therefore, also for standardization. The technical systems should become transparent and controllable by the human actors in order to keep them responsible. Thereby, this recommendation is in line with the idea of responsible research and innovation, a core idea of the EU's "Horizon 2020" research and innovation program: "[..] an inclusive approach to research and innovation (RI), to ensure that societal actors work together during the whole research and innovation process. It aims to better align both the process and outcomes of R&I with the values, needs and expectations of European society."[3]

The concept of responsible research and innovation is closely related to technology assessment, applied ethics, and STS research (Grunwald, 2011). A set of methods and requirements has been specified also by the engineers (VDI, 1991) for the ethical responsibilities of engineers (Hubig and Reidel, 2003), and specifically for human-to-human interoperability as a DIN PAS (Frey, 1995) for teaching and performing verbal human-to-human interaction and for an objective measurement of the quality of such interaction. These assessments ensure that technical systems serve the societal needs of humans.

However, for the need of standardization we also have to look the other way round: How can we get and describe the human in the loop of the technical system in order to get a comprehensive technical system including the human being.

HUMANS AND THEIR ROLES WITH RESPECT TO TECHNICAL SYSTEMS

Again: We need technology experts with multidisciplinary education and qualification and with some readiness to cross technological borders. Such personality issues must be clearly defined and, if at all feasible, formally classified in order to support the adequate assignment of these experts to different tasks during the life cycle of complex systems (cf. Figure 2). An (incomplete) set of such tasks in industrial as well as in socio-economic environments must include:

- Organizing *Technology Convergence*
- Organizing *knowledge management and learning*
- Establishing *competence management*
- Establishing *standardization management*
- Assigning *roles, tasks and responsibilities*, in particular for
- System design, engineering
- System operations, safety, emergency management
- Composing *teams and organizing change management*
- Organizing *cooperation with standardization bodies*

Obviously, these tasks relate to higher level human actions regarding justification and reasoning. With respect to the humans in the loop of the technical system we explicitly ask for the capability of changing perspectives, of changing one's mind according to current changes in context.

Considering knowledge, we must distinguish explicit knowledge (to be "programmed" into technical objects) from implicit and/or tacit knowledge, which is unique to humans. Implicit knowledge will be activated by human experts as soon as the programmed explicit knowledge of technical system components fails to keep the system status in, e.g., a pre-specified "safe" state space (i.e., to control the numerical values of a set of state variables in a way that the overall system behavior does not switch to, e.g., "dangerous" if and when automation loses control—"fail safe").

The introduction of adequate implicit and tacit knowledge into the knowledge base of a knowledge-driven technical system in the right place and the right time is one of the key reasons for putting the human expert in the loop.

This approach makes the human expert become a part of (an "object" within) the total system. Integrating and standardizing humans in the loop would—that's the idea—substantially improve coping with unexpected/critical system behavior. Extending such existing modeling methods and tools by a (semi-)formal description of the unique abilities of human beings will result in a consistent system description during all phases of the system life cycle, including the critical system design phase (cf. Figure 2). If possible at all, this additional human-type "object," carrying and introducing implicit (or "tacit") knowledge, must be integrated into the design as well as into the operational environment, enabling the system designers and operators to work in a homogenous environment of design and operating methods and tools, irrespective of dealing with technical system components or with human experts. Together with concepts and methods from ethical and legal reflection as introduced above, this will result in a smooth collaborative life cycle support for complex systems, comprising the unique abilities, properties and capabilities of human beings.

As an indispensable aspect, our approach includes responsibility of human experts in charge for the specification, design and operation of such systems throughout the system life cycle. In parallel, we want to discuss propositions of fairness and target-oriented acting as aspects of ethical reflection within the processes of standardization, system design, and system operation.

Integrating the human designer and operator in the system life cycle environment requires experienced human experts taking the roles and tasks of an ultimate authority, i.e., being responsible for an optimal design as well as for safe and efficient operations of complex and "smart" systems. In the following, we will further discuss both such roles of human experts with respect to the required standards for the design and engineering of converging technology systems. In doing so, we will consider the task and load of real-time decision making of humans in the case of exceptional system status situations.

For the reasons mentioned above, the application of human knowledge in decision making and emergency management is of crucial importance. We distinguish the following three types of "knowledge" to be considered here (cf. Figure 1 and Figure 3):

- **Explicit Knowledge:** Impersonal, algorithmic, teachable, "programmable" (if correct semantic interpretation by all actors involved can be guaranteed);
- **Implicit Knowledge:** Personal by education, teachable, access on demand, "programmable" (after successful teaching);
- **Tacit Knowledge:** Personal by experience, not teachable, access on stimulation.

In particular, tacit knowledge is a unique human ability. For this reason, tacit knowledge should be a very valuable source of support for complex decision making in a multi-technology environment. Tacit knowledge, however, may not be available or cannot be activated in case of, e.g., a safety-critical operational system status asking for fast and optimized reactions of the human operator. Nevertheless, tacit knowledge is among the key abilities of human experts and must be integrated into complex actions when, e.g., fighting exceptional system emergencies.

Figure 3. Types and applications of knowledge

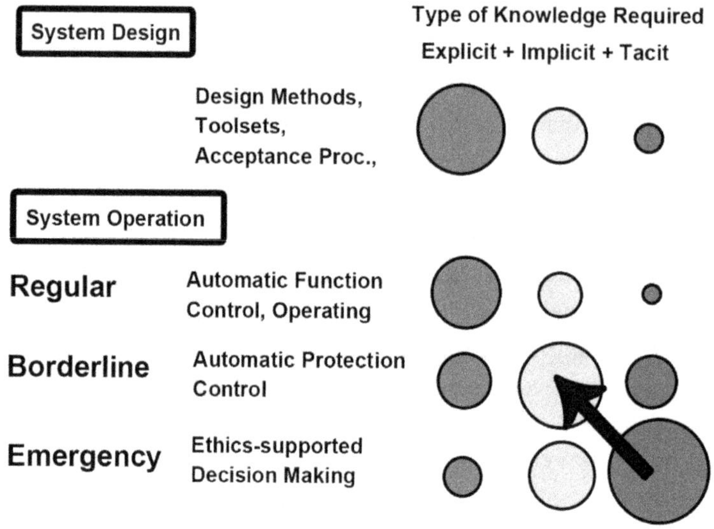

Figure 3 illustrates the use of these three types of knowledge; the diameters of the circles indicate the importance of the respective types of knowledge. During the design phase, the human experts are dealing with complex but deterministic system components. For this reason, explicit knowledge is the overwhelming choice of knowledge applied. Tacit knowledge is entering the stage during system operations as soon as the design flaws will pop up. Figure 3 shows three operational states where tacit knowledge becomes the outstanding choice when dealing with worst-case application problems. Figure 3 underlines an important knowledge-type transfer process: Any successful application of tacit knowledge will change the type of such item of knowledge into explicit. This is a valuable possibility of knowledge acquisition and must be integrated in the operating rules to be followed by the operators of such complex systems.

Which are the human properties, abilities and capabilities we can use for a "technical" description of humans? The "dense description" of a human being mentioned above could be taken as such a description but is obviously optimized to describe humans in human-human interaction. One can imagine a full set of properties and abilities with several hundred topics. In the following, we refer, as an example, to the categorization of which we found helpful as a first step to describe humans in terms that can be technically interpreted:

- **Properties:** Represent the biophysical, biochemical, ergonomic, sensoric, and motoric inventory of humans. Properties of a human are not "unique" since they are found, more or less distinctly, with animals as well. Their combination, however, is unique in comparison with machines. When discussing shop floor safety in the context of Industry 4.0 (BMWi 2013), we presume the implementation and consideration of all the existing standards with respect to, among others, MMI, life-long education, or physical and mental healthcare support.
- **Abilities:** Are those that are "unique" to humans only, and thus are an essential source of additional options for problem solving by humans in case of, e.g., faulty automation components where the human expert is the "last resort" for gaining back control (cf. Figure 1):

- *Free Will:* Supports the human expert in deciding and acting *whatever* the system status requires, *however* and *whenever* human interference is needed under real-time conditions. Free will can even override laws and regulations if required to avoid critical consequences of, e.g., the breakdown of automation functions within a complex technical or socio-economic system.
- *Creativity:* Of the human to go beyond limits when taking over control due to functional or resource deficits of automation functions. Creativity may, however, be limited by individual factors of the human problem solver, such as emotions, education, social position/barriers, self-consciousness, etc. Creativity must be controlled by optimization considerations assessing the consequences of free-will-based actions.
- *Prudence:* Free will and creativity need some control for achieving optimal results in case of challenging situations to be solved by the human in the loop under real-time conditions. There is a strong relationship between prudence and moral.
- *Cognition:* Is needed to capture and understand the environment and the consequences of human actions, resulting in an ongoing improvement of the explicit and the tacit knowledge inventory (cf. Figure 1) of the technical/societal system and the human experts involved. Cognition is a human-oriented ability including the combination of different sources of knowledge for determining and evaluating the system status.
- *Explicit Knowledge:* Implements support for problem solving by employing algorithmic situation-related standards of all kinds, which are an indispensable source of explicit knowledge and should be implemented and maintained by methods and tools of formal semantics.
- *Tacit Knowledge:* Supports problem solving by collecting and employing experience derived unconsciously from daily occurrences. Tacit knowledge helps coping with unexpected situations through finding analogies between such situations. It should be turned into explicit knowledge through, e.g., exploration events involving human experts with different backgrounds.
- *Abstraction:* Is a unique ability of human beings to model functionalities and structures in a generalized way, thus supporting the transfer and application of knowledge within and between communities of different technological and societal background. There are standards, directives and guidelines that can be introduced in organizations, aiming at the common understanding of abstractions (VDI/VDE, 2003).
- *Language:* Is the basic and atavistic ability of mankind to communicate everyday simple information as well as highly complex issues with other humans and/or machines under real-time conditions. Language can be a highly versatile communication carrier between and within international "communities" of humans (H2H) and machines (M2M, (H2M, M2H) based on standardized formal and informal semantic agreements. In order to minimize misunderstandings, a large number of language subsets and contexts (e.g., System Engineering and programming languages) have been developed and published, which, in the context of this book, will be further divided among human experts in their different roles and responsibilities. Ethics in this field must make sure that language-based communication is included in the education systems, providing an understanding and estimation of high communication quality through precise language use (Frey, 1995).

A role defines a set of tasks which the user, assigned to that role, is expected or allowed to perform. Capabilities describe the degree of *being capable to perform such tasks.* Both concepts are of very practical importance when deciding about the qualification of persons who are to take over the tasks indicated in, e.g., Fig.1. In our ethics-based concept this capability issue has a specific importance:

It is unethical as well as uneconomic to place responsibilities on persons which they cannot fulfill.

The performance of technical systems during their life cycle is as much determined by system design as it is by the quality of operators during the performance period of a plant. It is therefore important to carefully consider the individual capabilities of candidates before assigning a certain task. As an example, we should not task an extremely precise but time taking person to act as a trouble shooter under real-time pressure. Vice versa, precision in fulfilling System Engineering tasks is certainly the most effective way of avoiding production time problems. This means, however, that the concept of applying use cases for the definition and codification of ethical decision systems must be understood and supported by experts with a wide range of backgrounds and motivations.

The roles and responsibilities in the different phases of a common system life cycle are shown in Figure 2. Roles are part of the use case methodology and must be specified using the respective templates.

USE CASE (UC) METHOD TO IMPLEMENT THE HUMAN IN THE LOOP

It is essential to any kind of closed loop systems that all steps of this structure are implemented and followed up at any time and under the required real-time conditions. This real-time requirement is valid and mandatory for the human in the loop as well. If the situation of a production facility requires the take-over of control by human operators (human as the ultimate measure—"ultima ratio") due to their abilities discussed in section 3, the operators must be supported by a property-optimized environment, which means the proper and complete implementation of the related standards in the fields of human engineering (among others, the V-model).

As a second prerequisite, we need standardized methods for the definition of tasks and processes in all phases of a system's life cycle (LC). Figure 2 gives an overview of engineering requirements, the allocation of the resulting tasks and, consequently, the related responsibilities. A promising approach for the definition of human actions in cooperation with machine actions is the so-called "use case" (UC) method (DKE, 2014). Recent applications of use cases in the development of standards for smart grids have shown the capabilities of this method for specifying system functionalities (i.e., the set of actions) within complex systems with a great variety of application fields. Originally, this UC method was standardized for optimally specifying sequences of automatic machine actions without any intervention of humans for safety reasons.

However, the current German platform project called "Industry 4.0" (acatech, 2013) explicitly introduces physical human-machine interaction and cooperation under shop floor and real-time conditions. For this reason, we need standardized description methods and action models supporting the integration of the "*human being*" as a decider and action provider with very unique abilities and capabilities, along with the well understood and standardized *world of machine interactions*. The UC method offers the possibility to describe the human being in technical terms as decider and action provider.

The UC method has turned out to be a highly effective specification method for the functionality of complex, networked systems embracing components from different technology fields and, thus, implementing Technology Convergence. Standards for implementing and using the UC method are under way at CENELEC and IEC.

Experience with the application of the UC method and related toolsets (such as an UC repository) has shown that an efficient training towards an optimal use of the included flexibility concepts is essential. So far, there is not much literature in this field but some support is under way: DIN and the German Ministry of Economics and Energy (BMWi) have issued—under the research program called Innovation through Standardization (INS)—a project for introducing the UC method in various fields of technology and socio-economic issues.

The current UC methodology is not designed for describing the activities and actions of human beings. Human beings have social roots and behaviors of frequently unknown origin and extent. In addition, human beings have the ability to apply their abilities in a creative and adaptive manner according to his/her current intentions and wishes. As mentioned above, human beings have accumulated "tacit" knowledge enabling them to successfully deal with unanticipated situations where the programmed automatic behavior of machines cannot succeed due to the lack of appropriate intentions and/or actions implemented.

Two systematic ways out of this dilemma should be considered:

1. Extension of the current UC methodology in a way that expected human behavior can be included in the description of the complete technical system, including humans and machines and standardized interfaces. This would be an extension referring to the "technology push" idea, resulting in situation-specific use cases prepared in the design phase of a technical system.

Another approach could provide more flexible and effective support in dealing with such unexpected events in a more timely manner:

2. Description and storage of predefined and observed human behavior modules, administrated by means of an ethical decisions repository system. Such ethics-based "models" of human behavior could then be combined with adequate UC "models," describing topical system demands and actions to be carried out by humans and tracking data to locate adequate and available human resources which could be activated when necessary.

This concept of separating a human resources and capabilities pool management from the UC repository would provide higher flexibility in linking available human properties, abilities, and capabilities with topical demands for human interference in exceptional situations. As an example, the "system" would send a topical description of an unexpected situation to some dispatcher instance (human or machine), together with a request for human interference. The dispatcher would then search a "capability" database for appropriate and available human resources. This database could include predefined and already standardized moral rules.

The result of this transaction would then influence a decision-making activity towards either trying to fix the situation through an appropriate and available human expert or shutting down the technical system due to the unavailability of such an expert.

This scenario would make use of the descriptive capabilities of the systems, enabling the modeling of a wide range of technical as well as societal consequences and actions. Such actions then could be linked to adequate properties of available technical experts, resulting in a quick and effective formation of a problem-solving team.

Both ways of dealing with the requirement of cooperative behavior of humans and machines will require numerous standards for interfacing and for data, information and knowledge management.

EXAMPLE FROM THE SMART GRID: SUCCESSFUL ACTIONS OF HUMAN EXPERTS IN THE LOOP: PREVENTING AN ELECTRICAL NETWORK COLLAPSE

We resist the temptation of going into details of many prominent real-life accidents exemplifying the importance of ethics in support of our hypothesis of the human in the loop to be the last resort for tackling unexpected problems. Instead, we want to mention a highly interesting event from the electric power grid arena where the human in the loop was able to successfully stabilize the grid in a both complex and unusual situation. It was the combination of codified explicit technical knowledge, the application of the tacit knowledge and of the professional ethics of a team of humans in the loop which prevented a breakdown of large parts of the electric grid in Germany and in Europe. A detailed documentation is available from the European power grid authority.

The situation was not even completely unexpected, but the existing hierarchically organized automation systems were not able to handle it: It was December 26, 2009, a Europe-wide holiday, with clear sky and strong winds. These weather conditions caused a substantial generation of electric energy in photovoltaic installations and in wind turbines throughout northern Europe, which, due to the holiday, dramatically exceeded the electric power demands of industry and private households. The German energy law, however, assigns priority to the consumption of the so-called renewable energies. Since there was, and still is, no local capacity for storing such surplus energy, the first possibility of getting rid of it is selling it to the European grid, i.e., to the neighboring national grids, which, however, requires sufficient transportation capacities. These were—and still are—not available on an adequate level, neither to the southern part of the local German grid nor to the national grids of the neighboring national grids in the western parts of Europe. The only possible way out was to sell the surplus energy to the eastern European partner grids which, however, required transport capacities from western Europe to the east of Europe. Due to the political situation during the cold war until the beginning of the 1990s, the west-to-east electric energy transportation lines in Europe are still not as powerful as needed for flexible trading of energy in times of a surplus situation in western Europe. And, after all, December 26 is a trans-European holiday!

The situation was resolved without a black-out of the European electric power grid. A multinational team of humans in the loop was able to dispatch the surplus energy in unusually short sequences of operational events, making use of their tacit knowledge and following their professional ethics: The highest goal in operating the electric power grid is the availability of power at any time and with acceptable, standardized quality.

All in all, this was not only a demonstration of successful multinational collaboration. It was also a proof of our hypothesis of the positive effect of clear ethical guidelines to be followed by the individual. And last but not at all least, the activation of tacit human knowledge through discussions in an emergency team resulted in a step-by-step process of handling the situation.

This example of a joint application of ethics and technologies for optimizing the electric power system under the legal framework of the "Energiewende" (energy transition), clearly shows the need for cooperative standardization of technologies and ethical reflections.

CONCLUSION

This paper is a first attempt to combine the algorithmic view (object) of technical systems with an ethically reflected human-oriented view (subject) as parts of complex systems. The basic difficulty of this approach stems from the existence of unique abilities governing the human behavior but being less calculable compared with the algorithms of the technical world. The paper provides some ideas for research and for standardization in this heterogeneous plot. We presented a first description of human properties and abilities which could be used for a first "technical" description of humans in the loop of technical systems. We proposed the use case method as a first but promising approach for improving life cycle systems engineering by implementing and reflecting the role of the human in the system loop. Use cases for human actions and even use cases for ethical decision making could be developed and implemented in the use case schemes.

Especially with the latter, we argue in line with other scholars who ask for ethical decision making in technical systems (Wallach & Allen, 2008). Particularly in military contexts (Arkina, 2008; Sparrow, 2006) and for the use of autonomous cars, the implementation of ethically based decision making is (controversly) discussed. In their thoughts about moral reasoning and decision making, Van de Voort et al. (2015) also consider mediation between machines. Their arguments support our proposal to start with definable use cases in which also moral aspects can be described. Ethical rules for deciding on the moral aspects implemented in the system's life cycle can then solve potential moral conflicts.

Moreover, we argue that we need both perspectives of reflection. Describing humans in a technical way in order to implement them in a system life cycle approach supports system development and leads to better informed decision making by the human system designers and operators. General assessment of the technical systems with respect to their (un)intended effects, with respect to their (un)desired consequences, with respect to main and side-effects of technological processes, and with respect to related chances and risks for individuals and society is also a necessary task to be undertaken in order to support political decision making and standardization highly affecting our daily life.

Convergent technologies ask for highly sophisticated control systems in order to meet the high societal expectations.

This basic idea of an ethically—or finally multidisciplinary—extended system life cycle approach has gained some interest in the technical world of production enterprises as well as in industrial associations and, not at least, in standards developing organizations (SDOs).The reason for this very recent interest could be the advent of highly complex distributed and networked systems such as the "smart" systems (Smart Grid, Smart City, Smart Traffic, etc.), which obviously and in spite of advanced networks, powerful computation, cheap but smart sensors, and well advanced engineering need the human expert in order to cope with such complexity. The related novel tasks for the human experts and troubleshooters, to be carried out under real-time conditions and under stressing responsibilities assigned to these experts, may be the trigger for increased research into the potential of ethical reflection as discussed in this paper.

To put it in short: The cooperation of technology and engineering oriented persons with ethics experts will turn out to be a highly fruitful constellation when dealing with the complexity of the technical world and with the societal consequences of situations which are unavoidable due to the global growth of population and the scarcity of resources. As an expectation so far, ethics may help tackling this situation through extending their current duality of "YES—YOU MUST" vs. "NO—YOU MUST NOT" by "TAKE RESPONSIBILITY—TRY"!

REFERENCES

Acatech. (2013). *Securing the future of German manufacturing industry - Recommendations for implementing the strategic initiative INDUSTRIE 4.0.* Retrieved from http://www.acatechapterde/fileadmin/user_upload/Baumstruktur_nach_Website/Acatech/root/de/Material_fuer_Sonderseiten/Industrie_4.0/Final_report__Industrie_4.0_accessible.pdf

Arkina, R. C. (2008). Governing Lethal Behavior: Embedding Ethics in a Hybrid Deliberative/Reactive Robot Architecture. In *Proceedings of the First AGI Conference*. Amsterdam: IOS Press. doi:10.1145/1349822.1349839

BMWi. (2013). *E-Energy Abschlussbericht. Ergebnisse und Erkenntnisse aus der Evaluation der sechs Leuchtturmprojekte*. Retrieved from http://www.doc4web.de/download/3475350288573

Christaller, T., Decker, M., Gilsbach, J.-M., Hirzinger, G., Lauterbach, K., Schweighofer, E., & Sturma, D. et al. (2001). *Robotik. Perspektiven für menschliches Handeln in der zukünftigen Gesellschaft*. Berlin: Springer.

DKE. (2011). *Smart Grid. Deutsche Normungsroadmap "E-Energy/Smart Grids 2.0.* Retrieved from http://www.e-energy.de/documents/DKE_Roadmap_SmartGrid_230410_Engllish.pdf

DKE. (2012). *The German Standardization Roadmap for Electromobility—Version 2*. Retrieved from https://www.vde.com/en/dke/std/Documents/E-Mobility_Normungsroadmap_V2_EN.pdf

DKE. (2014). *DKE/K STD_1113 Use Cases: Methodiken und Referenzmodelle*. DKE.

Frey, C. (1995). *Die intrapersonelle Balance-Theorie im Führungsverhalten*. Berlin: Springer. doi:10.1007/978-3-322-97704-5

Gethmann, C. F., & Sander, T. (1999). Rechtfertigungsdiskurse. In A. Grunwald & S. Saupe (Eds.), *Ethik in der Technikgestaltung. Praktische Relevanz und Legitimation* (pp. 117–151). Berlin: Springer. doi:10.1007/978-3-642-60033-3_7

Grunwald, A. (2002). *Technikfolgenabschätzung—eine Einführung*. Berlin: Edition Sigma.

Grunwald, A. (2011). Responsible Innovation: Bringing together Technology Assessment, Applied Ethics, and STS researchapter *Enterprise and Work Innovation Studies, 7*, 9–31.

Grunwald, A. (2013). *Handbuch Technikethik*. Stuttgart: J.B. Metzler.

Hubig, C., & Reidel, J. (Eds.). (2003). Ethische Ingenieurverantwortung. Handlungsspielräume und Perspektiven der Kodifizierung. Berlin: Edition Sigma.

ISO. (2010). *Guidance on social responsibility – ISO 26000:2010*. ISO.

Maurer, M., Gerdes, J. C., Lenz, B., & Winner, H. (Eds.). (2015). *Autonomes Fahren. Technische, rechtliche und gesellschaftliche Aspekte*. Berlin: Springer.

Sellars, W. (1997). *Empiricism and the Philosophy of Mind*. Cambridge, MA: Harvard University Press.

Sparrow, R. (2007). Killer Robots. *Journal of Applied Philosophy*, *24*(1), 62–77. doi:10.1111/j.1468-5930.2007.00346.x

Van de Voort, M., Pieters, W., & Consoli, L. (2015). Refining the ethics of computer-made decisions: A classification of moral mediation by ubiquitous machines. *Ethics and Information Technology*, *17*(1), 41–56. doi:10.1007/s10676-015-9360-2

VDI. (1991). *Technikbewertung – Begriffe und Grundlagen. VDI-Richtlinie 3780*. Düsseldorf: VDI Verlag.

VDI/VDE. (2003). *Formalisierte Prozessbeschreibungen. VDI/VDE-Richtlinie 3682*. Düsseldorf: VDI Verlag.

Wallach, W., & Allen, C. (2008). *Moral machines: Teaching robots right from wrong*. Oxford, UK: Oxford University Press.

KEY TERMS AND DEFINITIONS

Ethics: Ethics refers to standards of right and wrong, which establish how humans should act or behave.

Human: In the context of this Chapter, a Human Being acting in direct cooperation with technical or societal systems and being the Ultimate Authority for the solution of technical and/or societal problems.

Multi-Disciplinarity: Integrating methods and concepts of a multiplicity of scientific fields in pursuing the design and operation of complex systems.

Real-Time Systems: Systems fulfilling their tasks within application dependent time constraints.

Reflection: Ethical reflection is the reflection on societal moral argumentation schemes by professional ethics in order to support human action.

System Engineering: Designing and implementing systems based on application related specifications.

Technology Convergence: Employing technologies of different kinds and origins for implementing the functions of complex distributed systems.

Unique Human Abilities/Features: Properties not found at other creatures with respect to technical and/or societal systems.

ENDNOTES

1 The example of self-defense can illustrate this point, because the action of "killing somebody" can be morally judged as an—ethically inacceptable—murder and as a self-defense situation in which the killing is ethically accepted.

2 We follow here the argumentation of Christaller et al. (2001, p. 116 ff.)

3 For more information see: https://ec.europa.eu/programmes/horizon2020/en/h2020-section/societal-challenges and http://ec.europa.eu/information_society/newsroom/cf/horizon2020/document.cfm?doc_id=3778 (last accessed May 2015).

Compilation of References

ABA Section of Antitrust Law. (2011). *Handbook on antitrust aspects of standard setting* (2nd ed.). Chicago, IL: ABA Publishing.

Abbott, A. (1988). *The system of profession: an essay on the division of expert labor*. University of Chicago Press.

Abdelkafi, N., & Makhotin, S. (2014). Seizing opportunities for the support of innovation through committee standards and standardization: Insights from German SMEs. *International Journal of IT Standards and Standardization Research, 12*(2), 38–56. doi:10.4018/ijitsr.2014070103

Abernethy, M. A., & Stoelwinder, J. U. (1995). The role of professional control in the management of complex organizations. *Accounting, Organizations and Society, 20*(1), 1–17. doi:10.1016/0361-3682(94)E0017-O

Acatech. (2013). *Securing the future of German manufacturing industry - Recommendations for implementing the strategic initiative INDUSTRIE 4.0*. Retrieved from http://www.acatechapterde/fileadmin/user_upload/Baumstruktur_nach_Website/Acatech/root/de/Material_fuer_Sonderseiten/Industrie_4.0/Final_report__Industrie_4.0_accessible.pdf

ACEA (2009). *Translation of MIIT document 'Administrative Rules on Access for New-energy Vehicle Manufacturers and Products' for NEV policies MNCs have to comply with the regulations and comments, No. (44 of GCY [2009]), (for internal use only)*. ACEA.

Acemoglu, D., Gancia, D., & Zilibotti, F. (2012). Competing engines of growth: Innovation and standardization. *Journal of Economic Theory, 147*(2), 570–601. doi:10.1016/j.jet.2010.09.001

ACWG. (2012). *Auto Components Working Group Position Paper, 144-145, by EUCCC*. Retrieved from http://www.europeanchamber.com.cn/en/publications-archive?categoryId=3

Adolphi, H. (1997). *Strategische Konzepte zur Organisation der betrieblichen Standardisierung. In DIN Normungskunde* (Vol. 38). Berlin: Beuth Verlag.

Adoption. (2011). In *Merriam-Webster*. Retrieved from http://www.merriam-webster.com/thesaurus/adopt

AFNOR. (1967). *La normalisation dans l'entreprise*. Paris: AFNOR.

Ajzen, I. (1991). The theory of planned behavior. *Organizational Behavior and Human Decision Processes, 50*(2), 179–211. doi:10.1016/0749-5978(91)90020-T

Akerlof, G. A. (1970). Market for Lemons - Quality Uncertainty and Market Mechanism. *The Quarterly Journal of Economics, 84*(3), 488–500. doi:10.2307/1879431

Allen, P. S. (Ed.). (1913). *Opus Epistolarum Des Erasmi Roterodami* (vol. 2). Oxford, UK: Oxford University Press.

Alvesson, M. (2001). Knowledge work: Ambiguity, image and identity. *Human Relations*, *54*(7), 863–886. doi:10.1177/0018726701547004

Amit, R., & Zott, C. (2012). Creating Value Through Business Model Innovation. *MIT Sloan Management Review*, *53*(3), 41–49.

Andersen, E., & Fjeldstad, Ø. D. (2003). Understanding interfirm relations in mediation industries with special reference to the Nordic mobile communication industry. *Industrial Marketing Management*, *32*(5), 397–408. doi:10.1016/S0019-8501(03)00013-0

Anh, V. T. (2007). An Introduction - The History of Standardization. In W. Hesser, A. Feilzer, & H. de Vries et al. (Eds.), *Standardisation in Companies and Markets* (2nd ed.; pp. 45–76). Hamburg, Germany: Helmut Schmidt University.

ANSI. (2015). *Essential Requirements: Due process requirements for American National Standards*. Retrieved from http://publicaa.ansi.org/sites/apdl/Documents/standards%20Activities/American%20National%20standards/Procedures,%20Guides,%20and%20Forms/2015_ANSI_Essential_Requirements.pdf

APCO. (2010). *China's 12th Five-Year Plan, "How it actually works and what's in store for the next five years", December 2010*. Retrieved from http://www.apcoworldwide.com/

Argyres, N., & Bigelow, L. (2010). Innovation, modularity, and vertical de-integration: Evidence from the early US auto industry. *Organization Science, 21*(4), 842-853.doi: 10.1287/orsc.1090.0493

Arkina, R. C. (2008). Governing Lethal Behavior: Embedding Ethics in a Hybrid Deliberative/Reactive Robot Architecture. In *Proceedings of the First AGI Conference*. Amsterdam: IOS Press. doi:10.1145/1349822.1349839

Arksey, H., & Knight, P. (1999). *Interviewing for Social Scientist*. London: Sage Publications.

Arthur, W. B. (1989). Competing Technologies, Increasing Returns, and Lock-in by Historical Events. *The Economic Journal*, *394*(99), 116–131. doi:10.2307/2234208

Arthur, W. B. (1996). Increasing returns and the new world of business. *Harvard Business Review*, *74*, 100–109. PMID:10158472

Atkinson, D. R. (2012). *Enough is enough: Confronting Chinese innovation Mercantilism*. The Information Technology & Innovation Foundation.

AXELOS. (2009). *Managing Successful Projects with PRINCE2, 2009 Edition*. Available from: https://www.axelos.com/best-practice-solutions/prince2/what-is-prince2

Axelrod, R. (1984). *The evolution of Cooperation*. Basic Books.

Backhouse, J., Hsu, C. W., & Silva, L. (2006). Circuits of Power in Creating de jure Standards: Shaping an International Information Systems Security Standard. *Management Information Systems Quarterly*, *30*(Special Issue), 413–438.

Bailyn, L. (1985). Autonomy in the R&D Lab. *Human Resource Management*, *24*(2), 129–146. doi:10.1002/hrm.3930240204

Barley, S. R., & Kunda, G. (2001). Bringing work back in. *Organization Science*, *12*(1), 76–95. doi:10.1287/orsc.12.1.76.10122

Barry, A., & Slater, D. (2002). Introduction: The technological economy. *Economy and Society*, *31*(2), 175–193. doi:10.1080/03085140220123117

Barzel, Y. (1982). Measurement Cost and the Organization of Markets. *The Journal of Law & Economics*, *25*(1), 27–48. doi:10.1086/467005

Basden, A. (2004). Dooyeweerd and Freedom. In *The Dooyeweerd Pages*. Retrieved January 7, 2010, from http://www.dooy.salford.ac.uk/freedom.html

Basden, A. (2010). *The Dooyeweerd Pages*. Retrieved January 7, 2010, from http://www.dooy.salford.ac.uk/index.html

Bauer, C.-O. (1980). Technische Normen – ihre Methoden und Wirkungen als Gegenstand von Forschung und Lehre. *DIN Mitteilung*, *59*, 329.

Bayus, B. L., Jain, S., & Rao, A. G. (2001). Truth or Consequences: An Analysis of Vaporware and New Product Announcements. *JMR, Journal of Marketing Research*, *38*(1), 3–13. doi:10.1509/jmkr.38.1.3.18834

Becker, M. C. (Ed.). (2008). *Handbook of organisational routines*. Cheltenham, UK: Edward Elgar. doi:10.4337/9781848442702

Bekkers, R., & Updegrove, A. (2012, September). *A Study of IPR Policies and Practices of aRepresentative Group of Standard Setting Organizations Worldwide*. Symposium conducted at a meeting of the National Academies of Science. Retrieved from http://sites.nationalacademies.org/xpedio/groups/pgasite/documents/webpage/pga_072197.pdf

Bekkers, R., Iversen, E., & Blind, K. (2006). *Patent pools and non-assertion agreements: coordination mechanisms for multi-party IPR holders in standardization*. Paper for the EASST 2006 Conference, Lausanne, Switzerland.

Bekkers, R. (2001). *Mobile Telecommunication Standards. GSM, UMTS, Tetra and Ermes*. Boston: Artech House.

Bekkers, R., Iversen, E., & Blind, K. (2012). Emerging ways to address the reemerging conflict between patenting and technological standardization. *Industrial and Corporate Change*, *21*(4), 901–931. doi:10.1093/icc/dtr067

Benner, M. J., & Tushman, M. L. (2003). Exploitation, exploration, and process management: The productivity dilemma revisited. *Academy of Management Review*, *28*, 238–256.

Benner, M. J., & Veloso, F. M. (2008). ISO 9000 practices and financial performance: A technology coherence perspective. *Journal of Operations Management*, *26*(5), 611–629. doi:10.1016/j.jom.2007.10.005

Bentham, J. (1838). Plan of the Panopticon. In J. Bowring (Ed.), *The Works of Jeremy Bentham* (vol. 4, pp. 172–173). Edinburgh, UK: William Tait.

Besen, S. M., & Farrell, J. (1991). The role of the ITU in standardization: Pre-eminence, impotence or rubber stamp? *Telecommunications Policy*, *15*(4), 311–321. doi:10.1016/0308-5961(91)90053-E

Besen, S. M., & Farrell, J. (1994). Choosing How to Compete: Strategies and Tactics in Standardization. *The Journal of Economic Perspectives*, *8*(2), 117–131. doi:10.1257/jep.8.2.117

Betancourt, D., & Walsh, R. (1995). The evolution of strategic standardization management. *StandardView*, *3*(3), 117–126. doi:10.1145/226191.226211

Beunza, D., & Stark, D. (2004). Tools of the trade: The socio-technology of arbitrage in a Wall Street trading room. *Industrial and Corporate Change*, *13*(2), 369–400. doi:10.1093/icc/dth015

Bevan, R. (2011). *Changemaking: Tactics and Resources for Managing Organizational Change*. Seattle, WA: ChangeStart Press.

Biddle, B., Curci, F. X., Haslach, T. F., Marchant, G. E., Askland, A., & Gaudet, L. (2012). The expanding role and importance of standards in the information and communications technology industry. *Jurimetrics*, *52*, 177–209.

Bird, G. B. (1998). The business benefit of standards. *StandardView*, *5*(2), 76–80. doi:10.1145/301688.301691

Blind, K. (2009). *Standardisation: a catalyst for innovation*. Retrieved from http://hdl.handle.net/1765/17558

Blind, K. (2013). *The Impact of Standardization and Standards on innovation, Compendium of Evidence on the Effectiveness of innovation Policy Intervention*. Retrieved from http://www.innovation-policy.org.uk/compendium/

Blind, K., & Großmann, A.-M. (2014). An Exploratory Analysis of Company Standards –Evidence from the German Standardization Panel. In I. Mijatovic & K. Jakobs (Eds.), *EURAS Proceedings 2014: Cooperation between Standardisation Organisations and the Scientific and Academic Community: EURAS Annual Standardisation Conference* (Vol. 19). Aachen, Germany: Verlagshaus Mainz GmbH Aachen.

Blind, K., & Grupp, H. (2000). *Standards Statistics as New Indicators for the Diffusion of Technology*. Paper presented to the International JA Schumpeter Conference, Manchester, UK.

Blind, K., Bekkers, R., Dietrich, Y., Iverson, E., Kohler, F., Muller, B., . . . Verweijen. (2011). *Study on the interplay between standards and intellectual property rights (IPRs) (final report)*. Berlin: Author. Retrieved from http://ec.europa.eu/enterprise/policies/european-standards/files/standards_policy/ipr-workshop/ipr_study_final_report_en.pdf

Blind, K., Bierhals, R., Iversen, E., Hossain, K., Rixius, B., Thumm, N., & van Reekum, R. (2004). Interaction between Standardisation and Intellectual Property Rights. European Commission Technical Report EUR 21074 EN, Seville.

Blind, K., Jungmittag, A., & Mangelsdorf, A. (2011). *The economic benefits of standardization: An update of the study carried out by DIN in 2000*. Retrieved September 26, 2014, from DIN German Institute for Standardization e. V.: http://www.din.de/sixcms_upload/media/2896/DIN_GNN_2011_engl_akt_neu.pdf

Blind, K., Thumm, N., Bierhals, R., Hossein, K., Iversen, E., van Reekum, R., & Rixius, B. (2002). *Study on the Interaction between Standardisation and Intellectual Property Rights*. Final Report to the DG Research of the European Commission (EC Contract No G6MA-CT-2000-02001).

Blind, S., & Bühring, J. (1996). Die ökonomische Theorie der Standards und ihre Anwendung auf den Medienbereich. *Homo Oeconomicus*, (13), 515–560.

Blind, K. (2002). Driving forces for Standardization at Standardization development organizations. *Applied Economics*, *34*(16), 1985–1998. doi:10.1080/00036840110111158

Blind, K. (2004). *The Economics of Standards – Theory, Evidence, Policy*. Cheltenham, UK: Edward Elgar.

Blind, K. (2004). *The Economics of Standards: Theory, Evidence, Policy*. Cheltenham, UK, Northampton, MA: Edward Elgar Publishing Limited.

Blind, K. (2004). *The economics of Standards: Theory, Evidence, Policy*. Northampton, MA: Edward Elgar Publishing.

Blind, K. (2011). An economics analysis of Standards competition: The example of the ISO ODF and OOXML Standards. *Telecommunications Policy*, *35*(4), 373–381. doi:10.1016/j.telpol.2011.02.007

Blind, K., & Gauch, S. (2009). Research and Standardisation in Nanotechnology: Evidence from Germany. *The Journal of Technology Transfer*, *34*(3), 320–342. doi:10.1007/s10961-008-9089-8

Blind, K., Gauch, S., & Hawkins, R. (2010). How stakeholders view the impacts of international ICT Standards. *Telecommunications Policy*, *34*(3), 162–174. doi:10.1016/j.telpol.2009.11.016

Blind, K., & Thumm, N. (2004). Intellectual property protection and standardization. *International Journal of IT Standards and Standardization Research*, *2*(2), 61–75. doi:10.4018/jitsr.2004070105

Blind, K., & Thumm, N. (2004). Interrelation between patenting and standardisation strategies: Empirical evidence and policy implications. *Research Policy*, *33*(10), 1583–1598. doi:10.1016/j.respol.2004.08.007

Blumberg, B., & Cooper, D. R. (2011). *Business Research Methods*. London: Mc-Graw Hill.

BMBF. (2011). *Fortschrittsbericht der Nationalen Plattform Elektromobilität (Dritter Bericht)*. Retrieved from http://www.bmbf.de/de/22327.php?hilite=elektromobilit%E4t

BMWi. (2013). *E-Energy Abschlussbericht. Ergebnisse und Erkenntnisse aus der Evaluation der sechs Leuchtturmprojekte*. Retrieved from http://www.doc4web.de/download/3475350288573

Bohannan, C., & Hovenkamp, H. (2012). *Creation without restraint: promoting liberty and rivalry in innovation*. Oxford, UK: Oxford University Press. doi:10.1093/acprof:oso/9780199738830.001.0001

Boh, W. F., Soh, C., & Yeo, S. (2007). Standards Development and diffusion: A case study of RosettaNet. *Communications of the ACM*, *50*(12), 57–62. doi:10.1145/1323688.1323695

Boh, W. F., & Yellin, D. (2007). Using Enterprise Architecture Standards in Managing Information Technology. *Journal of Management Information Systems*, *23*(3), 163–207. doi:10.2753/MIS0742-1222230307

Boiral, O. (2003). ISO 9000: Outside the iron cage. *Organization Science*, *14*(6), 720–737. doi:10.1287/orsc.14.6.720.24873

Boiral, O. (2003). ISO 9000: Outside the iron case. *Management Science*, *14*(6), 720–737.

Boiral, O. (2012). ISO certificates as organizational degrees? Beyond the rational myths of the certification process. *Organization Studies*, *33*(5-6), 633–654. doi:10.1177/0170840612443622

Boiral, O., & Roy, M. (2007). ISO 9000: Integration rationales and organisational impacts. *International Journal of Operations & Production Management*, *27*(2), 226–247. doi:10.1108/01443570710720630

Borneman, W. R. (2010). *Rival Rails: The Race to Build America's Greatest Transcontinental Railroad*. Old Saybrook, CT: Tantor.

Botzem, S., & Dobusch, L. (2012). Standardization cycles: A process perspective on the formation and diffusion of transnational Standards. *Organization Studies*, *33*(5-6), 737–762. doi:10.1177/0170840612443626

Bourgeois, L. J. III. (1985). Strategic goals, perceived uncertainty, and economic performance in Volatile Environments. *Academy of Management Journal*, *28*(3), 548–573. doi:10.2307/256113

Bousquet, F. (2003). Dealing with standardization: Do you need a guru? Paradoxes and tricks of standardization management. In *Proceedings of the 3rd Conference on Standardization and Innovation in Information Technology*. Piscataway, NJ: IEEE. doi:10.1109/SIIT.2003.1251194

Boxenbaum, E., & Jonsson, S. (2008). Isomorphism, diffusion and decoupling. In The Sage handbook of organizational institutionalism. Sage Publications.

Boynton, A. C., & Zmud, R. W. (1987). Information Technology Planning in the 1990s. *Management Information Systems Quarterly*, *11*(1), 59–71. doi:10.2307/248826

Brehm, S., & Brehm, J. W. (1981). *Psychological Reactance – a Theory of Freedom and Control*. New York: Academic Press.

Breznitz, D., & Murphree, M. (2013). *The Rise of China in Technology Standards: New Norms in Old Institutions*. Retrieved from http://origin.www.uscc.gov/sites/default/files/Research/RiseofChinainTechnologyStandards.pdf

Briscoe, J., Fawcett, S., & Tood, R. (2005). The Implementation and Impact of ISO 9000 among Small Manufacturing Enterprises. *Journal of Small Business Management*, *43*(3), 309–330. doi:10.1111/j.1540-627X.2005.00139.x

Brooks, R. G., & Geradin, D. (2011). Interpreting and enforcing the voluntary FRAND commitment. *International Journal of IT Standards and Standardization Research*, *9*(1), 1–23.

Brown, A., Van Der Wiele, T., & Loughton, K. (1998). 'Smaller enterprises' experiences with ISO 9000. *International Journal of Quality & Reliability Management*, *15*(3), 273–285. doi:10.1108/02656719810198935

Brown, J. S., & Duguid, P. (1991). Organizational learning and communities of practice: Toward a unified view of working, learning and innovation. *Organization Science*, *2*(1), 40–56. doi:10.1287/orsc.2.1.40

Brunson, N., & Jacobsson, B. (2000). The contemporary expansion of Standardization. In N. Brunson & B. Jacobsson (Eds.), *A world of Standards*. New York: Oxford University Press.

Brunsson, N. (2000). Standardization and fashion trends. In A world of standards. Oxford Univ. Press.

Brunsson, N., & Jacobsson, B. (2000a). The contemporary expansion of standardization. In A world of standards. Oxford Univ. Press.

Brunsson, N., & Jacobsson, B. (2000b). Following standards. In A world of standards. Oxford Univ. Press.

Brunsson, N. (2000). The Pros and Cons of Standardisation – An Epilogue. In *A World of Standards* (pp. 169–173). Oxford, UK: Oxford University Press.

Brunsson, N., & Jacobsson, B. (2000). Following Standards. In *A World of Standards* (pp. 125–137). Oxford, UK: Oxford University Press.

Brunsson, N., & Olsen, J. P. (1993). *The reforming organization*. New York: Routledge.

Brunsson, N., & Sahlin-Andersson, K. (2000). Constructing organizations: The example of public sector reform. *Organization Studies*, *21*(4), 721–746. doi:10.1177/0170840600214003

Buckler, B. (1996). A learning process mode to achieve continuous improvement and innovation. *The Learning Organization*, *3*(3), 31–39. doi:10.1108/09696479610119660

Budd, R. W., Throp, R. K., & Donohew, L. (1967). *Content analysis of communications*. New York: Macmillan.

Buffett, W. (2002). *Letter to Berkshire Stakeholders*.

Burkert, M., Davila, A., Mehta, K., & Oyon, D. (2014). Relating alternative forms of contingency fit to the appropriate methods to test them. *Management Accounting Research*, *25*(1), 6–29. doi:10.1016/j.mar.2013.07.008

Byrne, B. M., & Golder, P. A. (2002). The diffusion of anticipatory standards with particular reference to the ISO/IEC information resource dictionary system framework standard. *Computer Standards & Interfaces*, *24*(5), 369–379. doi:10.1016/S0920-5489(02)00057-0

Callon, M. (1998). Introduction: The embeddedness of economic markets in economics. In M. Callon (Ed.), *The Laws of the Markets*. Blackwell. doi:10.1111/j.1467-954X.1998.tb03468.x

Callon, M., & Muniesa, F. (2005). Economic markets as calculative collective devices. *Organization Studies*, *26*(8), 1129–1250. doi:10.1177/0170840605056393

Cannady, C. (2013). *Technology licensing and development agreements*. Oxford University Press.

Cao, C., Suttmeier, R. P., & Simon, D. F. (2006, December). China's 15-Year Science Plan: Mapping Research and innovation Strategies for the 21st Century. *Physics Today*, *59*(2), 38–43. doi:10.1063/1.2435680

Cargill, C. F. (1989). *Information Technology Standardization: Theory, Process, and Organizations*. Bedford, MA: Digital Press.

Carnall, C., & Todnem, R. (2007). *Managing Change in Organizations* (6th ed.). Harlow, UK: Pearson Education.

Carrier, M. (2009). *Innovation for the 21st Century – Harnessing the power of intellectual property and antitrust law*. Oxford University Press. doi:10.1093/acprof:oso/9780195342581.001.0001

Carruthers, B., & Stinchcombe, A. (1999). The social structure of liquidity: Flexibility, markets, and states. *Theory and Society*, *28*(3), 353–382. doi:10.1023/A:1006903103304

Cary, G. S., Nelson, M. W., Kaiser, S. J., & Sistla, A. R. (2011). The case for antitrust law to police the patent holdup problem in standard setting. *Antitrust Law Journal*, *77*(3), 913–945.

Casadesús, M., & Karapetrovic, S. (2005). Has ISO 9000 lost some of its lustre? *International Journal of Operations & Production Management*, *25*(6), 580–596. doi:10.1108/01443570510599737

CEC. (1996). *Communication from the commission to the council and the parliament on 'standardization and the global information society: The European approach'*. COM(96) 359 final. Retrieved from http://eur-lex.europa.eu/LexUriServ/LexUriServ.do?uri=COM:1996:0359:FIN:EN:PDF

CESIP, China Europe Standardization Information Platform. (2013). Retrieved from http://www.eu-china-standards.eu/

Chandler, A. D. Jr. (1977). *The Visible Hand: The Managerial Revolution in American Business*. Belknap Press.

Chesbrough, H. W. (2003). *Open innovation: The new imperative for creating and profiting from technology*. Boston: Harvard Business School Press.

Chevalier, F. (2012). *Programme d'amélioration des Processus de gestion de projets adaptés à l'envergure des projets*. École de technologie supérieure. (in French)

Child, J. (1972). Organizational structure, environment and performance: The role of strategic choice. *Sociology*, *6*(1), 1–22. doi:10.1177/003803857200600101

China Daily. (2012). *Milestone investment agreement is signed from China Daily*. Retrieved from http://www.chinadaily.com.cn/business/2012- 05/14/content_15282436.htm

Choi, B., Raghu, T. S., & Vinze, A. (2004). Addressing a standards creation process: A focus on ebXML. *International Journal of Human-Computer Studies*, *61*(5), 627–648. doi:10.1016/j.ijhcs.2004.04.002

Choi, D. G., Lee, H., & Sung, T.-. (2011). Research profiling for 'standardization and innovation'. *Scientometrics*, *88*(1), 259–278. doi:10.1007/s11192-011-0344-7

Christaller, T., Decker, M., Gilsbach, J.-M., Hirzinger, G., Lauterbach, K., Schweighofer, E., & Sturma, D. et al. (2001). *Robotik. Perspektiven für menschliches Handeln in der zukünftigen Gesellschaft*. Berlin: Springer.

Christensen, C. M., Grossman, J. H., & Hwang, J. (2009). *The innovator's prescription: A disruptive solution for health-care*. New York: McGraw-Hill.

Ciborra, C. (2007). Digital Technologies and Risk: A Critical Review. In Hanseth & Ciborra (Eds.), Risk, Complexity and ICT. Edward Elgar Publishing. doi:10.4337/9781847207005.00007

Cohen, M. D., Burkhart, R., Dosi, G., Egidi, M., Marengo, L., Warglien, M., & Winter, S. (1996). Routines and Other Recurrent Action Patterns of Organizations: Contemporary Research Issues. *Industrial and Corporate Change*, *5*(3), 653–698. doi:10.1093/icc/5.3.653

Cohen, W. M., & Levinthal, D. A. (1990). Absorptive capacity: A new perspective on learning and innovation. *Administrative Science Quarterly*, *35*(1), 128–152. doi:10.2307/2393553

Collis, C., & Donnelly, T. (2012). Joint Ventures and the Development of the Chinese Automotive. *International Journal of Automotive Technology and Management*, *12*(4), 318–319. doi:10.1504/IJATM.2012.051358

Collis, D. J., & Rukstad, M. G. (2008). Can you say what your strategy is? *Harvard Business Review*, *86*(4), 82–90. PMID:18435009

COM I. (2011). *311*. Communication from the Commission to the European Parliament, the Council, and the European Economic and Social Committee on 'A strategic vision for European standards: Moving forward to enhance and accelerate the sustainable growth of the European economy by 2020', final. Retrieved from http://eur-lex.europa.eu/LexUriServ/LexUriServ.do?uri=COM:2011:0311:FIN:EN:PDF

COM. (2007). *581 final*. Communication from the Commission to the European Parliament, the Council, the European Economic and Social Committee and the Committee of the Regions, The European Interest: Succeeding in the age of globalisation. Retrieved from http://eur-lex.europa.eu/Result.do?T1=V5&T2=2007&T3=581&RechType=RECH_naturel&Submit=Search

COM. (2008a). *133 final*. Communication from the Commission to the Council, the European Parliament and the European Economic and Social Committee, Towards an increased contribution from standardisation to innovation in Europe. Retrieved from http://ec.europa.eu/enterprise/policies/european-standards/standardisation-policy/policy-activities/innovation/

COM. (2008b). *465 final*. Communication from the Commission to the European Parliament, the Council and the European Economic and Social Committee, An Industrial Property Rights Strategy for Europe. Retrieved from http://ec.europa.eu/enterprise/policies/innovation/policy/intellectual-property/index_en.htm

COM. (2010). *546 final*. Communication from the Commission to the European Parliament, The Council, The European Economic and Social Committee and The Committee of the Regions Europe 2020 Flagship Initiative Innovation Union. Retrieved from http://ec.europa.eu/research/innovation-union/index_en.cfm?pg=keydocs

Connor, P. E. (1997). Total Quality Management: A selective commentary on its human dimensions, with special reference to its downside. *Public Administration Review*, *57*(6), 501–509. doi:10.2307/976961

Contreras, J. L. (2015b). A market reliance theory for FRAND commitments and other patent pledges. *Utah Law Review*, *2015*(2), 479-558.

Contreras, J. L. (2013). Fixing FRAND: A pseudo-pool approach to standards-based patent licensing. *Antitrust Law Journal*, *79*(1), 47–97.

Contreras, J. L. (2015a). Patent pledges. *Arizona State Law Journal*, 47(3).

Contreras, J. L. (2015c). A brief history of FRAND: Analyzing current debates in standard-setting and antitrust through a historical lens. *Antitrust Law Journal*, 30(1), 39-120.

Contreras, J. L. (Ed.). (2007). *Standards development patent policy manual*. Chicago, IL: ABA Publishing.

Contreras, J. L., & Hernacki, A. T. (2014). Copyright termination and technical standards. *University of Baltimore Law Review*, *43*, 221–254.

Contreras, J. L., & Newman, D. L. (2014). Developing a framework for arbitrating standards-essential patent disputes. *Journal of Dispute Resolution*, *2014*, 23–50.

Contreras, J. L., Roth, H., & Lewis, M. (2011). Toward a rational framework for sustainable building materials. *Standards Engineering*, *63*(5), 1–17.

Corbett, C. J., Montes-Sancho, M. J., & Kitsch, D. A. (2005). The Financial impact of ISO 9000 Certification in the United States: An Empirical Analysis. *Management Science*, *51*(7), 1046–1059. doi:10.1287/mnsc.1040.0358

Cotter, T. F. (2014). Comparative law and economics of standard-essential patents and FRAND royalties. *Texas Intellectual Property Law Journal*, *22*, 311–363.

Cox, W. (2012). *Reducing Greenhouse Gases from Personal Mobility: Opportunities and Possibilities*. Available at http://reason.org/files/reducing_greenhouse_gases_mobility_development.pdf

Creswell, J. W. (2009). *Research Design - Qualitative, Quantitative, and Mixed Methods Approaches*. Los Angeles, CA: Sage.

Cummings, J., & DiPasquale, J. (1997). *A primer on the secondary mortgage market*. Boston: City Research.

Cusumano, M. (2010). Technology Strategy and Management – The Evolution of Platform Thinking. *Communications of the ACM*, *53*(1), 32–34. doi:10.1145/1629175.1629189

Cyert, R. M., & March, J. G. (1963). *A behavioral theory of the firm*. Prentice Hall.

Dacin, M. T., Ventresca, M. J., & Beal, B. D. (1999). The embeddedness of organizations: Dialogue & directions. *Journal of Management*, *25*(3), 317–356. doi:10.1177/014920639902500304

Daidj, N., Grazia, C., & Hammoudi, A. (2010). Introduction to the Non-Cooperative Approach to Coalition Formation: The Case of the Blu-Ray/HD-DVD Standards' War. *Journal of Media Economics*, *23*(4), 192–215. doi:10.1080/08997764.2010.527206

Dale, B. G. (2007). Quality Management Systems. In B. G. Dale, T. Van der Wiele, & J. Van Iwaarden (Eds.), *Managing Quality* (pp. 279–309). Malden, MA: Blackwell Publishing.

David, P. A. (1987). Some new Standards for the economics of Standardization in the information age. In P. Dasgupta & P. Stoneman (Eds.), Economic policy and technological performance (pp. 206-239). doi:10.1017/CBO9780511559938.010

David, P. A. (1995). Standardisation policies for network technologies: the flux between freedom and order revisited. In Standards, Innovation and Competitiveness – The Politics and Economics of Standards in Natural and Technical Environments (pp. 15-35). Aldershot, UK: Edward Elgar.

David, P. A. (1995). Standardization policies for network technologies: The flux between freedom and order revisited. In R. Hawkins, R. Mansell, & J. Skea (Eds.), *Standards, Innovation and Competitiveness: The Politics and Economics of Standards in natural and Technical Environments*. Cheltenham, UK: Edward Elgar.

David, P. A., & Steinmueller, W. E. (1994). Economics of compatibility standards and competition in telecommunication networks. *Information Economics and Policy*, *6*(3-4), 217–241. doi:10.1016/0167-6245(94)90003-5

Davis, F. D. (1989). Perceived usefulness, perceive ideas of use, and user acceptance of information technology. *Management Information Systems Quarterly*, *13*(3), 319–340. doi:10.2307/249008

Davis, F. D., Bagozzi, R. P., & Warsaw, P. R. (1989). User acceptance of computer technology: A comparison of two theoretical models. *Management Science*, *35*(8), 982–1003. doi:10.1287/mnsc.35.8.982

De Vries, H. J. (2006). *Standards for business - How companies benefit from participation in international standards setting*. Retrieved August 12, 2014, from http://www.iec.ch/about/globalreach/academia/pdf/vries-1.pdf

De Vries, H.J. (1997) Standardization – What's in a name? *Terminology – International Journal of Theoretical and Applied Issues in Specialized Communication*, *4*(1), 55-83.

de Vries, H. (2007). Fundamentals of Standards and Standardization. In W. Hesser, A. Feilzer, & H. de Vries et al. (Eds.), *Standardisation in Companies and Markets* (2nd ed.; pp. 1–44). Hamburg, Germany: Helmut Schmidt University.

De Vries, H. J. (1999). *Standardization – A Business Approach to the Role of National Standardization Organisations*. Boston: Kluwer Academic Publishers. doi:10.1007/978-1-4757-3042-5

de Vries, H. J. (2006). The Paradox of Standardization and Innovation. *ISO Focus*, *3*(February), 40–43.

De Vries, H. J. (2015). Standardisation – A Developing Field of Research. In P. Delimatsis (Ed.), *The Law, Economics and Politics of International Standardsation*, (pp.15-41). Cambridge, UK: Cambridge University Press.

De Vries, H. J., Slob, F. J., & Zuid-Holland, V. G. (2006). Best practice in company standardization. *International Journal of IT Standards and Standardization Research*, *4*(1), 62–85. doi:10.4018/jitsr.2006010104

De Vries, H. J., & Van Delden, M. (2006). How to integrate standardisation in knowledge management. *ISO Focus*, *3*(3), 37–40.

den Uijl, S., de Vries, H. J., & Bayramoglu, D. (2013). The rise of MP3 as the market standard: How compressed audio files became the dominant music format. *International Journal of IT Standards and Standardization Research*, *12*(1), 1–26. doi:10.4018/jitsr.2013010101

Denrell, J., Fang, C. & Winter, S. G. (2003). The Economics of Strategic Opportunity. *Strategic MGMT. J.*, *24*, 977-78.

Deutsches Institut für Normung e. V. (DIN). (Ed.). (2000). Gesamtwirtschaftlicher Nutzen der Normung: Volkswirtschaftlicher Nutzen. Der Zusammenhang zwischen Normung und technischem Wandel, ihr Einfluss auf den Außenhandel und die Gesamtwirtschaft. Berlin: Beuth Verlag GmbH.

Dey, P. K., Clegg, B., & Bennett, D. (2009). Managing implementation issues in enterprise resource planning projects. In M. H. Sherif (Ed.), *Handbook of enterprise integration*. Boca Raton, FL: Auerbach. doi:10.1201/9781420078220-c24

Dick, G. P. M., Heras, I., & Casedesús, M. (2008). Shedding light on causation between ISO 9001 and improved business performance. *International Journal of Operations & Production Management*, *28*(9), 687–708. doi:10.1108/01443570810881811

DIN 820-1. (2014). Berlin: Beuth Verlag GmbH.

DIN EN 45020. (2007). Berlin: Beuth Verlag GmbH.

DIN. (2000). *Economic benefits of standardization: Summary of results*. Beuth Verlag GmbH and Deutsches Institut für Normung. Retrieved from http://www.din.de/sixcms_upload/media/2896/economic_benefits_standardization.pdf

DIN. (2000). *Economic benefits of standardization: Summary of results*. Final report and practical examples. Retrieved September 26, 2014, from DIN German Institute for Standardization e. V.: http://www.din.de/sixcms_upload/media/2896/economic_benefits_standardization.pdf

DIN. (2004). *Economic benefits of standardization: Summary of results*. Berlin: Beuth Verlag GmbH and Deutsches Institut für Normung.

Dixit, A. K., & Stiglitz, J. E. (1977). Monopolistic Competition and Optimum Product Diversity. *The American Economic Review*, *67*(3), 297–308.

Djelic, M., & Sahlin-Andersson, K. (2006). *Transnational governance: Institutional dynamics of regulation*. Cambridge University Press. doi:10.1017/CBO9780511488665

DKE. (2011). *Smart Grid. Deutsche Normungsroadmap "E-Energy/Smart Grids 2.0.* Retrieved from http://www.e-energy.de/documents/DKE_Roadmap_SmartGrid_230410_Engllish.pdf

DKE. (2012). *The German Standardization Roadmap for Electromobility—Version 2.* Retrieved from https://www.vde.com/en/dke/std/Documents/E-Mobility_Normungsroadmap_V2_EN.pdf

DKE. (2014). *DKE/K STD_1113 Use Cases: Methodiken und Referenzmodelle.* DKE.

DNP. (2014). *Indikatorenbericht 2014: Deutsches Normungspanel.* Retrieved June 10th, 2015, from projects.inno.tu-berlin.de/DNP/Bericht_2014.pdf

Dokko, G., & Rosenkopf, L. (2010). Social capital for hire? Mobility of technical professionals and firm influence in wireless Standards committees. *Organization Science, 21*(3), 677–695. doi:10.1287/orsc.1090.0470

Dölle, J. E. (2013). *Lieferantenmanagement in der Automobilindustrie. Struktur und Entwicklung der Lieferantenbeziehungen von Automobilherstellern. Schriften zur Unternehmensentwicklung.* Wiesbaden, Germany: Springer Gabler Verlag.

Donaldson, L. (2001). *The contingency theory of organizations. Foundations for organizational science.* Thousand Oaks, CA: SAGE Publications. doi:10.4135/9781452229249

Dooyeweerd, H. (1955). A new critique of theoretical thought. In The general theory of the modal spheres (vol. 2). Philadelphia: The Presbyterian and Reformed Publishing Company.

Dooyeweerd, H. (1957). A new critique of theoretical thought. In The structures of individuality of temporal reality (vol. 3). Philadelphia: The Presbyterian and Reformed Publishing Company.

Dosi, G. (1982). Technological paradigms and technological trajectories: A suggested interpretation of the determinants and directions of technical change. *Research Policy, 11*(3), 147–162. doi:10.1016/0048-7333(82)90016-6

Dranove, D., & Gandal, N. (2003). The DVD-vs.-DIVX standard war: Empirical evidence of network effects and preannouncement effects. *Journal of Economics & Management Strategy, 12*(3), 363–386. doi:10.1162/105864003322309518

Drazin, R., & Van de Ven, A. H. (1985). Alternative forms of fit in Contingency Theory. *Administrative Science Quarterly, 30*(4), 514–539. doi:10.2307/2392695

Drezner, D. W. (2002). *Who rules? The Regulation of Globalization, Previous versions of the paper were presented at the 2001 American Political Science Association and other events.* Retrieved from www.danieldrezner.com/research/whorules.pdf

Duffie, D. (2008). *Innovations in Credit Risk Transfer: Implications for Financial Stability.* BIS Working Papers, No. 255.

Dul, J., & Hak, T. (2008). *Case study methodology in business research.* Oxford, UK: Butterworth-Heinemann/Elsevier.

Duncan, R. B. (1972). Characteristics of organizational environments and perceived environmental uncertainty. *Administrative Science Quarterly, 17*(3), 313–327. doi:10.2307/2392145

Dutch Standardisation Forum. (2012). *BOMOS2i: The practical approach.* Author.

Earley R., Kang, L. &Green-Weiskel, L. (2011). *Electric vehicles in the context of sustainable development in China.* The innovation Center for Energy and Transportation (iCET), Beijing Los Angeles New York, Background Paper No.9, CSD19/2011/BP9, 7-8.

Ebers, M., & Oerlemans, L. (2013). The variety of governance structures beyond market and hierarchy. *Journal of Management.* doi: .10.1177/0149206313506938

EC. (2009). *Modernising ICT Standardisation in the EU - The Way Forward.* Brussels: European Commission.

EC. (2011). *A strategic vision for European standards: Moving forward to enhance and accelerate the sustainable growth of the European economy by 2020*. Brussels: European Commission.

Eckartz, S., & Folmer, E. (2015). *BOMOD, Management and Development Model for Open Data*. TNO Publication.

Economides, N. (1996). *The Economics of Networks. International Journal of Industrial Organization*, *14*(6), 673–699. doi:10.1016/0167-7187(96)01015-6

Economides, N., & Katsamakas, E. (2006). Two-sided competition of proprietary vs. open source technology platforms and the implications for the software industry. *Management Science*, *52*(7), 1057–1071. doi:10.1287/mnsc.1060.0549

EEAS. (2013). Retrieved from http://eeas.europa.eu/china/

Egyedi, T. (2014). The Impact of Competing Standards: On Innovation and Interoperability for E-government. *PIK - Praxis der Informationsverarbeitung und Kommunikation, 37*(3), 211-215.

Egyedi, T. M. (2003). Consortium problem redefined: Negotioting 'democracy' in the actor network on Standardization. *Journal of IT Standards and Standardization Research,* (2), 22-38.

Egyedi, T. M. (2001). Why JavaTM was -not- standardized twice. *Computer Standards & Interfaces*, *23*(4), 253–265. doi:10.1016/S0920-5489(01)00078-2

Egyedi, T. M., & Blind, K. (2008). *The dynamics of standards*. Cheltenham, UK: Edward Elgar.

Egyedi, T. M., & Hawkins, R. (2010). Standards Consortia. In W. Hesser, A. J. Feilzer, & H. J. de Vries (Eds.), *Standardisation in Companies and Markets* (3rd ed.; pp. 443–490). Hamburg, Germany: Helmut Schmidt University.

Egyedi, T., & Sherif, M. (2010). Standards dynamics through an innovation lens: Next-generation ethernet networks. *IEEE Communications Magazine*, *48*(10), 166–171. doi:10.1109/MCOM.2010.5594692

Eisenhardt, K. M. (1989). Building theories from case study research. *Academy of Management Review*, *14*, 488–511.

Eisenhardt, K. M., & Graebner, M. E. (2007). Theory building from cases: Opportunities and challenges. *Academy of Management Journal*, *50*(1), 25–32. doi:10.5465/AMJ.2007.24160888

Eisenmann, T., Parker, G., & Van Alstyne, M. W. (2006). Strategies for two-sided markets. *Harvard Business Review*, *84*(10), 92–101. PMID:16649701

Ellul, J. (1970). *The meaning of the city*. Grand Rapids, MI: Eerdmans.

EPO (European Patent Office). (2007). *Scenarios for the Future, March 2007*. Retrieved from http://www.epo.org/topics/patent system/scenarios-for-the-future/

Ernst, D. (2012). American's voluntary standards system – A "best practice" model for innovation policy? *East-West Center Working Papers, Economics Series*, *128*, 1-57.

Ernst, D. (2011).*Indigenous innovation and Globalization: The Challenge for China's Standardization Strategy*. La Jolla, CA: UC Institute on Global Conflict and Cooperation.

Espeland, W. N. (1998). *The struggle for water*. University of Chicago Press.

EU Mobility and Transport Website. (2013). Retrieved from http://ec.europa.eu/transport/themes/urban/vehicles/road/electric_en.htm

EUCCC Study. (2012). *Executive summary*. Retrieved from http://www.europeanchamber.com.cn/en/publications-position-paper

European Commission Directorate-General for Enterprise and Industry (EC). (2014). *Patents and Standards: A Modern Framework for IPR-Based Standardization*. Author.

Farrell, J., Hayes, J., Shapiro, C., & Sullivan, T. (2007). Standard setting, patents, and hold-up. *Antitrust Law Journal*, *74*(3), 603–670.

Farrell, J., & Saloner, G. (1985). Standardization, compatibility, and innovation. *The Rand Journal of Economics*, *16*(1), 70–83. doi:10.2307/2555589

Farrell, J., & Saloner, G. (1986). Installed base and compatibility: Innovation, Product Preannouncements, and predation. *The American Economic Review*, *76*, 943–954.

Farrell, J., & Saloner, G. (1988). Coordination through Committees and Markets. *The Rand Journal of Economics*, *19*(2), 235–252. doi:10.2307/2555702

Farrell, J., & Saloner, G. (1992). Converters, Compatibility, and the Control of Interfaces. *The Journal of Industrial Economics*, *40*(1), 9–35. doi:10.2307/2950625

Farrell, J., & Shapiro, C. (1988). Dynamic Competition with Switching Costs. *The Rand Journal of Economics*, *19*(1), 123–137. doi:10.2307/2555402

Faull, J., Nikpay, A., & Taylor, D. (2014). *The EU law of competition*. Oxford, UK: Oxford University Press.

Felin, T., & Foss, N. J. (2005). Strategic organization: A field in search of micro-foundations. *Strategic Organization*, *3*(4), 441–455. doi:10.1177/1476127005055796

Felin, T., & Foss, N. J. (2009). Experience and Repetition as Antecedents of Organizational Routines and Capabilities: A Critique of Behaviorist and Empiricist Approaches. *SMG WP*, *9*, 2009.

Feuchter, H. (1996). *Werknormung. Normungskunde* (Vol. 37). Berlin: Beuth Verlag GmbH.

FIA. (2011). *Federation Internationale de l'Automobile*. Brussels: Towards E-Mobility the Challenges Ahead.

Filipovic, E. (2013). *Mapping Electric-Mobility: Standards Infrastructure for Market uptake*. Paper presented at SIIT 2013 Conference, Sophia-Antipolis, France. Retrieved from http://ieee-siit.org/program/

Fincham, R., Fleck, J., Proctor, R., Scarbrough, H., Tierney, M., & Williams, R. (1994). *Expertise and Innovation: Information Technology Strategies in the Financial Services Sector*. Oxford, UK: Clarendon Press.

Fiss, P. C., & Zajac, E. (2006). The symbolic management of strategic change: Sensegiving via framing and decoupling. *Academy of Management Journal*, *49*(6), 1173–1193. doi:10.5465/AMJ.2006.23478255

Flaherty, T. (1986). Coordinating International Manufacturing and Technology. In M. Porter (Ed.), *Competition in Global Industries* (pp. 83–110). Boston: Harvard Business School Press.

Fligstein, N., & Goldstein, A. (2010). The anatomy of the mortgage securitization crisis. In M. Lounsbury & P. M. Hirsch (Eds.), Markets on trial: The economic sociology of the U.S. financial crisis: Part A (Research in the Sociology of Organizations, Volume 30) (pp. 29-70). Emerald Group Publishing Limited. doi:10.1108/S0733-558X(2010)000030A006

Fojcik, T. M., & Proff, H. (2014). *Accelerating market diffusion of battery electric vehicles through alternative concepts*. Paper presented at the 20th GERPISA International Colloquium: Structuring new automotive industries, restructuring old automotive industries and the new geopolitics of the global automotive sector, Krakow, Poland.

Folmer, E. & Verhoosel, J. (2011). *State of the Art on Semantic IS Standardization, Interoperability & Quality*. Enschede, The Netherlands: TNO, University of Twente, CTIT, NOiV.

Folmer, E. (2012). *Quality of Semantic Standards*. Enschede, The Netherlands: TNO, University of Twente, CTIT, NOiV.

Folmer, E., & Berends, W. (2009). Top IS research on quality of transaction standards, a structured literature review to identify a research gap. In *Proceedings of 6th International Conference on Standardization and Innovation in Information Technology (SIIT 2009)*. Verlag Mainz.

Folmer, E., Heddier, M., Matzner, M., Räckers, M., & Becker, J. (2014). A Method for Managing IT-Based Boundary Objects: Design and Application in the Public Sector. In *Proceedings of the European Conference on Information Systems (ECIS 2014)*. Tel Aviv, Israel: ECIS.

Folmer, E., & Punter, M. (2011). *Management and Development Model for Open standards (BOMOS) version 2*. The Hague, The Netherlands: Netherlands Open in Connection.

Folmer, E., & Zsolt Kulcsar, I. (2013). Instant Standard Concept for Data standards Development.*Americas Conference on Information Systems (AMCIS)*, Chicago, IL.

Fomin, V. V., & de Vries, H. J. (2009). How Balanced is Balanced Enough? Case Studies of Stakeholders' (Under-) Representation in Standardization Process. In *Proceedings of the 6th International Conference on Standardization and innovation in Information Technology SIIT 2009*. Aachen: Wissenschaftsverlag Mainz.

Fomin, V.V., Su, J., & Gao, P. (2011). Indigenous standard development in the presence of dominant international standards: the case of the AVS standard in China. *Technology Analysis & Strategic Management*, *23*(7), 748.

Forselius, R. M. (1998). *Strategic standardization management and competitive advantage*. (Unpublished Doctoral Dissertation). University of New Haven, New Haven, CT.

Foss, K. (1996). *A transaction cost perspective on the influence of Standards on product development: Examples for the fruit and vegetable market*. Working paper No 96-9. DRUID Working Paper Series, Copenhagen.

Foucault, M. (1976). *Histoire de la sexualité I – La volonté de savoir*. Paris: Éditions Gallimard.

Foucault, M. (1977). *Discipline and Punish; Birth of the Prison* (A. Sheridan, Trans.). London: Penguin. (Original work published 1975)

Foucault, M. (1982). Afterword. The subject and power. In H. L. Dreyfus & P. Rabinov (Eds.), *Michel Foucault. Beyond structuralism and hermeneutics* (pp. 208–226). Brighton, UK: University of Chicago Press.

Franklin, A., & Gale, D. (1994). *Financial innovations and risk sharing*. Cambridge, MA: MIT Press.

Frey, C. (1995). *Die intrapersonelle Balance-Theorie im Führungsverhalten*. Berlin: Springer. doi:10.1007/978-3-322-97704-5

Freyschuss, A. (2015). Så blev en svensk koncern världsledande med hjälp av Standarder. *SIS*. Retrieved from http://www.sis.se/Global/Om%20SIS/kundseminarium%202011/camfil_ppt.pdf

Furusten, S. (2000). The Knowledge Base of Standards. In *A World of Standards* (pp. 71–84). Oxford, UK: Oxford University Press.

Gallagher, S. (2007). The complementary role of dominant designs and industry Standards. *IEEE Transactions on Engineering Management*, *54*(2), 371–379. doi:10.1109/TEM.2007.893991

Gallagher, S. R. (2012). The battle of the blue laser DVDs: The significance of corporate strategy in standards battles. *Technovation*, *32*(2), 90–98. doi:10.1016/j.technovation.2011.10.004

Garcia, L., Laporte, C. Y., Arteaga, J., & Bruggmann, M. (2015). Implementation and Certification of ISO/IEC 29110 in an IT Startup in Peru. Software Quality Professional Journal, 17(2).

Garcia, S. (2005). How standards enable adoption of project management practice. *IEEE Software, 22*(5), 22–29. doi:10.1109/MS.2005.122

Garcia, S., & Turner, R. (2006). *CMMI Survival Guide: Just Enough Process Improvement*. Addison Wesley Professional.

Gauch, S. (2008). + vs. –, Dynamics and effects of competing standards of recordable DVD media. In T. M. Egyedi & K. Blind (Eds.), The dynamics of standards (pp. 47-67). Cheltenham, UK: Edward Elgar.

Gauch, S., & Blind, K. (2014). Technological convergence and the absorptive capacity of standardisation. *Technological Forecasting and Social Change*, 1–14.

Gawer, A., & Cusumano, M. A. (2008). How Companies Become Platform Leaders. *MIT Sloan Management Review, 49*(2), 29–30.

Gerdin, J., & Greve, J. (2004). Forms of contingency fit in management accounting research—a critical review. *Accounting, Organizations and Society, 29*(3-4), 303–326. doi:10.1016/S0361-3682(02)00096-X

Gereffi, G., & Lee, J. (2012). Why the World Suddenly Cares About Global Supply Chains. *Journal of Supply Chain Management, 48*(3), 24–32. doi:10.1111/j.1745-493X.2012.03271.x

Gerst, M. (2010). *CNCA CCC IP Support Study commissioned by the IPR2 project of the European Commission: IPR and Trade Secrecy related issues in the Chinese CCC certification process*. (not published).

Gerst, M. (2013). *IP Issues in Technology Standards in a Multi-Stakeholder Environment – Analysing New Energy Vehicle Standardization in Europe and China*. (Unpublished MS Thesis).

Gerst, M., & Xudong, G. (2013b). Innovation Management of Electrical Vehicle Charging. Infrastructure Standards in the Sino-European Context. In *Proceedings of the 2013 ITU Kaleidoscope Academic Conference. Building Sustainable Communities*. Kyoto, Japan: ITU. Retrieved from http://www.itu.int/en/ITU-T/academia/kaleidoscope/2013/Documents/K-2013_Proceedings.pdf

Gerst, M., & Gao, X. (2013a). Electric Vehicle Standardization Management. In M. Zelm et al. (Eds.), *Enterprise Interoperability: I-ESA'12 Proceedings*. John Wiley & Sons, Inc. doi:10.1002/9781118561942.ch63

Gerundino, D., & Hilb, M. (2010, June). *The ISO Methodology - Assessing the economic benefits of standards*. ISO Focus, International Organization for Standardization.

Gethmann, C. F., & Sander, T. (1999). Rechtfertigungsdiskurse. In A. Grunwald & S. Saupe (Eds.), *Ethik in der Technikgestaltung. Praktische Relevanz und Legitimation* (pp. 117–151). Berlin: Springer. doi:10.1007/978-3-642-60033-3_7

Ghaziani, A., & Ventresca, M. J. (2005). Keywords and Cultural Change: Frame Analysis of Business Model Public Talk, 1975–2000. *Sociological Forum, 20*(4), 523–559. doi:10.1007/s11206-005-9057-0

Ghemawat, P. (2002). Competition and Business Strategy in Historical Perspective. *Business History Review, 76*(37), 37–38. doi:10.2307/4127751

Gibbs, G. (2007). *Analyzing qualitative data*. Los Angeles, CA: SAGE Publications.

Gibson, C. S. (2007). Globalization and the Technology Standards Game: Balancing Concerns of Protectionism and Intellectual Property in International Standards. Suffolk University Law School Legal Studies Research Paper No. 07-39, 2007. *Berkeley Technology Law Journal, 22*, 6.

Gilbert, R. J. (2011). Deal or no deal? Licensing negotiations in standard-setting organizations. *Antitrust Law Journal*, *77*(3), 855–888.

Ginsberg, A., & Venkatraman, N. (1985). Contingency perspectives of organizational strategy: A critical review of the empirical research. *Academy of Management Review*, *10*(3), 421–434.

Goudzwaard, B. (1979). *Capitalism and Progress: A Diagnosis of Western Society*. Toronto: Wedge.

Gouldner, A. W. (1954). *Patterns of industrial Bureaucracy – A case study of modern factory administration*. New York: The Free Press.

Grant, R. M. (2004). *Contemporary strategy analysis: concepts, techniques, applications* (5th ed.). Oxford, UK: Blackwell Publishing.

Graubner, M. (2006). *Task, firm size, and organizational structure in management consulting: An empirical analysis from a contingency perspective*. Wiesbaden, Germany: Deutscher Universitats-Verlag.

Greenstein, S., & Stango, V. (2006). *Standards and Public Policy*. Cambridge University Press. doi:10.1017/CBO9780511493249

Grewal, D. S. (2008). *Network power: The social dynamics of globalization*. New Haven, CT: Yale University Press.

Grimes, K. R. (2003). *ISO 9001:2000: a practical quality manual explained*. Milwaukee, WI: ASQ Quality Press.

Grossmann, A.-M. (2013). Suppliers' motives for applying company Standards – How company Standards can affect the buyer-supplier relationship. In *EURAS Proceedings 2013: The Euras Board Series*. Verlagshaus Mainz.

Großmann, A.-M., & von Gruben, P.-V. (2014). The Role of Company Standards in Supply Chains - The Case of the German Automotive Industry. In T. Blecker (Ed.), Innovative Methods in Logistics and Supply Chain Management: Current Issues and Emerging Practices (pp. 99–121). Berlin: epubli GmbH.

Grunwald, A. (2002). *Technikfolgenabschätzung—eine Einführung*. Berlin: Edition Sigma.

Grunwald, A. (2011). Responsible Innovation: Bringing together Technology Assessment, Applied Ethics, and STS researchapter *Enterprise and Work Innovation Studies*, *7*, 9–31.

Grunwald, A. (2013). *Handbuch Technikethik*. Stuttgart: J.B. Metzler.

Guler, I., Guillén, M., Macpherson, J., & Guillen, M. F. (2002). Global competition, institutions, and the diffusion of organizational practices: The international spread of ISO 9000 quality certificates. *Administrative Science Quarterly*, *47*(2), 207–232. doi:10.2307/3094804

Haan, M. A. (2003). Vaporware as a Means of Entry Deterrence. *The Journal of Industrial Economics*, *3*(3), 345–358. doi:10.1111/1467-6451.00204

Habermas, J. (1987). The Theory of Communicative Action: Lifeworld and System: A Critique of Functionalist Reason (vol. 2; T. McCarth, Trans.). Boston: Beacon. (Original work published 1981)

Hallett, T. (2010). The myth incarnate: Recoupling processes, turmoil, and inhabited institutions in an urban elementary school. *American Sociological Review*, *75*(1), 52–74. doi:10.1177/0003122409357044

Hamel, G., & Prahalad, G. K. (1996). *Competing for the future*. Boston: Harvard Business School Press.

Hanseth, O., Monteiro, E., & Hatling, M. (1996). Developing information infrastructure: The tension between standardization and flexibility. *Science, Technology & Human Values*, *21*(4), 407–426. doi:10.1177/016224399602100402

Hargrave, T. J., & Van De Ven, A. H. (2006). A collective action model of institutional innovation. *Academy of Management Review*, *31*(4), 864–888. doi:10.5465/AMR.2006.22527458

Harvey, M. G., & Novicevic, M. M. (2006). The World is Flat: A Perfect Storm for Global Business? *Organizational Dynamics*, *35*(3), 207–219. doi:10.1016/j.orgdyn.2006.05.002

Hauschildt, J. (1997). Innovationsmanagement: Vahlens Handbücher der Wirtschafts- und Sozialwissenschaften (2nd ed.). Munich: Vahlen.

Haverkamp, A. (2004). *Bezint eer gij begint – Standaardisatie van werkmethoden bij Polynorm Automotive – Onderzoek op basis van een christelijk-filosofische analyse*. Rotterdam: Erasmus Universiteit, Faculteit Bedrijfskunde.

Haverkamp, A. (2006). Een veelzijdig kwaliteitsverbeteringsproject. *Synaps*, *4*(20), 21–26.

Helfat, C. E. (1997). Know-how and asset complementarity and dynamic capability accumulation: The case of R&D. *Strategic Management Journal*, *18*(5), 339–360. doi:10.1002/(SICI)1097-0266(199705)18:5<339::AID-SMJ883>3.0.CO;2-7

Henson, S. J., & Hooker, N. H. (2001). Private Sector Management of Food Safety: Public Regulation and the Role of Private Controls. *International Food and Agribusiness Management Review*, *4*(1), 7–17. doi:10.1016/S1096-7508(01)00067-2

Herman, M. K. (2010). Negotiating standards-related patent licenses: How the deal is done (Parts 1 & 2). *Landslide*, *3*(2-3), 31-41.

Hesser, W., & Meyer, R. (1993). Parameter der Wirtschaftlichkeit von Normungsvorhaben – Das Wachstum der Typenvielfalt als überbetrieblich verwendbarer Parameter. *DIN Mitteilung*, (72), 349–354.

Hesser, W. (2010). Standardisation within a company – A strategic perspective. In W. Hesser (Ed.), *Standardisation in companies and markets* (3rd ed.; pp. 215–258). Hamburg, Germany: Helmut Schmidt University.

Hesser, W. (2010). Standardisation within a Company – a Strategic Perspective. In W. Hesser (Ed.), *Standardisation in Companies and Markets* (3rd ed.; pp. 215–258). Hamburg.

Hesser, W., Czaya, A., & Riemer, N. (2007). Development of Standards. In W. Hesser, A. Feilzer, & H. de Vries et al. (Eds.), *Standardisation in Companies and Markets* (2nd ed.; pp. 123–170). Hamburg, Germany: Helmut Schmidt University.

Hesser, W., & Inklaar, A. (1997). *An introduction to Standards and Standardization. In DIN-Normungskunde* (Vol. 36). Berlin: Beuth Verlag GmbH.

Hevner, A. R., & March, S. T. (2004). Design science in information systems research. *MIS Quarterly: Management Information Systems*, (1), 75–105.

Hevner, A., & Chatterjee, S. (2010). *Design Research in Information Systems - Theory and Practice*. New York: Springer. doi:10.1007/978-1-4419-5653-8

Hill, P. (1999). Tangibles, intangibles and services: A new taxonomy for classification of output. *The Canadian Journal of Economics. Revue Canadienne d'Economique*, *32*(2), 426–447. doi:10.2307/136430

Hofstede, G. (1978). The Poverty of Management Control Philosophy. *Academy of Management Review*, *3*(3), 450–461.

Hölk, S. (2012). Die deutsche Normungs-Roadmap Elektromobilität – Version 2. Arbeitsgruppe 4 „Normung, Standardisierung und Zertifizierung" der Nationalen Plattform Elektromobilität (NPE), Gemeinsame Geschäftsstelle Elektromobilität der Bundesregierung (Hrsg.). Berlin.

Holler, M. J. (1996). Die Rationalität strategischer Normung in Europa. In K.-E. Schenk, D. Schmidtchen, & M. E. Streit (Eds.), Jahrbuch für Neue Politische Ökonomie (pp. 137–153). Tübingen: Mohr-Siebeck.

Holmstrom, B. (1999). The Firm as a Subeconomy. *Journal of Law Economics and Organization, 15*(1), 74–102. doi:10.1093/jleo/15.1.74

Hubig, C., & Reidel, J. (Eds.). (2003). Ethische Ingenieurverantwortung. Handlungsspielräume und Perspektiven der Kodifizierung. Berlin: Edition Sigma.

Hudson, J., & Jones, P. (2001). Measuring the efficiency of stochastic signals of product quality. *Information Economics and Policy, 13*(1), 35–49. doi:10.1016/S0167-6245(00)00029-9

Iacono, S., & Kling, R. (2001). Computerization movements: The rise of the internet and distant forms of work. In J. Yates & J. Van Maanen (Eds.), *Information technology and organizational transformation: History, rhetoric, and practice* (pp. 93–136). Thousand Oaks, CA: Sage Publications. doi:10.4135/9781452231266.n4

Imai, M. (1986). *Kaizen – the key to Japan's competitive success*. New York: Random House.

IMF. (2006). *Global financial stability report: Market development and issues, April 2006*. IMF.

Interest. (2005). *D02 – Report on the results of the survey among researchers*. Retrieved from http://www.nets.rwth-aachen.de/Interest/D02.pdf

ISI. (2006). *INTEREST Integrating research and standardisation: D04 - Report on Case Studies*. Retrieved October 11, 2014, from http://www-i4.informatik.rwth-aachen.de/Interest/documents.html

ISI. (2007). *INTEREST Integrating research and standardisation: D 10 – Final Report*. Retrieved October 11, 2014, from http://www-i4.informatik.rwth-aachen.de/Interest/documents.html

ISO. (2005, May). *ISO/IEC JTC1/SC7 N3288, New Work Item Proposal – Software Life Cycles for Very Small Enterprises*. Retrieved from http://www.jtc1-sc7.org/

ISO. (2006). *ISO/IEC/IEEE 16085:2006 - Systems and software engineering — Lifecycle Processes — Risk management*. Geneva, Switzerland: International Organization for Standardization/International Electrotechnical Commission.

ISO. (2008). *ISO 9001:2008 Quality Management systems – Requirements*. Geneva: International Organisation for Standardisation.

ISO. (2008a). *ISO/IEC/IEEE 12207:2008, Systems and software engineering– Software life cycle Processes*. Geneva, Switzerland: International Organization for Standardization/International Electrotechnical Commission.

ISO. (2008b). *ISO/IEC/IEEE 15288:2008, Systems and software engineering– System life cycle Processes*. Geneva, Switzerland: International Organization for Standardization/International Electrotechnical Commission.

ISO. (2009). *ISO/IEC/IEEE 16326:2009 - Systems and software engineering — Lifecycle Processes — Project management*. Geneva, Switzerland: International Organization for Standardization/International Electrotechnical Commission.

ISO. (2009). *ISO/TS 16949:2009 Quality Management systems – Particular requirements for the application of ISO 9001:2008 for automotive production and relevant service part organisations*. Geneva: International Organisation for Standardisation.

ISO. (2010). *Economic Benefits of Standards: Implementation Guide*. International Organization for Standardization. Retrieved from http://www.standardsinfo.net/info/benefits/benefits_s1.html

ISO. (2010). *Guidance on social responsibility – ISO 26000:2010*. ISO.

ISO. (2011). *Economic benefits of standards - International case studies - Volume 1*. Retrieved March 13, 2015, from International Organization for Standardization: http://www.iso.org/iso/benefits_of_standards.html

ISO. (2011). *The ISO Survey of Management Standard Certifications 2011*. Geneva: ISO Central Secretariat.

ISO. (2011a). *ISO/IEC/IEEE 15289:2011, Systems and software engineering - Content of systems and software life cycle Process information products (Documentation)*. Geneva, Switzerland: International Organization for Standardization/International Electrotechnical Commission.

ISO. (2011b). *ISO/IEC TR 29110-5-1-2:2011, Software Engineering - Lifecycle Profiles for Very Small Entities (VSEs) - Part 5-1-2: Management and engineering guide - Generic profile group: Basic profile*. International Organization for Standardization/International Electrotechnical Commission. Available at no cost from ISO at: http://standards.iso.org/ittf/PubliclyAvailableStandards/c051153_ISO_IEC_TR_29110-5-1_2011.zip

ISO. (2012). *Economic benefits of standards - International case studies - Volume 2*. Retrieved March 13, 2015, from International Organization for Standardization: http://www.iso.org/iso/benefits_of_standards.html

ISO. (2012). *ISO/IEC TR 29110-5-1-1:2012, Software Engineering - Lifecycle Profiles for Very Small Entities (VSEs) - Part 5-1-1: Management and engineering guide - Generic profile group: Entry profile*. International Organization for Standardization/International Electrotechnical Commission. Available at no cost from ISO at: http://standards.iso.org/ittf/PubliclyAvailableStandards/

ISO. (2014). *ISO/IEC TR 29110-5-6-2:2014 - Systems Engineering – Lifecycle Profiles for Very Small Entities (VSEs) - Management and engineering guide: Generic profile group: Basic profile*. International Organization for Standardization/International Electrotechnical Commission. Available at no cost from ISO at: http://standards.iso.org/ittf/PubliclyAvailableStandards/index.html

ISO/IEC Guide 2. (2004). Geneva: ISO/IEC.

Iversen, E. (1999). Standardization and intellectual property rights: conflicts between innovation and diffusion in new telecommunications systems. In K. Jakobs (Ed.), *Standards and Standardization: A Global Perspective* (pp. 80–101). Idea Group Publishing.

Iversen, E. J., Oversjoen, E., & Lie, H. T. (2004). Standardization, innovation and IPR. *Teletronikk*, *2*, 65–79.

Jacobides, M. G. (2005). Industry change through vertical disintegration: How and why markets emerged in mortgage banking. *Academy of Management Journal*, *48*(3), 465–498. doi:10.5465/AMJ.2005.17407912

Jacobsson, B. (2000). Standardization and expert knowledge. In A World of Standards. Oxford Univ. Press.

Jain, S. (2012). Pragmatic agency in technology standards setting: The case of Ethernet. *Research Policy*, *41*(9), 1643–1654. doi:10.1016/j.respol.2012.03.025

Jakobs, K. & Kritzner, J. (2009). How to Select the Best Platform for ICT Standards Development. In *Information Communication Technology Standardization for E-Business Sectors: Integrating Supply and Demand Factors*. Hershey, PA: Information Science Reference.

Jakobs, K. (2000). Standardization Processes in IT: Impact: Problems and Benefits of User Participation. Wiesbaden, Germany: Vieweg+Teubner.

Jakobs, K. (2015). Corporate ICT standardisation management – Lessons from the literature and from case studies. In *Proceedings of the 7th Annual Conference of the EuroMed Academy of Business*. Nicosis, Cyprus: EuroMed Press.

Jakobs, K., & Gerst, M. (2013). *How to Perceive the (Future) Role of China in ICT Standardisation – A European Perspective*. Paper presented at the 8th IEEE Conference on Standardisation and innovation in Information Technology. Retrieved from http://ieee-siit.org/program/

Jakobs, K. (2006). Shaping User-side Innovation Through Standardisation – The Example of ICT. *Technological Forecasting and Social Change*, *73*(1), 27–40. doi:10.1016/j.techfore.2005.06.007

Jakobs, K. (2008). The IEEE 802.11 WLAN installation at RWTH Aachen University: A case of voluntary vendor lock-. In T. M. Egyedi & K. Blind (Eds.), *The dynamics of standards* (pp. 99–116). Cheltenham, UK: Edward Elgar.

Jakobs, K. (2011). How people and stakeholders shape standards – The case of IEEE 802.11. In J. Filipe & J. Cordeiro (Eds.), *Web information systems and technologies 2010* (pp. 1–13). Heidelberg, Germany: Springer. doi:10.1007/978-3-642-22810-0_1

Jakobs, K. (2011).The European Union's ICT StandardizationPolicy - Changes Ahead!? In *New Applications in IT Standards: Developments and Progress*. IGI Global.

Jakobs, K., & Blind, K. (2011). The European ICT Standardization Policy – Recent Developments, Current Situation, and a Brief Outlook. In *Proceedings of the 16th EURAS Conference. EURAS contributions to standardization research* (vol. 5). EURAS.

Jakobs, K., Procter, R., & Williams, R. (2001). The making of standards. *IEEE Communications Magazine*, *39*(4), 102–107. doi:10.1109/35.917511

Johansson, M., & Jonsson, A. (2012). The package logic: A study on value creation and knowledge flows. *European Management Journal*, *30*(6), 535–551. doi:10.1016/j.emj.2012.04.003

Johnson, G. & Scholes, K. (2005). *Exploring corporate strategy: text and cases.* Academic Press.

Kalpakjian, S. (1995). *Manufacturing Engineering and Technology* (3rd ed.). Reading, MA: Addison-Wesley.

Kalsbeek, L. (1975). *Contours of a Christian Philosophy: An Introduction to Herman Dooyeweerd's Thought*. Toronto: Wedge.

Kang, B., & Motohashi, K. (2015). Essential intellectual property rights and inventors' involvement in standardization. *Research Policy*, *44*(2), 483–492. doi:10.1016/j.respol.2014.10.012

Kaniadakis, A., Constantinides P. (2014). Innovating Financial Information Infrastructures: the transition of legacy assets to the securitization market. *Journal of the Association of Information Systems, 15*(5).

Kast, F. E., & Rosenzweig, J. E. (Eds.). (1973). *Contingency views of organization and management*. Chicago: Science Research Assoc.

Katz, M. L., & Shapiro, C. (1985). Network externalities, competition and compatibility. *The American Economic Review*, *75*, 424–440.

Katz, M., & Shapiro, C. (1985). Network Externalities, Competition, and Compatibility. *The American Economic Review*, *75*, 424–440.

Katz, M., & Shapiro, C. (1994). Systems competition and Network effects. *The Journal of Economic Perspectives*, *8*(2), 93–115. doi:10.1257/jep.8.2.93

Kayworth, T., & Sambamurthy, V. (2000). Facilitating localized exploitation and enterprise-wide integration in the use of IT infrastructures: The role of PC/LAN infrastructure standard. *The Data Base for Advances in Information Systems*, *31*(4), 54–81. doi:10.1145/506760.506765

Kelly, M. (1994). Foucault, Habermas and the Self-Referentiality of Critique. In M. Kelly (Ed.), *Critique and Power – Recasting the Foucault/Habermas Debate*. Cambridge, MA: The MIT Press.

Kennedy, S., Suttmeier, R.P. & Su, J. (2008). *Standards, Stakeholders, and innovation: China's Evolving Role in the Global Knowledge Economy*. NBR Reports, 4.

Kesan, J. P., & Hayes, C. M. (2013). FRAND's forever: Standards, patent transfers, and Licensing Commitments. *Indiana Law Journal (Indianapolis, Ind.)*, *89*, 231–314.

Kindleberger, C. P. (1983). Standards as public, collective and private goods. *Kyklos*, *36*(3), 377–396. doi:10.1111/j.1467-6435.1983.tb02705.x

Kogut, B. (2000). The network as knowledge: Generative rules and the emergence of structure. *Strategic Management Journal*, *21*(3), 405–425. doi:10.1002/(SICI)1097-0266(200003)21:3<405::AID-SMJ103>3.0.CO;2-5

Kondo, Y. (1996). Are creative ability and work standardisation in contradictory relationship? *Training for Quality*, *4*(3), 35–39. doi:10.1108/09684879610125734

Kondo, Y. (2000). Innovation versus Standardisation. *The TQM Magazine*, *12*(1), 6–10. doi:10.1108/09544780010287177

Kortum, S., & Lerner, J. (1999). What is behind the recent surge in patenting? *Research Policy*, *28*(1), 1–22. doi:10.1016/S0048-7333(98)00082-1

Kovacevicj, A. (2009). *Google's Approach to Competition*. Google Public Policy Blog.

Krasner, H. (1998, November). Using the cost of quality approach for software. *Crosstalk*, *11*(November), 6–11.

Krechmer K. (2000). *The Fundamental Nature of Standards: Economics Perspective*. Paper presented at the International JA Schumpeter Conference, Manchester, UK.

Krechmer, K. (2005). *Standards Mark the Course of Economic Progress*. Unpublished Paper.

Krechmer, K. (2009). Open standards: A Call for Change. *IEEE Communications Magazine*, *47*(May), 88–94. doi:10.1109/MCOM.2009.4939282

Kroes, N. (2010). *How to get more Interoperability in Europe: Openness at the heart of the EU Digital Agenda*. Retrieved 06-17, 2010, from http://europa.eu/rapid/pressReleasesAction.do?reference=SPEECH/10/300&format=HTML

Kroll, P., & Kruchten, P. (2003). *The Rational unified Process made easy: A practitioner's guide to the RUP*. Boston: Addison-Wesley Longman Publishing Co.

Krugman, P. (2007, December 3). Innovating Our Way to Financial Crisis. *New York Times*.

Kunda, G. (1992). *Engineering culture: culture and control in a high-tech organization*. Philadelphia: Temple Univ. Press.

Kwaka, J., Leeb, H., & Fomin, V. V. (2011). Government coordination of conflicting interests in standardisation: Case studies of indigenous ICT standards in China and South Korea. *Technology Analysis & Strategic Management, 23*(7), 789-806.

Lagrosen, Y., & Lagrosen, S. (2005). The effects of Quality Management – A survey of Swedish quality professionals. *International Journal of Operations & Production Management*, *25*(10), 940–952. doi:10.1108/01443570510619464

Land, S. K. (1997). *Results of the IEEE Survey of Software Engineering Standards Users*. Software Engineering Standards Symposium and Forum, 1997. Emerging International Standards, Walnut Creek, CA. doi:10.1109/SESS.1997.595987

Laporte, C. Y. (2009). *Contributions to software engineering and the development and deployment of international software engineering standards for very small entities*. (Doctoral dissertation). Université de Bretagne Occidentale, Brest.

Laporte, C. Y., Alexandre, S., & O'Connor, R. (2008b). A Software Engineering Lifecycle Standard for Very Small Enterprises. In R.V. O'Connor et al. (Eds.), EuroSPI 2008. Dublin, Ireland: CCIS 16.

Laporte, C. Y., Chevalier, F., & Maurice, J.-C. (2013c, February). *Improving Project Management for Small Projects.* ISO Focus+, International Organization for Standardization. Retrieved from http://www.iso.org/iso/home/news_index/iso_magazines/isofocusplus_index/isofocusplus_2013/isofocusplus_2013-02.htm

Laporte, C. Y., Hébert, C., & Mineau, C. (2014b). Development of a Social Network Website Using the New ISO/IEC 29110 Standard Developed Specifically for Very Small Entities. *Software Quality Professional Journal, 16*(4), 4-25. Available at not cost at: http://asq.org/pub/sqp/past/vol16_issue4/index.html

Laporte, C. Y., Houde, R., & Marvin, J. (2014c). *Systems Engineering International Standards and Support Tools for Very Small Enterprises.* Paper to be presented at the 24th Annual International Symposium of INCOSE (International Council on Systems Engineering), Las Vegas, NV. doi:10.1002/j.2334-5837.2014.tb03167.x

Laporte, C. Y., Séguin, N., & Villas Boas, G. (2013a, February). *Seizing the benefits of software and systems engineering standards.* ISO Focus, International Organization for Standardization. Retrieved from http://www.iso.org/iso/home/news_index/iso_magazines/isofocusplus_index/isofocusplus_2013/isofocusplus_2013-02.htm

Laporte, C. Y., Alexandre, S., & Renault, A. (2008c, June). The Application of International Software Engineering Standards in Very Small Enterprises. *Software Quality Professional Journal, ASQ, 10*(3), 4–11.

Laporte, C. Y., & O'Connor, R. V. (2014b). A Systems Process Lifecycle Standard for Very Small Entities: Development and Pilot Trials. In B. Barafort, R. V. O'Connor, & R. Messnarz (Eds.), *Systems, Software and Services Process Improvement, CCIS* (Vol. 425). Springer-Verlag. doi:10.1007/978-3-662-43896-1_2

Laporte, C. Y., & O'Connor, R. (2014a). Systems and Software Engineering Standards for Very Small Entities - Implementation and Initial Results. QUATIC'2014. *9th International Conference on the Quality of Information and Communications Technology*, Guimarães, Portugal.

Laporte, C. Y., O'Connor, R., & Fanmuy, G. (2013b, May/June). International Systems and Software Engineering Standards for Very Small Entities. *Crosstalk, 26*(3), 28–33.

Laporte, C. Y., & Palza Vargas, E. (2012). The Development of International Standards to facilitate Process Improvements for Very Small Enterprises. In *Software Process Improvement and Management: Approaches and Tools for Practical Development* (pp. 34–61). Hershey, PA: IGI Global. doi:10.4018/978-1-61350-141-2.ch003

Laporte, C. Y., Renault, A., & Alexandre, S. (2008a). Applying ISO/IEC Software Engineering Standards in Very Small Enterprises. In *Software Process Improvement for Small and Medium Enterprises: Techniques and Case Studies* (pp. 42–70). Hershey, PA: Idea Group Inc. doi:10.4018/978-1-59904-906-9.ch002

Latour, B. (1987). *Science in Action.* Cambridge, MA: Harvard University Press.

Laursen, K., & Salter, A. J. (2014). The paradox of openness: Appropriability, external search and collaboration. *Research Policy, 43*(5), 867–878. doi:10.1016/j.respol.2013.10.004

Layne-Farrar, A. (2010). Nondiscriminatory pricing: Is standard setting different? *Journal of Competition Law & Economics, 6*(4), 811–838. doi:10.1093/joclec/nhq009

Layne-Farrar, A., Llobet, G., & Padilla, J. (2014). Payments and participation: The incentives to join cooperative standards-setting efforts. *Journal of Economics & Management Strategy, 23*(1), 24–49. doi:10.1111/jems.12046

Layne-Farrar, A., Padilla, A. J., & Schmalensee, R. (2007). Pricing patents for licensing in standard-setting organizations: Making sense of FRAND commitments. *Antitrust Law Journal, 74*(3), 671–706.

Lazaric, N., & Denis, B. (1995). Routinization and memorization of tasks in a workshop: The case of the introduction of ISO norms. *Industrial and Corporate Change*, *15*(5), 873–896.

Lazina, L. K. (2014). Standardization Landscape of Company Standards at Volkswagen AG: Short Version. In I. Mijatovic & K. Jakobs (Eds.), *EURAS Proceedings 2014: Cooperation between Standardisation Organisations and the Scientific and Academic Community: EURAS Annual Standardisation Conference* (Vol. 19). Aachen, Germany: Verlagshaus Mainz GMbH Aachen.

Lea, G., & Hall, P. (2004). Standards and intellectual property rights: An economic and legal perspective Information. *Economic Policy*, *16*, 71.

Lee, H., & Oh, S. (2006). A standards war waged by a developing country: Understanding international standard setting from the actor-network perspective. *The Journal of Strategic Information Systems*, *15*(3), 177–195. doi:10.1016/j.jsis.2005.10.002

Leiponen, A. E. (2008). Competing through cooperation: The organization of Standard setting in wireless telecommunications. *Management Science*, *54*(11), 1904–1919. doi:10.1287/mnsc.1080.0912

Leland, H. (1979). Quacks, lemons, and licensing: A theory of minimum quality standards. *Journal of Political Economy*, *87*(6), 1328–1346. doi:10.1086/260838

Lelong, B., & Mallard, A. (1995). Presentation. *Reseaux*, *18*(102), 9–34.

Lemley, M. (2002). Intellectual Property Rights and Standard Setting Organizations. In *California Law Review (1889 2002), 1900: While standardization has great economic value in many markets, group standard setting also poses some potential threats to competition.* Retrieved from http://papers.ssrn.com/sol3/papers.cfm?abstract_id=310122

Lemley, M. A. (2002). Intellectual property rights and standard-setting organizations. *California Law Review*, *90*(6), 1889–1980. doi:10.2307/3481437

Lemley, M. A., & Shapiro, C. (2007). Patent holdup and royalty stacking. *Texas Law Review*, *85*(7), 1991–2049.

Lemley, M. A., & Shapiro, C. (2013). A simple approach to setting reasonable royalties for standard-essential patents. *Berkeley Technology Law Journal*, *28*, 1135–1166.

Lerner, J., & Tirole, J. (2014). A Better Route to Tech Standards. *Science*, *343*(6174), 972–973. doi:10.1126/science.1246439 PMID:24578566

Levin, A. (2010). *Bridging the Standardization Gap: China and the ITU-T*. Retrieved from http://www.china-eu-standards.org/dissemination/slides/Levin.pdf

Leyshon, A., & Thrift, N. (2008). The capitalism of almost everything: The future of finance and capitalism. *Theory, Society and Culture*, *24*(7-8), 97–115. doi:10.1177/0263276407084699

Lichtman, D. (2010). Understanding the RAND commitment. *Houston Law Review*, *47*, 1023–1050.

Lin, Ch.-I., & Jang, W.-Y. (2008). Successful ISO 9000 implementation in Taiwan: How can we achieve it, and what does it mean? *International Journal of Productivity and Performance Management*, *57*(8), 600–622. doi:10.1108/17410400810916044

Lindkvist, K. (1981). Approaches to textual analysis. In K. E. Rosengren (Ed.), *Advances in content analysis*. Beverly Hills, CA: Sage.

Li, X., & Chen, Y. (2012). Corporate IT Standardization: Product Compatibility, Exclusive Purchase Commitment, and Competition Effects. *Information Systems Research*, *23*(4), 1158–1174. doi:10.1287/isre.1110.0412

Loch, C. H., & Huberman, B. A. (1999). A punctuated-equilibrium model of technology diffusion. *Management Science, 45*(2), 160–177. doi:10.1287/mnsc.45.2.160

Lo, L. K., & Chang, D. S. (2007). The difference in the perceived benefits between firms that maintain ISO certification and those that do not. *International Journal of Production Research, 48*(5), 1881–1897. doi:10.1080/00207540600733709

Longenecker, J. G., & Pringle, C. D. (1978). The illusion of Contingency Theory as a general theory. *Academy of Management Review, 3*(3), 679–683.

Louhiala-Salminen, L., & Kankaanranta, A. (2012). Language as an issue in international internal communication: English or local language? If English, what English? *Public Relations Review, 38*(2), 262–269. doi:10.1016/j.pubrev.2011.12.021

Löwer, U. M. (2005). *Interorganisational Standards - Managing Web Services Specifications for Flexible Supply Chains.* Heidelberg, Germany: Physica-Verlag - A Springer Company.

Lundqvist, B. (2014). *Standardization under EU competition rules and US antitrust laws: The rise and limits of self-regulation.* Edward Elgar Publishing Limited. doi:10.4337/9781781954867

MacKenzie, D. (2011). The credit crisis as a problem in the sociology of knowledge. *American Journal of Sociology,* (May): 2011.

MacQueen, H. (2008). *Contemporary Intellectual Property: Law and Policy* (2nd ed.). Oxford University Press.

Magd, H. A. E. (2008). ISO 9001:2000 in the Egyptian manufacturing sector: perceptions and perspectives. *International Journal of Quality & Reliability Management, 25*(2), 173–200. doi:10.1108/02656710810846934

Mahadevappa, B., & Kotreshwar, G. (2004). Quality Management Practices in Indian ISO 9000 Certified Companies: An Empirical Evaluation. *Total Quality Management, 15*(3), 295–305. doi:10.1080/1478336042000183389

Mair, C. (2012). *Openness, intellectual property and standardization in the European ICT sector.* Retrieved from https://papers.ssrn.com/sol3/papers.cfm?abstract_id=2065271

Malisius, G., & Weidner, T. (1998). Die Normenabteilung – fit für die Zukunft. In *DIN Mitteilung,* (pp. 272-275). Academic Press.

Manders, N. B. (2015). Implementation and Impact of ISO 9001. ERIM Ph.D. Series Research in Management. Rotterdam, The Netherlands: Erasmus Research Institute of Management (ERIM).

Manders, B. (2015). *Implementation and Impact of ISO 9001. ERIM PhD Series in Research in Management 337.* Rotterdam: Erasmus Research Institute of Management – ERIM.

Manders, B. (2015). *Implementation and Impact of ISO 9001. ERIM PhD Series Research in Management.* Rotterdam: Erasmus Research Institute of Management – ERIM.

Manyika, J., Remes, J., & Woetzel, J. (2014). A productivity perspective on the future of growth. *The McKinsey Quarterly.*

March, J. G. (1991). Exploration and Exploiting in Organizational Learning. *Organization Science, 2*(1), 71–87. doi:10.1287/orsc.2.1.71

March, J. G., & Simon, H. A. (1958). *Organizations.* Cambridge, MA: Blackwell publishers.

Markus, M. L., Dutta, A., Steinfield, C. W., & Wigand, R. T. (2008). The Computerization Movement in the US Home Mortgage Industry: Automated Underwriting from 1980 to 2004. In K. Kraemer & M. Elliott (Eds.), *Computerization Movements and Technology Diffusion: From Mainframes to Ubiquitous Computing* (pp. 115–144). Medford, NY: Information Today.

Markus, M. L., & Steinfield, C. W. (2006). Industry-wide Information Systems Standardization as collective action: The case of U.S. residential mortgage industry. *Management Information Systems Quarterly*, 439–465.

Markus, M. L., Steinfield, C. W., Wigand, R. T., & Minton, G. (2006). Industry-wide IS standardization as collective action: The case of the US residential mortgage industry. *Management Information Systems Quarterly*, *30*(Special Issue), 439–465.

Martínez-Costa, M., Choi, T. Y., Martinez, J. A., & Martínez-Lorente, A. R. (2009). ISO 9000/1994, ISO 9001/2000 and TQM: The performance debate revisited. *Journal of Operations Management*, *27*(6), 495–511. doi:10.1016/j.jom.2009.04.002

Maskus, K., & Merrill, S. A. (Eds.). (2013). Intellectual property challenges for standard-setting in the global economy. Washington, DC: National Research Council (NRC).

Masten, S. E. (1991). A legal basis for the firm. In O. E. Williamson & S. G. Winter (Eds.), *The Nature of the Firm: Origins, Evolution and Development* (pp. 196–212). New York: Oxford University Press.

Matutes, C., & Regibeau, P. (1996). A selective review of the economics of Standardization. Entry deterrence, technological progress and international competition. *European Journal of Political Economy*, *12*(2), 183–209. doi:10.1016/0176-2680(95)00013-5

Matyusz, Z. (2012). *The effect of contingency factors on the use of manufacturing practices and operations performance.* (PhD thesis). Cornvinus University of Budapest.

Maurer, M., Gerdes, J. C., Lenz, B., & Winner, H. (Eds.). (2015). *Autonomes Fahren. Technische, rechtliche und gesellschaftliche Aspekte*. Berlin: Springer.

Maxwell, J. A. (2005). *Qualitative Research Design - An Interactive Approach.* Thousand Oaks, CA: Sage.

Mayring, P. (2010). *Qualitative Inhaltsanalyse: Grundlagen und Techniken* (11th ed.). Weinheim: Beltz. doi:10.1007/978-3-531-92052-8_42

McGregor, J. (2012). *China's Drive for 'Indigenous innovation' A web of industrial policies*. Global Regulatory Cooperation Project for US Chamber of Commerce. Retrieved from http://www.uschamber.com/reports/chinas-drive-indigenous-innovation-web-industrial-policies

McKinnon, R. (2013). *Consent of the Networked: The Worldwide Struggle For Internet Freedom*. New York, NY: Basic Books.

Mendoza, R. A., & Ravichandran, T. (2011). An Exploratory Analysis of the Relationship Between Organizational and Institutional Factors Shaping the Assimilation of Vertical Standards. *International Journal of IT Standards and Standardization Research*, *9*(1), 24–51. doi:10.4018/jitsr.2011010102

Meyer, N. (2012). *Public intervention in private rule-making: the role of the European Commission in industry standardization*. (PhD thesis). The London School of Economics and Political Science. Retrieved from http://etheses.lse.ac.uk/236/

Meyer, J. W., & Rowan, B. (1977). Institutionalized Organizations: Formal Structure as Myth and Ceremony. *American Journal of Sociology*, *83*(2), 340. doi:10.1086/226550

Meyer, R. (1995). *Parameter der Wirksamkeit von typenreduzierenden Normungsvorhaben. DIN-Normungskunde* (Vol. 34). Berlin: Beuth Verlag.

Michel, S. (2011). Bargaining for RAND royalties in the shadow of patent remedies law. *Antitrust Law Journal*, *77*(3), 889–911.

Miles, M. B., & Huberman, A. M. (1984). *Qualitative data analysis: A sourcebook of new methods*. Sage Publications.

Miles, R. E., Snow, C. C., & Meyer, A. D., & Coleman, H. J., Jr. (1978). Organizational strategy, structure, and process STOR. *Academy of Management Review*, *3*(3), 546–562. PMID:10238389

Miller, D. (1996). Configurations revisited. *Strategic Management Journal*, *17*(7), 505–512. doi:10.1002/(SICI)1097-0266(199607)17:7<505::AID-SMJ852>3.0.CO;2-I

Miller, J. S. (2007). Standard setting, patents, and access lock-in: RAND licensing and the theory of the firm. *Indiana Law Review*, *40*, 351–396.

Mintzberg, H. (1979). *The structuring of organizations*. Prentice-Hall Inc.

Mione, A. (2015). The value of intangibles in a situation of innovation: Questions raised by the case of Standards. *Journal of Innovation Economics & Management*, *2*(17), 49–68. doi:10.3917/jie.017.0049

Miotti, H. (2009, June). The economic impact of standardization technological change. In *Standards growth in France*. AFNOR.

Modi, S. B., & Mabert, V. A. (2007). Supplier development: Improving supplier performance through knowledge transfer. *Journal of Operations Management*, *25*(1), 42–64. doi:10.1016/j.jom.2006.02.001

Moll, R. (2013, February). *Being prepared – A bird's eye view of SMEs and risk management*. ISO Focus +. Geneva, Switzerland: International Organization for Standardization.

Mom, T. J. M., Van den Bosch, F. A. J., & Volberda, H. W. (2002). Exploratie én exploitatie van kennis - Het managen van de determinanten van horizontale kennisuitwisseling. *Management en Organisatie*, *56*(5), 23–43.

MST. (2012). *12th Five-Year Special Plan on Technical Standards – Draft for Comments*. Obtained privately.

Mueller, J. M. (2002). *Patent Misuse through the capture of industry standards*. Retrieved from http://papers.ssrn.com/sol3/papers.cfm?abstract_id=1346546

Mueller, T., Dittes, S., Ahlemann, F., Urbach, N., & Smolnik, S. (2015). Because Everybody is Different Towards Understanding the Acceptance of Organizational IT Standards. In *Proceedings of the 48th Hawaii International Conference on System Sciences* (HICSS). Kauai, HI: HICSS. doi:10.1109/HICSS.2015.487

Muffatto, M. (1999). Introducing a platform strategy in product development. *International Journal of Production Economics*, *60-1*, 145–153. doi:10.1016/S0925-5273(98)00173-X

Myers, M. D. (1999). Investigating Information Systems with Ethnographic Research. *Communications of the Associations for Information Systems, 2*(23), 1–20.

Nair, A., & Prajogo, D. (2009). Internalisation of ISO 9000 standards: The antecedents role of functionalist and institutionalist drivers and performance implications. *International Journal of Production Research*, *47*(16), 4545–4568. doi:10.1080/00207540701871069

Nakamura, S. (2000). *The new Standardisation: Keystone of continuous improvement in manufacturing*. Portland, OR: Productivity Press.

Narayanan, V., & Chen, T. (2012). Research on technology Standards: Accomplishment and challenges. *Research Policy*, *41*(8), 1375–1406. doi:10.1016/j.respol.2012.02.006

Naveh, E., & Erez, M. (2004). Innovation and Attention to Detail in the Quality Improvement Paradigm. *Management Science*, *50*(11), 1576–1586. doi:10.1287/mnsc.1040.0272

Naveh, E., Marcus, A., & Moon, H. K. (2004). Implementing ISO 9000: Performance improvement by first or second movers. *International Journal of Production Research*, *42*(9), 1843–1863. doi:10.1080/00207540410001662912

Nelson, R. R., & Winter, S. G. (1982). *An evolutionary theory of economic change*. Cambridge, MA: Belknap press/ Harvard University Press.

Nickerson, J. V., & zur Muehlen, M. (2006). The ecology of standards processes: Insights from Internet standards making. *Management Information Systems Quarterly*, *30*(1), 476–488.

Nicolau, J. L., & Sellers, R. (2002). The stock market's reaction to quality certification: Empirical evidence from Spain. *European Journal of Operational Research*, *142*(3), 632–641. doi:10.1016/S0377-2217(01)00312-5

Nicoloso, E. (2007). Bridging the gap between Quality Management systems and product quality. *ISO Management Systems*, *7*(1), 7–11.

Nonaka, I. (1994). A Dynamic Theory of Organisational Knowledge Creation. *Organization Science*, *5*(1), 14–37. doi:10.1287/orsc.5.1.14

Nonaka, I., & Takeuchi, H. (1995). *The knowledge creating company: how Japanese companies create the dynamics of innovation*. New York: Oxford University Press.

Normann, R., & Ramirez, R. (1993). From value chain to value constellation: Designing interactive strategy. *Harvard Business Review*, *71*, 65–78. PMID:10127040

Normann, R., & Ramirez, R. (1994). *Designing interactive strategy – from value chain to value constellation*. Chichester, UK: Wiley & Sons.

O'Connor, R. V., & Laporte, C. Y. (2010). Towards the Provision of Assistance for Very Small Entities in Deploying Software Lifecycle Standards. In *Product Focused Software Development and Process Improvement* (PROFES 2010). PROFES.

OECD & Eurostat. (2005). Oslo manual: Guidelines for collecting and interpreting innovation data (3rd ed.). Paris: Organisation for Economic Co-operation and Development; Statistical Office of the European Communities.

OECD. (2008). *Review of innovation Policy China*. Retrieved from http://www.oecd.org/sti/inno/oecdreviewsofinnovationpolicychina.htm

Ogilvie, S. (2004). Guilds, efficiency, and social capital: Evidence from German proto-industry. *The Economic History Review*, *57*(2), 286–333. doi:10.1111/j.1468-0289.2004.00279.x

Okhomatovskiy, I., & David, R. J. (2012). Setting your own Standards: Internal corporate governance codes as a response to institutional pressure. *Organization Science*, *23*(1), 155–176. doi:10.1287/orsc.1100.0642

Okun, D. T. (2010). China: Intellectual Property Infringement, Indigenous innovation Policies, and Frameworks for Measuring the Effects on the U.S. Economy. Investigation No. 332-514. U.S. International Trade Commission, USITC Publication 4199 (amended), November 2010.

Oliver, C. (1991). Strategic Responses to Institutional Processes. *Academy of Management Review*, *16*(1), 145–179.

Oshri, I., & Weeber, C. (2006). Cooperation and competition standards-setting activities in the digitization era: The case of wireless information services. *Technology Analysis and Strategic Management*, *18*(2), 265–283. doi:10.1080/09537320600624196

OTA. (1992). *Global standards: Building blocks for the future*. TCT-512. Washington, DC: Congress of the United States, Office of Technology Assessment. Retrieved from http://www.princeton.edu/~ota/disk1/1992/9220/9220.PDF

Ouchi, W. G. (1977). The relationship between organizational structure and organizational control. *Administrative Science Quarterly*, *22*(1), 95–113. doi:10.2307/2391748

Padma, P., Ganesh, L. S., & Rajendran, C. (2008). A study on the critical factors of ISO 9001:2000 and organisational performance of Indian manufacturing firms. *International Journal of Production Research*, *46*(18), 4981–5011. doi:10.1080/00207540600793299

Parasuraman, A., Zeithaml, V. A., & Berry, L. L. (1985). A conceptual model of service quality and its implications for future research. *Journal of Marketing*, *49*(4), 41–50. doi:10.2307/1251430

Parker, G., & Van Alstyne, M. W. (2003). Two sided networks: A theory of information product design. *Management Science*, *51*(10), 1494–1504. doi:10.1287/mnsc.1050.0400

Perez-Aleman, P. (2011). Collective learning in global diffusion: Spreading quality Standards in a developing country cluster. *Organization Science*, *22*(1), 173–189. doi:10.1287/orsc.1090.0514

Petroski, H. (1992). *To engineer is human: The role of failure in successful design*. Vintage.

Pfaffenberger, B. (2000). The Rhetoric of Dread: Fear, Uncertainty, and Doubt (FUD) in Information Technology Marketing. *Knowledge, Technology & Policy*, *13*(3), 78–92. doi:10.1007/s12130-000-1022-x

Pfeffer, J. (1982). *Organizations and Organization Theory*. Marshfield, MA: Pitman.

PMI. (2013a). *A Guide to the Project Management Body of Knowledge (PMBOK® Guide)* (5th ed.). Newtown Square, PA: Project Management Institute.

PMI. (2013b). *Organizational Project Management Maturity Model (OPM3®)* (3rd ed.). Newtown Square, PA: Project Management Institute.

Poba-Nzaou, P., & Raymond, L. (2011). Managing ERP system risk in SMEs: A multiple case study. *Journal of Information Technology*, *26*(3), 170–192. doi:10.1057/jit.2010.34

Pollock, N., & Williams, R. (2011). Who decides the shape of product markets? The knowledge institutions that name and categorize new technologies. *Information and Organization*, *21*(4), 194–217. doi:10.1016/j.infoandorg.2011.08.001

Poon, M. (2009). From New Deal Institutions to Capital Markets: Commercial consumer risk scores and the making of subprime mortgage finance. *Accounting, Organizations and Society*, *34*(5), 654–674. doi:10.1016/j.aos.2009.02.003

Porter, M. E. (1985). Competitive advantage: Creating and sustaining superior performance. New York: Free Press.

Porter, M. (1980). *Competitive strategy*. New York: Free Press.

Porter, M. E. (1985). *Competitive advantage: creating and sustaining superior performance*. New York: Free Press.

Porter, M. E. (1996). What is strategy? *Harvard Business Review*, *74*(6), 61–78. PMID:10158474

Porter, M. E. (2004). *Competitive strategy: techniques for analyzing industries and competitors*. New York: Free Press.

Porter, M. E. (2008). *On Competition, Updated and expanded*. Boston: Harvard Business School Publication.

Posner, R. (2013). *Antitrust Law* (9th ed.). University of Chicago Press.

Potter, N., & Sakry, M. (2002). *Making Process Improvement Work*. Addison-Wesley – Pearson Education.

Pound, J., Lehn, K. & G. Jarel. (1986). Are Takeovers Hostile to Economic Performance? *Regulation, 10*, 25-30.

Powell, W., & Grodal, S. (2005). Networks for Innovators. In The Oxford Handbook of innovation. Oxford.

Preda, A. (2006). Sociotechnical agency in financial markets: The case of the stock ticker. *Social Studies of Science*, *36*(5), 753–782. doi:10.1177/0306312706059543

Prudhomme, D. (2012). *Dulling the Cutting-Edge: How Patent-Related Policies and Practices Hamper innovation in China, study of the European Chamber of Commerce in China*. Academic Press.

Prusa, T., & Schnitz, J. Jr. (1991). Are New Firms an Important Source of Innovation? Evidence from the PC Software Industry. *Economics Letters*, *35*(3), 39–342. doi:10.1016/0165-1765(91)90155-E

Puffert, D. (2009). *Tracks across Continents, Paths through History: The Economic Dynamics of Standardization in Railway Gauge*. Chicago: University of Chicago Press.

Pugh, D. S., Hickson, D. J., Hinnings, C. R., & Turner, C. (1969). The context of organization structures. *Administrative Science Quarterly*, *14*(1), 91–114. doi:10.2307/2391366

Quinn, S. (2009). *Towards a market for mortgage-backed securities: Credit lending, the Federal budget, and the politics of debt management*. University of California Group of Economic History, Working Paper.

Rada, R., & Craparo, J. (2000). Standardizing Software Projects. *Communications of the ACM*, *43*(12), 21–25. doi:10.1145/355112.355117

Rada, R., & Ketchell, J. (2000). Sharing standards: Standardizing the European information society. *Communications of the ACM*, (3): 21–25.

Radice, R. A., Roth, N. K., O'Hara, A. C., & Ciarfella, W. A. (1985). A Programming Process Architecture. *IBM Systems Journal*, *24*(2), 79–90. doi:10.1147/sj.242.0079

Rammert, W. (2003). Two Paradoxes of Fragmented Knowledge Production: Combining Heterogeneous and Cultivating Non-Explicit Knowledge. *SASE Annual Conference*, Aix-en-Provence, France.

Rathmell, J. M. (1966). What is meant by services? *Journal of Marketing*, *4*(4), 32–36. doi:10.2307/1249496

Ray, G., Barey, J. B., & Muhanna, W. A. (2004). Capabilities, business processes, and competitive advantage: Choosing the dependent variable in empirical tests of the resources based view. *Strategic Management Journal*, *25*(1), 23–38. doi:10.1002/smj.366

RaymondE. (2009). http://www.catb.org/~esr/jargon/html/F/FUD.html

Reichwald, R., & Piller, F. T. (2009). Interaktive Wertschöpfung: Open Innovation, Individualisierung und neue Formen der Arbeitsteilung (2nd ed.). Wiesbaden: Gabler.

Reimers, K., & Li, M. (2005). Effectiveness of governance structures for 3G Standardization and implications for China's 3G policy. In P. Cunningham & M. Cunningham (Eds.), eAdoption and the knowledge economy: Issues, applications, case studies (pp. 256-263). IOS Press.

Reuters. (2014). Retrieved from http://www.reuters.com/article/2014/10/25/us-china-autos-idUSKCN0IE04M20141025

Reuters. (2015). Retrieved from http://www.reuters.com/article/2015/01/20/us-china-economy-idUSKBN0KT04920150120

Riddiough, T. J. (2011). Can securitization work? Economic, Structural and Policy considerations. *Journal of Portfolio Management*, *37*(5), 24–38. doi:10.3905/jpm.2011.37.5.024

Riillo, C. A. F. (2013). The engagement in stadarization activities: A firm level analyse of formal and company standardization. In K. Jakobs, H. J. De Vries, A. Genesh, A. Gulacsi, & I. Soetert (Eds.), *EURAS Proceedings 2013. Standards: Boosting European Competitiveness: EURAS Annual Standardisation Conference* (vol. 18, pp. 325–339). Aachen, Germany: Verlagshaus Mainz GMbH Aachen.

Rochet, J.-C., & Tirole, J. (2003). Platform competition in two-sided markets. *Journal of the European Economic Association, 1*(4), 990–1029. doi:10.1162/154247603322493212

Rogers, E. M. (2003). *Diffusion of Innovations* (5th ed.). New York: Free Press.

Rogers, E. M. (2003). *Diffusion of innovations* (5th ed.). New York: Simon & Schuster International.

Rombach, B. (1993). Success at the expense of control. In The Reforming Organization. Routledge.

Rongping, M., & Zhuoliang, W. (2005). *The Role of Standards in National Technology Policy in China*. Retrieved from http://strategicstandards.com/files/China.pdf

Rosenkopf, L., Metiu, A., & George, V. P. (2001). From the Bottom Up? Technical Committee Activity and Alliance Formation. *Administrative Science Quarterly, 46*(4), 748–772. doi:10.2307/3094830

Roth, K., & Morrison, A. J. (1992). Business-level competitive strategy: A contingency link to internationalization. *Journal of Management, 18*(3), 473–487. doi:10.1177/014920639201800303

Ruef, M., & Scott, W. R. (1998). A Multidimensional Model of Organizational Legitimacy: Hospital Survival in Changing Institutional Environments. *Administrative Science Quarterly, 43*(4), 877–904. doi:10.2307/2393619

Rumelt, R. P., Schendel, D., & Teece, D. J. (1991). Strategic Management and Economics. *Strategic Management Journal, 12*(S2), 5–29. doi:10.1002/smj.4250121003

Salop, S. C., & Scheffman, D. T. (1983). Raising Rivals' Costs. *The American Economic Review*, (73), 267–271.

Salop, S.C., & Sheffman, D.T. (1983). Raising Rivals' Costs. *The American Economic Review, 73*(2), 267-271.

Salzberger, E. M. (2012). *Law and economics of innovation*. Cheltenham, UK: Edward Elgar. doi:10.4337/9781784714345

Samuelson, P. (2007). Questioning copyrights in standards. *Boston College Law Review, 48*, 193–224.

Sanchez, R., & Mahoney, J. T. (1996). Modularity, flexibility, and knowledge management in product and organizational design. *Strategic Management Journal, 17*(S2), 63–76. doi:10.1002/smj.4250171107

Sandholtz, K. W. (2012). Making Standards stick: A theory of coupled vs. Decoupled compliance. *Organization Studies, 33*(5-6), 655–679. doi:10.1177/0170840612443623

Santos, F. M., & Eisenhardt, K. M. (2005). Organizational Boundaries and Theories of Organization. *Organization Science, 16*(5), 491–508. doi:10.1287/orsc.1050.0152

Sauder, M., & Espeland, W. N. (2009). The discipline of rankings: Tight coupling and organizational change. *American Sociological Review, 74*(1), 63–82. doi:10.1177/000312240907400104

Schaefer, M. (2011). *Grusswort zur "Sino-German Electric Vehicles Science and Technology Cooperation" am 24*. Retrieved from http://www.china.diplo.de/Vertretung/china/de/__pr/2011/reden-bo-2011/111024-emobilitaet-ps.html

Schengen. (1990). The Schengen acquis - Convention implementing the Schengen Agreement of 14 June 1985 between the Governments of the States of the Benelux Economic Union, the Federal Republic of Germany and the French Republic on the gradual abolition of checks at their common borders. *Official Journal L 239*, 22/09/2000 P. 0019 – 0062.

Schilling, M. (1999). Winning the Standards race: Building installed base and the availability of complementary goods. *European Management Journal*, *17*(3), 265–274. doi:10.1016/S0263-2373(99)00005-5

Schoechle, T. (2009). *Standardization and Digital Enclosure: The Privatization of Standards, Knowledge, and Policy in the Age of Global Information Technology*. Hershey, PA: IGI Global. doi:10.4018/978-1-60566-334-0

Schreier, M. (2012). *Qualitative content analysis in practice*. Thousand Oaks, CA: SAGE Publications.

Schumpeter, J. (1934). *The theory of economic development*. London: Transaction publishers.

Schumpeter, J. (1943). *Capitalism, Socialism and Democracy*. Unwin University Books.

Scott, W. R., & Meyer, J. W. (1991). The organization of societal sectors: Propositions and early evidence. In The new institutionalism in organizational analysis. Univ. of Chicago Press.

SEI. (2010). *Software Engineering Institute. (2010). CMMI for Development, Version 1.3*. Pittsburgh, PA: Carnegie Mellon University. CMU/SEI.

Seidl, D. (2007). Standard setting and following in corporate governance: An observation-theoretical study of the effectiveness of governance codes. *Organization*, *14*(5), 705–728. doi:10.1177/1350508407080316

Sellars, W. (1997). *Empiricism and the Philosophy of Mind*. Cambridge, MA: Harvard University Press.

SESEC Project. (2013). Retrieved from http://quality-partnerships.cn/standards-in-china/sesec-project/

Shapiro, C., & Varian, H. (1999). Information Rules: A Strategic Guide to the Network Economy. Harvard Business Press.

Shapiro, C., & Varian, H. R. (1998). *Information rules: A strategic guide to the network economy*. Harvard Business Review Press.

Shapiro, C., & Varian, H. R. (1999). *Information rules - A Strategic Guide to the Network Economy*. Boston: Harvard Business School Press.

Shapiro, C., & Varian, H. R. (1999). The art of Standards wars. *California Management Review*, *41*(2), 8–32. doi:10.2307/41165984

Sheehan, N. T., & Foss, N. J. (2007). Enhancing the prescriptiveness of the resource-based view through Porterian activity analysis. *Management Decision*, *2*(3), 240–260.

Sherif, M. H. (2003). When Is Standardization Slow?. *International Journal of IT Standards and Standardization Research*, *1*(1), 19–32. doi:10.4018/jitsr.2003010102

Shiller, R. J. (2003). *The New Financial Order: Risk in the 21st Century*. Princeton, NJ: Princeton University Press.

Shin, D.-H., Kim, H., & Hwang, J. (2015). Standardization revisited: A critical literature review on standards and innovation. *Computer Standards & Interfaces*, *38*, 152–157. doi:10.1016/j.csi.2014.09.002

Sidak, J. G. (2013). The meaning of FRAND, part I: Royalties. *Journal of Competition Law & Economics*, *9*(4), 931–1055. doi:10.1093/joclec/nht040

Sidak, J. G. (2015). The meaning of FRAND, part II: Injunctions. *Journal of Competition Law & Economics*, *11*(1), 201–269. doi:10.1093/joclec/nhv005

Simons, C. A. J., & De Vries, H. J. (2002). *Standaard of maatwerk – Bedrijfskeuzes tussen uniformiteit en verscheidenheid*. Schoonhoven, The Netherlands: Academic Service.

Singh, R. M., & Dahlin, K. B. (2009). *The winner does not take it all: Survival of niche standards.* Paper presented at the EGOS 2009 "The Social Dynamics of Standardization", Barcelona.

SIPO. (2013). Retrieved from http://english.sipo.gov.cn/news/official/201303/t20130326_789188.html

Skitol, R. A. (2005). Concerted buying power: Its potential for addressing the patent holdup problem in standard setting. *Antitrust Law Journal, 72*(2), 727–744.

Slob, F. J. C., & De Vries, H. J. (2001). Wie die (petro)chemische Industrie von Normen profitiert. *DIN-Miteilungen, 80*(6), 440–442.

SLoCAT@ Rio + 20. (2012). *Expanding the Use of electric mobility: Options for sustainable urban transport.* UN Department of Economic and Social Affairs (DESA). Available at http://www.slocat.net/content/expanding-use-electric-mobility-options-sustainable-urban-transport

Smit, J. (2004). *Het drama Ahold.* Amsterdam: Balans.

Sousa, R., & Voss, C. A. (2008). Contingency research in operations management practices. *Journal of Operations Management, 26*(6), 697–713. doi:10.1016/j.jom.2008.06.001

Sparrow, R. (2007). Killer Robots. *Journal of Applied Philosophy, 24*(1), 62–77. doi:10.1111/j.1468-5930.2007.00346.x

Spence, A. M. (1984). Cost Reduction, Competition, and Industry Performance. *Econometrica, 52*(1), 101–112. doi:10.2307/1911463

Spivak, S. M., & Brenner, F. C. (2001). *Standardization essentials: principles and practice.* New York: Dekker.

Sprenger, C. C., & Ten Have, S. (1996). Kennismanagement als motor van de lerende organisatie. *Holland Management Review, 51*, 69–76.

Spring, M. B., & Weiss, M. B. (1995). Financing the standards development process. In B. Kahin & J. Abbate (Eds.), *Standards policy for information infrastructure* (pp. 289–320). Boston: MIT Press.

Stabell, C. B., & Fjeldstad, Ø. D. (1998). Configuring value for competitive advantage: On chains, shops, and networks. *Strategic Management Journal, 19*(5), 413–437. doi:10.1002/(SICI)1097-0266(199805)19:5<413::AID-SMJ946>3.0.CO;2-C

Steinfield, C. W., & Markus, M. L. (2011). Cooperative Advantage and Vertical Information System Standards: An Automotive Supply Chain Case Study.*44th Hawaii International Conference on System Sciences (HICSS)*, Hawaii. doi:10.1109/HICSS.2011.130

Stevens, J. M., Steensma, H. K., Harrison, D. A., & Cochran, P. L. (2005). Symbolic or substantive document? The influence of ethics codes on financial executives' decisions. *Strategic Management Journal, 26*(2), 181–195. doi:10.1002/smj.440

Stevenson-Yang, A., & DeWoskin, K. (2005, March). China Destroys the IP Paradigm. *Far Eastern Economic Review*, 9-18.

Stoneman, P., & Diederen, P. (1994). Technology diffusion and public policy. *The Economic Journal, 105*(425), 918–930. doi:10.2307/2234987

Storz, C. (2007). Compliance with international standards: The EDIFACT and ISO 9000 standards in Japan. *Social Science Japan Journal, 10*(2), 217–241. doi:10.1093/ssjj/jym048

Suarez, F. F. (2004). Battles for technological dominance: An integrative framework. *Research Policy, 33*(2), 271–286. doi:10.1016/j.respol.2003.07.001

Suárez, F. F., & Utterback, J. M. (1995). Dominant designs and the survival of firms. *Strategic Management Journal, 16*(6), 415–430. doi:10.1002/smj.4250160602

Sullivan, L. H. (1896). The Tall Office Building Artistically Considered. *Lippincott's Magazine*, *57*, 403–409.

Suttmeier, P., & Yao, X. (2011). *China's IP Transition: Rethinking Intellectual Property Rights in a Rising China*. NBR Special Report, 7.

Suttmeier, R. P., Yao, X., & Tan, A. Z. (2006). *Standards of Power? Technology, Institutions, and Politics in the Development of China's National Standards Strategy*. Retrieved from http://kms2.isn.ethz.ch/serviceengine/Files/RESSpecNet/106854/ipublicationdocument_singledocument/DEC48CA8-82BF-4097-AEBF-3C47AA85F045/en/10_Apr06.pdf

Suttmeier, R.P. & Xiangkui, Y. (2004). *China's Post-WTO Technology Policy – Standards, Software, and the Changing Nature of Techno-Nationalism*. NBR Special report no. 7.

Swaminathan, J. M. (2001). Using Standardized Operations. *California Management Review*, *43*(3), 125–135. doi:10.2307/41166092

Swan, J., & Scarbrough, H. (2005). The Politics of Networked Innovation. *Human Relations*, *58*(7), 913–943. doi:10.1177/0018726705057811

Swann, G. P. (2000). The Economics of Standardization: Final Report for Standards and Technical Regulations Directorate Department of Trade and Industry. Manchester Business School, University of Manchester.

Swann, G.M.P. (2010). *The economics of standardization: An update*. Report for the UK Department of BIS.

Swann, P. (2000). *The economics of standardization: Final report for Standards and Technical Regulations Directorate Department of Trade and Industry*. Retrieved July 17, 2014, from https://www.gov.uk/government/uploads/system/uploads/attachment_data/file/16506/The_Economics_of_Standardization_-_in_English.pdf

Swann, P. (2010). *The economics of standardization: An update*. Retrieved May 17, 2015, from https://www.gov.uk/government/uploads/system/uploads/attachment_data/file/16509/The_Economics_of_Standardization_-_an_update_.pdf

Swann, G. M. P., & Lambert, R. (2010). Why do standards enable and constrain innovation? In J. C. Gratz & K. Jakobs (Eds.), *Services standardisation: EURAS contributions to standardisation research* (Vol. 2, pp. 357–376). Aachen, Germany: Mainz Academic Publishers.

Swann, G. P. (2000). *The economics of Standardization*. Manchester, UK: University of Manchester.

Swann, G. P. (2010). *The economics of Standardization: an update. Report for the UK Department of Business, Innovation and Skills*. BIS.

Swann, P. (2000). *The Economics of Standardization. Final Report for Standards and Technical Regulations*. Manchester, UK: Directorate Department of Trade and Industry, University of Manchester.

Swanson, D. G., & Baumol, W. J. (2005). Reasonable and nondiscriminatory (RAND) royalties, standards selection, and control of market power. *Antitrust Law Journal*, *73*(1), 1–58.

Swanson, E. B., & Ramiller, N. C. (1997). The Organising Vision in Information Systems Innovation. *Organization Science*, *8*(5), 5. doi:10.1287/orsc.8.5.458

SWD. (2012). *458 final*. Commission Staff Working Document, Towards enhanced patent valorisation for growth and jobs. Retrieved from http://ec.europa.eu/enterprise/search/index_en.htm?q=SWD%282012%29+458+final

Szilagyi, J. A., & Wallace, M. (1980). *Organizational behavior and performance*. Santa Monica, CA: Goodyear.

Tagscherer, U. (2012). Electric mobility in China – A policy review. Fraunhofer ISI Discussion Papers innovation Systems and Policy Analysis No. 30, Karlsruhe, February 2012, 8.

Tamm Hallström, K. (2000). Organizing the process of standardization. In A World of Standards. Oxford Univ. Press.

Tamm Hallström, K. (2004). *Organizing international standardization: ISO and the IASC in quest for authority*. Northampton, MA: Edward Elgar.

Tamm-Halström, K. (2004). *Organizing International Standardisation – ISO and the IASC in Quest of Authority*. Cheltenham, UK: Edward Elgar.

Tang, R. (2009). *The Rise of China's Auto Industry and Its Impact on the U.S. Motor Vehicle Industry*. Report for Congress Prepared for Members and Committees of Congress.

Tassey, G. (1995). The Roles of Standards as Technology Infrastructure. In Standards, Innovation and Competitiveness (pp. 161–71). Camberley, UK: Edward Elgar.

Tassey, G. (1982). The role of government in supporting measurement standards for high-technology industries'. *Research Policy*, *11*(5), 311–320. doi:10.1016/0048-7333(82)90053-1

Tassey, G. (2000). Standardization in technology-based markets. *Research Policy*, *29*(4-5), 587–602. doi:10.1016/S0048-7333(99)00091-8

Taylor, F. W. (1993). *The Principles of Scientific Management, and Shop Management*. London: Routledge/Thoemmes Press. (Original work published 1903)

TCEITSRP. (2012). *Deliverable 3: Review of State of Art: China and EU*. Retrieved from http://www.china-eu-standards.org/deliverables/index.htm

Teece, D. J. (1986). Profiting from technological innovation: Implications for integration, collaboration, licensing and public policy. *Research Policy*, *15*(6), 285–305. doi:10.1016/0048-7333(86)90027-2

Teece, D. J. (2007). Explicating dynamic capabilities: The nature and microfoundations of (sustainable) enterprise performance. *Strategic Management Journal*, *28*(13), 1319–1350. doi:10.1002/smj.640

Teece, D. J. (2010). Business Models, Business Strategy and Innovation. *Long Range Planning*, *43*(2-3), 172–194. doi:10.1016/j.lrp.2009.07.003

Teichmann, H. (2010). *Global Standardization: Coping with Multilingualism. In EURAS contributions to standardisation research 3*. Aachen, Germany: Verlag Mainz.

Terziovski, M., Power, D., & Sohal, A. S. (2003). The longitudinal effects of the ISO 9000 certification process on business performance. *European Journal of Operational Research*, *146*(3), 580–595. doi:10.1016/S0377-2217(02)00252-7

Terziovski, M., Samson, D., & Dow, D. (1997). The business value of Quality Management systems certification – Evidence from Australia and New Zealand. *Journal of Operations Management*, *15*(1), 1–18. doi:10.1016/S0272-6963(96)00103-9

Tether, B. S., Hipp, C., & Miles, I. (2001). Standardisation and particularization in services: Evidence from Germany. *Research Policy*, *30*(7), 1115–1138. doi:10.1016/S0048-7333(00)00133-5

Thévenot, L. (2009). Governing life by standards: A view from engagements. *Social Studies of Science*, *39*(5), 793–813. doi:10.1177/0306312709338767

Thompson, J. (1967). *Organizations in Action: Social Science Bases of Administrative Theory*. McGraw-Hill.

Thompson, J. D. (1967). *Organizations in action*. New York: McGraw-Hill.

Tidd, J., & Bessant, J. R. (2009). *Managing innovation: Integrating technological, market and organizational change* (4th ed.). Chichester, UK: Wiley.

Timmermans, S., & Epstein, S. (2010). A world of standards but not a standard world: Toward a sociology of standards and standardization. *Annual Review of Sociology*, *36*(1), 69–89. doi:10.1146/annurev.soc.012809.102629

Toth, R. B. (1996). *Standards Activities of Organizations in the United States*. NIST Special Publication 806. Retrieved from http://gsi.nist.gov/global/docs/SP%20806.pdf

Trade. (2013). Retrieved from http://ec.europa.eu/trade/policy/countries-and-regions/countries/china/

Tremblay, N. (2012). *Implantation de la norme d'ingénierie système ISO/IEC 29110 chez CSinTrans (Rapport de maitrise)*. École de Technologie Supérieure. (in French)

Tu, Q., Vonderembse, M. A., Ragu-Nathan, T. S., & Ragu-Nathan, B. (2004). Measuring modularity-based manufacturing practices and their impact on mass customization capability: A customer-driven perspective. *Decision Sciences*, *35*(2), 147–168. doi:10.1111/j.00117315.2004.02663.x

Tushman, M. (1979). Work characteristics and subunit communication structure: A contingency analysis. *Administrative Science Quarterly*, *24*(1), 82–98. doi:10.2307/2989877

Tzelepis, D., Tsekjouras, K., Skuras, D., & Dimara, E. (2006). The effects of ISO 9001 on firms' productive efficiency. *International Journal of Operations & Production Management*, *26*(10), 1146–1165. doi:10.1108/01443570610691111

U.S. Dep't of Justice & Fed. Trade Comm'n. (2007). *Antitrust enforcement and intellectual property rights: Promoting innovation and competition.* Retrieved from www.usdoj.gov/atr/public/hearings/ip/222655.pdf

Ulrich, K. (1995). The role of product architecture in the manufacturing firm. *Research Policy*, *24*(3), 419–440. doi:10.1016/0048-7333(94)00775-3

Updegrove, A. (2006). Standards stakeholders: Who should, and who does, set standards? *Consortium Standards Bulletin, 5*(4). Retrieved from http://www.consortiuminfo.org/bulletins/apr06.php#feature

Updegrove, A. (2007). *Government Policy and "Standards – Based Neo-Colonialism"*. Retrieved from www.consortiuminfo.org

Updegrove, A. (2009). A concise introduction to free and open software. *Standards Today*, *8*(5). Retrieved from http://www.consortiuminfo.org/bulletins/aug09.php#feature

Utterback, J. M., & Abernathy, W. J. (1975). A dynamic model of process and product innovation. *Omega*, *3*(6), 639–656. doi:10.1016/0305-0483(75)90068-7

Vallas, S. P. (2003). Why teamwork fails: Obstacles to workplace change in four manufacturing plants. *American Sociological Review*, *68*(2), 223–250. doi:10.2307/1519767

Van de Kaa, G., De Vries, H. J., Van Heck, E., & Van den Ende, J. (2007). The Emergence of Standards: a Meta-analysis. In *Proceedings of the 40th Hawaii International Conference on Systems Sciences*. IEEE.

Van de Kaa, G., Van den Ende, J., De Vries, H. J., & Van Heck, E. (2011). Factors for winning interface format battles: A review and synthesis of the literature. *Technological Forecasting and Social Change*, *78*(8), 1397–1411. doi:10.1016/j.techfore.2011.03.011

Van de Voort, M., Pieters, W., & Consoli, L. (2015). Refining the ethics of computer-made decisions: A classification of moral mediation by ubiquitous machines. *Ethics and Information Technology*, *17*(1), 41–56. doi:10.1007/s10676-015-9360-2

Van den Bossche, P. (2003). *The electric vehicle: raising the standards*. (PhD dissertation). Vrije Universiteit Brussel.

Van Kemenade, E. A., Hardjono, T. W., & De Vries, H. J. (2011). The willingness of professionals to contribute to their organisation's certification. *International Journal of Quality & Reliability Management, 28*(1), 27–42. doi:10.1108/02656711111097535

Van Veldhuisen, A. (1996). *De Macht van het Systeem – Een studie naar kwaliteitssystemen gebaseerd op het werk van Michael Foucault*. Rotterdam: Erasmus Universiteit, Faculteit Bedrijfskunde.

Van Wegberg, M. (2004). Standardization and Competing Consortia: The Trade-off Between Speed and Compatibility. *International Journal of IT Standards and Standardization Research, 2*(2), 18–33. doi:10.4018/jitsr.2004070102

van Wessel, R. (2010). Toward Corporate IT Standardization Management: Frameworks and Solutions. Hershey, PA: IGI Global.

Van Wessel, R. (2010). *Toward Corporate IT Standardization Management: Frameworks and Solutions*. Hershey, PA: IGI Global. doi:10.4018/978-1-61520-759-6

van Wessel, R., Ribbers, P., & de Vries, H. (2006). Effects of IS Standardization on Business Process Performance: A Case in HR IS Company Standardization. In *Proceedings of the 39th Annual Hawaii International Conference on System Sciences*. HICSS. doi:10.1109/HICSS.2006.141

Vandemoortele, T., & Deconinck, K. (2013). When are private Standards more Stringent than Public Standards? *American Journal of Agricultural Economics, 96*(1), 154–171. doi:10.1093/ajae/aat064

VDI. (1991). *Technikbewertung – Begriffe und Grundlagen. VDI-Richtlinie 3780*. Düsseldorf: VDI Verlag.

VDI/VDE. (2003). *Formalisierte Prozessbeschreibungen. VDI/VDE-Richtlinie 3682*. Düsseldorf: VDI Verlag.

Venkatraman, N., & Camillus, J. C. (1984). Exploring the concept of "Fit" in Strategic Management. *Academy of Management Review, 9*(3), 513–525.

Venkatraman, N., & Prescott, J. (1990). Environment-strategy coalignment: An empirical test of its performance implications. *Strategic Management Journal, 11*(1), 1–23. doi:10.1002/smj.4250110102

Venkatraman, S., Bala, H., Venkatesh, V., & Bates, J. (2008). Six strategies for electronic medical records systems. *Communications of the ACM, 51*(11), 140–144. doi:10.1145/1400214.1400243

Vetter, G. R. (2007). Open source licensing and scattering opportunism in software standards. *Boston College Law Review, 48*, 225–250.

Volberda, H. W., & Weerdt, N. (2012). Contingency fit, institutional fit, and firm performance: A metafit approach to organization–environment relationships. *Organization Science, 23*(4), 1040–1054. doi:10.1287/orsc.1110.0687

Vries, H. J. (2006). Competing E-Purse Systems: A Standards Battle. *Journal of Cases on Information Technology, 8*(1), 1–15. doi:10.4018/jcit.2006010101

Wainwright, T. (2009). Laying the foundations for a crisis: Mapping the historico-geographical construction of residential mortgage-backed securitization in the UK. *International Journal of Urban and Regional Research, 33*(2), 372–388. doi:10.1111/j.1468-2427.2009.00876.x

Walgenbach, P. (2001). The production of distrust by means of producing trust. *Organization Studies, 22*(4), 693–714. doi:10.1177/0170840601224006

Walgenbach, P. (2007). Façade and means of control: The use of ISO 9000 standards. In L. Theuvsen, A. Spiller, M. Peupert, & G. Jahn (Eds.), *Quality Management in Food Chains* (pp. 29–43). Wageningen: Wageningen Academic Publishers.

Wallach, W., & Allen, C. (2008). *Moral machines: Teaching robots right from wrong*. Oxford, UK: Oxford University Press.

Wang, P., Wang, Y., & Hill, J. (2010). *Standardization Strategy of China: Achievements and Challenges*. East-West Center Working Paper, Economics Series No. 107.

Wang, A., Liao, W., & Hein, A. P. (2014). *Bigger, better, broader: A perspective on China's auto market in 2020*. McKinsey & Company.

Wang, P., & Ramiller, N. C. (2009). Community learning in information technology innovation. *Management Information Systems Quarterly*, *33*(4), 709–734.

Warner, A., Fairbank, J., & Steensma, H. (2006). Managing uncertainty in a formal Standards-based industry: A real options perspective on acquisition timing. *Journal of Management*, *32*(2), 279–298. doi:10.1177/0149206305280108

Weber, M. (1922). *Wirtschaft und Gesellschaft*. Tübingen: J.C.B. Mohr.

Week, B. (1998). *A $2.5 trillion market you hardly know*. Available at: http://www.bloomberg.com/bw/stories/1998-10-25/a-2-dot-5-trillion-market-you-hardly-know

Wegberg, M. (2004). Standardization Process of Systems Technologies: Creating a Balance between Competition and Cooperation. *Technology Analysis and Strategic Management*, *16*(4), 457–478. doi:10.1080/0953732042000295784

Weill, P., & Olson, M. H. (1989). An assessment of the contingency theory of management information systems. *Journal of Management Information Systems*, *6*(1), 59–85.

Weill, P., & Ross, J. (2004). *IT Governance – How Top Performers Manage IT decision Rights for Superior Results*. Boston, MA: Harvard Business School Press.

Weissinger, R. (2013). Economic benefits of standards - Results and future perspectives of the "ISO Methodology". In K. Jakobs, H. J. de Vries, A. Genesh, A. Gulacsi, & I. Soetert (Eds.), *EURAS Proceedings 2013. Standards: Boosting European Competitiveness: EURAS Annual Standardisation Conference* (vol. 18, pp. 385–395). Aachen, Germany: Verlagshaus Mainz GmbH Aachen.

Weitzel, T. (2003). *A Network ROI, MISQ Special Issue Workshop Standard Making: A Critical Research Frontier for Information Systems*. Paper presented at Pre-Conference Workshop, International Conference on Information Systems, Seattle, WA.

Wenström, H., Ollner, J., & Wenström, J. (2010). *Focus on Industry and Standards – An Integrated Approach*. Stockholm: SIS.

Werle, R. and E. J. Iversen (2006, March). Promoting Legitimacy in Technical Standardization. *Science, Technology & Innovation Studies*, 19-39.

Werr, A., Stjernberg, T., & Docherty, P. (1997). The functions of methods of change in management consulting. *Journal of Organizational Change Management*, *10*(4), 288–307. doi:10.1108/09534819710175839

West, J. (2003). How open is open enough? Melding proprietary and open source platform strategies. *Research Policy*, *32*(7), 1259–1285. doi:10.1016/S0048-7333(03)00052-0

West, J. (2007). *The economic realities of Open standards: Black, white, and many shades of gray. In Standards and Public Policy* (pp. 87–122). Cambridge, UK: Cambridge University Press.

West, J., & Dedrick, J. (2006). Scope and Timing of Deployment: Moderators of Organizational Adoption of the Linux Server Platform. *International Journal of IT Standards and Standardization Research*, *4*(2), 1–23. doi:10.4018/jitsr.2006070101

Westphal, J. D., & Zajac, E. J. (2001). Decoupling policy from practice: The case of stock repurchase programs. *Administrative Science Quarterly*, *46*(2), 202–228. doi:10.2307/2667086

Williams, R. (2009). *Common ICT Standards or Divided Markets: The EU and China*. China-Eu Standards Project Workshop held in conjunction with the I-ESA Conference.

Williams, R., & Edge, D. (1992). The Social Shaping of Technology: a Review of UK Research Concepts, Findings, Programmes and Centres. In New Technology at the Outset. Campus-Verlag & Westview.

Williamson, O. (1979). Transactions-Cost Economics: The Governance of Contractual Relations. *The Journal of Law & Economics*, *22*(2), 233–262. doi:10.1086/466942

Williamson, O. E. (1985). *The economic institutions of capitalism: firms, markets, relational contracting*. New York: Free Press.

Wissema, J. G., Messer, H. M., & Wijers, G. J. (1991). *Angst voor veranderen? Een mythe! – Of: Hoe u veranderingsbereidheid op de werkvloer vergroot*. Assen: Van Gorcum.

Woodward, J. (1965). *Industrial organization: Theory and practice*. London: Oxford University Press.

Wooton, L. M. (1977). The mixed blessings of contingency management. *Academy of Management Review*, *2*(3), 431–441.

Wright, C., Sturdy, A., & Wylie, N. (2012). Management innovation through Standardization: Consultants as Standardizers or organizational practice. *Research Policy*, *41*(3), 652–662. doi:10.1016/j.respol.2011.12.004

WTO. (2005). *World Trade Report 2005: Standards, 'offshoring' and air transport*. Geneva: World Trade Organization.

Wu, Y., Balasubramanian, S., & Mahajan, V. (2004). When is a Preannounced New Product Likely to be Delayed? *Journal of Marketing*, *68*(2), 101–113. doi:10.1509/jmkg.68.2.101.27792

Yeung, A. C. L., Lee, T. S., & Chan, L. Y. (2003). Senior management perspectives and ISO 9000 effectiveness: An empirical research. *International Journal of Production Research*, *41*(3), 545–569. doi:10.1080/0020754021000033896

Yin, R. K. (2003). *Case study research: Design and methods* (3rd ed.). Thousand Oaks, CA: SAGE Publications.

Yin, R. K. (2009). *Case Study Research – Design and Methods* (4th ed.). Thousand Oaks, CA: SAGE.

Yin, R. K. (2009). *Case Study Research: Design and Methods* (4th ed.). London: Sage.

Zbaracki, M. (1998). The rhetoric and reality of Total Quality Management. *Administrative Science Quarterly*, *43*(3), 602–636. doi:10.2307/2393677

Zhao, K., Xia, M., & Shaw, M. (2005). Vertical e-business standards and standards developing organizations: A conceptual framework. *Electronic Markets*, *15*(4), 289–300. doi:10.1080/10196780500302690

Zhao, K., Xia, M., & Shaw, M. (2007). An integrated model of consortium-based e-business Standardization: Collaborative Development and adoption with network externalities. *Journal of Management Information Systems*, *23*(4), 247–271. doi:10.2753/MIS0742-1222230411

Ziegler, K., & Gerst, M. (2012). *Sino-German Cooperation on Standardisation and Conformity assessment for New Energy Vehicles*. Feasibility Study for BMWi (German Ministry of Economy and Technology) and CNCA.

Zott, C., Amit, R., & Massa, L. (2011). The Business Model: Recent developments and future research. *Journal of Management*, *37*(4), 1019–1042. doi:10.1177/0149206311406265

About the Contributors

Kai Jakobs joined Aachen University's (RWTH) Computer Science Department as a member of the technical staff in 1985. Over the years, his research interests moved away from the technical nuts and bolts of communication systems to socio-economic aspects with a focus on IT standardization. He is the co-author of a text book on data communications and more recently, five books on standards and standardization in IT. He has been on the program committees of numerous international conferences, and has also served as an external expert on evaluation panels of various European R&D programs, on both technical and socio-econimic issues. He holds a PhD in computer science from the University of Edinburgh.

* * *

Nizar Abdelkafi is head of the research unit "Business Model Engineering and Innovation" at Fraunhofer MOEZ and senior researcher at the department of Innovation Management and Innovation Economics at the University of Leipzig. He holds an Industrial Engineering diploma from the National Engineering School Tunis, a Master degree in Business Administration from the Technische Universität München, and a PhD degree from the Hamburg University of Technology. Nizar Abdelkafi is interested in innovative business models, along with innovation and srtandardization and has published his research in two books, several international journals such as International Journal of Innovation Management and IEEE transactions on Engineering Management, as well as over 30 conference papers and book chapters.

Carl Cargill is the Standards Principal at Adobe Systems, Advanced Technology Labs, where he works on structuring Adobe's standardization activities for the next decade, as well as attempting to provide standardization theory that applies in the "massively connected world." Carl has been a leader in standardization—both in practice and in theory—for over 25 years. His writings include two books on the subject, Information Technology Standardization: Theory, Process, and Organizations and Open Systems Standardization: A Business Approach, and scores of articles on the subject of standardization and its practical applications. He has testified several times before Congress, and has been on both Office of Technology Assessment and General Accounting Office panels as an expert on standardization. Prior to joining Adobe, Carl was director of corporate standards at Sun Microsystems, where he managed Sun's standardization strategies, activities, and portfolios. He has served on the boards of W3C, Object Management Group, Open Mobile Alliance, The Open GIS Consortium, The Open Group, Enterprise Grid Alliance, ECMA, and OSGi. During the rest of his career, has was a product strategist, marketing manager, pricing manager, program manager, and at one time an Air Force intelligence officer. His interests include medieval history and the study of magic (as distinguished from quantum mechanics). He holds a Bachelor in Arts from the University of Colorado in Medieval European History (1969) and a Masters in the Science of Administration (Management Engineering) from the George Washington University (1975).

Frédéric Chevalier is an electrical engineer and project leader. He is the change agent of the improvement program that consisted in defining and implementing project management processes for an engineering company. He obtained a Master degree in engineering, with a concentration in engineering project management, from the École de technologie supérieure in 2012. He is a member of the professional association of engineers of the Province of Québec (Ordre des ingénieurs du Québec) and of PMI.

Jorge L. Contreras is an Associate Professor of Law at the University of Utah, and a Senior Policy Fellow at American University Washington College of Law. He writes and speaks frequently on topics including technical standards, patent litigation and antitrust law. He serves as Co-Chair of the American Bar Association's Technical Standardization Committee, a member of the American National Standards Institute (ANSI) IPR Policy Committee, and the Advisory Council of NIH's National Center for Advancing Translational Science (NCATS). He previously served on the National Academy of Science's Committee on IP Management in Standard-Setting Processes and as Co-Chair of the National Conference of Lawyers and Scientists. He edited the Technical Standards Patent Policy Manual (ABA Publishing: Chicago, 2007) and has written numerous articles, book chapters and commentaries relating to intellectual property and standards development. His work has appeared in scholarly publications including Science, Nature, Telecommunications Policy, Standards Engineering, IEEE Internet Computing, Jurimetrics, American University Law Review, Antitrust Law Journal, Berkeley Technology Law Journal, and Harvard Journal of Law and Technology. Professor Contreras is the founding editor of SSRN's Law, Policy and Economics of Technical Standards eJournal, and was the winner of the Standards Engineering Society's (SES) 2011 and (with co-author Andrew Updegrove), 2015 scholarly paper competitions. His work has been supported, among others, by the U.S. National Institute of Standards and Technology (NIST) and a daVinci Fellowship awarded by the Center for the Protection of Intellectual Property.

Alice de Casanove is Chairperson of ISO TC 279 on Innovation Management, and Telecom Faculty Leader with Airbus Defence and Space Engineering Academy. She has 10 years of experience in innovation management; definition of standards in participating in various committees at ISO and ETSI as technical redactor and chair, survey of its impact on the company strategy; and intellectual property right management as part of her various positions in start-up and large groups for the consumer electronics, multimedia and space telecommunication industries. She graduated in 2004 from Telecom ParisTech, with major in digital telecommunication and signal processing; and specialized training on IPR from University of Orsay.

Henk J. de Vries is Associate Professor of Standardisation at the Rotterdam School of Management, Erasmus University, Department of Technology and Operations Management. His research and teaching focus on standardisation from a business perspective. Henk is President of the European Academy for Standardisation EURAS. He is author and co-author of more than 300 publications in the field of standardisation. See http://www.rsm.nl/hdevries.

Michael Decker is Professor for Technology Assessment (TA) at the Institute for Philosophy of the Karlsruhe Institute of Technology (KIT) in Germany and director of the Institute for Technology Assessment and System Analysis (ITAS) at the KIT. He is the spokesperson of the German Speaking Network of TA, spokesperson for the Topic "Key Technologies and Innovation Processes" in the Helmholtz Program "Technology, Innovation and Society," and leads the advisory board "Society and Technology" of the Association of German Engineers (VDI). He did his PhD in physics at the University of Heidelberg and a habilitation in TA at the University of Freiburg. Before coming to the KIT, Decker worked as scientist at the German Aerospace Center (DLR) in Stuttgart and at the Europäische Akademie GmbH in Ahrweiler. Main research areas include: Theory and methodology of TA; TA of new and emerging sciences and technologies; epistemology of inter- and transdisciplinary knowledge.

Tineke M. Egyedi, founder of DIRoS, is policy advisor and senior researcher Standardization with a focus on policy, processes and impact issues. She has done studies and projects for the EU, standards bodies, science foundations, government agencies and international companies. She publishes in journals ranging from IEEE Communications Magazine to International Journal of Hydrogen Energy. Based on the book "Inverse Infrastructures: Disrupting Networks from Below" (2012, Edward Elgar; co-edited by Donna Mehos) and her work on 'Standards & Flexible Infrastructures' for the Next Generation Infrastructures Foundation, she was invited to develop NGI MOOC modules (2014). She initiated the game Setting Standards used by universities internationally (multiple disciplines), NIST (US policy makers) and NEN (Chinese ISO standardizers). She was president of the European Academy for Standardization (2005-2011) and chair of the International Cooperation for Education about Standardization (2009-2010). She is expert to the UNECE WP6 on 'Regulatory Cooperation and Standardization Policies' and member of IEEE committees (e.g. co-chair TPC SIIT2015). She regularly lectures at home and abroad.

Erwin Folmer, in 1999, started working for KPN Research as an innovator, and was especially involved in designing order entry and billing systems for new KPN services such as ADSL. From 2001 Erwin joined TNO to lead the European OpenXchange project focusing on business to business integration. Since then he has been involved in standardization and interoperability research with a special interest in semantic standards. He is involved in a number of standard domains such as temporary staffing industry, waste, invoices, transportation and education, amongst others. Erwin is the founder of the SETU Timecard standard, which is used nowadays by the whole Dutch staffing industry. He chaired many standardization workgroups, including HR-XML SIDES, SETU and STOSAG. Currently he chairs the Standards steering board for the Dutch Semantic Invoice Model. As a standardization expert he was a member of the Netherlands Open in Connection (2009-2011) project team to improve the adoption of open standards (and open source software) in the Netherlands. As part of this role, Erwin authored the BOMOS (management and development of open standards) book that many standardization organizations use as a guidance when organizing the management of standards. In 2009 he joined the University of Twente part-time to embark on a PhD research program on standardization, while continuing his work for TNO untill 2015. From 2015 onwards Erwin left TNO and joined the Kadaster, the Dutch land registry. As well as being involved in teaching activities, Erwin has also presented his work at many international scientific conferences and in 2010 participated in the doctoral consortium of the American Conference on Information Systems. Erwin's research has led to many publications in books, magazines, conference proceedings, and journals; Electronic Markets and Information Systems and E-Business Management, amongst others.

Amalia Foukaki holds a B.Sc. in Business Administration from Athens University of Economics and Business (2010). During her undergraduate studies she worked for a global student-run organization, AIESEC Hellas, leading and managing national and international projects. She obtained remarkable international experience through interaction and collaborations with Katholieke Universiteit of Leuven (Belgium), European Voluntary Service (Romania), and R.O.C. Council for Cultural Affairs (Taiwan). She joined the Institute of Economic Research in 2012, right after obtaining a M.Sc. in Corporate and Financial Management from School of Economics and Management, Lund University.

Xudong Gao is faculty of Department of Innovation, Entrepreneurship. He received his Bachelor in Industrial Management Engineering from Harbin Institute of Technology in 1988, and Master in Industrial Economics from Renmin University of China in 1991, Ph.D. in Management from MIT Sloan School of Management in 2003. He teaches Strategic Management, Technology Strategy, High Tech Entrepreneurship. His main Research Interests include Competitive Strategy, Technology Strategy, Management of Technological Innovation. Since 2005, he has been the project leader or key member for 15 research projects funded by National natural science foundation of china, China National Petroleum Corporation, etc. He has published more than 30 papers during last five years in Research Policy, Research-Technology Management, Science of Science and Management of S & T, Mobile Communications, Int. J. Business and Systems Research, International Journal of Technology Management, etc.

Martina Gerst is an expert in innovation, ICT standardisation, IPR and Supply Chain Management and has extensive experience in these areas in Europe, US and China. She has been working as Senior Manager for leading international companies like Daimler, Siemens or KPMG. Currently, Dr Gerst is working as a Senior Expert in Innovation, Technology and IPR for numerous European organisations and companies. Besides, she is affiliated with the Tsinghua University, School of Economics and Management and the University of Edinburgh, Institute for the Study of Science, Technology and Innovation researching New Electrical Vehicles technologies, IPR in ICT and ICT standards. Dr Gerst obtained her PhD from the University of Edinburgh on research of socio-technical dynamics of ICT Innovations. Before, she graduated in Business Informatics from the University of Saarbruecken/Germany and the Grand Ecole Institut Commerciale de Nancy/France.

Anne-Marie Großmann completed her Bachelor in Philosophy and Economics from the London School of Economics in 2009. 2010 she joined the Berlin Doctoral Programme in Economics and Management science and received a Master in Economics and Management Science from the Humboldt Universität of Berlin in 2012. Since 2012 Anne-Marie is a research fellow at the Chair of Innovation Economics at the Technische Universität Berlin. She received her PhD for her thesis with the title "The Microeconomics of Standards: Five Essays on the Relation of Standards to Innovation and Inter-Firm Relationships" (Technische Universität Berlin, 2015). Her research interest is primarily on company standards and the economic aspects of standards.

Ries Haverkamp is Director of European Quality, Environment, Safety and Health at Interface European Manufacturing. Previous jobs include Operations Manager at Gezu, Quality Engineer at DHL Exel Supply Chain, and Manager of the Quality Department at Mitsubishi Caterpillar Forklift Europe. His research interests include standardisation, quality management and the multi-disciplinarity of business administration. He conducted the research work for this chapter as an external PhD-candidate at the Rotterdam School of Management, Erasmus University Rotterdam, The Netherlands.

Anique Hommels is associate professor at the Maastricht University research group on Science, Technology & Society (MUSTS). Her research focuses on the vulnerability of critical infrastructures, including emergency communication. Recent publications include Vulnerability in Technological Cultures, 2014, MIT Press (co-edited with Wiebe Bijker and Jessica Mesman) and The Making of Europe's Critical Infrastructure. Common Connections and Shared Vulnerabilities, 2013, Palgrave (co-edited with Per Hogselius, Arne Kaijser and Erik van der Vleuten).

Magnus Johansson received his PhD in Business Administration from the School of Economics and Management at Lund University in 2008. After the dissertation, Magnus Johansson spent more than four years in the telecom and automotive industries working with strategic analysis, planning and product portfolio management. As of 2012 he holds a postdoc position at the Standardization Research Centre (SRC), Lund Institute of Economic Research. His research is focused on business model and value creation strategies as well as competence and knowledge management in association with standards, innovation and network effects. He has also done research within the field of pricing and the interrelation between value creation and value capture strategies.

Antonios Kaniadakis is a Lecturer in IT Management at the School of Electronic Engineering and Computer Science, Queen Mary, University of London. His research focuses on social aspects of ICT innovation, predominantly in the finance sector and in social media. He holds a PhD in Technology and Innovation Studies from Edinburgh University.

Matts Kärreman received his PhD in Business Administration from the School of Economics and Management at Lund University. He is an Associate Professor at the the Department of Business Administration, Lund University, and has a position at the Standardization Research Centre (SRC), Lund Institute of Economic Research. His research is focused on decision making in the context of strategic management, corporate governance, as well as competence and knowledge management in association with standards, innovation and network effects.

Claude Y. Laporte has been a professor since 2000 at the École de technologie supérieure (ÉTS), an 8,000-student engineering school, where he teaches software engineering. His research interests include software process improvement in small and very small enterprises, as well as software quality assurance. He has worked in defense and transportation enterprises for over 20 years. He received a Master's degree in Physics from the Université de Montréal, a Master's degree in Applied Sciences from the École Polytechnique de Montréal and a Ph.D. from the Université de Bretagne Occidentale (France). In addition, he was awarded an honorary doctorate by the Universidad de San Martin de Porres (Peru) in 2013. He is the Editor of ISO/IEC JTC1 SC7 Working Group 24, tasked to develop ISO/IEC 29110 life cycle standards and guides for Very Small Entities. He is the Co-chair of the INCOSE Systems Engineering for Very Small Entities WG. He is a member of INCOSE, IEEE, PMI and a member of the professional association of engineers of the Province of Québec (Ordre des ingénieurs du Québec). He is the co-author of two French books on software quality assurance published in 2011 by Hermes Science-Lavoisier and one English textbook, on the same topic, published by John Wiley and Sons in 2016.

Luisa Kim Lazina completed her Bachelor in Economics and holds her Master in Industrial and Network Economics. She is a Phd candidate at the Chair of Innovation Economics at Technische Universität (TU) Berlin. Since October 2013 she is working in the standardization department of a German automotive manufacturer. 2004 till 2007 she worked for a large German logistics company, where she conducted an apprenticeship in industrial management. Her research interest is primarily on company standards and standardization strategies in the automotive industry.

Sergiy Makhotin studied Industrial Engineering and Management at the University of Bremen, with Special Focus on Innovation, Project Management and Process Engineering. Since December 2011 he has been working as research fellow at Fraunhofer MOEZ. He is also a PhD Student at the University of Leipzig, at the department of Innovation Management and Innovation Economics. Sergiy Makhotin is interested in the Research Areas of standardization, Business Models and Electric Mobility. Sergiy He co-authored many papers published in journals such as the International Journal of Innovation Management and International Journal of IT Standards and Standardization Research, and in several conference proceedings.

Justin Pierce studied law in England, where he was later called to the Bar of England and Wales. He is employed as an in-house lawyer as well as having a position in academia. Academically, he has lectured in a number of institutions and has presented at numerous conferences alongside other academic writings in the field of EU competition law and standardisation. He is currently undertaking a PhD in EU competition law and standardisation.

Piet M. A. Ribbers was full professor of Information Management, at Tilburg University from 1987 to 2014 where he chaired the Department of Information Systems and Management. He currently is Emeritus Professor at Tilburg University. At Tias Business School, the school for executive education of Tilburg University, he is the Academic Director of the Executive MSc in Information Management (MIM), the professional program on Business and Enterprise Architecture (BEA), and of the Masterclass Information Security Management (ISM). His research interests relate to the interaction between the 'business domain' and 'IS domain'. Examples of specific research themes are: IT Governance, Alignment (including business case analysis), Sourcing and Information Quality.

Jasper Roes works for TNO as a business consultant in the field of interoperability and standardization. In this function he advises organizations in different sectors and branches that are working with or on digital collaboration. Next to advising organizations Jasper is also involved in project that design and create (open) standards that can be used to exchange information between organizations. Jasper focuses on using ICT to make collaborations between companies and devices smarter and more profitable for all organizations involved. Jasper has a special interest in the energy sector.

Kurt Sandholtz is an Assistant Professor of Organizational Behavior and director of the Management Communications program in the Marriott School of Management at Brigham Young University in Provo, Utah, USA. He earned his Ph.D. in Management Science and Engineering at Stanford University. Prior to his academic career, Kurt spent 15 years in management consulting. He has worked with a diverse portfolio of clients such as Qualcomm, Coca-Cola, ConocoPhillips, Nokia, Amgen, DuPont, and Novartis. Kurt's scholarly work explores issues at the intersection of institutions and occupations. Among the occupations he has studied are HR professionals, design engineers, and data scientists. His research has been published in Organization Studies and Strategic Entrepreneurship Journal. He also has published numerous practitioner-oriented articles, and is the co-author of two books: Beyond Juggling: Rebalancing Your Busy Life (2002) and HR Competencies: Mastery at the Intersection of People and Business (2008). The field work that led to the chapter in this volume was funded by National Science Foundation grant # ITR 0427173.

Hartwig Steusloff was born 1937 in Gelsenkirchen, Germany. 1963 he received the degree of a "Diplom-Ingenieur" in communication technology from the Technical University of München, Germany. He joined Teldix (Telefunken Bendix) military aircraft systems as a development engineer for sensor systems. 1968 he started his scientific career at the Fraunhofer Institute of Vibration Sensor Systems in Tübingen, Germany which moved to Karlsruhe, Germany, in 1968. Hartwig Steusloff received his PHD degree (Dr. rer.nat.) in Computer Science in year 1977 from the Tchnical University of Karlsruhe. His thesis defined the Programming Language PEARL (Process and Experiment Automation Real-time Language) for distributed computer control systems. 1984 he was appointed Head of Fraunhofer Institute of Information and Data Processing in technical and biological application fields which he held until his retirement in 2004. Hartwig's main professional interests still are dealing with complex distributed systems with particular applications in Smart Energy Grids, in transportation and traffic systems and in natural resources management with a special focus on water resources. His additional dedication is concerned with international standardization. From 1991 until 2014 he was deputy chairman of the German Commission Electrotechnology Electronics and Information Technology (DKE). At present he is carrying through a project on the inclusion of ethics in system control and with the related standards. Hartwig is also the chairman of DIN FOCUS.ICT focusing at the consistent inclusion of ICT in standardization projects.

Andrew Updegrove is a founding partner of Gesmer Updegrove LLP, a Boston-based technology law firm. He has a broad range of experience in representing both mature and emerging high technology companies of all types in all aspects of their legal affairs. Since 1988, he has also represented and helped structure more than 135 worldwide standard setting, open source, research and development, promotional and advocacy consortia, including some of the largest standard setting and open source development organizations in the world. He spends a significant part of his time giving strategic advice to clients of the firm. His leadership in collaboration related matters is widely recognized. In 2004, he was the sole representative of the consortium community to be appointed as a member of the United States National Standards Strategy Committee. He has served on the Boards of Directors and Executive Committee of the American National Standards Institute (ANSI), the Boards of Directors of the Free Standards Group (FSG) and the Linux Foundation and the Boards of Advisors of HL7, an ANSI accredited developer of electronic health standards for clinical and administrative data and of Open Source for America, a group promoting the use of open source software by government, and a charter Fellow of the OpenForum Academy, a sister organization to OpenForum Europe, an organization promoting the use of open source and standards throughout Europe. In 2011, he was appointed to a National Academies of Science Committee commissioned by the U.S. Patent and Trademark Office to advise it on intellectual property policies around the world. He has been retained by many of the largest corporations in the world to assist in setting up international consortia, and by both multinational companies as well as government agencies to advise them in setting their standards-related policies and goals. He has also provided testimony to the Department of Justice and Federal Trade Commission on standard setting and intellectual property rights, and written and filed pro bono "friend of the court" briefs in major standards-related litigation before the Federal Circuit Court, the Supreme Court, and the Federal Trade Commission. In 2010, he provided testimony on the future of the National Institute of Science and Technology (NIST) to the House Committee on Science and Technology's Subcommittee on Technology and Innovation. In May of 2002, he conceived and launched ConsortiumInfo.org, an extensive website intended to provide the most comprehensive and detailed source of news and information on the Internet on standard-setting and forming and maintaining consortia. The site serves up to a million page views a month to a global audience, including up to 50,000 visitors a month from China. In December of 2002, he conceived and launched the Consortium Standards Bulletin (now Standards Today), an eJournal of news, ideas and analysis on standard setting. Standards Today is sent to a large and rapidly growing list of standards professionals and other subscribers at major corporations, government agencies and universities throughout the world. Other sections of the site include the Standards MetaLibrary, with over 1,900 categorized and abstracted articles, and the Standards Blog, which attracts up to 100,000 visitors a month. Besides his efforts at the ConsortiumInfo.org Website, he writes and speaks frequently both domestically and internationally on topics involving standards, open source software and consortia. In 2005 he was selected to receive the ANSI President's Award for Journalism for his work at ConsortiumInfo.org and Standards Today. Mr. Updegrove is a graduate of Yale University and the Cornell University Law School.

Robert M. van Wessel holds a Master in Electrical Engineering from Twente University and a PhD in Business Administration from Tilburg University (Department of Information Systems and Management). He is associated with Rotterdam School of Management, Erasmus University. Robert's research interests relate to the interaction of Business and Information Technology, in particular Business performance and the Value of IT, Enterprise Architecture, Data Management, IT Governance, Portfolio Management, Information Security Management and IT Standardisation and Standards. Robert is currently Business Architect at a financial services enterprise and founder of ApexIS, a firm providing consultancy services to companies in his areas of research.

Paul-Vincent von Gruben attained his Bachelor Degree in Industrial Engineering at the Technische Universität Berlin in 2012. Afterwards, he worked in a manufacturing company on projects in Germany and the UK. In 2014, he completed his Master of Science in Management at the University of Edinburgh Business School. After his graduation, he joined the Chair of Innovation Economics at the Technische Universität Berlin and is currently a Ph.D. Student while employed at an electronic component producer. His research focus is on the economics of standardization, especially on the microeconomic level.

Index

A

B

C

M

N

O

P

Q

R

S

T

U

V

W

www.ingramcontent.com/pod-product-compliance
Lightning Source LLC
LaVergne TN
LVHW060359300926
849162LV00006B/1308
* 9 7 8 1 4 6 6 6 9 7 3 7 9 *